Nanostructure Science and Technology

For other titles published in this series, go to
www.springer.com/series/6331

James H. Dickerson • Aldo R. Boccaccini
Editors

Electrophoretic Deposition of Nanomaterials

Springer

Editors
James H. Dickerson
Vanderbilt University
Nashville, TN, USA
james.h.dickerson@vanderbilt.edu

Aldo R. Boccaccini
University of Erlangen-Nuremberg
Erlangen, Germany
aldo.boccaccini@ww.uni-erlangen.de

ISSN 1571-5744
ISBN 978-1-4419-9690-9 e-ISBN 978-1-4419-9730-2
DOI 10.1007/978-1-4419-9730-2
Springer New York Dordrecht Heidelberg London

Library of Congress Control Number: 2011932793

Printed on acid-free paper

Springer is part of Springer Science+Business Media (www.springer.com)

Preface

Electrophoretic Deposition of Nanomaterials

Electrophoresis and dielectrophoresis have been used extensively in biology, chemistry, materials science, and bioengineering for the manipulation and processing of a variety of biological materials (proteins, cells, etc.), colloids, and other liquid phase objects. In parallel, electric field assisted deposition schemes, such as electrodeposition, electrochemical deposition, and electroplating, have been employed for a number of years in materials science, industrial materials processing, and thin film applications involving precious metal coatings, composite ceramic formation, and bio-active materials, various paints and dyes.

Recent research interest in nanoscience and nanotechnology has focused on discovering facile techniques to manipulate materials that historically have shown to be difficult to handle because of their size. In particular, nanomaterials pose a substantial challenge in their efficient handling due to their diminished size. Of the multitude of available techniques to control the distribution of these materials, the electrophoretic deposition (EPD) of nanomaterials appears to be ideally suited for the distribution and deposition of nanomaterials, providing a facile, expeditious means to produce tightly packed films of nanoparticle, nanotubes, and other nanostructured materials. EPD combines aspects of electrophoresis—the translation of charged particles, suspended in a solution, due to an ambient, direct current (dc) electric field—and dielectrophoresis—the locomotion of dipolar, polarizable, or charged particles, also in solution, due to alternating current (ac) or gradient electric fields—to deposit nanostructures onto conducting electrodes. In traditional EPD, a dc voltage is applied across the cell, thereby creating an electric field that transports charged particles to the electrodes where they deposit to form a cast film. The primary advantages of electrophoretic deposition as a technique to distribute, to aggregate, and to compose films of nanomaterials include site-selectivity, dense packing of the nanomaterials, size-scalability of the film, and marked control over the deposition thickness of the film. Thus, EPD can rapidly fabricate films comprising multiple monolayers of tightly packed nanomaterials, with short-range van der Waals interactions a stabilizing influence.

Electrophoretic deposition, first investigated in depth by Hamaker and Koelmans in the 1940s and 1950s, has been applied to cast uniform layers of particles on conducting surfaces. By applying a potential across the two electrodes, charged particles are transported to the appropriately biased electrode (e.g., negatively charged particles migrate to the anode) where they accumulate to form a cast film. Compact casts of nanocrystals, nanotubes, or nanoparticles have been fabricated by EPD from suspensions that contain polar and non-polar organic solvents, as well as water. An essential advantage of EPD is its significantly shorter cycle times relative to other wet-casting processes. Most large-scale deposition schemes, including Langmuir-Blodgett or evaporative self-assembly, are at least an order of magnitude slower than EPD. The ability to cast films at higher rates, as well as the potential to control nanocrystal mobility through the applied electric field, presents compelling advantages of EPD relative to the other techniques. For thin films of nanomaterials to be competitive with and to supplant bulk crystalline materials in optical, electronic, and magnetic applications, the facile and rapid production of ordered, homogeneous, densely packed, and topographically smooth nanomaterial films must be realized. The inability to cast ordered defect-free films at an industrial scale and at short cycle times remains a significant obstacle to the commercialization of nanocrystalline films.

EPD presents several additional advantages over other casting processes, such as evaporative self-assembly and Langmuir-Blodgett casting. EPD is *scalable*, as demonstrated by applications in the ceramics and coatings industries, where films are deposited onto substrates as large as automotive bodies (e.g., primer coating) and as small as nanoscale electrodes. The structure and properties of the cast film can be tuned to targeted values by manipulating process variables, such as applied dc and ac voltages, frequency, and nanocrystal surface chemistry, thereby affirming the flexibility of the process. Finally, by preparing suitable templates as deposition electrodes, *patterned* films can be cast. These combined characteristics make EPD the ideal deposition scheme to produce robust nanoparticle thin films.

The rapidly emerging nanomaterials market has motivated sectors of the metals, ceramics, electronics, and other industries to consider the incorporation of nanotechnology and/or nanomaterials into their products, techniques, and protocols. Commercial and industrial markets in Europe, Asia, and North America have demonstrated interests in using nanomaterials in device deliverables and other systems. Burgeoning research investment and developing markets in South America, Australia, and Africa are following these trends.

The ultimate goal of this book is to provide a comprehensive, integrated view of the basic research, materials science, and engineering of the deposition of nanomaterials via electrophoresis and a view of commercial and industrial applications associated with this technique. Further, this book will provide an invaluable, contemporary reference for the development of fundamental theory and experiment, advanced experimental and manufacturing techniques, and industrial applications of electrophoretic deposition of nanomaterials. This monograph represents contributions from a large breadth of the science and technology that are involved in this versatile deposition process, disciplines that include electrochemistry, materials science, physics, chemistry, chemical engineering, ceramics engineering, bioengineering, and electrical engineering, among others. This volume represents the initial foray into illustrating a variety of characteristic components of this rapidly emerging field of nanoscience and nanotechnology.

This monograph begins with two chapters that overview the fundamental concepts, methodologies, equations, terms, and phenomena associated with electric field control and manipulation of colloidal particles (Paul J. Sides) and, more specifically, with the electrophoretic deposition of nanoparticles in polar solvents (Rodrigo Moreno and Begoña Ferrari). Subsequent chapters discuss various applications of electrophoretic deposition of nanoparticles, focusing on: non-polar solvent-based nanocrystal deposition (James H. Dickerson); carbon nanotube films and carbon nanotube-based composites (Aldo R. Boccaccini, Milo S. P. Shaffer, and Cengiz Kaya); advanced ceramics applications (Partho Sarkar, Debnath De, Tetsuo Uchikoshi, and Laxmidhar Besra; Rolf Clasen; and Saša Novak, Katja König, and Aljaž Ivekovič); solid state lighting and display devices (Jan B. Talbot); and electroactive materials applications (Li Tao, Chen Yanhong and Ma Jan).

Contributors

Laxmidhar Besra Colloids & Materials Chemistry Department, Institute of Minerals & Materials Technology (IMMT), Bhubaneswar, Orissa 751013, India

Aldo R. Boccaccini Department of Materials Science and Engineering, University of Erlangen-Nuremberg, 91058 Erlangen, Germany
e-mail: aldo.boccaccini@ww.uni-erlangen.de

Rolf Clasen Saarland University, Saarbrücken, Germany
e-mail: r.clasen@nanotech.uni-saarland.de

Debnath De 161 Sagewood Drive, Malvern, PA 19355, USA
e-mail: ded@jmusa.com

James H. Dickerson Department of Physics and Astronomy, Vanderbilt University, Station B #351807, 2301 Vanderbilt Place, Nashville, TN 37235-1807, USA
e-mail: james.h.dickerson@vanderbilt.edu

Begoña Ferrari Instituto de Cerámica y Vidrio, CSIC, Kelsen 5, 28049 Madrid, Spain
e-mail: bferrari@icv.csic.es

Jan Ma School of Materials Science and Engineering, Nanyang Technological University, 639798 Singapore, Singapore
e-mail: asjma@ntu.edu.sg

Cengiz Kaya Department of Metallurgical and Materials Engineering, Yildiz Technical University, Istanbul, Turkey
e-mail: cengizk@yildiz.edu.tr

Katja König Department for Nanostructured Materials, Jožef Stefan Institute, Jamova c. 39, SI-1000, Ljubljana, Slovenia

Aljaž Ivekovič Department for Nanostructured Materials, Jožef Stefan Institute, Jamova c. 39, SI-1000, Ljubljana, Slovenia

Rodrigo Moreno Instituto de Cerámica y Vidrio, CSIC, Kelsen 5, 28049 Madrid, Spain
e-mail: rmoreno@icv.csic.es

Saša Novak Department for Nanostructured Materials, Jožef Stefan Institute, Jamova c. 39, SI-1000, Ljubljana, Slovenia
e-mail: sasa.novak@ijs.si

Dennis C. Prieve Department of Chemical Engineering, Carnegie Mellon University, Pittsburgh, PA, USA

Partho Sarkar Carbon & Energy Management, Alberta Innovates-Technology Futures, 250 Karl Clark Road, Edmonton, AB T6N 1E4, Canada
e-mail: Partha.Sarkar@albertainnovates.ca

Milo S. P. Shaffer Department of Chemistry, Imperial College London, London, UK
e-mail: m.shaffer@imperial.ac.uk

Paul J. Sides Department of Chemical Engineering Carnegie Mellon University Pittsburgh, PA 15213, USA
e-mail: ps7r@andrew.cmu.edu

Jan B. Talbot Department of NanoEngineering, University of California, San Diego, 9500 Gilman Drive, La Jolla, CA 92093-0448, USA
e-mail: jtalbot@ucsd.edu

Tao Li School of Materials Science and Engineering, Nanyang Technological University, 639798 Singapore, Singapore

Tetsuo Uchikochi Nano Ceramics Centre, Fine Particle Processing Group, National Institute for Materials Science (NIMS), 1-2-1 Sengen, 305-0047, Tsukuba, Ibaraki, Japan

Christopher L. Wirth Department of Chemical Engineering, Carnegie Mellon University, Pittsburgh, PA 15213, USA

Chen Yanhong School of Materials Science and Engineering, Nanyang Technological University, 639798 Singapore, Singapore

Contents

Part I
Fundamentals of Electrophoretic Deposition

Chapter 1
Mechanisms of Directed Assembly of Colloidal Particles in Two Dimensions by Application of Electric Fields

Paul J. Sides, Christopher L. Wirth and Dennis C. Prieve

1.1 Introduction

When electric fields interact with particles immersed in liquids and levitated near electrodes, the particles assemble into patterns such as ordered arrays or chains. For example, direct electric current flowing through an aqueous solution held between two parallel-plate electrodes produces arrays of colloidal particles near one of the electrodes, as in Fig. 1.1 [1]. Electric fields imposed in-plane, by contrast, cause 1-D aggregation in the form of chained gold nanoparticles stretching from one electrode to the other, as in Fig. 1.2 [2]. Figures 1.1 and 1.2 demonstrate that purposeful application of electric fields can manipulate particles to form repeating structures. Naturally, these phenomena interested scientists and engineers. Investigations into the translation of particles laterally along surfaces when electric fields were applied normally to those surfaces, and into why particles organized themselves into wires when electric fields were applied, led to proposed mechanisms by which colloidal particles move relative to the nearby surface and relative to each other; these mechanisms are the main topics of this account.

1.1.1 Mechanisms of Electrically Driven Particle and Fluid Motion

The mechanisms of electrically directed assembly of colloidal particles are electrophoresis, electroosmosis, electrohydrodynamics, dielectrophoresis, and induced dipole interactions. *Electrophoresis* (EP) is translation of a colloid when an imposed electric field moves excess charge in the diffuse layer surrounding the particle, which produces slip at the interface between the particle and the liquid; con-

P. J. Sides (✉)
Department of Chemical Engineering Carnegie Mellon University Pittsburgh, PA 15213, USA
e-mail: ps7r@andrew.cmu.edu

J. H. Dickerson, A. R. Boccaccini (eds.), *Electrophoretic Deposition of Nanomaterials*, Nanostructure Science and Technology,
DOI 10.1007/978-1-4419-9730-2_1,

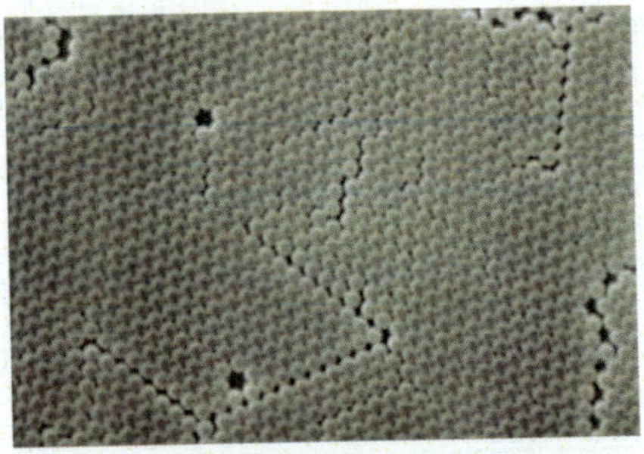

Fig. 1.1 Polystyrene particles, 4 μm diameter, deposited electrophoretically at 2 V applied across a 6 mm gap containing 0.01 mM KCl for 1,800 s, then 6 V for 30 s, and dried [1]. This figure gives an idea of the ordering possible. (Reprinted in part with permission from Bohmer [1], © 1996 American Chemical Society)

sequently, the particle translates in the direction opposite the overall slip velocity. The other mechanisms act in the scenarios shown in Fig. 1.3.The false color of the images is proportional to the absolute value of the local electric field.

Electroosmosis (EO) is electrically driven slip between a charged immobile solid and a liquid, which produces flow with respect to fixed coordinates. In Fig. 1.3a gravity, double layer repulsion, and the electrophoretic force hold a dielectric colloid in a force balance that continually restores the particle to its most probable height above the electrode after Brownian excursions. Electric current between two plane-parallel electrodes establishes an electric field with magnitude E_∞ and direction normal to the electrodes. The particle disturbs the otherwise constant (yellow) electric field far from the particle. The strong tangential electric field at the particle moves mobile unbalanced charge associated with the *diffuse* part of the electric double layer at its interface. EO is occurring at the particle and also at the electrode if there are potential gradients along it arising from the disturbance of the imposed electric field by the particle. Although an isolated particle merely stirs the fluid in its vicinity, neighboring particles entrain or repel each other; EO flow thereby causes lateral motion of the particles.

We define *electrohydrodynamic flow* (EHD), by contrast, as fluid motion arising from the action of electric fields on small charge imbalances that exist in the diffusion layer, i.e. *outside* the diffuse part of the double layer. The gradient of electric field strength normal to the lower electrode in Fig. 1.3b arises from concentration variations necessary to generate unbalanced charge outside the diffuse layer. Lateral electric field components exert force on the unbalanced charge in the *diffusion* layer, thereby moving fluid along the particle and the electrode. Again, mutual entrainment causes lateral motion of neighboring particles.

Dielectrophoresis (DEP), appearing in Fig. 1.3c is particle migration that occurs when a particle with an induced dipole moment is in a region where the gradient of the electric field is nonzero. If the two in-plane (blue) electrodes of Fig. 1.3c are polarized oppositely, lines of current with position dependent spacing, and hence

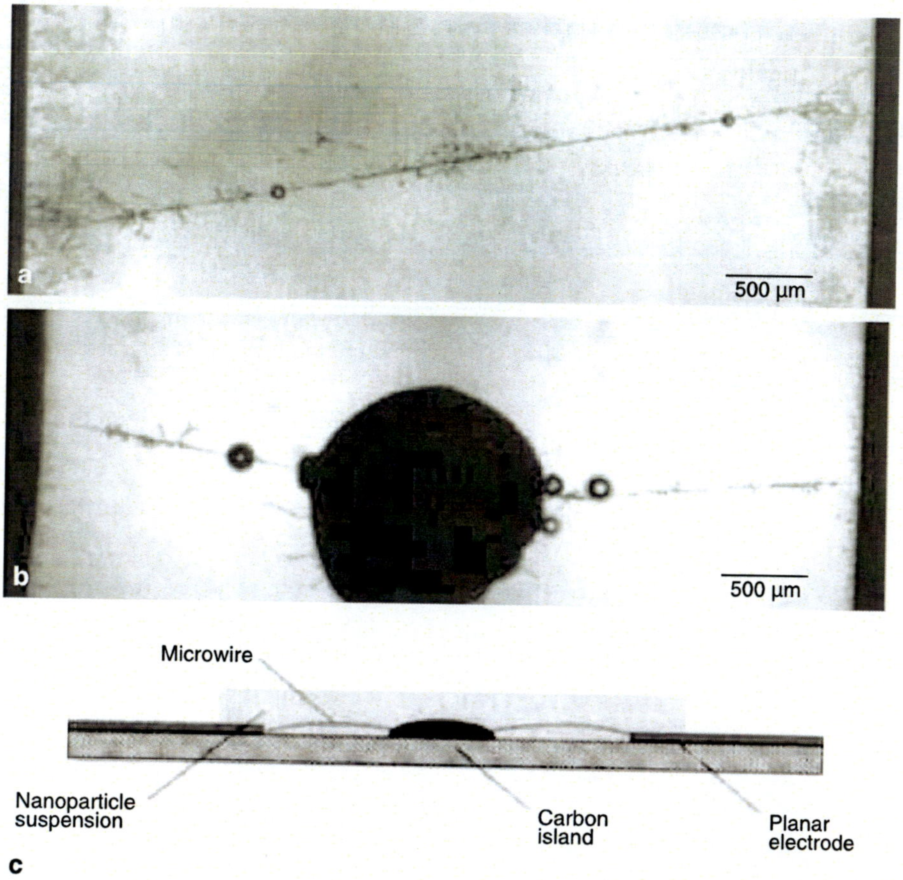

Fig. 1.2 Figure from Hermanson et al. [2] showing a microwire made of gold nanoparticles (15–30 nm) and formed between in-plane electrodes. **a** The microwire connected two gold electrodes 5 mm apart. **b** The microwire connected to a carbon spot between the gold electrodes. **c** The wire traveled through the solution. The wires exhibited ohmic behavior with resistivity much higher than bulk gold. (From Hermanson et al. [2]. Reprinted with permission from AAAS)

gradients, connect the two electrodes. The particle of Fig. 1.3c migrates in the direction of the electric field gradient by DEP and along the field lines connecting the two conductors by EP. EO is also occurring along the surface between the electrodes.

Induced dipole interaction, ID in Fig. 1.3d, is the force between two particles that arises from an imposed electric field; the field polarizes each particle and the induced charges interact at a distance. Two particles having centers connected by a line parallel to the imposed field attract. Two particles having centers connected by a line perpendicular to the imposed field repel each other.

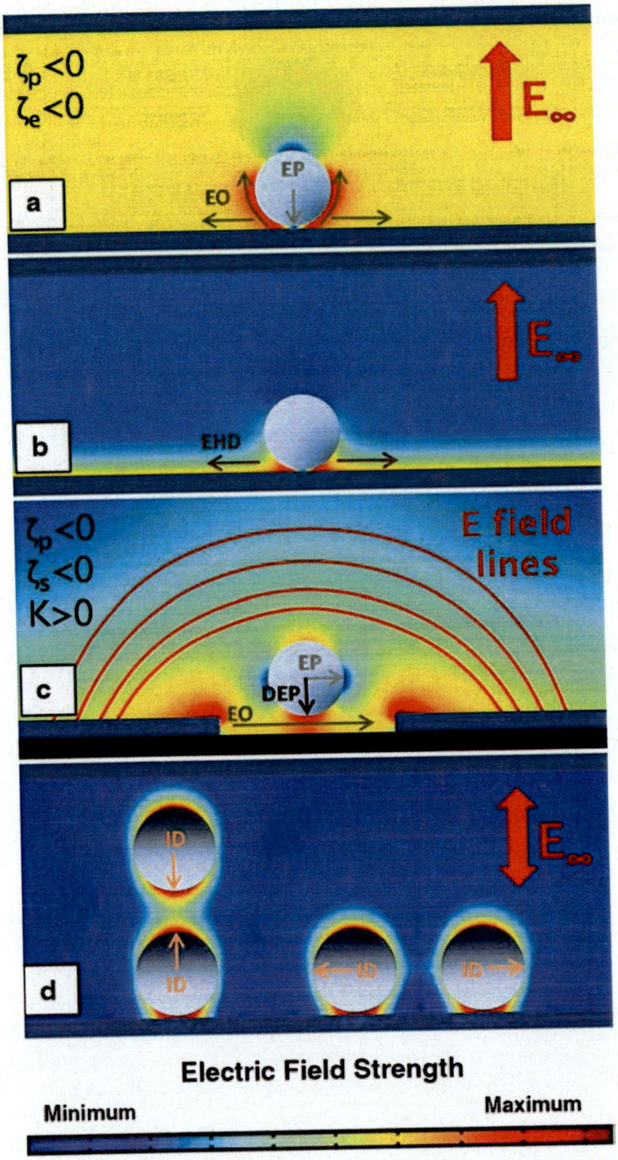

Fig. 1.3 Four scenarios illustrate aspects of the mechanisms to be discussed. The false color indicates the absolute value of the local electric field. **a** The electrodes face each other across the electrolyte. The dielectric particle experiences an electrophoretic force in addition to gravity and double layer repulsion. Interaction of the imposed electric field drives electroosmotic slip along the particle and on the electrode. **b** Faradaic reactions generate concentration gradients in the vicinity of the particle, which causes a small amount of unbalanced charge to appear in the *diffusion* layer in addition to the charge in the *diffuse* layer. The electric field interacts with the charge to produce an electrohydrodynamic body force, causing the fluid to move. **c** Electrodes coated on the same insulating surface are polarized with respect to each other. Lines of current/electric field arc

We present the theory and experimental observations of these mechanisms in action. The variety of the experimental results reflects the simultaneous action and competition among mechanisms.

1.1.2 Prior Reviews

H. A. Pohl coined the term "dielectrophoresis" and wrote the book on the subject [3, 4]. Many subsequent investigators of dielectrophoresis and induced dipole effects cite Jones's *Electromechanics of Particles* [5]. Restricting his attention to particles having diameters in the range from 1 μm to 1 mm, Jones discussed the effective moment of particles and then used that formulation to develop topics within dielectrophoresis and induced dipole interactions. He treated particles with multiple shells having different dielectric properties, levitation of particles by dielectrophoresis, chaining, and electrostatic adhesion of particles to surfaces. Green et al. [6] surveyed dynamics of particles with a principal focus on dielectrophoresis. Velev and co-workers, in a series of reviews, [7–10] summarized the use of directed assembly of colloidal particles to form functional elements. Emphasizing the building of electrical function from colloidal particles, Velev [7] described the use of electric fields to micro-manipulate particles to form wires, sensors, and displays. Velev and Bhatt [8] surveyed electrically directed colloidal assembly for on-chip devices. They gave background on the mechanisms discussed in Sect. 1.1.1 and summarized instances where these mechanisms have been used to control particles for practical purposes such as sensing, μTAS applications, and gathering particles. Gupta and Velev [9] addressed the use of colloidal assembly as a "bottom up" process, in contrast to techniques based on, for example, lithography. They categorized the materials produced from the point of view of dimensionality (1-D–3-D) and catalogued multiple self-assembly techniques including directed assembly with the use of electric fields, putting the methods that are the subject of this review into context with other methods. They concluded that the capabilities are impressive, but the large scale applications to this point have been few. The fourth review [10] traces the historical development and describes current directions in the use of *dc* and *ac* electric fields to manipulate particles, in this case treating both observed phenomena and models based on the mechanisms. The primary subject was dielectrophoresis and induced dipole interactions, but they also described induced charge electroosmosis of particles having non-uniform electrical characteristics, such as Janus particles.

General discussions of electrophoresis and electroosmosis are found in textbooks and other extended treatments [11, 12]. Prieve et al. [13] concentrated on the

from one conductor to the other. Particles move along the lines of current by electrophoresis and move in the direction of electric field gradient by dielectrophoresis. Electroosmotic slip occurs on the surface between electrodes. **d** Induced dipoles repel each other when a line joining their centers is perpendicular to the electric field; they attract each other when the line joining their centers is parallel to the electric field experimental evidence and models for directed assembly of particles where electric fields are applied in the normal direction between two equivalent uniform conductors facing each other across electrolyte. Their treatment included a discussion of electrochemical fundamentals, a crucial component in the description of directed assembly with the aid of electric fields.

1.1.3 Purpose and Outline of This Contribution

The primary purpose of the present contribution is to survey mechanisms reported when electric fields interact with particles near surfaces. The theory of each mechanism of Sect. 1.1.1 is described. The secondary purpose is to focus in more detail on assembly mechanisms involving electrohydrodynamics (EHD) and electroosmosis (EO), which have received relatively less critical attention [13] than assembly involving dielectrophoresis and induced dipole interactions [4–10]. This review, therefore, concentrates on EHD and EO effects in 2-D directed assembly of particles with the aid of electric fields. Both of these mechanisms involve mutual entrainment of particles in fluid flow resulting from the action of the applied electric field—distorted by the particles themselves—on charged fluid elements. EHD and EO are associated with three different regions of charge: (1) the diffuse layer on the particle, (2) the diffuse layer on the electrode and (3) the bulk fluid outside the diffuse layers. The differences among the models stem mainly from which fluid elements are charged and whether that charge is present without the applied electric field or induced by it.

This survey begins with a timeline of observations and proposed mechanisms related to directed assembly with opposed planar electrodes in Sect. 1.2. Electrophoresis is briefly mentioned in Sect. 1.3 and a formula for calculating electrophoretic "force" is derived. EHD in its circumscribed definition is discussed in Sect. 1.4. The treatment of the role of electroosmosis in the case of two opposed planar electrodes is broken into two sections. Section 1.5 is about equilibrium charge electroosmosis, and Sect. 1.6 is about induced charge electroosmosis. We highlight mechanisms that have been developed in sufficient detail to make quantitative estimates of the aggregation or disaggregation rates. Section 1.7 is a presentation of the theory of dielectrophoresis and induced dipole interactions with electric field gradients and with each other. The survey continues with Sect. 1.8, a summary of demonstrated capabilities and one projected application not previously considered, to our knowledge. Concluding remarks are provided in Sect. 1.9.

1.2 Electrohydrodynamics and Electroosmosis in Directed Assembly: The Uniform Field Limit

The following is a presentation of observations and suggested mechanisms with a tight focus on phenomena encountered when voltage was applied across two opposing electrodes with electrolyte in between and particles near one of them, the geometry of Fig. 1.3a and b; the response of nominally homogeneous particle pairs as well as ensembles was investigated in a number of laboratories. The purpose of this chronology is to provide a context for Sects. 1.3, 1.4, 1.5, and 1.6.

1.2.1 Phenomena Observed in a Uniform Field Normal to the Electrode

1984 Richetti [14] reported electrically induced 2-D crystallization of colloidal particles in de-ionized water. Applying 1.7 V *ac* at 1 kHz between two parallel-plate electrodes separated by 15 μm (nominal 110 kV/m), they observed hexagonal arrays of 2 μm polyvinyl toluene particles, appearing in Fig. 1.4. When the cell voltage was reduced to 0.8 V, the arrays "diffused back to the bulk of the system." Noting "the attraction between the spheres in the two-dimensional regime is much less transparent" because one expects repulsion from induced side-by-side dipoles, they recommended a "complete electrohydrodynamical treatment." At 20 V *ac* peak to peak, the ordering was lost above 5 kHz, but the particles formed columns perpendicular to the electrode at these frequencies. The equilibrium distance between particles was appreciably larger than the radius.

1993 Giersig and Mulvaney [15] electrodeposited 14 nm gold particles onto a graphite electrode by applying 50 mV *dc* between the graphite and an Al-foil counter electrode located about 2 mm away. Figure 1.5 is an image from their work. They also observed the formation of large hexagonal close-packed arrays on the electrode. Reversing the polarity of the electric field removed the deposit, indicating that particles in the 2-D array were not permanently adsorbed. A variety of thiol stabilizers were used for the Au sols [16]. The average gap between adjacent particles was directly proportional to, but somewhat less than, twice the calculated chain length of the stabilizer used for the Au sol, which suggests attraction between

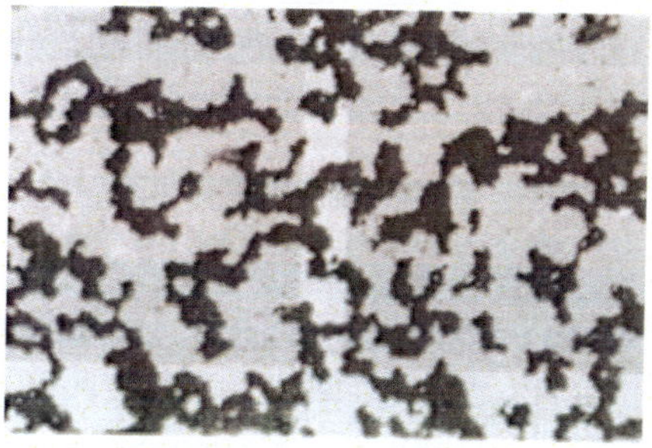

Fig. 1.4 Particles, 2 μm in diameter, aggregated on a conductive transparent electrode. Magnification 500×: approximately 100 μm in view. The particles were randomly distributed before the field was applied. The nominal imposed electric field was 100 kV/m. From Richetti et al. 1984 [14], perhaps, the earliest observation of 2-D directed assembly of colloidal particles in the plane parallel geometry. (Reprinted by permission from Richetti et al. [14], © 1984 EDP Sciences)

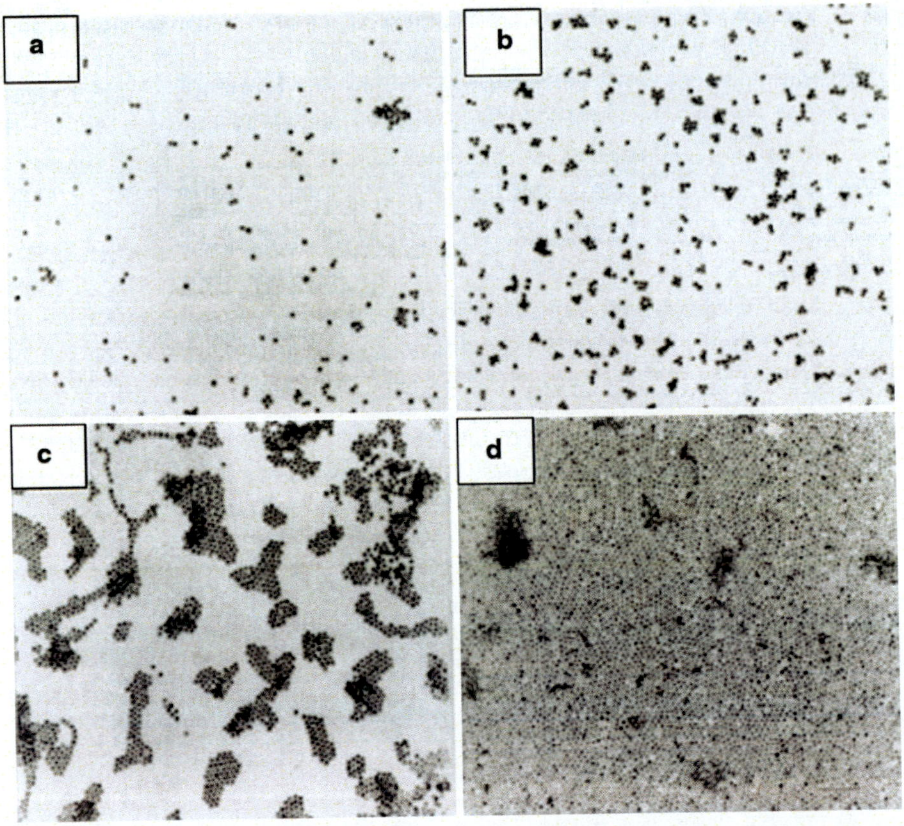

Fig. 1.5 Gold nanoparticles 14 nm in diameter aggregated on carbon coated copper microscope grids at 0.1 kV/m. for 5, 10, 15, and 35 s for photographs (**a–d**), respectively [16]. The sample in each case was removed from solution with leads attached, then dried. Trisodium citrate ion was used as a stabilizer. The total applied dc voltage was 200 mV, which was too small to pass continuous faradaic current if water splitting was occurring, but the authors performed control experiments that indicated the effect was not chemical. (Reprinted in part with permission from Giersig and Mulaney [16], © 1993 American Chemical Society)

particles despite their like charge. Although electrophoretic deposition of particles usually produces multilayer deposits, [17] the duration of the experiment might have been insufficient to saturate a monolayer, or electrostatic repulsion between the robustly stable gold particles was stronger than the repulsion between one particle and the electrode. They reported multilayer deposits when the electric field or duration of the experiment was increased.

1996 Trau et al. [18], applying approximately 10 kV/m *dc* between ITO-coated glass and a polished brass counter electrode, observed aggregation of 2 μm latex particles, Fig. 1.6. "Coagulation" was also observed with alternating current at 1 kHz and the voltage amplitude raised to approximately 20 kV/m. An electrohy-

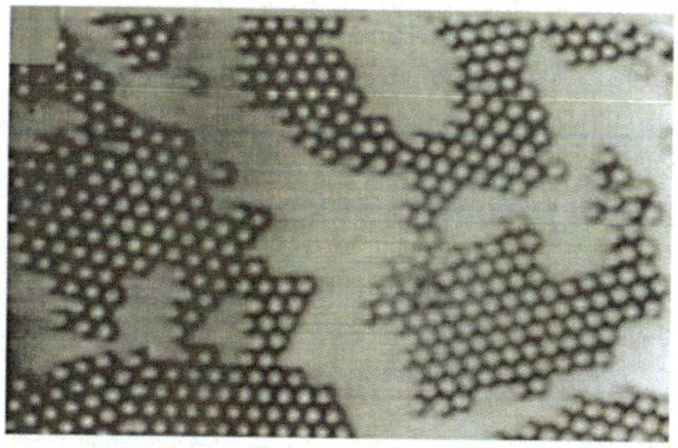

Fig. 1.6 Aggregated particles 2 μm in diameter after approximately one minute of exposure to 2 V *dc* dropping across 200 μm, for a nominal field of 10 kV/m [18]. (From Trau et al. [18]. Reprinted with permission from AAAS)

drodynamic mechanism both for *ac* and *dc* experiments, based on departure from electroneutrality *outside* the diffuse layer, was suggested.

Böhmer [1] independently reported *in situ* observations of the aggregation of 4 μm latex particles using video microscopy. A nominal field 0.5 kV/m was applied between ITO-coated glass and a Pt counter electrode about 4 mm away. Many 2-D clusters of close-packed particles were observed. Reversing the electric field drove particles apart; turning off the electric field allowed the aggregated particles to diffuse apart on a longer time scale. An image from this work, before the sample dried as in Fig. 1.1, appears in Fig. 1.7. Böhmer, after conversations with Yuri Solomentsev and John Anderson at Carnegie Mellon, suggested that bulk flow driven by electroosmotic slip along the particles mutually entrained or dispersed them [1].

1997 Yeh et al. [19] observing aggregation of 2 μm latex particles caused by a nominal 25 kV/m *ac* electric field at 1 kHz, proposed that the mechanism was electroosmotic flow generated by induced *diffuse layer* charge on the electrode. An image from their report appears in Fig. 1.8. In a separate experiment they observed steady toroidal circulation of 1 μm particles near a 20 μm bead; the steady flow was remarkable because the voltage was oscillating. The flow was toward the large bead along the surface of the electrode and away from the bead far from the electrode.

Solomentsev et al. [20] following Bohmer [1], reported experiments in which negative or positive *dc* electric fields of approximately 0.1 kV/m were imposed on ITO facing a Pt counter electrode. 10 μm spheres aggregated or dispersed when the voltage was applied and reversed. The rate of aggregation supplied data for their mechanism based on electroosmotic slip on the particles. The particles clearly re-

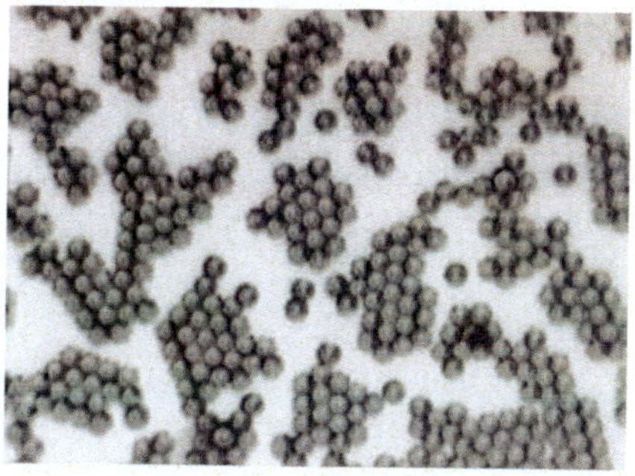

Fig. 1.7 Particles 10 μm in diameter aggregated in 0.01 mM KCl during *dc* polarization [1]. (Reprinted in part with permission from Bohmer [1], © 1996 American Chemical Society)

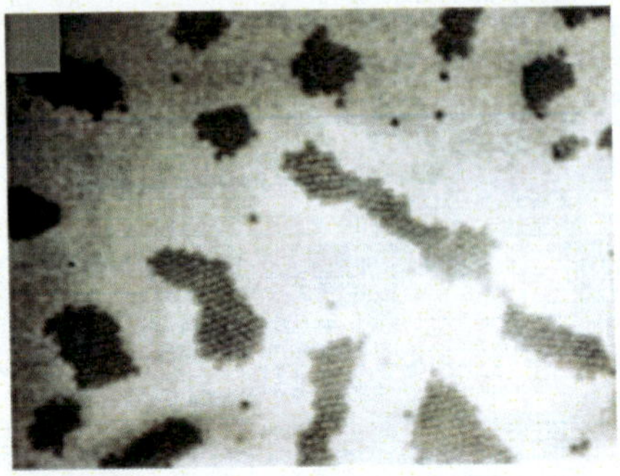

Fig. 1.8 Aggregation of carboxylated polystyrene beads 2 μm in diameter during *ac* polarization, 1.5 V and 1 kHz applied across 60 μm of sodium azide solution, for a nominal field of 25 kV/m [19]. (Reprinted by permission from Macmillan Publishers Ltd: Yeh et al. [19], © 1997)

mained mobile on the electrode during the application of the voltage; the mobility was essential to achieving a highly ordered layer. Particles can remain mobile even when a reversed voltage is applied if the equilibrium zeta potential of the electrode dominates its induced charge.

Trau et al. [21] also provided theory that showed how lateral electric field variations could drive electrohydrodynamic flow.

2000 Solomentsev et al. [22] and Guelcher et al. [23] reported observations and quantitative measurements of relative motion when *dc* electric fields were imposed on particle pairs adjacent to an ITO electrode in Na_2CO_3 and $HClO_4$ solutions. The positively charged particles were 2.5 and 10 μm in diameter. The particles aggregated when the field was negative and dispersed when the field was positive as shown in Fig. 1.9. They combined [22] the particle motion studies with particle elevation measurements in support of their model based on mutual entrainment in electroosmotic flows driven by the slip at the particles; the model agreed well with their data when lateral electrophoresis and wall hindrance were included but did not capture increased velocities at small interparticle separations.

2001 Gong et al. [24] confined polystyrene particles between closely spaced ITO electrodes. For example, particles 3 μm in diameter were confined between electrodes separated by 3.5 μm. The electrolyte was a 1:1 combination of de-ionized water and glycerine with Triton X-100 added as a stabilizer. The nominal fields were *ac* 1,000–3,000 kV/m at 100 Hz. Particles aggregated, as found by previous investigators, [14, 18, 19] when not confined but formed expanded ordered arrays, showing substantial repulsion, when the particles were closely confined. Figure 1.10, taken from a later report [25], illustrates an essential finding. Relatively modest changes in the degree of confinement (electrode separation 23% larger than the 4.2 μm particle diameter reduced to 3%, with a concomitant increase of electric field) switched the dominant interaction between particles from attractive to repulsive. They attributed the change to an increased importance of induced dipole repulsion because strict confinement inhibited electroosmotic flow. Field strength and particle radius also were important. Reducing the field strength from 3,000 kV/m to 600 kV/m at constant strict confinement and particle radius produced the same observed changes as varying confinement; the electric field strength required to order the particles depended on the inverse cube of the particle radius, in support of the induced dipole effect.

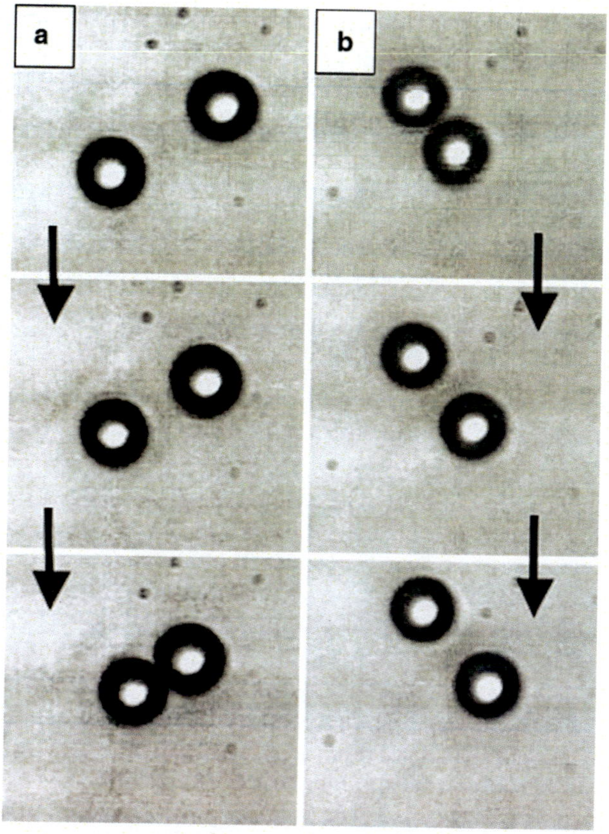

Fig 1.9 Negatively charged particles 9.7 μm in diameter near indium tin oxide electrodes first **a** aggregating during *dc* anodic polarization and then **b** moving apart during *dc* cathodic polarization [23]. The elapsed time in (a) was 30 s and in (b) was 21 s. During anodic polarization the direction of the slip velocity on the particle was away from the electrode, which drew fluid from infinity along the electrode toward the particle and entrained the neighboring particle. The flow was reversed during cathodic polarization. This figure is strong evidence for the mechanism based on equilibrium charge electroosmosis at the particle as presented by Solomentsev et al. [22]. (Reprinted from Guelcher et al. [23], © 2000, with permission from Elsevier)

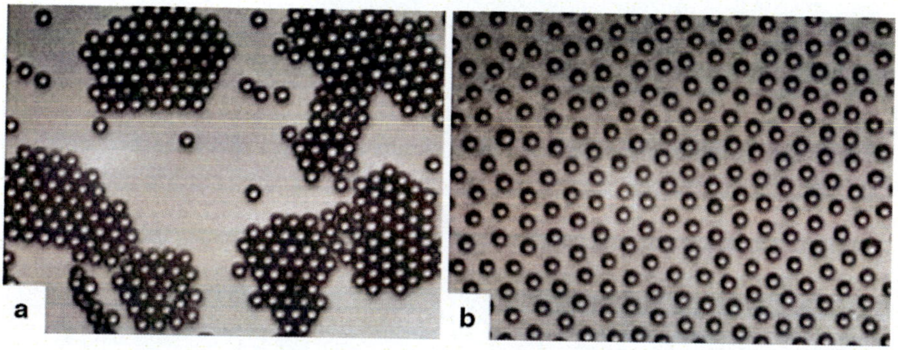

Fig. 1.10 The effect of strict confinement on the interaction between particles [25]. The particles were 3 μm in diameter aggregated at 100 Hz ac polarization. **a** Aggregated particles where the ratio of the cell gap to the particle diameter was 1.23. **b** The same cell, different location, where the same ratio of gap to particle diameter was 1.03. The particles repelled each other. The polarizing voltage was the same, so the nominal electric field was also 20% stronger. The authors concluded that close confinement weakened electroosmotic flows that normally aggregate particles. (Reprinted in part with permission from Gong et al. [25], © 2002 American Chemical Society)

Sides [26] (also in [27]) elaborating the electrohydrodynamic theory of Trau et al. [21], identified a transport parameter that predicted dependence of the direction of EHD flow on the transport properties of the ions present and pointed out that the current density under small particles might be nearly uniform because electrode kinetic limitations might override ohmic hindrance to the flow of current beneath particles near electrodes.

2002 Nadal et al. [28] observed frequency-dependent attraction or repulsion between particles of various sizes in NaOH on ITO in fields around 20 kV/m. Particles 1.5 μm in diameter formed tight clusters in NaOH at 1 kHz, but formed expanded arrays at 2 kHz. They identified a critical frequency that they argued marked the balance point between dipole repulsion and flow of an "electrohydrodynamic nature" that depended principally on particle radius. For example, they reported an "explosive" dispersion of 1.5 μm particles aggregated below the critical frequency when the frequency was raised above the critical frequency. They demonstrated this repulsion in experiments where an optical trap held particles between two electrodes, and they presented evidence of radial flow of a tracer particle toward a larger particle.

Kim et al. [29] reported measurements on particle doublets under *ac* polarization of 3 kV/m, in $NaHCO_3$. As in Nadal et al. [28] aggregation or separation of particle doublets depended on frequency in some cases. For example, 9.7 μm particles aggregated at frequencies below 500 Hz and separated at 1 kHz. This observation echoed the identification of a critical frequency. The separation rate at 1 kHz, however, was much lower than the aggregation rate at 100 and 200 Hz; the particle pair experiment did not exhibit the explosive repulsion of Nadal [28], but the electric fields Kim et al. used were substantially lower. Kim et al. [30] noted a stationary gap between the particles aggregated at lower frequencies, probably due to induced dipole repulsion. More interesting was that two 9.7 μm diameter particles immersed in NaOH *separated* at 100 Hz and 3.5 kV/m, as shown in Fig. 1.11, in clear contrast to the response of particles from the same batch under equivalent conditions in $NaHCO_3$. They attributed the reversal to the large mobility of the hydroxyl ion, which the model of Sides [26] predicted would affect the direction of motion. The aggregation rate of particles in bicarbonate at 100 Hz [30] was insensitive to the zeta potential; unmodified particles having a zeta potential of −53 mV aggregated in bicarbonate at 100 Hz at approximately the same rate as particles having a zeta potential of −15 mV after treating the original particles with the neutral triblock copolymer poly(ethylene glycol)-poly-(propylene glycol)-poly(ethylene glycol). In *dc* conditions in the same solution, the untreated particles aggregated as expected, but the treated particles with the smaller zeta potential did not aggregate, presumably because the electroosmotic slip on the particle was reduced. The same was true for the experiments in KOH. The *dc* results were affected when the particle zeta was changed from −57 to −25 mV, but the *ac* results were not affected.

Fagan et al. [31] reported the first measurements of particle motion normal to the electrode, shown in Fig. 1.12. Use of Total Internal Reflection Microscopy (TIRM) [32] resolved the nanometric motion of particles in response to the imposed electric field. The particles jigged randomly due to Brownian motion, but averaging the

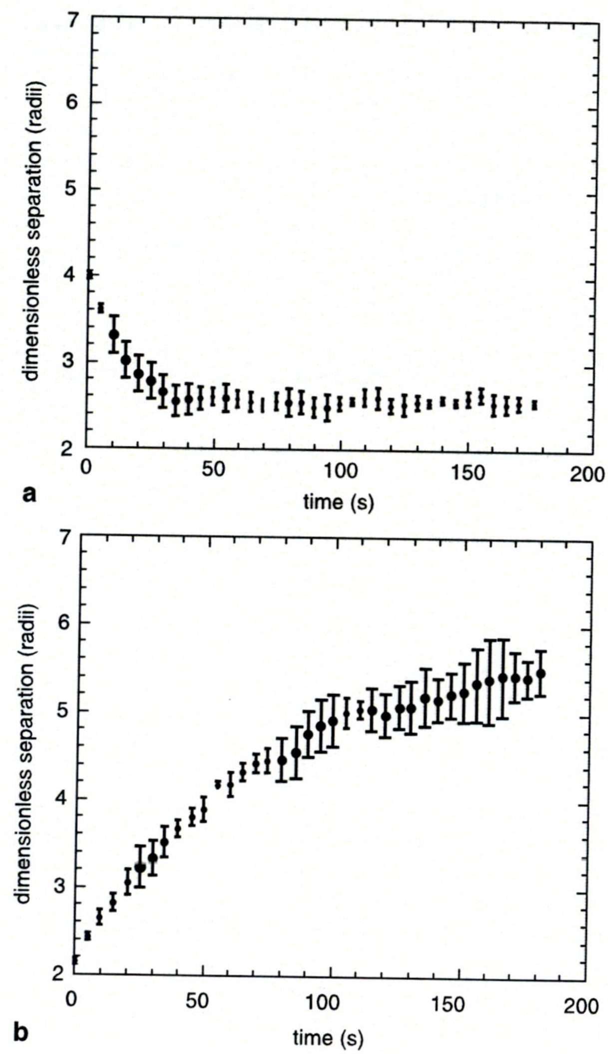

Fig. 1.11 Results from Kim et al. [30] demonstrating the effect of electrolyte in otherwise identical circumstances of ac polarization. The ordinates are dimensionless distance between two particles of a pair (normalization by radius). **a** Aggregation in bicarbonate. **b** repulsion in KOH at the same ionic strength. The frequency was 100 Hz. The critical range for comparison is dimensionless distances between 2 and 4. Particles moved somewhat faster during aggregation in bicarbonate than in KOH. Also, note the steady separation >2 radii in bicarbonate, presumed to indicate induced dipole repulsion. (Reprinted in part with permission from Kim et al. [30], © 2002 American Chemical Society)

measured particle heights revealed that individual particles moved toward or away from the electrode in a repeating deterministic fashion at the frequency of the electric field. The particle changed direction more sharply at its maximum height than it did at its minimum height, [31] which reflected increased hydrodynamic hindrance to motion as the particle approached the electrode. Particles in KOH were pulled toward the electrode during polarization, on average, while particles in $NaHCO_3$ were pushed away from the electrode during polarization. This observation marked a second difference between hydroxide and carbonate solutions, which echoed the two opposite aggregation behaviors reported by Kim et al. [30]. Fagan et al. [31] identified an unexpected phase angle between the scattered intensity and the elec-

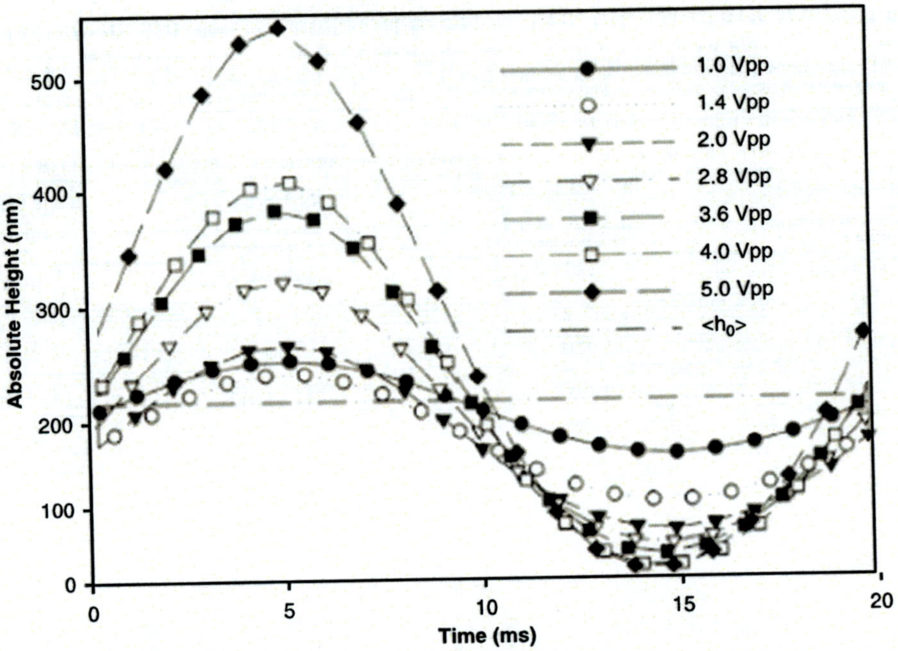

Fig. 1.12 Fagan et al. measured the absolute height of particles during ac polarization in KOH, 50 Hz [31]. The amplitude of oscillation was not symmetric during the two halves of the cycle, at least in part because the particle experienced strong hindrance to motion toward the electrode while it was being driven closer. The frequency of oscillation was the same as the driving frequency. This was the first time the detailed motion of the particles in the direction normal to the electrode was documented. (Reprinted in part with permission from Fagan et al. [31], © 2002 American Chemical Society)

tric current. One expects a 90° phase angle from simple arguments, but Fagan et al. found a phase angle different from this expected quarter cycle.

Gong et al. [25] followed their 2001 report with observations of two dimensional phase behavior demonstrated by confining particles in cells with a wedge geometry that produced different degrees of confinement in a single experiment. They demonstrated a dramatic change from closely aggregated particle clusters to an expanded order as in their previous work. In this case the nominal field varied between approximately 6,000 kV/m for the expanded ordering and 5,000 V/m for the cluster formation. They also reported wormlike ordering and honeycomb ordering in their confined geometry.

Brisson and Tilton [33] demonstrated that yeast cells aggregated in the presence of *ac* polarization, just as particles aggregated.

2004 Ristenpart et al. [34] experimented with 2.7 μm polystyrene particles in 1 mM KCl at 500 Hz and 6 V falling across an ITO counter-electrode and Pt film working electrode. Since the spacer was 0.5 mm, the nominal electric field amplitude was 12 kV/m. The particles, randomly distributed on the Pt electrode before the experiment, aggregated into clusters after 30 s. They interpreted the results with a mass action model of particle aggregation on the electrode, and a scaling law deduced on the basis of induced charge on the electrode, as discussed later in this contribution.

Liu [35] reported on crystallization/nucleation experiments with 1 μm particles in deionized water contained between ITO electrodes 120 μm apart. Peaks in the radial distribution function corresponded to ordering as the nominal field strength rose from 7.5 to 2,300 kV/m at a frequency of 600 Hz. The position of the first peak was 20% larger than the particle radius. Frequency variation from 450 Hz to 1.5 kHz did not change the radial distribution function significantly.

Fagan et al. [36], continuing TIRM measurements, reported that the unexpected deviation of a phase angle from 90° during ac polarization, previously reported as being *less* than 90° [31] in KOH, also appeared in bicarbonate but was *greater* than 90°.

2005 Fagan et al. [37] demonstrated a variety of effects of *ac* polarization on the behavior of individual particles. 6.2 μm particles displayed similar stochastic behavior at zero frequency and 2 kHz, while the particle motion at 400 Hz was clearly a mixture of deterministic and stochastic motion. Fagan et al. took advantage of this to investigate the influence of the imposed electric field at 10 kHz; he obtained the potential energy profiles from the distributions of elevations sampled by Brownian motion [32]. Results from these experiments appear in Fig. 1.13. The analysis of the potential energy curves of particles in KOH, HNO_3, and $NaHCO_3$ at 10 kHz revealed: (1) Particles in KOH experienced an increased downward force as shown in the increase slope of the potential energy profile, which agreed with the tendency for particles in KOH to be drawn closer to the electrode observed at 100 Hz; (2) The same particles in $NaHCO_3$ experienced an upward force at 10 kHz; (3) Particles in HNO_3 experienced an upward force at 10 kHz, which agreed with a change from downward force to upward force above 300 Hz found earlier.

2006 Liu et al. [38] generated radial distribution functions to study the closeness of the packing in ordered arrays as a function of frequency at constant electric field strength. Liu et al. identified upper and lower critical frequencies where liquid-like to crystalline and crystalline to liquid-like transitions occurred. For a fixed field strength of 15 kV/m, a crystal to liquid-like transition was observed above 500 Hz for 5 μm particles, with some variation introduced by salt concentration. The range of crystalline order was several kHz for 3 μm diameter particles. Fagan et al. [39] proposed a mechanism of rectified motion that depended on differing hindrance to lateral motion as a function of particle height above the electrode. This mechanism is discussed in detail later.

Zhou et al. [40] observed complicated behavior of mixtures of particles and yeast cells, which they explained with a mixture of mechanisms including electroosmosis and dielectrophoresis. Figure 1.14 shows microphase separation of particles into a halo around yeast cells. Figure 1.15 shows a bizarre streaming of cells and particles between two agglomerations.

Santana-Solano et al. [41] reported the rotation of particles during directed assembly of spheres in the 0.1–0.5 kHz frequency range of *ac* ECEO in the geometry of Fig. 1.3a. Particles, separated initially by several radii, began rotating on an axis

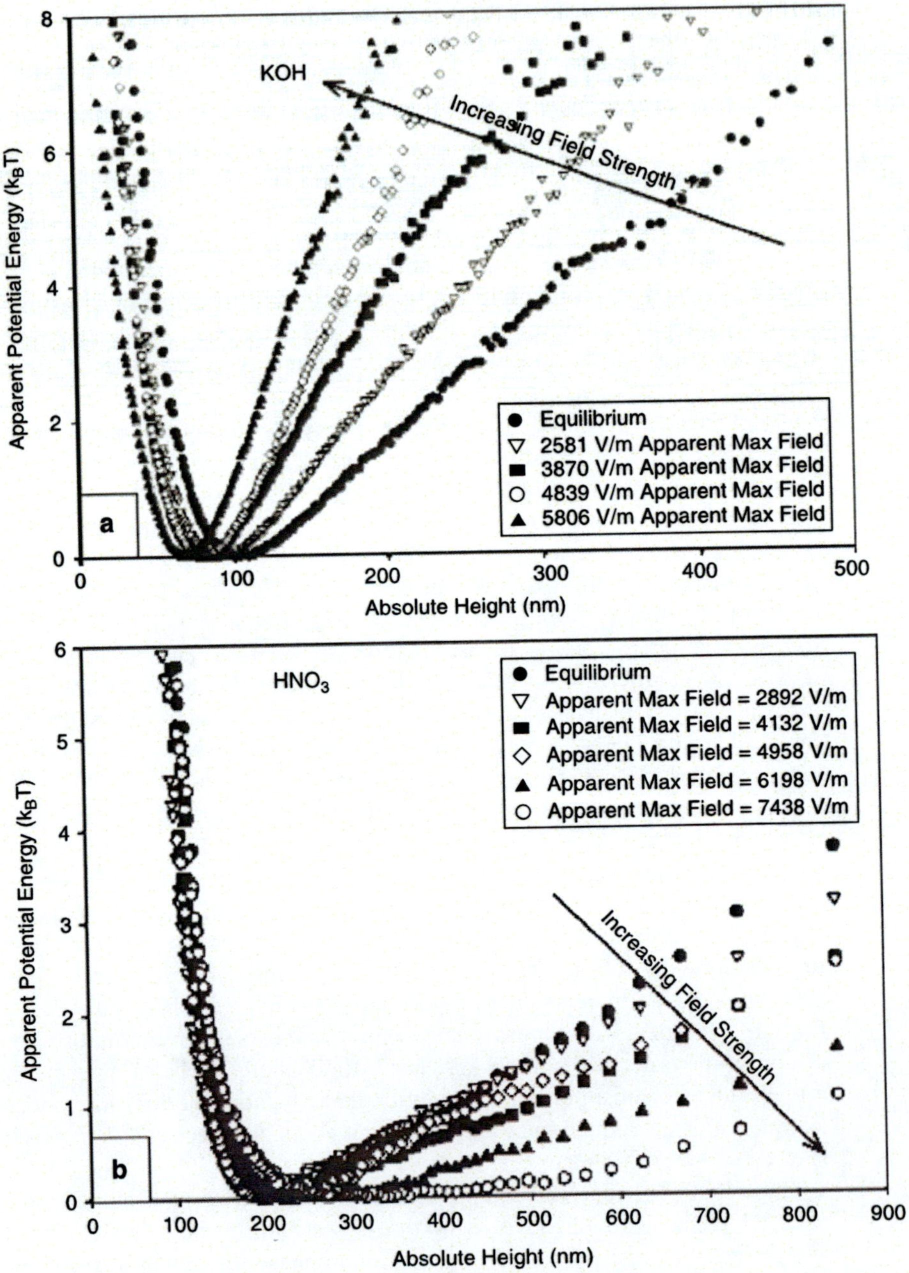

Fig 1.13 [37] Potential energy curves deduced at frequencies sufficiently high (and amplitude of oscillation sufficiently low) that usual TIRM methods [32] could be used. During polarization at 10 kHz, particles in KOH appeared heavier as the electric field strength increased, while particles in HNO_3 appeared more buoyant. Fagan et al. thereby demonstrated that the effect of differing ions, which leads to attraction or repulsion at lower frequencies, appears also at high frequencies. (Reprinted in part with permission from Fagan et al. [37], © 2005 American Chemical Society)

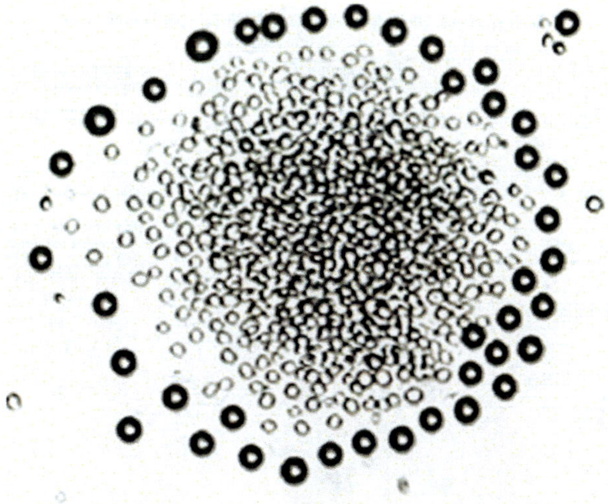

Fig. 1.14 Demonstration of the separation of yeast (6 μm, −24 mV) from polystyrene particles, (9.6 μm, −43 mV) by 20 min at 60 Hz and 1 kV/m followed by 5 min at 5 Hz and 1 kV/m. The PS particles rose above the yeast and migrated outward to create a halo of PS particles around the yeast cells [40]. (Reprinted from Zhou et al. [40], © 2006 with permission from Elsevier)

in a manner consistent with aggregative EHD flow that pushes them toward each other. If viewing a pair of nearby radii from the side, the particle on the left rotated clockwise and the particle on the right rotated counterclockwise as they translated toward each other. The particles translated toward each other much more slowly than coupling the rotation and translation rates would predict [41]. When the particles met, they continued rotating. When a cluster had formed, the particles at the periphery rotated in place but the particles at the center of the cluster ceased rotating. The frequency of rotation was proportional to the square of the amplitude of the applied potential.

2007 Ristenpart et al. [42] followed tracer particles (300 nm) moving in the vicinity of a 50 μm particle in order to probe angularly symmetric flow in the larger particle's vicinity. The nominal applied fields were 16, 24, and 36 kV/m, and the frequency was 750 Hz. Experiments at 24 kV/m and frequencies of 500, 1,000, and 1,500 Hz also were performed. The tracer particles moved along the electrode toward the target particle, disappeared under it, and reappeared briefly when the focal plane was moved from near the electrode toward the upper surface of the particle. Ristenpart et al. [43] also attempted to account for *dc* effects by modeling the electroosmotic flow resulting from induced charge on the electrode.

Hoggard et al. [44] experimentally explored Fagan's hypothesis that the phase angle between the current and the particle height could be used to predict the direction of pairwise particle motion. They tracked particle motion and electric current through the cell at a frequency of 100 Hz and nominal electric field 1,800 V/m. The phase angle between the current and the TIRM signal discussed previously was a reliable predictor for pairwise aggregation or separation of particles for a variety of electrode and electrolyte combinations.

Following Ristenpart's model [35], Liu et al. [45] determined aggregation rate constants and equilibrium particle separations. The aggregation coefficient in-

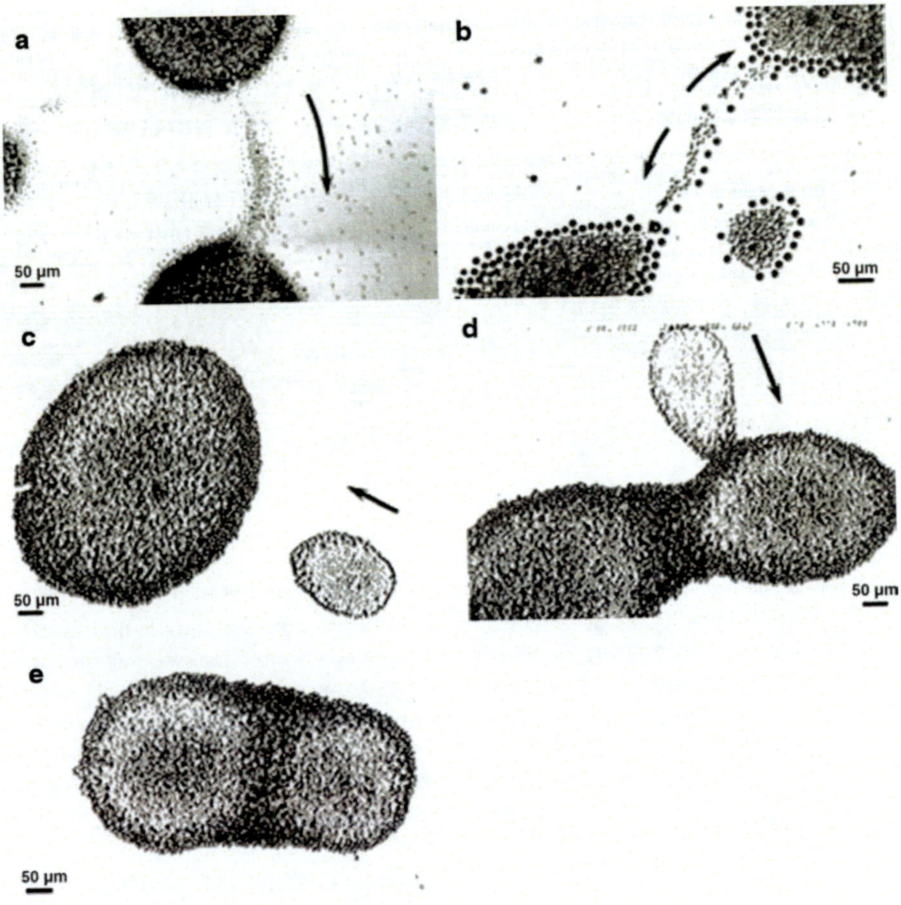

Fig. 1.15 Phenomena observed during polarization of yeast cell/PS particle mixtures (see Fig. 1.14) at 5 Hz and 1 kV/m. Clusters transfer yeast cells one to the other, move toward each other, and merge [40]. (Reprinted from Zhou et al. [40], © 2006 with permission from Elsevier)

creased with frequency to about 400 Hz and then decreased. The equilibrium interparticle separation varied from a few percent to 20% of the particle diameter depending on frequency and particle size. Zhang and Liu [46] reported on the effect of particle size on the frequency window within which particles can be ordered. Liu et al. [47] also explored the effect of temperature; higher temperatures expanded the frequency range over which ordering occurred. Different combinations of temperature, field strength, and frequency allowed fine tuning of the interparticle separation from near zero to over one particle diameter, as shown in Fig. 1.16. The three articles by Liu and co-workers provided useful information on the effects of the main operating parameters: field, particle size, frequency, and temperature.

2008 Hoggard et al. [48] investigated the pertinence of pairwise particle behavior to the behavior of particle ensembles. Prior experiments had established that in

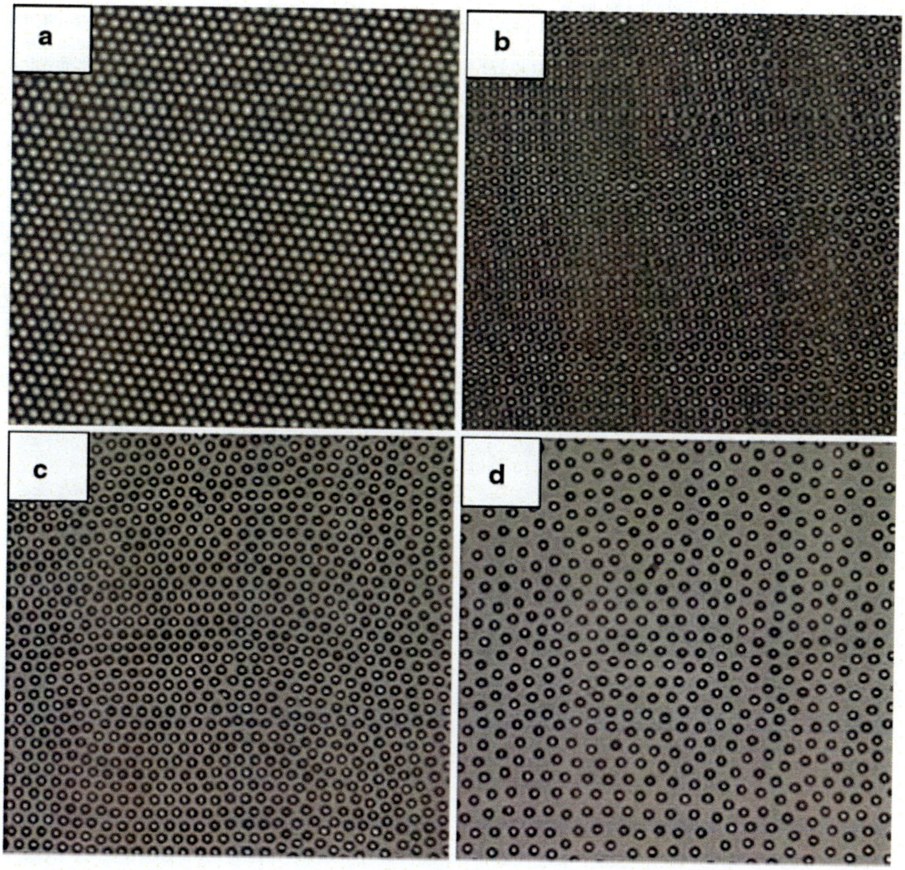

Fig. 1.16 Demonstration of tuning the spacing of particles (polystyrene, 3 μm, −86 mV) in ordered arrays by increasing the electric field strength from **a** 58 kV/m to **d** 83 kV/m while holding the frequency and temperature constant at 278 K and 40 Hz, respectively [47]. (Reprinted with permission from Liu et al. [47], © 2007 American Institute of Physics)

KOH at 100 Hz, particle pairs separated, while in $NaHCO_3$ particle pairs aggregated. These tendencies were observed in ensembles of particles also (10 kV/m). An ensemble of particles in bicarbonate formed ordered arrays where the separation between particles was less than the radius of the particles while ensembles of particles in KOH formed ordered arrays with particle spacings much greater than the particle radii, indicating repulsion-dominated ordering. These effects appear in Fig. 1.17. Hoggard et al. also observed interesting behavior in KOH when a large anomalous, but reproducible, voltage spike was present in the system due to equipment malfunction. Particles clustered in bicarbonate as before, but much more quickly. Concentrated initial ensembles of particles in KOH, however, were driven apart and formed 3d particle "worms" extending out from the electrode. The worms decomposed when the power was terminated. This observation echoed Richetti's

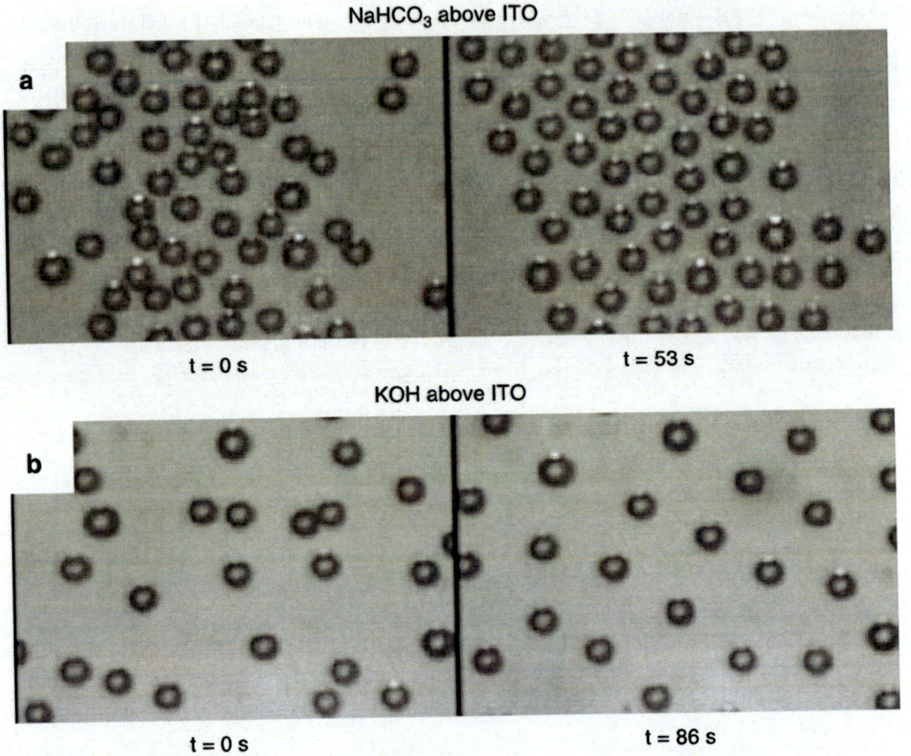

Fig. 1.17 Demonstration that the same particles at the same frequency and same applied voltage behave oppositely in KOH and bicarbonate solution, **a** aggregating in bicarbonate and **b** repelling each other in KOH [48]. (Reprinted in part with permission from Hoggard et al. [48], © 2008 American Chemical Society)

original observation [1] of formations perpendicular to the plane of the electrode. Hoggard et al. [48] also observed that large particles shepherded smaller particles, as shown in Fig. 1.18, drawing them into their vicinity when the power was turned on and releasing them when the power was off. Xie and Liu [49] used a line of colloidal particles as a template from which to aggregate epitaxially and order particles in an out-of-plane geometry. After the particles were initially aggregated, the fixed line of colloidal particles helped anneal the assembly.

2010 Yariv [50], using an asymptotic analysis taking advantage of the three different length scales in the problem shown in Fig 1.3a (Debye length, gap between the particle and the electrode, and particle radius), compared the strength of the electroosmotic flow driven by slip on the particle to the strength of the electroosmotic flow driven by slip at an adjacent electrode. For equal zeta potentials, electroosmotic slip on the particle dominated the flow in the liquid in the vicinity of the particle by approximately an order of magnitude, which justifies Solomentsev's neglect of the flow due to the electrode [20].

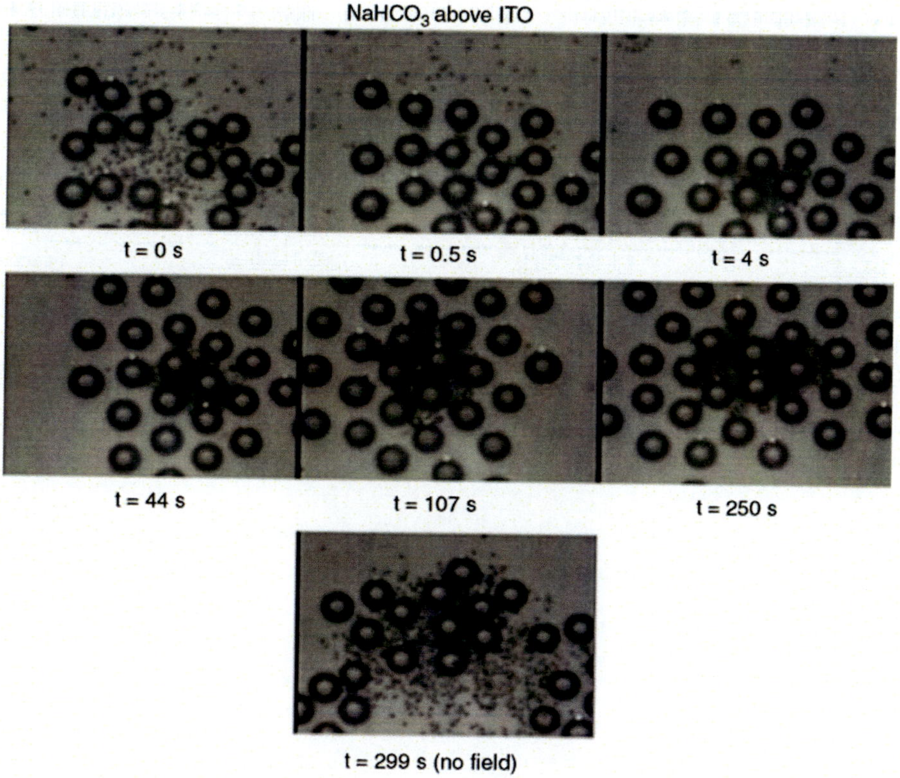

Fig. 1.18 Demonstration of the gathering of small particles (1 μm) by large particles (10 μm) in bicarbonate solution. Polarization at 100 Hz with a compound oscillating voltage [48]. (Reprinted in part with permission from Hoggard et al. [48], © 2008 American Chemical Society)

1.2.2 Summary

Experiments in labs around the globe have established that electric fields, applied in the normal direction to particles levitated very near one of two plane parallel electrodes, cause particles to move laterally and either form arrays with gaps between them smaller than their diameters or form ordered "repulsive" arrays with gaps between them several diameters wide. The effect is observed in both *ac* and *dc* polarization. The mechanisms described in Sect. 1.1.1 have all been invoked to explain the effects. Two mechanisms are attractive and relatively simple. In *dc*, the mutual entrainment in flow arising from electroosmosis on the particle captures the salient effects. In *ac*, imbalance between the aggregation driven by electroosmosis of induced charge on the electrode and induced dipole repulsion plausibly leads to either mutual entrainment or net repulsion. One problem is that neither of these two effects accounts for the different responses in bicarbonate and hydroxide systems. Furthermore, other paradoxes arise from this model, but these issues are best discussed in the context of explicit descriptions of the models based on the mechanisms, which is the purpose of the next four sections.

1.3 Electrophoresis

1.3.1 Electrophoretic Velocity

Electrophoresis is the motion of charged particles in an electric field. Equation 1.1 gives the particle velocity v_p for particles having zeta potential ζ_p when the ratio of the particle radius to the Debye length (κ^{-1}) is infinite.

$$v_p = \frac{\varepsilon \zeta_p E_\infty}{\eta} \quad \text{for} \quad \kappa a \to \infty. \tag{1.1}$$

Here ε is the dielectric permittivity of the electrolyte, η is the viscosity, and E_∞ is the uniform electric field as in Fig. 1.3a. If the ratio of the particle radius to the Debye length is zero, the result is

$$v_p = \frac{2\varepsilon \zeta_p E_\infty}{3\eta} \quad \text{for} \quad \kappa a \to 0. \tag{1.2}$$

When the ratio of the particle radius to the Debye length is not infinitely small or infinitely large, the surface conductivity of the diffuse layer and transport of charge in an out of the diffuse layer complicate the calculations. O'Brien and White [51] treated these complications; Khair and Squires [52] recently showed that plotting the electrophoretic mobility against dimensionless surface conductivity collapsed results at large κa onto a single curve. For present purposes we refer the reader to these treatments and reference works, [11, 12] with the exception of the following perspective.

1.3.2 Electrophoretic "Force"

Effects in addition to those mentioned above further complicate the electrophoresis of particles near electrodes. Examining mechanisms of particle motion near electrodes up to the point of deposition, we must consider electrophoresis when particles are near electrodes. When far from electrodes, particles are force-free and move by electroosmotic slip in the direction of the potential gradient. If the particle moves by electrophoresis parallel or perpendicular to the electrode, wall hindrance retards its motion. For example, Solomentsev et al. [22] included hindrance of the motion of particles moving by electrophoresis parallel to an electrode. When near an electrode, however, particles moving normal to it encounter a suite of forces including double layer electrostatic attraction or repulsion and van der Waals attraction; if the particle is sedimenting, gravity is also present. The following treatment provides some guidance in the rendering of the electrophoretic contribution in this situation. Consider a dilute ensemble of particles near an electrode but not being deposited on it. The flux statement for particles is

$$J = 0 = -q_{NE} D_\infty \frac{dC}{dz} + [q_{NE}(h) v_{NE}(h) + q_{EP}(h) v_{EP}]C, \tag{1.3}$$

where J is the particle flux to the electrode (zero for non-depositing particles), EP denotes electrophoretic velocity and NE denotes velocity arising from all other forces on the particle. D_∞ is the Einstein diffusion coefficient and C is the concentration of particles. The quantities q_i are hindrance factors that are known functions of h, the gap between the particle and the electrode, in the absence of slip on the electrode. The hindrance coefficient is different for electrophoretic motion because of the slip boundary condition on the particle. Rewriting Eq. 1.3 as

$$\frac{d \ln C}{dz} = \frac{q_{NE}(h) v_{NE}(h) + q_{EP}(h) v_{EP}}{q_{NE} D_\infty} = \frac{F(h)}{kT} + \frac{q_{EP}(h)}{q_{NE}(h)} \frac{v_{EP}}{D_\infty}, \tag{1.4}$$

where $F(h)$ is the force due to non-electrophoretic effects, one establishes correspondence between other forces and electrophoresis, which yields an expression appropriate for the electrophoretic "force."

$$F_{EP} \equiv \frac{q_{EP}(h_m)}{q_{NE}(h_m)} \frac{kT}{D_\infty} v_{EP} = q(h) 6\pi a \varepsilon \zeta_p E_\infty \quad \text{where } q(h) \cong \frac{h + 1.544}{h + 0.3}. \tag{1.5}$$

Solomentsev et al. [20] provided the approximation in Eq. 1.5 for the ratio of the hindrance experienced by a particle moving by electrophoresis toward an electrode [53] to the hindrance experienced by a particle with zero zeta potential; [54] $q(h)$ becomes constant at zero gap h and infinitely thin diffuse layers. Equation 1.5 is useful in force balances on mobile particles near electrodes.

1.3.3 Electrophoresis and Nanoparticles

For the electroosmotic flows to cause assembly by the mechanisms discussed here, the particles must reside in the same plane parallel to the electrode face. In particular their centers should be about one particle radius away from the electrode. For particles several microns in diameter, that can be achieved by gravitational sedimentation before the electroosmosis begins. In this section, we consider achieving this goal for much smaller particles using electorphoresis to drive the particles to the surface. Writing Eq. 1.3 for only diffusion of particles and electrophoretic flux, one obtains

$$J = 0 = -D_\infty \frac{dC}{dz} + C\frac{\varepsilon\zeta}{\eta}E_\infty. \tag{1.6}$$

Continuity in 1-D requires this flux to be independent of z. Scaling the equation to the particle radius, in order to reflect the degree of confinement to the electrode, and using the Einstein diffusivity of the particle, one deduces a characteristic electric field.

$$E_\infty = \frac{kT}{6\pi\varepsilon\zeta_p a^2} \tag{1.7}$$

Thus, if particles do not sediment and they bear the same sign of charge as the electrode, one must deliver them to the surface with an electric field that depends on the inverse square of the radius. For $\zeta = -100$ mV, and $a = 10$ nm, the required electric field is 30 kV/m.

1.4 Electrohydrodynamics

The term "electrohydrodynamics" is often used in the context of directed particle assembly. For the purposes of this review, we restrict the definition of *electrohydrodynamics* (*EHD*) to the fluid motion that ensues when electric fields interact with small charge imbalances in the liquid *outside* the electric double layer. This definition highlights the distinction between this bulk effect and electroosmosis, an electrokinetic effect. When an electric field causes ions to move in an otherwise electroneutral solution, unbalanced charge appears in regions of concentration gradients. The electric field acts on this unbalanced charge outside of the region normally associated with the electric double layer. Thus, we distinguish electrohydrodynamic flow from electroosmosis because the Smoluchowski equation does not describe EHD. Describing electrohydrodynamics theoretically means solving equations governing the transport of ions simultaneously with the electrostatic and hydrodynamic aspects of the problem.

1.4.1 Theory of EHD

The starting points for EHD flow, as for many of the mechanisms described in this contribution, are Eqs. 1.8–1.9 representing the flux of species i and Poisson's equation.

$$\mathbf{N}_i = -D_i \nabla c_i - \frac{ez_i D_i}{kT} c_i \nabla \phi \tag{1.8}$$

$$-\varepsilon\nabla^2\phi = e\sum_i \nu_i c_i \tag{1.9}$$

If the fluxes of all species are specified, Eqs. 1.8 and 1.9 provide a sufficient number of relationships to close the problem. Material balances given by

$$\frac{\partial c_i}{\partial t} = -\nabla\cdot\mathbf{N}_i \tag{1.10}$$

provide the extra relationships when the flux of reactant or current is not specified *a priori.* Typically one solves Eqs. 1.8–1.10 or a subset thereof to obtain the electric field as a function of time and position and then solves the fluid dynamic equations

$$0 = -\nabla P + \eta\nabla^2\mathbf{v} + \rho_e\mathbf{E}, \tag{1.11}$$

and

$$\nabla\cdot\mathbf{v} = 0 \tag{1.12}$$

for the flow pattern, where η is the liquid viscosity and P denotes pressure. Inertia is unimportant for the viscous flow typical in this context. In the case of EHD, the boundary conditions are taken as no-slip at all solid/liquid interfaces.

When a varying potential is specified, the problem must include the material balance Eq. 1.10. The simplest version of this approach takes a binary electrolyte with one reacting ion as the basis. The first thrust is calculation of the electrostatic potential and the ion concentrations simultaneously with cognizance of continuity and electroneutrality at every point [55, 56]. For a binary equi-valent electrolyte, one replaces Poisson's equation by electroneutrality and inserts flux statements for the two ions into their respective continuity equations, which allows elimination of the electric potential from the ion continuity equations and leaves

$$\frac{\partial c}{\partial t} + \mathbf{v}\cdot\nabla c - D\nabla^2 c = 0 \quad \text{where } D \equiv \frac{2D_+D_-}{D_+ + D_-}, \tag{1.13}$$

$c=c_+=c_-$ is the local number density of cations or anions, and $D_\pm$ are the ion diffusion coefficients. Boundary conditions such as specified flux or concentrations constrain the solution to Eq. 1.13. Equation 1.14 governs the relationship between the concentration profile and the electric field.

$$\nabla\cdot(c\mathbf{E}) = \frac{kT}{ze}\frac{D_+ - D_-}{D_+ + D_-}\nabla^2 c \tag{1.14}$$

Poisson's equation is applied to calculate the space charge density ρ. The induced space charge is given by [55]

$$\frac{\rho_e}{zec_\infty} = 2\kappa^{-2}\left[\frac{D_+ - D_-}{D_+ + D_-}\nabla^2\ln c - \frac{\mathbf{i}\cdot\nabla\ln c}{(D_+ + D_-)zec}\right] \tag{1.15}$$

where

$$\mathbf{i} = -ze(D_+ - D_-)\nabla c + (D_+ + D_-)\frac{(ze)^2}{kT}c\mathbf{E} \tag{1.16}$$

is the local current density. Equation 1.15 reveals that the ratio of the unbalanced charge density to the ion concentration roughly equals the squared ratio of the Debye length to a length scale characteristic of the region over which the concentration changes from its bulk value to its value just outside the diffuse layer. Furthermore, the coefficient multiplying the Laplacian reveals the dependence of the phenomenon on the asymmetry between the mobilities of the anion and the cation, identified by Sides [26, 27] as discussed below. If the Debye length is 10 nm and the diffusion layer thickness is 10 μm, for example, the ratio is 1 ppm. Nonetheless, the bulk fluid indeed contains unbalanced charge and EHD at some level exists as a consequence of ionic concentration gradients associated with faradaic reactions in the electrochemical cell. The unknowns are the fluid velocity, ion concentration, electric field, electric current, and charge density. The desired result is the local unbalanced charge ρ_e which is then used in Eqs. 1.11 and 1.12 to find the fluid velocity.

1.4.2 Comparison of Theory and Experiment

Richetti [14] speculated that a net positive cloud of charge between the particles could cause attractive force. The above "electrohydrodynamical treatment" was not available at the time. Trying to understand Richetti's observations and their own data, and following up the suggestion that electrohydrodynamics might play a role, Trau et al. [21] demonstrated EHD flow near an electrode. Using a perturbation method and a stream function, they solved Eqs. 1.8–1.10, and 1.12 by specifying a spatially varying current on an electrode; a corresponding system of flow cells with toroidal streamlines in each cell arose from the relationships. They argued that particles would be entrained in these flow cells and would aggregate, but there was no quantitative comparison to theory.

Sides [26, 27] used a simplified form of the model represented by Eqs. 1.13–1.16 to calculate the EHD around an isolated dielectric sphere where faradaic current and phase relationships introduced by double layer charging were included. A sinusoidally varying voltage was applied to an electrochemical system with linear electrode kinetics. Sides introduced the logic governing the distribution of current beneath spherical particles near electrodes, showing that the current is distributed uniformly, as opposed to the potential, for small spheres. Equation 1.14 becomes Laplace's equation for vanishing concentration gradients. Sides solved for the potential distribution in this limit. Although Laplace's equation does not apply in regions where concentration gradients are nonzero as in EHD, the assumption of small departures from the bulk concentration mitigated this error. In addition to exhibiting the expected rectification of flow, the analysis of Sides [27, 26] agreed with some experimental observations. The theory revealed that the sign of the coefficient of the Laplacian in Eq. 1.15, a transport parameter called out in Eq. 1.17, governed whether particle pairs aggregate or separate. Here t_i is the transference number of a given ion. The transference number indicates the fraction of current carried by that ion in regions of uniform concentration.

$$\hat{t} \equiv \frac{t_+}{z_+} + \frac{t_-}{z_-} \quad \text{where } t_+ \equiv \frac{z_+ D_+}{z_+ D_+ - z_- D_-},\; t_- \equiv \frac{z_- D_-}{z_+ D_+ - z_- D_-} \tag{1.17}$$

The sign of the parameter depends on the relative individual ionic diffusion coefficients of the two ions involved; it is positive for cases where the cation is more mobile than the anion and negative otherwise. Likewise, the EHD theory predicted the existence of a critical frequency above which the sign of the expected particle motion changed. Realizing that differences of ion mobility should cause particles in hydroxide solution to repel each other, Sides suggested that a co-worker try experiments in potassium hydroxide solution. Experiments where alternating current was passed between two plane parallel electrodes demonstrated that the pairwise interaction of particles in KOH was repulsive (see Fig. 1.11), while the interaction of particles from the same batch in bicarbonate was attractive [30].

The EHD theory also predicted the existence of a critical frequency above which the sign of the attraction changed, which agreed with experimental findings of Nadal et al. [28]. The critical frequency was proportional to the reciprocal of the relaxation time for transport of the neutral salt on the length scale of the particle, which echoed the original comments of Richetti et al. [14] regarding the importance of the mobility of protons. The observations of Liu et al. [46] likewise showed that higher temperature raised the critical frequency, in agreement with this result, because ion diffusivity is proportional to temperature directly and exponentially proportional to temperature through the liquid viscosity. The pairwise particle attraction in KCl, [34] which has nearly equal ionic mobilities, revealed limitations of the theory; even in cases where aggregation was observed, the EHD theory predicted velocities lower than observed experimentally. Later, Fagan et al. [36] determined that EHD was unlikely to exert sufficient force to move particles as strongly as observed experimentally. While the predictions of the theory are elegant and in part found in experiment, the low unbalanced charge density produced outside the diffuse layer and the fact that particles aggregate in KCl calls the importance of the EHD mechanism (as defined here *not* including electroosmosis) into question.

1.5 Equilibrium Charge Electroosmosis (ECEO)

The flow that ensues when an imposed electric field interacts with mobile unbalanced equilibrium charge adjacent to particles or electrodes is equilibrium charge electroosmosis (ECEO). Charge generally exists at solid liquid interfaces, whether the liquid is polar or nonpolar, due to differential adsorption or ionization in the absence of current. Even though the interface is electroneutral, charge is separated between immobile charge on the electrode and mobile ions in the liquid. Unlike solution outside the diffuse layer where the space charge density is usually negligible, the concentration of space charge within the diffuse layer is on the same scale as the ionic strength. An imposed electric field exerts a force on the mobile charged species, which drags the liquid along with them. The net effect is a slip velocity at the solid given by Smoluchowski's formula Eq. 1.18, where ε is the dielectric permittivity, ζ is the zeta potential of the surface (whether electrode or particle), η is viscosity of the liquid, φ is electric potential, and the subscript s denotes the surface gradient.

$$\mathbf{v}_{slip} = \frac{\varepsilon\zeta}{\eta}\nabla_s\phi \tag{1.18}$$

The imposed field can either be constant or alternating. Figure 1.3a illustrates a system, in the context of this contribution, where electroosmotic slip drives bulk flow near particles and electrodes. The overall electric field is imposed normally to the electrode, but it engenders electroosmotic slip along both the particle and the electrode because both have equilibrium mobile diffuse-layer charge adjacent to them. The electroosmotic slip drives bulk flow that exerts viscous stress on any neighboring particle, leading to mutual entrainment in the plane of the electrode.

1.5.1 dc ECEO

1.5.1.1 Theory

Calculation of the flow and resulting particle velocities require solving a series of problems. First, one needs the electric field in the immediate vicinity of the particle. Neglecting ionic concentration gradients everywhere for simplicity, one solves Laplace's equation in the electrolyte surrounding a dielectric particle such as the particle shown in Fig. 1.3a.

$$\nabla^2\phi = 0 \tag{1.19}$$

The boundary condition far from the particle is a linearly varying potential proportional to a constant uniform electric field and the boundary condition on the particle surface is zero radial component of the electric field. The electrode boundary condition requires some discussion because the distribution of electric current in the vicinity of the particle depends on a trade-off between the ohmic disturbance of potential caused by the particle and the kinetics of the electrode reaction. The extreme cases are constant current at the electrode or constant potential [55]. In the case represented in Fig. 1.3a, one applies the following condition at the electrode when only charge transfer kinetics limits the reaction rate.

$$-\sigma_e \nabla\phi|_{z=0}\cdot\mathbf{n} = \mathbf{i}\cdot\mathbf{n} = i_o\left(e^{\frac{\alpha_a F\eta_s}{RT}} - e^{\frac{-\alpha_c F\eta_s}{RT}}\right), \quad \eta_s \equiv \phi_{WE} - \phi|_{z=0} \tag{1.20}$$

Here σ_e is the electrolyte conductivity, i_o is a parameter called the exchange current density that is characteristic of the intrinsic reaction rate, α_i are symmetry coefficients that assign amounts of the surface overpotential η_s to the anodic and cathodic partial currents. φ_{WE} indicates the potential of the working electrode, φ at $z=0$ is the potential in solution just outside the diffuse layer, and $\mathbf{n}$ denotes the unit vector normal to the electrode.

In the case of Fig. 1.3a, scaling of the problem reveals that the smaller the particle, the more uniform the current distribution [26]. Newman [55] identified two relevant dimensionless groups, one a dimensionless reaction rate coefficient and the other a dimensionless reaction rate.

$$J \equiv \frac{\alpha i_o Fa}{\sigma_e RT}, \qquad \delta \equiv \frac{\alpha \left| i_{avg} \right| Fa}{\sigma_e RT} \tag{1.21}$$

(Two are necessary because of the nonlinearity of the Butler Volmer equation.) J gives the ratio of the ohmic impedance to the intrinsic impedance of the charge transfer reaction for a particle of radius a, while δ is the same ratio for the reaction at the actual average current density. When current densities are small with respect to the exchange current density i_o, J governs the distribution, otherwise δ does. Low values of the parameters mean that the electrolyte is sufficiently conductive or that the reactions are sufficiently slow that current flows into the interstices between the particles and the electrode to make the current density uniform. For example, the value of J is less than 0.1 for a 1 μm radius, an exchange current density of 10 A/m^2 (reasonably fast), and a kinetic symmetry coefficient α of unity. The current density under the particle likely is uniform, which means the potential is nonuniform, and lateral gradients of potential along the electrode attain their maximum value; in this case electroosmotic slip exists at both the particle surface and the electrode. This conclusion might change for much larger particles.

The solution of Eq. 1.19, subject to its boundary conditions, gives the electric potential at all locations. Hence, the electric field adjacent to the electrode and the particle can be determined and supplied to Eq. 1.18, which provides a slip boundary condition at the particle and/or the electrode to the Stokes equations

$$0 = -\nabla P + \eta \nabla^2 \mathbf{v}, \quad \nabla \cdot \mathbf{v} = 0. \tag{1.22}$$

Solving Eq. 1.22 with appropriate boundary conditions on the particle and electrode gives the flow field in the vicinity of the particle. Figure 1.19 shows a calculation of the flow field around a single particle near an electrode in an unbounded space under various assumptions about the zeta potentials of the particle and electrode. Figure 1.19a is the case where the particle zeta potential is nonzero and the electrode zeta potential is zero. Figure 1.19b shows the other case where flow is only due to ECEO on the electrode. Figure 1.19a and b show that the flow intensity due to ECEO on the particle is distributed away from the particle while the ECEO flow intensity based on the electrode's equilibrium charge is concentrated under the particle. The flow is an order of magnitude stronger for ECEO based on the particle. (In Fig. 1.19a and b the val-

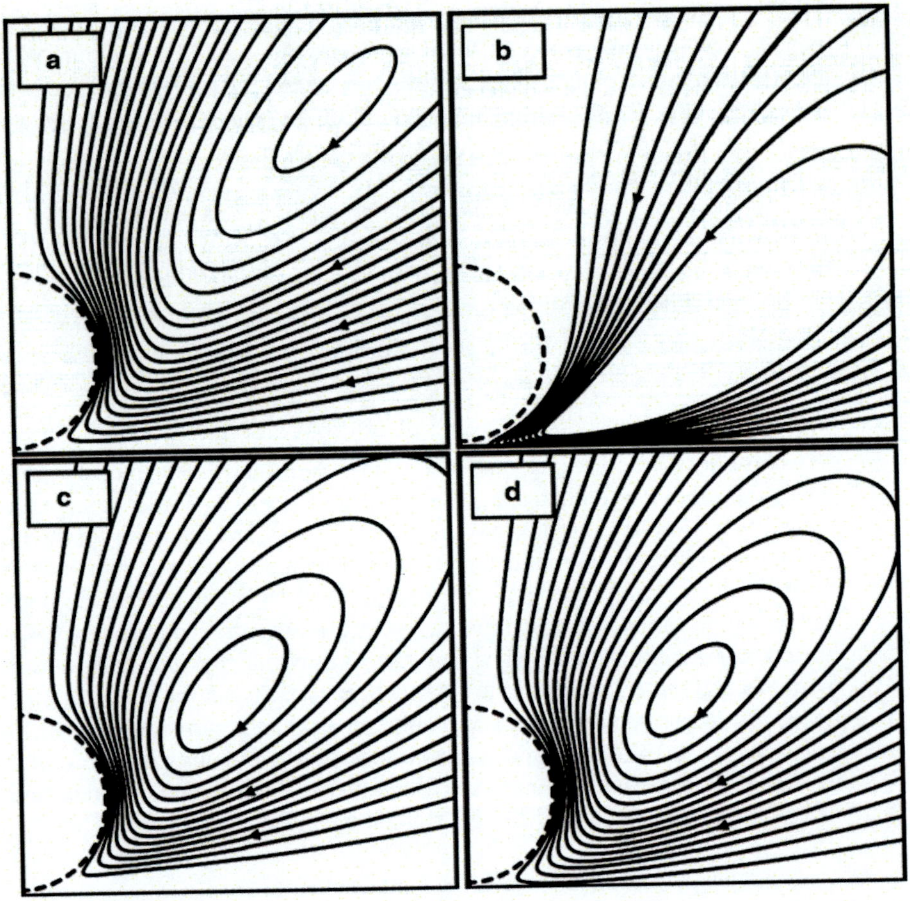

Fig 1.19 Streamlines of electroosmotic flow for equal electric fields imposed far from the particles: **a** slip only on the particle; **b** slip only on the electrode; **c** slip on both at the same same zeta potentials; **d** slip on both at zeta potentials of opposite sign. The stream function increment between lines was the same for all cases but (b), which was one order of magnitude lower. Slip on the particle dominates the flow pattern even when the zeta potentials are opposite in polarity (comparing (c) and (d))

ues of the stream functions were scaled differently to reveal the patterns.) To illustrate this point, Fig. 1.19c and d show cases where the zeta potential of the electrode was flipped from being the same as the zeta potential of the particle to having the opposite sign. The overall direction of flow was undisturbed, and the pattern more closely resembled Fig. 1.19a than b. Thus, mutual entrainment of particles in ECEO based on the particle is stronger than entrainment in ECEO based on the electrode for equivalent zeta potentials. ECEO driven by the particle depends on the normal component of the imposed electric field, while ECEO based on the electrode depends on secondary lateral components arising from the particle's disturbance of the primary electric field. The previous comments apply to the laterally directed mutual entrainment of particles; Fagan et al. [36] showed that the normal force exerted by ECEO associated with the electrode can be substantial because its effect is concentrated under the particle. Thus, ECEO might contribute substantially to the out-of-plane motion of the particle.

The resulting electroosmotic flow either pumps liquid into the space between particles, causing them to separate, or pumps liquid out from between the particles, causing them to aggregate. The direction and magnitude depend on both the particle zeta potential and the direction of the uniform electric field far from the electrode. One must calculate the rate of aggregation or separation by integrating the viscous stresses over the particle surfaces and balancing them with viscous resistance to lateral motion. We write the interparticle velocity as

$$\frac{dr}{dt} = U_{ECEO} \tag{1.23}$$

where U_{ECEO} is the net lateral translation of the particle as a consequence of mutual entrainment of particles in ECEO flow.

Solomentsev et al. [20] proposed that aggregation in *dc* experiments arises from mutual entrainment of particles, as described above. They solved Eq. 1.19 with a boundary condition of uniform potential on the electrode, which meant that electroosmotic slip at the electrode vanished. Far from the electrode and the particle, the fluid remained undisturbed by the electroosmotic flow, $\mathbf{r} \to \infty \;\Rightarrow\; \mathbf{v} \to 0,\; P \to 0.$ The solution of these equations yielded closed streamlines resembling those in Fig. 1.19a. Solomentsev et al. [20] assumed that a second colloidal particle near the electrode becomes entrained in the electroosmotic flow created by the first; they calculated the ECEO (entrainment velocity) by means of Faxen's law

$$\mathbf{u} = \mathbf{u}_{ECEO} + \mathbf{u}_{EP} = \mathbf{v} + \frac{1}{6}a^2\nabla^2\mathbf{v} + \frac{\varepsilon\zeta}{\mu}\mathbf{E} \tag{1.24}$$

where all terms are evaluated at the center of the second sphere ignoring the presence of the second sphere. The radial component of the second term is about 30% of the first and acts in the opposite direction [20]. Assuming the two spheres have identical ζ_p, the electrophoretic contribution acts in the opposite direction, has a maximum contribution of 20% of the Faxen result at contact, and becomes less important as separation increases. All three terms scale with $\varepsilon\zeta E_\infty/\eta = u_0$. Solomentsev et al. evaluated u_r in the plane $z = 1.05a$, which contains the center of the first sphere. They showed that the results are insensitive to z in the range $1 \le z/a \le 1.1$.

Since the presence of the nearby wall (the surface of the electrode) retards the motion of the second particle, Solomentsev et al. [22] used a factor q_w to correct for wall hindrance. The basic equation for calculating the interparticle separation between doublets was

$$\frac{dr}{dt} = U_{ECEO} + U_{EP} \cong 2q_w(h)\left(u_{ECEO}(r) + u_{EP}(r)\right) \tag{1.25}$$

where r was the center-to-center distance between the two spheres and $h = z - a$ was the gap between the bottom of the sphere and the electrode. The factor of 2 meant the entrainment was mutual. Solomentsev et al. [22] established a value for the hindrance factor q_w appearing in Eq. 1.25 by measuring the hindered diffusion of isolated particles near the electrode in the absence of the applied field.

1.5.1.2 Comparison Between Theory and Experiment

Guelcher observed the rate of aggregation of particle pairs in *dc* fields [23]. The doublets aggregated at different rates owing to Brownian motion. Averaging the interparticle distances as a function of time, Guelcher observed that reversing the polarity of the electric field, in turn, reversed the direction of interparticle motion, but the particles' relative velocity was unchanged. Varying the electric field from 20–100 V/m, he observed direct proportionality between the rate of aggregation and the electric field. Finally, aggregation of positively charged particles required the opposite electrode polarity to cause motion in the same direction for negatively charged particles. These predictions are consistent with the ECEO model in which the speed of aggregation is proportional to the particle zeta potential and to the first power of the electric field appearing in Eq. 1.18.

Figure 1.20 [22] shows a serious quantitative comparison between the *dc* ECEO model and pairwise aggregation, including wall hindrance, ECEO, and EP effects.

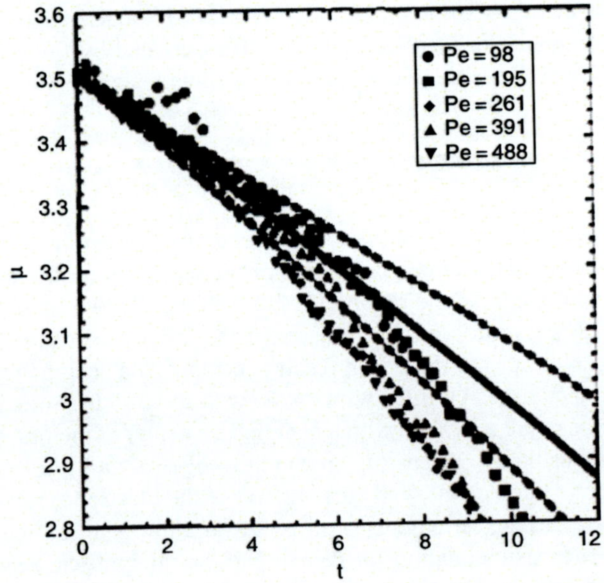

Fig. 1.20 Comparison of the electroosmotic flow model (particle-based) to experimental data for aggregation of particles during *dc* polarization, including lateral electrophoresis and wall hindrance; all parameters used in the calculation were determined independently [22]. The ordinate is the mean separation between particle centers, normalized by radius. The agreement is good at short times. Note that particles moved faster at longer dimensionless time because intensification of the electric field between the particles strongly increases the slip velocity. (Reprinted in part with permission from Solomentsev et al. [22], © 2000 American Chemical Society)

At a dimensionless center-to-center separation (μ=r/a) of 3.5<μ<3.2 particle radii, the observed average rate of aggregation in *dc* fields was within experimental error of the rate predicted by Eq. 1.25 (See Fig. 1.20); however, the particles moved more quickly relative to each other for smaller separations (See 3.2<μ<2.8 on the same figure.). The most likely explanation for the deviation between the predicted trajectory and averages of the observed trajectories was the influence of two-body intensification of the electric field between the particles as they approached each other. Use of the factor of two in Eq. 1.25 essentially implies linear superposition of the single particle flows and electric field. At large separations, deviations from linear superposition would be within the noise, but the slope of the lines in Fig. 1.20 clearly showed increase of the interparticle speed at smaller separations where the electric field between the particles was strong. Solomentsev et al. [22] mentioned this possibility but only explored the effect of Brownian motion. Inclusion of induced dipole effects in Eq. 1.25 would have made the agreement worse as the particles approached other and probably was not significant because the fields used were much smaller than used in *ac* experiments. Figure 1.20 represents a substantial body of work aimed at comparing a subtle theory, with no parameters other than those measured independently or found in handbooks, to experimental observations of a complex mixed deterministic and stochastic phenomenon. The agreement, subject to the limitations due to the assumed superposition described above, was good.

1.5.2 *ac ECEO*

Directed assembly with direct current acting on the particle's equilibrium charge is effective, but direct current means passing faradaic current with associated unwanted effects, such as alteration of the electrode or the solution properties (pH, concentration), or both. When it was established [14, 18, 19] that directed assembly with *ac* current was possible, and that different phenomena were involved, most attention shifted to that mode. Any mechanism depending purely on the first power of the electric field cannot move two particles together or apart when their relative motion is integrated over one symmetrical cycle. For example, in quasi-steady flow, a sinusoidal electric field $E_\infty \cos\omega t$ causes $u_r(r, t)$ to vary like $(\delta u)\cos\omega t$, where $\delta u(r)$ is the amplitude of oscillation in u_r. Integrating Eq. 1.25 over a cycle, one obtains

$$\left\langle \frac{dr}{dt} \right\rangle = 2q_w\,(\delta u)\,\frac{\omega}{2\pi}\int_0^{2\pi/\omega} \cos\omega t\; dt = 0. \tag{1.26}$$

Net drift of two particles toward each other or away, therefore, requires breaking of the antisymmetry of the two halves of the cycle. ECEO, whether based on the particle or the electrode, should be insignificant for directed assembly with *ac* current. Equation 1.25, however, contains the product of two factors, the particle speed and the hindrance factor q_w. If q_w is likewise an oscillating quantity, the antisymmetry can break, and net interparticle motion can occur. Fagan et al. [39] suggested that such a break arises from oscillations in the elevation h of spheres above the electrode because q_w is a function of h. Consider a sphere oscillating with a small amplitude normal to the electrode in an electric field $E_\infty \cos\omega t$; [31] the particle's height above the electrode behaves like $\cos(\omega t + \theta)$ where θ represents a phase angle between the field and the particle height. Then, the oscillations in elevations follow

$$h(t) = h_0 + (\delta h)\cos(\omega t + \theta) \tag{1.27}$$

and the corresponding oscillation in the wall correction becomes $q_w(t) = q_{w0} + (\delta q_w)\cos(\omega t + \theta)$.

Now, Eq. 1.26 predicts

$$\begin{aligned}\left\langle \frac{dr}{dt} \right\rangle &= 2(\delta u)\frac{\omega}{2\pi}\int_0^{2\pi/\omega} [q_{w0} + (\delta q_w)\cos(\omega t + \theta)]\cos\omega t\, dt \\ &= (\delta u)(\delta q_w)\cos\theta\end{aligned} \tag{1.28}$$

In other words, if the particle height and electric field are out of phase by an angle differing from 90°, the interparticle velocity integrates to a nonzero value; hence, two nearby particles experience a tendency to aggregate or separate by this mechanism.

1.5.2.1 Theory

Particle motion during *ac* polarization arises from a superposition of entrainment in flow and induced dipole repulsion,

$$\frac{dr}{dt} = U_{ECEO} + U_{EP} + U_{DD} \tag{1.29}$$

where r is the center to center distance between particles, U_{ECEO} is the net drift velocity arising from mutual entrainment, U_{EP} is lateral electrophoresis, and U_{DD} is migration caused by induced dipole repulsion between the particles. The first two of these mechanisms were discussed in Sect. 1.5.1. They remain in force under *ac* polarization; the difference is that the net lateral motion integrates to zero each cycle unless the phase angle between the particle height and the current is different from $\pi/2$, as shown above. U_{DD} arises from identical dipoles, induced in each particle and normal to a line connecting their centers. The net drift velocity driven by dipole repulsion can be calculated from [39]

$$U_{DD} = \frac{q}{6\pi\mu a}\left[\frac{3}{4\pi\varepsilon}\frac{(\alpha E_\infty)^2}{r^4}\right] \tag{1.30}$$

where α is the polarizability of the particle and αE_∞ is the magnitude of the induced dipole moment. (See Sect. 1.7 for more details on induced dipole effects.) The induced dipole interaction is always repulsive between identical particles arranged side by side with respect to the electric field.

1.5.2.2 Comparison of Theory to Experiment

The work of Prieve and co-workers provides the evidence for *ac* ECEO; they [31, 36, 37, 39, 44] studied the response of single particles to *ac* polarization with both video microscopy and TIRM, [32] which allowed measurement of the instantaneous height of the particle above a polarized electrode in addition to lateral motion. Particles having a diameter of a few microns and levitated by surface forces move toward the surface or away from it in response to the imposed electric field. Figure 1.21 [44] shows a typical measurement of particle height as a function of time (5.7 μm nominal particle diameter). Figure 1.21a shows the raw signal, including jumps due to Brownian motion, while Fig. 1.21b shows the averaged signal after conversion to height *h*. The frequency was 100 Hz. The amplitude of *h* was somewhat more than

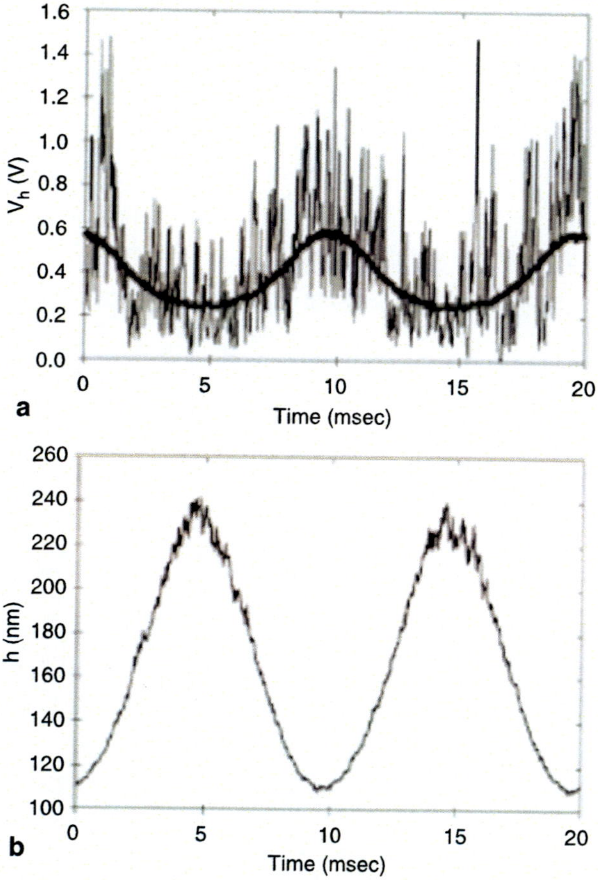

Fig. 1.21 Raw scattering data and conversion to height of a particle (5.7 μm diameter) in 0.15 mM KOH subjected to *ac* polarization at 100 Hz on Pt [44]. **a** Scattering intensity measured instantaneously and averaged over a thousand cycles. Brownian motion causes the scattering intensity to fluctuate stochastically. **b** The absolute particle height after conversion. Note the blunting of the peak when the particle is closest to the electrode as in Fig. 1.12. The maximum of intensity corresponded to a minimum of height. (Reprinted in part with permission from Hoggard et al. [44], © 2007 American Chemical Society)

100 nm, around 2% of the nominal particle diameter. The imposed potential varied sinusoidally, but h was more rounded at the minimum height compared to its shape at the maximum height. Extra hindrance due to the particle's proximity to the wall explains the blunting of h at the minimum height; the extra hindrance is proportional to the ratio of particle diameter to the gap between the particle and the wall, for small separations. The value of h had only one maximum and one minimum value per cycle, which was expected if the mechanism driving the particle's motion was proportional to the first power of the electric field, as in the ECEO mechanism.

A crucial observation obtained from curves of the height function, as in Fig. 1.21, appears in Fig. 1.22 [44]. The current through the cell is included in the graph with the scattered signal. Fagan discovered [31] and Hoggard [44] confirmed that the

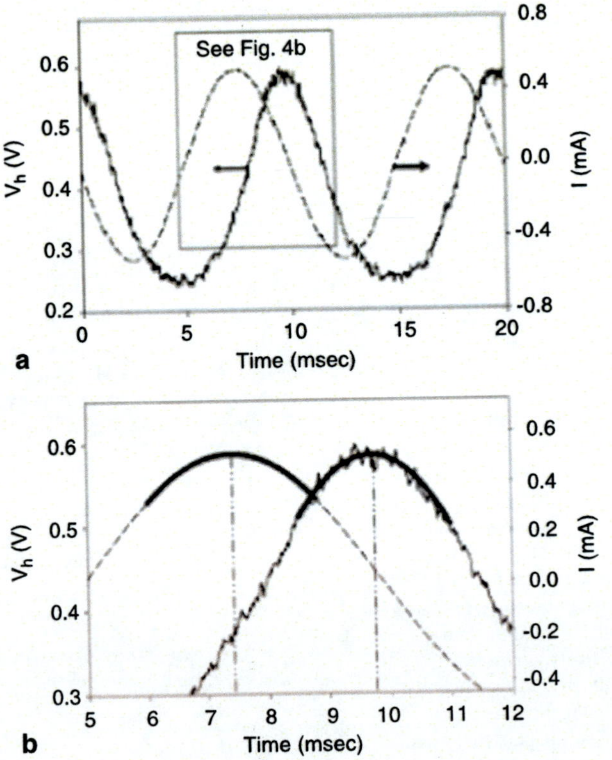

Fig. 1.22 Measured scattering intensity and current in a TIRM experiment as a function of time during ac polarization at 100 Hz [44]. The intensity of scattered light, related to the particle's height (see Fig. 1.21), should be 90° out of phase with the current density, if all forces other than the electrophoretic force are independent of height. **a** The current density and the scattering intensity. **b** The time difference between peaks can be converted to phase angle, which shows that the current leads the scattering intensity by 81.4°. A phase angle different from 90° causes two particles to move laterally relative to each other, as discussed in Sect. 1.5.2. The dependence of double layer repulsion on particle height leads to a phase angle different from 90°, as discussed in the text. (Reprinted in part with permission from Hoggard et al. [44], © 2007 American Chemical Society)

phase angle between the current and the scattering intensity were not 90° out of phase. In other words, the phase angle θ of Eqs. 1.27 and 1.28 was not $\pi/2$; hence, the condition necessary for a net drift of particles toward each other or away from each other was fulfilled (Eq. 1.28).

To translate these observations into particle motion, Fagan et al. [39] extended the model for *dc* ECEO to alternating fields by assuming that Eq. 1.31 applies in a quasi-steady fashion at each instant of time:

$$U_r = 2q_w(h)u_r(r). \tag{1.31}$$

Anti-symmetry (over one period of oscillation in the electric field) is broken by prescribing oscillations in the elevation according to Eq. 1.27. The particle trajectory $r(t)$ was predicted by substituting Eq. 1.30 and Eq. 1.31 into Eq. 1.29 and integrating numerically. Measured values for h_0, δh, θ and ζ were used. Comparing theoretical predictions of aggregation or separation rates to the data of Kim et al. [29], based on experimentally observed height functions and current, Fagan et al. [39] obtained excellent agreement between the *ac*-ECEO model for separation of particle doublets in KOH; the model under-predicted the aggregation rate for bicarbonate by a factor of 4. The phase angle between the current and the particle height was a predictor of whether particle pairs would aggregate or separate, as confirmed by Hoggard et al. [44] who explored other electrolyte and electrode combinations. In all cases, the observed fact of the phase angle being greater or less than $\pi/2$ predicted the correct direction of aggregation or separation of particle doublets. Figure 1.23 shows the agreement obtained with no adjustable parameters for various combinations of electrodes and electrolytes. The model typically under-predicted the rate, not surprisingly, because it used the same linear superposition of flow fields and electric fields as Solomentsev et al. [22] as discussed in Sect. 1.5.1.

One possible source of a phase angle has been identified, theoretically, and awaits experimental confirmation. The particle behaves as a damped harmonic oscillator. Writing an expression for the velocity of a particle in the direction normal to the electrode, one obtains

$$\frac{1}{\mu_D(h)}\frac{dh}{dt} = \frac{B}{\lambda}e^{-h/\lambda} - G + q_w(h)\frac{6\pi a\varepsilon\zeta_p}{\sigma_{el}}\bar{i} \tag{1.32}$$

where μ_D is the hindered diffusion mobility [54], B is the constant related to diffuse layer repulsion in the DLVO theory, λ is the Debye length, and G is the particle weight. Let

$$h = h_m + \bar{\delta}e^{j(\omega t+\theta)}, \quad i = \bar{i}e^{j\omega t}, \quad \mu_D(h) \to \mu_{Dm}, \quad q_w(h) \to q_w(0) \tag{1.33}$$

Where h_m is the particle's most probable height, $\bar{\delta}$ is a small amplitude of oscillation, and μ_{Dm} is the hindered diffusion mobility evaluated at h_m, for simplicity. Now we rewrite Eq. 1.32 in terms of δ. After canceling the terms at the most probable height, one can solve for the complex amplitude and phase angle of the response of the particle height to current density.

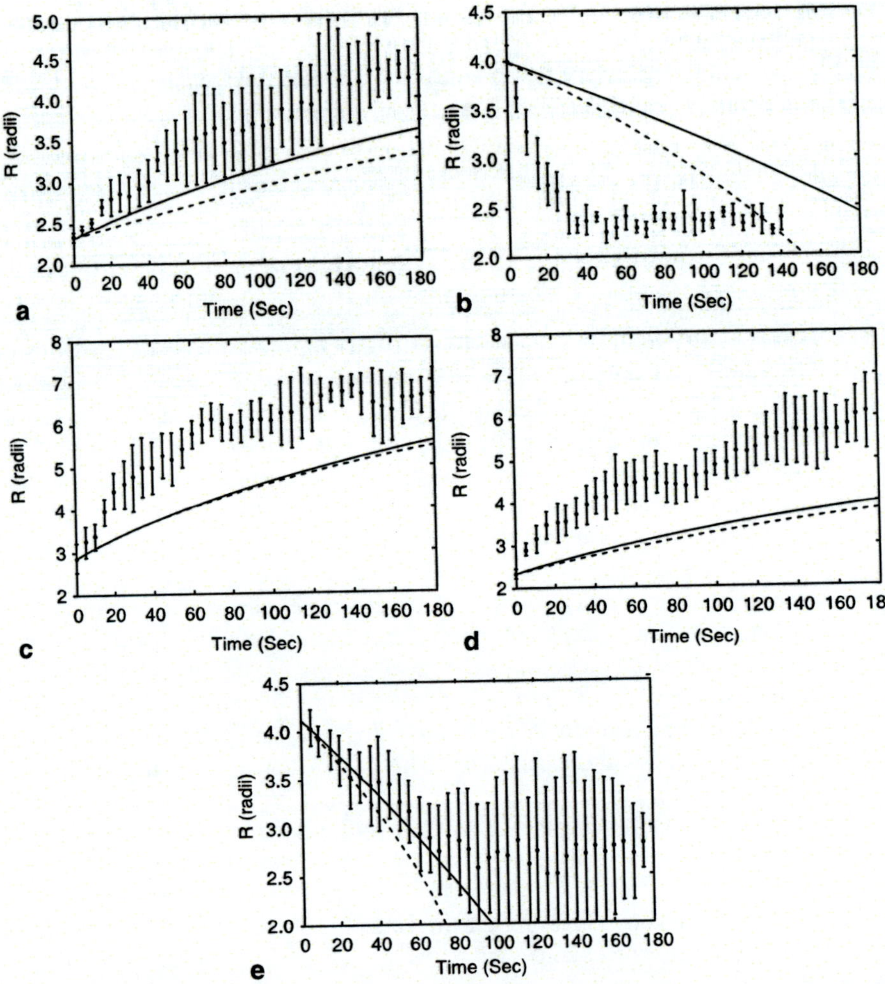

Fig. 1.23 Measured interparticle distances, R(t), compared to predictions of the ECEO model involving a phase angle between the particle height and current density [44]. The solid lines are predictions including ECEO at the particle, lateral electrophoresis, and induced dipole repulsions. The dotted lines neglect the dipole repulsion. **a** H_2CO_3-Pt, **b** KCl-Pt, **c** KOH-ITO, **d** KOH-Pt, and **e** $NaHCO_3$-Pt. The model generally underestimated the aggregation or separation rate because the superposition of the combined effect of the two particles neglected the intensification of the electric field between them. (Reprinted in part with permission from Hoggard et al. [44], © 2007 American Chemical Society)

$$\bar{\delta}e^{j\theta} = \frac{A\bar{i}}{\frac{1}{\mu_{Dm}}j\omega + \frac{B}{\lambda^2}e^{-h_m/\lambda}} \tag{1.34}$$

Multiplying the numerator and denominator by the complex conjugate and solving for θ, one obtains the phase angle

$$\theta = \tan^{-1} \frac{\omega \lambda^2}{\mu_{Dm} B e^{-h_m/\lambda}}. \tag{1.35}$$

The angle θ varies from zero to $\pi/2$ as the frequency increases, which shows that a phase angle between the scattering intensity and the current density, consistent with the observed response of particles in KOH, where the corresponding phase angle was 81°, can be understood as a consequence of the diffuse layer repulsion even without invoking the possible effect of faradaic reactions on the phase. This effect does not explain the observed electrolyte dependence of the phase angle, but it demonstrates that damped harmonic oscillation of the particle in response to the imposed voltage leads to a phase angle.

1.5.2.3 Summary of *ac* ECEO

Prieve and co-workers have provided evidence and a model for the behavior of particles in the geometry of Fig. 1.3a. The phase angle between the current and the particle height was a predictor of whether particle pairs would aggregate or separate. The phase angle is not distinct within the noise of the measurements above 1 kHz, but Fagan et al. [37] found evidence of electrolyte dependence even at 10 kHz as described previously (see Fig. 1.13). Fagan et al. [39] also showed that the aggregation or separation rate can appear to be invariant with particle zeta potential because the phase angle is not only a function of frequency but also of particle zeta potential. This might explain the zeta potential paradox of Kim et al. [30] who found that aggregation rates in *ac* polarization did not vary when the zeta potential of the particles was reduced. (See [39] for extensive discussion of this point.)

1.5.3 Summary of ECEO Investigations

Quantitative agreement between theory and experiments, based on independently measured parameters, has established the importance of mechanisms based on ECEO both on the particle and the electrode. The case for *dc* ECEO seems closed. The arguments and results of Sect. 1.5.2 put forward by Prieve and co-workers are evidence that *ac* ECEO based on slip at the particle's surface can predict aggregation or repulsion. The mechanism in its current form requires measurement of a phase angle between the particle height and the current density, a capability possessed only by Prieve et al. A complete explanation of the origin of the phase angle would be useful and theory showing one origin of a phase angle was given. The sharpest contrast in the dependence of aggregation on the electrolyte is between bicarbonate solution and alkaline solutions involving sodium or potassium hydroxide. We stress that careful purging of carbon dioxide from alkaline solutions is vital to observe this difference because carbonate quickly forms by the reaction between carbonic acid and the dilute alkaline solutions that allow particle levitation above electrodes. One widely reported phenomenon that the *ac* ECEO mechanism does not seem to predict is the steady radial movement of particles during *ac* polarization. The pure *ac* ECEO mechanism does not predict any steady motion of the fluid, just the particles toward each other or away. Arguments presented in the next section, however, demonstrate that steady radial flow during *ac* polarization is expected if the potential distribution of the potential under the particles is nonuniform.

1.6 Induced Charge Electroosmosis (ICEO)

Applying a voltage across electrodes induces a space charge in the diffuse layers adjacent to each electrode. In electrochemical terms, the applied potential charges the double layer even under conditions of no specific adsorption and no equilibrium charge. Any component of the electric field parallel to either electrode can drive the mobile space charge of the diffuse layer into electroosmotic flow. The overall schematic remains the same as in Fig. 1.3a, with charge both on the particle and charge on the electrode, but here the charge on the electrode is taken as potential dependent. ICEO arises because the same applied potential that induced the charge provides lateral gradients of potential along the electrode to produce the flow. The existence of lateral electric field gradients immediately adjacent to the electrode stems from the arguments presented in Sect. 1.5.1 about the current distribution underneath the particle. If the current distribution is uniform, such as under small particles, lateral gradients of potential at the electrode exist. ICEO is often cited as a mechanism underlying particle aggregation for normally directed electric fields [34, 38, 42, 43, 45–47].

1.6.1 dc ICEO

1.6.1.1 Theory

ICEO relies on a functional relationship between electrode charge and the imposed electric field, while ECEO ignores the induced charge in favor of equilibrium charge. The theory of ICEO requires, therefore, simultaneous solution of the fundamental equations governing the electric potential and concentrations everywhere outside the Outer Helmholtz Plane adjacent to each electrode. Equations 1.36–1.38 represent the flux of species i, Poisson's equation, and species conservation at steady state.

$$\mathbf{N}_i = -D_i \nabla c_i - \frac{e z_i D_i}{kT} c_i \nabla \phi \tag{1.36}$$

$$-\varepsilon \nabla^2 \phi = e \sum_i z_i c_i \tag{1.37}$$

$$\nabla \cdot \mathbf{N}_i = 0 \tag{1.38}$$

A general statement of the governing equations for *dc* includes steady or quasi-steady viscous flow with an electrically driven body force.

$$0 = -\nabla P - e\nabla\phi \sum_i z_i c_i + \eta \nabla^2 \mathbf{v} \tag{1.39}$$

$$\nabla \cdot \mathbf{v} = 0 \tag{1.40}$$

Flow boundary conditions depend on assumptions about the problem. In the absence of mass transfer limitations, the Butler Volmer equation (Eq. 1.20) relates the current flowing through the electronic conductor to the potential drop across the interface, which is necessary in the *dc* case because the double layer charging current is zero.

As with ECEO, the *dc* ICEO problem has been approached by solving the electrical and concentration problem (Eqs. 1.18–1.20) first and using the electric field components along the particle and surface and a thin double layer approximation to drive electrosmostic flow. Ristenpart et al. [43] solved the above equations for a ternary univalent electrolyte in which protons were reduced and molecules were oxidized to produce a specified current and corresponding proton flux, while the other two ions were regarded as indifferent. The potential drop was taken as small in order to linearize the problem, and the equilibrium charge of the electrode was taken as zero. Their result for the profile of electrostatic potential ϕ as a function of position x was

$$\phi(x) = \frac{\Delta\phi}{2}\left(1 - \frac{\sinh \kappa x}{\sinh \kappa l}\right) - \frac{Il}{\sigma_e}\left(\frac{x}{l} - \frac{\sinh \kappa x}{\sinh \kappa l}\right) \tag{1.41}$$

where κ is the reciprocal Debye length, l is one-half the electrode separation, I is the specified current, σ_e is the electrolyte conductivity, and $\Delta\phi$ is the applied cell potential. Despite the substantial assumptions listed above, Eq. 1.41 is notable because it describes a continuous change of potential from the OHP of one electrode to the OHP of the other, which is different from the typical decomposition of the problem into bulk and double layer regions. The charge densities q_{ECEO} induced in the diffuse layer on either electrode are equal in magnitude, opposite in sign, and obtained from Eq. 1.41 using Gauss's law.

$$q_{ICEO} = \left(\frac{\Delta\phi}{2} - \frac{Il}{\sigma_e}\right)\kappa\varepsilon \tag{1.42}$$

Equation 1.42 is expected; with linearization, the induced charge is proportional to the potential drop from the cell's midplane to the OHP after subtraction of ohmic loss. Ristenpart et al. [43] neglected the second term in the parentheses of this expression because the ohmic loss for thin gaps is often much smaller than the polarization of the diffuse layer, leaving

$$q_{ECEO} = \frac{\Delta\phi}{2}\kappa\varepsilon \tag{1.43}$$

Equation 1.43 is precisely what one obtains by linearizing Gouy Chapman theory and substituting half the cell potential for the potential at the OHP. Ristenpart et al.'s model for the electrostatic potential and electroosmotic flow outside the diffuse layer parallels the *dc* ECEO model discussed in Sect. 1.5.1, except that surface conduction was added to the slip boundary condition on the particle,

$$\sigma_e(\mathbf{n} \cdot \nabla\phi) + \sigma_s \nabla_s^2 \phi = 0, \tag{1.44}$$

where the subscript s denotes surface conductivity. Following the argument of Sides [26] about the respective importance of ohmic and charge transfer resistances, Ristenpart et al. [43] assumed a uniform current density on the electrode, which engendered a tangential electric field along the surface of the electrode. Once the potential profile had been determined, flow Eqs. 1.39, 1.40 were solved subject to the far field condition. The tangential components of the electric field acted on the induced charge of Eq. 1.43 in the slip equation

$$v_t = \frac{\varepsilon\Delta\phi}{2\eta}E_t. \tag{1.45}$$

The streamlines of the electroosmotic flow based on induced charge resemble the pattern shown in Fig. 1.19b for flow based on equilibrium charge on the elecrode. From there Ristenpart et al. used the approach of Solomentsev et al. already described in Sect. 1.5.1 to calculate the rate of mutual entrainment.

1.6.1.2 Comparison Between Theory and Experiment

Ristenpart [43] used the parameters of Solomentsev et al. [22] to generate *dc* ICEO flow and resulting interparticle velocities of doublets for comparison to Solomentsev's data. Since Guelcher [22, 23] did not report the total applied voltage in his experiments, Ristenpart et al. [43] assumed values of $\Delta\phi/2=0.25$ and 0.5 V for the potential drop induced across the diffuse cloud on either electrode and then calculated the induced surface charge density from Eq. 1.43. Ristenpart's calculation of the rate of aggregation with ECEO without ICEO confirmed Solomentsev's calculations and averaged experimental observations already shown in Fig. 1.20. The inclusion of ICEO increased the predicted rate of aggregation at a separation of 3.5 radii. The amount of the increase in rate for $\Delta\phi=1$ V seemed to explain experiments where the initial rate was larger than the mean. Ristenpart et al. [43] argued that *dc* ICEO flow was comparable in importance to *dc* ECEO based on the particle alone. Several issues, however, call this conclusion into question.

The total potential drop and the current density were treated as independent parameters in this model; in fact; however, both Ohm's Law and equations, such as the Butler-Volmer relation Eq. 1.20, bind them. This omission arises from the conception of the problem as hermetic from the outer Helmholtz plane (OHP) of one electrode to the OHP of the other electrode. Ristenpart et al. [43] did not include the compact layer between each electrode and its OHP. Treating the overall double layer as a series combination of compact layer capacitance C_C and diffuse layer capacitance, one can show that the ratio of the potential difference across the diffuse layer $\Delta\phi_D$ to the potential difference across the compact layer $\Delta\phi_C$ can be estimated by

$$\frac{\Delta\phi_D}{\Delta\phi_C} = \frac{C_C}{\kappa\varepsilon \cosh\left(\frac{e\Delta\phi_D}{2kT}\right)}. \tag{1.46}$$

Taking a typical capacitance of the compact layer to be 20 μf/cm² and an ionic strength of 1 mM, one concludes that a 250 mV overall potential difference means the potential difference across the diffuse layer would be approximately 100 mV; at 500 mV overall potential difference, the potential difference across the diffuse layer would be 140 mV. At the large overall potential differences assumed, most of the potential drop included in the calculations falls across the compact layer and not across the diffuse layer. In the electrochemical literature, this difference is known as the "Frumkin correction" because A. N. Frumkin pointed out that the potential difference that matters in electrode reactions spans the compact layer. Since direct measurement of potential between the OHP and the electrode is impractical, the Frumkin correction was proposed in point of this distinction. Breiter et al. [56] discussed this principle and provided algebraic equations for calculating its magnitude, while Hu et al. [57] measured the diffuse layer potential during polarization. Both estimates and measurements indicate that the effective induced zeta potential is approximately a factor of ten smaller than the values used to reach the conclusions of Ristenpart et al. [43]. As a result, the possible effect of *dc* ICEO on aggregation was overestimated by the assumption of unrealistically large values for the potential drop across the diffuse layer on the electrode. Using an induced zeta potential on the electrode comparable to the zeta potential of the particle, as seems most appropriate, one sees little effect of slip at the electrode in the outer flow field most important for mutual entrainment (see Fig. 1.19). Yariv [50] recently emphasized this point.

Other issues also reduce the strength of the arguments for *dc* ICEO. The most compelling contradiction is that the interparticle motion due to ICEO, whether *dc* or *ac,* invariably is aggregative. Data of Guelcher et al. [23] demonstrate in Fig. 1.9, however, that reversing the sign of this zeta potential reversed the motion from aggregation to separation.

1.6.2 *ac* ICEO

ICEO associated with *ac* polarization is another mechanism that might explain the experimentally observed aggregation with alternating polarization. ICEO is the result of the imposed electric field interacting with the induced charge on the electrode. As with *dc* ICEO, both the induced charge and electric field that drives the flow are proportional to the imposed field. Their product does not integrate to zero over each cycle; we classify the flow, therefore as *ac* ICEO. Since the product of lateral electric fields and unbalanced charge density in the diffuse layer drives electroosmotic flow, and since the induced charge and the associated electric field components switch sign in each half of the cycle, fluid flows along the electrode toward an isolated particle and upward away from it.

1.6.2.1 Theory

The theory remains much the same as for *dc* ICEO with addition of transient terms related to steady oscillation of potential with zero net current and incompressible oscillatory flow with a finite net circulation. Only Eq. 1.38 becomes

$$\frac{\partial c_i}{\partial t} = -\nabla \cdot \mathbf{N}_i \tag{1.47}$$

Ristenpart et al. [34, 42] modeled particle motion based on *ac* ICEO in two contributions where solution of the electrical aspect of the problem included oscillation of the applied potential. Ristenpart et al. used the solution by Hollingsworth and Saville [58] for a 1:1 electrolyte with zero flux at two opposing electrodes across a gap *2h*. Their result for small oscillations of the potential and concentration was

$$\phi(z,t) = \frac{\Delta\phi}{2}\left\{1 - \frac{\sinh\left(\frac{\gamma z}{h}\right)\operatorname{csch}(\gamma) - j\frac{z}{h}\gamma\nu^2\coth(\gamma)}{1 - j\gamma\nu^2\coth(\gamma)}\right\} e^{-j\omega t} \tag{1.48}$$

where $j = \sqrt{-1}$, $\gamma^2 \equiv (\kappa l)^2\left(1 - j\nu^2\right)$, $\nu^2 \equiv \omega/(\kappa^2 D)$ and $D=D_+=D_-$.

Allowed oscillations of potential, in this case, are much less than kT/e, which is about 25 mV. When the Debye length is much less than the electrode spacing and the frequencies are low enough that the concentration profiles of the diffuse layers are quasi-steady, the bulk of the fluid outside the diffuse layers is electrically neutral and the ion concentrations remain unperturbed. The electric field in the bulk is uniform and given by

$$E_x(x,\ t) = \frac{\Delta\psi}{2h}\frac{\alpha}{\sqrt{1+\alpha^2}}e^{j(\omega t+\theta_{bulk})} \tag{1.49}$$

where $\alpha \equiv \omega l/\kappa D$ and $\theta_{bulk} \equiv \tan^{-1}(1/\alpha)$ is the phase angle. The charge densities induced in the diffuse layer on either electrode are equal in magnitude, opposite in sign, and deduced by applying Gauss's law as in the *dc* ICEO case.

$$q_{ICEO}(t) = \frac{\Delta\phi}{2\sqrt{1+\alpha^2}}\kappa\varepsilon e^{j(\omega t+\theta_{surf})} \tag{1.50}$$

where

$$\theta_{surf} = \tan^{-1}\left(\frac{-\alpha}{1+\nu^2\alpha}\right).$$

These results are the oscillatory equivalent of Eq. 1.43; they reduce to the *dc* case at zero frequency.

Electric field components tangent to the surface, arising from the disturbance of the field by the particle, drive electroosmotic flow along the electrode surface. In the earlier of two contributions, Ristenpart et al. [34] estimated the tangential component of the electric field by treating the dielectric particle as a point dipole $\boldsymbol{p}$ embedded in an infinite dielectric material having permittivity ε. The expressions are

$$\phi(\mathbf{r}) = \frac{\mathbf{r}\cdot\mathbf{p}}{\varepsilon|\mathbf{r}|^3} \tag{1.51}$$

$$\mathbf{p} = C_0\varepsilon a^3\mathbf{E}_\infty \tag{1.52}$$

$$C_0 = \frac{2\lambda-1}{2\lambda+2} \tag{1.53}$$

where

$$\lambda = \frac{\sigma_s}{a(\sigma_e - i\varepsilon\omega)}$$

on the electrode; E_∞ is $\Delta\phi/2h$. Once the electric field is known everywhere, Stokes equations are solved, subject to the far field condition and no slip on the particle. The electroosmotic flow is driven by the action of the electric field on induced surface charge of the electrode:

$$\text{at } z=0: \quad \mathbf{v} = \frac{q_{ICEO}}{\eta\kappa}\nabla_s\phi \tag{1.54}$$

Ristenpart et al. [34] used Eqs. 1.50 and 1.51 in Eq. 1.54 to deduce a scaled tangential velocity characteristic of the problem.

$$u_r \sim \frac{3\varepsilon}{\eta\kappa}E_\infty^2\left(C_0' + \frac{D\kappa^2}{\omega}C_0''\right) \tag{1.55}$$

where C_0' and C_0'' are the real and imaginary parts of the dipole coefficient. Ristenpart et al. [42] subsequently reported the results of a finite element analysis of the charge transport around the particle, including investigation of the effect of the electrode and surface charge transport on the particle.

1.6.2.2 Comparison Between Theory and Experiment

The primary experimental evidence for this model is based on video microscopy from several groups. In most cases, the connection between observation and theory is qualitative. Yeh et al. [19] reported aggregation of 2 μm polystyrene particles during polarization at 1 kHz and proposed the above mechanism in general terms. Nadal et al. [28] quantified observed velocity of tracer particles toward a target particle, but did not develop theory required to match the observations with experiments. Detecting no apparent influence of zeta potential of the particles, they endorsed the mechanism of Yeh et al. [19] balanced with induced dipole repulsion, as the two opposing mechanisms that underlie the existence of a critical frequency.

Ristenpart et al. [34] investigating the aggregation of 2.7 μm polystyrene in 1 mM KCl at 500 Hz, attributed the toroidal flow mentioned by Yeh et al. [19] and the radial flow described by Nadal et al. [28] to the steady component of the ICEO velocity (see Eq. 1.55). They joined experimental data and the ICEO mechanism presented in Eq. 1.55 by measuring initial aggregation rates of large ensembles as a function of E_∞ and ω and matching the measurements to a two-body model where the rate constant was assumed to scale like u_r. The measured rate constants were proportional to $\boldsymbol{E}_\infty^2$ and ω^{-1}, which agreed with their scaling in u_r as shown in Fig. 1.24. The agreement was somewhat surprising because the applied voltages of the experiments exceeded the limits of linear behavior on which the theory was based by a factor of a hundred and because extracting a lateral electric field component from an isolated dipole approximation neglected the effect of the nearby electrode. The interplay among the electrode kinetics, electrolyte conductivity, and particle size determine whether lateral electric field gradients even exist. Nevertheless, the demonstration that the *ac* ICEO model generated a steady radial velocity and a tendency, thereby, to aggregate nearby particles during *ac* polarization, was encouraging evidence in favor of the *ac* ICEO mechanism.

Santana-Solano [41] added more evidence by demonstrating that the frequency of rotation of spheres during aggregation was steady and proportional to $\boldsymbol{E}_\infty^2$, consistent with the *ac* ICEO mechanism. To explain the observation that particle pairs moved toward each other more slowly than their rotation would suggest, they invoked dipole repulsion but did not develop the model sufficiently to show that quantitative agreement between their observations and the hypothesized mechanisms. They concluded that their observations strongly support [41] the *ac* ICEO model as opposed to *ac* ECEO flow. Certainly their results are consistent with numerous observations of steady flow toward particles in this regard; however, we point out that *ac* ECEO would *not* be observable in their experiment because it is reversing at the frequency of the applied voltage, 100 Hz and above. One cannot see flow evidence of *ac* ECEO at the fundamental frequency unless the experiment can probe time at intervals much shorter than the period of oscillation and can capture particle movement within that period with a resolution much less than the particle radius. One cannot expect to observe the doubled frequency component of *ac* ICEO flow without special equipment, for the same reason.

Liu et al. [38] attempted quantitative reconciliation between their observations of the steady separation distance between aggregated particles and the ICEO theory. The calculated gap compared favorably to their predictions based on the ICEO theory, but did not capture a frequency above which particles were farther apart

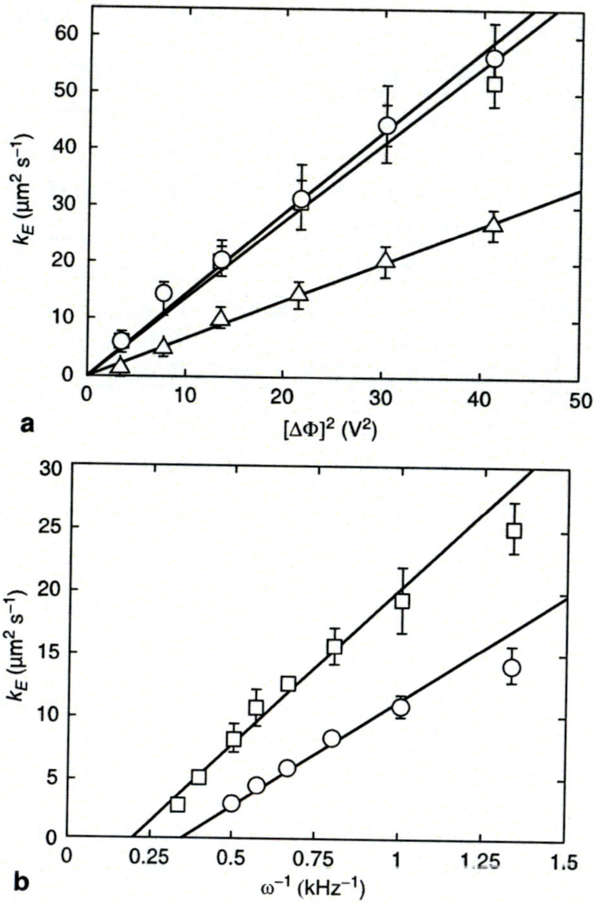

Fig. 1.24 **a** Aggregation rate constant k_E [34]. Particles were 2.7 μm polystyrene in 1.0 mM KCl. *Circles* and *squares:* two different particle mobilities, 500 Hz. *Triangles:* 1,000 Hz. *Lines* were fit to the data. The expected linear relationship between k_E and the square of the electric field (constant electrode separation) was obtained. Also, the rate constant was halved when the frequency was doubled, as predicted. **b** The aggregation rate constant k_E was inversely proportional to frequency above 1 kHz. The *circles* and *squares* corresponded to two different applied voltages. The solid lines were fit for frequencies of 1 kHz and larger, frequencies ten times larger than the minimum valid frequency according to the scaling. The value of k_E was a maximum at 500 Hz and much less than would be obtained by extrapolation of these graphs to a reciprocal frequency value of 2 kHz^{-1}. (Reprinted with permission from Ristenpart et al. [34], © 2004 American Institute of Physics)

even when dipole repulsion effects were included. Liu [45] reported on aggregation coefficients found in their own experiments based on the approach of Ristenpart [34]. Extending their frequency downward to 70 Hz, they reported again a sharply lower aggregation rate below 300 Hz and a gently declining rate above 300 Hz, as shown in Fig. 1.25. The decline of k_E with increasing frequency is consistent

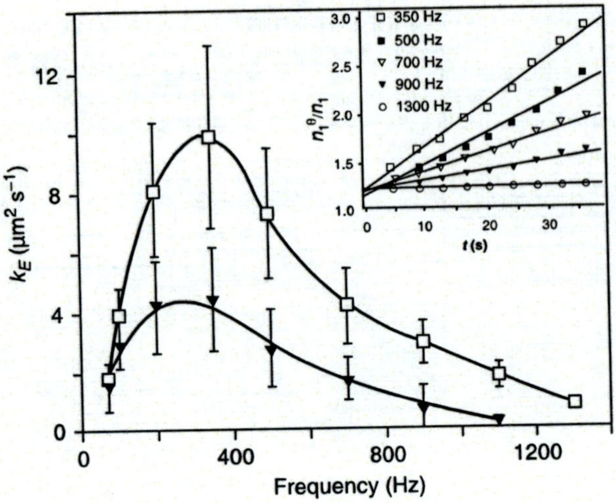

Fig. 1.25 Dependence of the aggregation coefficient kE on frequency, showing the decline at higher frequencies and a maximum around 300 Hz [45]. The aggregation coefficient as defined in [34] (see also Fig. 1.24). (Reprinted in part with permission from Liu et al. [45], © 2007 American Chemical Society)

with the predictions of the *ac* ICEO model in Eq. 1.55, but no model predicts both a lower and an upper extinction of the tendency to aggregate. Indeed, the work of Prieve and co-workers at 100 Hz [39, 44, 48] indicates that aggregation in chloride solution occurs at 100 Hz, which seems like a contradiction of a lower extinction frequency. The variation of the aggregation coefficient with particle radius agreed with theory, but the dependence of the tightness of the aggregate was not monotonic with the square of electric field.

Recognizing the inadequacy of the dipole approximation for calculating tangential electric fields at the electrode in the vicinity of a particle, Ristenpart et al. [42] included the influence of the particle in a subsequent comparison of theory and experiment. They reported agreement between particle tracking experiments and particle velocities measured experimentally in the vicinity of a 50 μm particle when they used surface conductivity as a fitting parameter. This value was then held constant, and the frequency was varied. The results appear in Fig. 1.26a for fixed frequencies and varying potential. Although the model fit the data well for the fixed frequency, it did not capture the dependence on frequency very well, as shown in Fig. 1.26b. This disagreement might stem from the dependence of the current distribution below the particle on frequency. The uniformity of capacitive current density along the surface of the electrode, which was addressed using numerical methods in Ristenpart et al. [42], can also be understood with the aid of a ratio of ohmic resistance introduced by the particle to capacitive impedance associated with the electrode, analogously with charge transfer impedance.

For a dielectric particle having negligible surface conductivity and $h/a \ll 1$, the following ratio characterizes the distribution of current on an ideally polarizable electrode adjacent to the particle. Here, the subscript "*ip*" indicates that the electrode is ideally polarizable, σ_e is the electrolyte conductivity, and C_{tot} is the overall double layer capacitance.

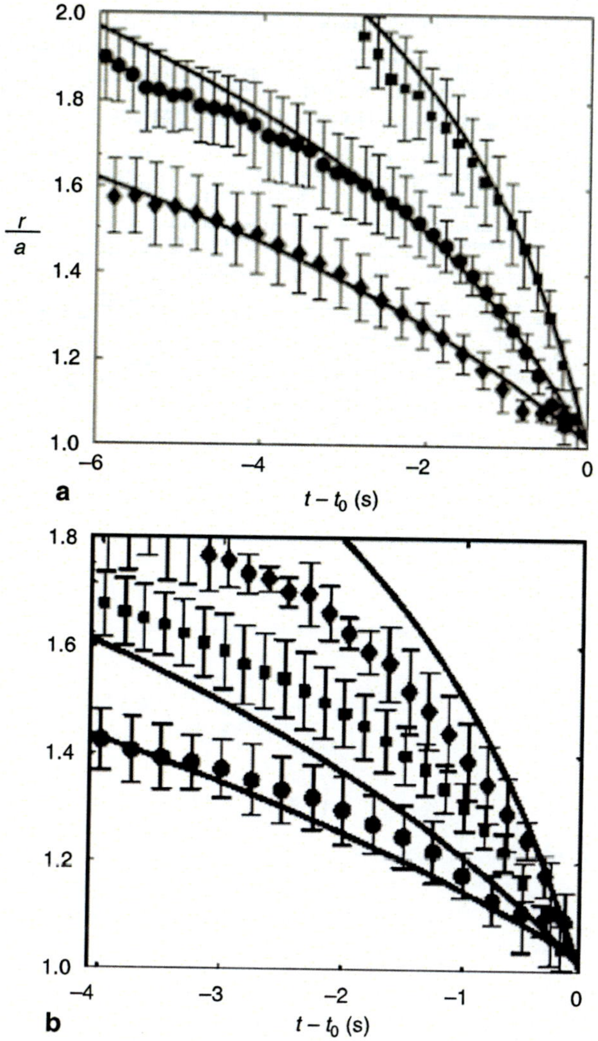

Fig. 1.26 Progress of 300 nm tracer particles toward a 50 μm target particle along the electrode, starting from time measured before the particle obscured the tracer [42]. **a** 750 Hz, potentials: 4, 6, 9 V. Model fit to data by adjusting the surface conductivity. **b** Frequencies: 0.5, 1, 1.5 kHz. Model results calculated using the surface conductivity found in (a). Part of the problem in the test of the model might have been that the large diameter of the particle, while useful for visualization, made the assumption of uniform current density less valid. (Ristenpart et al. [42],

$$J_{ip} \equiv \frac{Z_p}{Z_{tot}} = \frac{a\omega C_{tot}}{\sigma_e} \tag{1.56}$$

As J_{ip} decreases, the current distribution on the electrode beneath the particle becomes more uniform. In other words, for conductive fluid media and low frequencies (limited by the assumption of an ideally polarizable electrode), the impedance of the electrode dominates the current distribution and straightens the lines of current at the electrode. The smaller the fluid conductivity, the higher the frequency, and the larger the particle, however, the more the particle dictates the current distribution in its vicinity. Ristenpart et al. [42] stated that as particle radius increases, the assumption of uniform current distribution at the electrode becomes stronger because the diffuse layer collapses onto the electrode. Yet, this does not take into account the additional ohmic resistance presented by a larger particle, as shown above. Applying this perspective to Ristenpart's [42] calculations, one finds that J_{ip} was 0.17 for the calculations where the particle radius was 500 nm but was 17 for the 50 μm particle of the experiment. The dependence of the current distribution on frequency might explain the weaker observed dependence; as frequency increased, the current distribution shifted from dominance by electrode impedance to control by ohmic resistance, which mitigated the frequency dependence. The effect of frequency on *ac* ICEO flow, therefore, is at least twofold. First, the *ac* ICEO velocity of Eq. 1.55 depends explicitly on the reciprocal of the frequency, meaning that the velocity decreases as frequency increases. Second, increasing frequency short-circuits the impedance of the double layer, leading to ohmic control of the current density distribution and associated uniformity of the potential distribution beneath each particle. ICEO vanishes as the potential distribution becomes uniform.

One last consideration introduces a mystery into the ICEO mechanism of aggregation. Ristenpart et al. [34] noted "the electric stress (along the electrode) has a steady component … along with an oscillation at twice the imposed frequency." The doubled frequency ensues directly from the product of charge and electric field oscillating at the same frequency. Thus, the full expression of Eq. 1.55, obtained by taking the product of the real parts of the charge and electric field functions, is

$$u_r \sim \frac{3ar\,\varepsilon}{\mu\kappa} E_\infty^2 \left\{ \left(C_0' + \frac{D\kappa^2}{\omega} C_0'' \right) + \left(C_0' - \frac{D\kappa^2}{\omega} C_0'' \right) \cos 2\omega t + \left(\frac{D\kappa^2}{\omega} C_0' + C_0'' \right) \sin 2\omega t \right\} \tag{1.57}$$

The doubled frequency of oscillation of the velocity articulated in Eq. 1.57 is a fundamental feature of the ICEO theory; it stems directly from the product of the oscillating induced charge and the oscillating electric field. Observing the flow through a microscope with tracer particles, one expects to see only the net flow unless the framing rate of the camera and the magnification are such that the oscillatory motion of tracer particles could be resolved both in time and space, which would be difficult. The detailed vertical motion of a probe particle in response to ICEO with appropriately fast resolution in the time domain, however, should exhibit a doubled frequency because the inward aggregating flow must exert a normal force underneath each particle, as Fagan et al. [36] and recently Yariv [50] demonstrated. Fagan et al. [31] and Hoggard et al. [44] have observed and reported on the detailed response of single particles located near electrodes and exposed to *ac* fields. TIRM allowed acquisition of the particle motion with a resolution of one nanometer. Figures 1.12, 1.21, 1.22, and 1.27 are examples from their work at frequencies from 50 to 500 Hz. The particle height in these figures clearly oscillates *at the fundamental frequency;* there is no hint that the frequency of vertical oscillation of the particles is doubled. This lack of evidence of the ICEO mechanism at least requires an explanation; it undermines the importance of ICEO for aggregation.

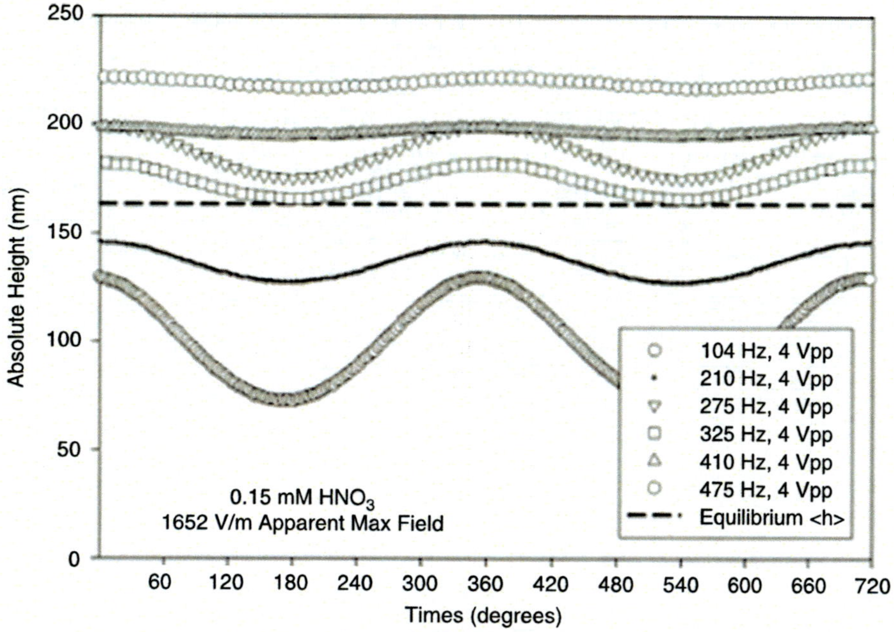

Fig. 1.27 The oscillation of a particle (polystyrenesulfonate, 4.6 μm nominal diameter, surfactant-free, polystyrene-amine subjected to a 1.652 kV/m nominal electric field in 0.15 mM nitric acid, at varying frequencies [37]. There is no evidence of a doubled frequency up to 475 Hz, the maximum frequency tested. The average height of the particle was lower than the equilibrium height below 275 Hz and above the equilibrium height at 275 Hz and above. (Reprinted in part with permission from Fagan et al. [37], © 2005 American Chemical Society)

1.6.2.3 Summary

The *ac* ICEO mechanism shows how an oscillating induced charge density and radial electric field components generate a steady flow that, in principle, can aggregate particles. Its strength is clear logic underpinning this rectification; experimental observations of different but related effects in several laboratories confirm this constant flow [28, 41, 42]. To our knowledge, however, no quantitative calculation based on fundamental theory and using independently measured parameters has accurately predicted the rate of lateral particle movement in the geometry of Fig. 1.3a to the accuracy that Prieve and co-workers [39, 44] have demonstrated with the *ac* ECEO mechanism of Sect. 1.5. Furthermore, experiments, in which the instantaneous movements of a probe particle were recorded, did not reveal the frequency doubling implicit in the model. The *ac* ICEO mechanism of particle aggregation is in a state analogous to the EHD theory of Sect. 1.4. Both mechanisms predict steady flow and declining strength of the flow as frequency increases, but neither has been shown to predict quantitatively (with independently determined parameters) observed pairwise motion; both are missing something. The EHD mechanism predicts that particles should not aggregate in KCl, but they clearly do aggregate. The *ac*

ICEO mechanism predicts that particles should oscillate at twice the frequency of the imposed voltage, but there is no evidence of doubled frequency in experiments able to observe that effect if it was significant [31, 44]. Finally, nothing in the *ac* ICEO mechanism accounts for the repulsion observed in KOH unless induced dipole repulsion dominates *ac* ICEO in KOH but not in KCl or $NaCO_3$ at the same ionic strength.

1.7 Directed Assembly Based on Induced Dipoles

Two mechanisms of particle aggregation require induced dipole moments on particles. Particles move by dielectrophoresis when the electric field gradient is nonzero; the velocity is proportional to the gradient of the electric field intensity [3]. The direction of motion is independent of the electric field's polarity but dependent on the sign of the Clausius-Mossotti function K that depends on the electrical properties of the particles. The direction of the velocity is often not the same direction as the direction of the electric field. Two clear cases where the field vector and the gradient are coincident are the case between two long concentric cylinder conductors and the case between two concentric spheres; the directions of dielectrophoresis and electrophoresis in these cases are identical. The field between two infinite flat plates, except for the narrow regions of the diffuse layers, and possibly mass transfer diffusion layers (for binary electrolyte), is uniform and no dielectrophoresis is expected. The gradients at the edges of parallel flat plates and between non-parallel flat plates usually propel particles along directions different from the electric field vector. Figure 1.3c shows this effect as the particle moves normal to the horizontal plane while the electric field lines follow curves from electrode to electrode. Induced dipoles interact with electric field gradients and with each other, either in attraction or repulsion. The three topics of this section, therefore, are basic theory of the induced dipole moment of particles, dielectrophoresis, and dipole-dipole interactions.

1.7.1 The Dipole Moment of a Spherical Particle

The dipole moment of a spherical particle is a topic both fundamental and complicated. The dipole moment appears both in the context of this section, dielectrophoresis, and in the context of dipole attraction and repulsion, the next section. The general form of the dipole moment of a spherical particle immersed in a medium is [5]

$$\tilde{\mathbf{p}}_{eff}(t) = 4\pi\varepsilon_m \tilde{K}(\omega) a^3 \mathbf{E}(t) \tag{1.58}$$

where p_{eff} is the dipole moment, ε_m is the dielectric permittivity of the continuous medium, K is the dipole coefficient, *a* is the particle radius, and $E(t)=E_\infty\ exp(-j\omega t)$

is the electric field. The tilde indicates a complex quantity. Two brief developments aid understanding of this form and the dipole coefficient. The potential around two equal point charges of opposite sign and separated by a distance d is given by an expansion of the superimposed potentials arising from each charge [5]

$$\phi(r,\theta) = \frac{q}{4\pi\varepsilon_m r_+{}^2} - \frac{q}{4\pi\varepsilon_m r_-{}^2} = \frac{qd\cos\theta}{4\pi\varepsilon_m r^2} + \frac{qd^3(3\cos^2\theta - 1)}{8\pi\varepsilon_m r^4} + \cdots. \tag{1.59}$$

The subscripts + and − indicate that distances from the positive and negative charges, respectively. One obtains a point dipole in the limit of small d; the first term expresses the value of the potential and the product qd becomes p_{eff}.

$$\phi(r,\theta) = \frac{p_{eff}\cos\theta}{4\pi\varepsilon_m r^2} \tag{1.60}$$

The potentials within and outside a homogeneous dielectric spherical particle p immersed in an infinite dielectric medium m, subject to an imposed electric field E_∞, and obtained by solving Laplace's equation in spherical coordinates, are

$$\phi_m(r,\theta) = -E_\infty r\cos\theta + \frac{A\cos\theta}{r^2} \quad \text{and} \quad \phi_p(r,\theta) = -Br\cos\theta \tag{1.61}$$

where $r>R$ for the potential ϕ_m and $r<R$ for potential ϕ_p. The quantities A and B are constants to be determined at the boundary of the particle. Equality of the potentials $\phi_m(R,\theta)=\phi_p(R,\theta)$ and continuity of the normal components of the electric displacement vectors at $r=R$ give

$$\varepsilon_m \left.\frac{\partial\phi_m}{\partial r}\right|_R = \varepsilon_p \left.\frac{\partial\phi_p}{\partial r}\right|_R \tag{1.62}$$

Applying these conditions to the equations for the potential, one finds a value for the constant A such that the potential in the medium is [5]

$$\phi_m(r,\theta) = -E_\infty r\cos\theta + \frac{\varepsilon_p - \varepsilon_m}{\varepsilon_2 + 2\varepsilon_p} R^3 E_\infty \frac{\cos\theta}{r^2}, \tag{1.63}$$

which allows equating the disturbance term in the exact result for ϕ_m, Eq. 1.63 to the point dipole relation Eq. 1.60.

$$\frac{p_{eff}\cos\theta}{4\pi\varepsilon_m r^2} = \frac{\varepsilon_p - \varepsilon_m}{\varepsilon_p + 2\varepsilon_m} R^3 E_\infty \frac{\cos\theta}{r^2} \Rightarrow p_{eff} = 4\pi\varepsilon_m K R^3 E_\infty \quad \text{where} \quad K \equiv \frac{\varepsilon_p - \varepsilon_m}{\varepsilon_p + 2\varepsilon_m} \tag{1.64}$$

This form of K also is appropriate for much more complicated situations, such as conducting particles and media in *ac* electric fields,

$$\tilde{p}_{eff} = 4\pi\varepsilon_m \tilde{K}(\omega) a^3 E_\infty \quad \text{where} \quad \tilde{K}(\omega) = \frac{\tilde{\varepsilon}_p - \tilde{\varepsilon}_m}{\tilde{\varepsilon}_p + 2\tilde{\varepsilon}_m} \text{ and } \tilde{\varepsilon}_i \equiv \varepsilon_i + j\frac{\sigma_i}{\omega}, \tag{1.65}$$

where σ is the electrical conductivity. The dielectric constants of the particle and the liquid, and the conductivity of the electrolyte in the context of aqueous salt solutions, are common physical properties. O'Konski [59] showed that a particle having both bulk conductivity and surface conductivity evinces Maxwell-Wagner polarization if one writes

$$\sigma_p = \sigma_p^b + \frac{2\sigma_p^s}{a} \tag{1.66}$$

where b denotes "bulk" and s denotes "surface." Green and Morgan [60] measured the dielectrophoretic response of submicron particles, varying conductivity, frequency, and particle size. They compared their results to predictions of Maxwell-Wagner theory for particles ranging from 557 nm down to 93 nm, taking surface conductivity into account. The agreement was best for the largest particles. The theory did not adequately explain the behavior of colloidal particles for all sizes. Saville et al. [61] argued that frequencies much higher than the reciprocal of the relaxation time for diffusion of ions over a particle radius leave mobile ions near the particle essentially in place. They used Bikerman's expression

$$\sigma_p^s = \lambda\sigma_e(1+3m)[e^{|e\zeta/2kT|} - 1] \quad \text{where } m \equiv \frac{2\varepsilon(kT)^2}{3\mu e^2 D_e} \tag{1.67}$$

to express the surface conductivity. Saville et al. [61] described this treatment as an extended Maxwell-Wagner limit of the dipole coefficient. They cited an estimate of the frequency at which this treatment should be accurate as σ_m/ε_m, which takes the value of 1 MHz for a 0.3 mM solution of KCl.

Zhou et al. [62] solving the Poisson-Boltzmann mean-field equations of colloidal electrostatics, calculated the polarizabilities of spherical particles at *dc* and over a range of frequencies, zeta potentials, and ionic strengths. Figure 1.28a is a plot of results for zero frequency and Fig. 1.28b is a sample plot over a frequency spectrum. Their results largely agreed with the extended Maxwell-Wagner theory of Saville et al. [61] at high frequencies where its assumptions were valid; at frequencies higher than σ_e/ε_e, the extended Maxwell-Wagner treatment provides a relatively simple estimate of the dipole coefficient. At low frequencies, the polarizability depended on zeta potential and the ionic strength; extrapolation of high frequency theories to low frequency can produce coefficients of the wrong sign.

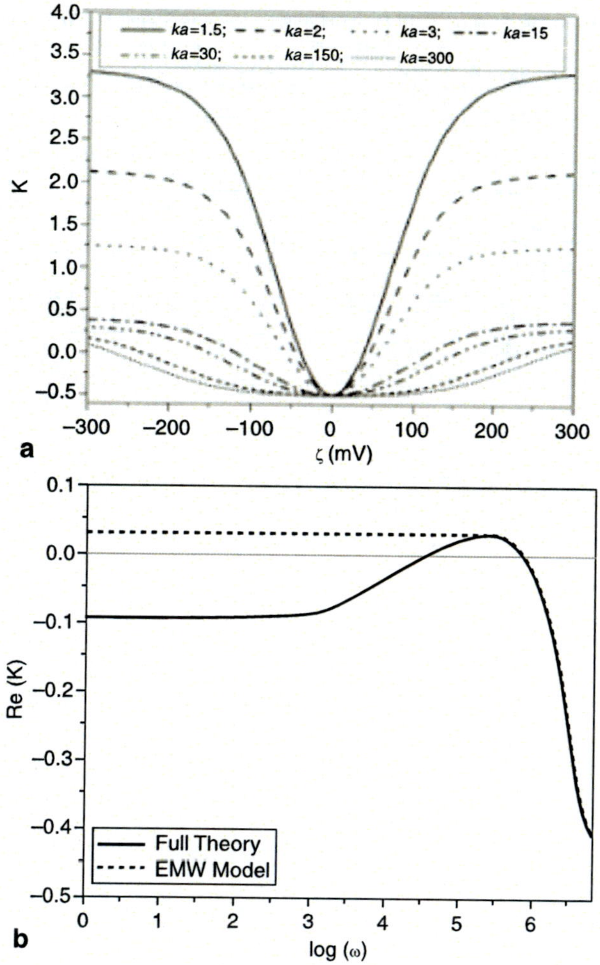

Fig 1.28 **a** Values of K (*top*) for static fields and **b** Re(K) (*bottom*) in *ac* fields for induced dipoles of charged dielectric particles [62]. In the static case, the dependence on the charge of the particle is strongest when the Debye length is comparable to the particle radius. $\kappa a = 150$ is still not perfectly thin. For *ac* polarization at high frequency, the model of Saville et al. [61] works well, but can give the wrong sign if used outside its range of validity. (Reprinted from Zhou et al. [62], © 2005 with permission from Elsevier)

1.7.2 Dielectrophoresis (DEP)

As mentioned previously, dielectrophoresis is particle migration arising from the interaction of a dipole with a gradient of electric field strength. Consider two opposite and equal charges q_1, q_2 separated by a distance s_x aligned with the x direction. If the associated electric field components in the x direction are $E_{1,x}$ and $E_{2,x}$, then the total force F_x is given by

$$F_x = q_1 E_{1,x} + q_2 E_{2,x} \cong -qE_{1,x} + qE_{1,x} + qs_x \frac{\partial E_x}{\partial x} = qs_x \frac{\partial E_x}{\partial x} = p_x \frac{\partial E_x}{\partial x}, \tag{1.68}$$

where p_x is a dipole moment. Equation 1.68 illustrates the origin of the force as the product of a dipole moment and a gradient of an electric field. More generally, the time-varying force on the particle is $\mathbf{F}(t) = \mathrm{Re}[\tilde{\mathbf{p}}_{eff}(t)] \cdot \nabla \mathrm{Re}[\mathbf{E}(t)]$, the steady part of which is [5]

$$\langle \mathbf{F}_{DEP}(t) \rangle = \frac{1}{2} \mathrm{Re}[\tilde{\mathbf{p}}_{eff}(t) \cdot \nabla \tilde{\mathbf{E}}(t)^*] = 2\pi \varepsilon_m a^3 \mathrm{Re}[\tilde{K}(\omega)] \nabla E_{rms}^2 \tag{1.69}$$

in which the asterisk designates the complex conjugate of the electric field.

Equation 1.69 expresses the essential features of dielectrophoresis, as summarized by Jones [5]. F_{DEP} is proportional to the volume of the particle and the dielectric permittivity of the liquid. Its force vector parallels the gradient of the electric field intensity and not the electric field vector except in special cases. Finally, the sign of the force depends on the sign of the frequency-dependent dipole coefficient K. Positive dielectrophoresis occurs when $K>0$ and negative dielectrophoresis occurs when $K<0$. A simple example illustrates this definition most clearly. Consider a long conductive rod concentric with a long conductive hollow cylinder and a dielectric fluid separating them. Solution of Laplace's equation in this geometry, when the outer cylinder is polarized positively with respect to the grounded rod, gives $\phi(r) \sim V \ln r$, where r is radial position measured from the axis and V is the applied voltage. Hence, $E_r \nabla E_r \sim -V^2 r^{-3}$ and F_{DEP} is negative for positive K and positive for negative K. The particle with positive K and, hence, positive dielectrophoresis moves toward the rod where the electric field intensity is higher. Under the opposite polarization, $\phi(r) \sim -V \ln r$, but $E_r \nabla E_r \sim -V^2 r^{-3}$, again. Thus, the direction of DEP is invariant with the sign of the polarization. This example also illustrates a scaling principle of dielectrophesis, that the dielectrophoretic force is sensitive to the square of the applied voltage and the reciprocal cube of the significant length of the electrode [63]. The steady part of the velocity of a particle in an electric field gradient is

$$\mathbf{v}_{DEP} = \frac{a^2 \varepsilon_m}{3\mu} \mathrm{Re}[\tilde{K}(\omega)] \nabla E_{rms}^2 \tag{1.70}$$

and is superimposed on components oscillating at twice the frequency. The complexities of effects governing the coefficient K were described in Sect. 1.7.1. The limits of the quantity in brackets are 1 and −1/2.

1.7.3 Dipole-Dipole Interactions

Consider first the force of interaction between two identical uncharged particles whose line of centers given by r_{12} is parallel to the imposed electric field. Jones [5] showed that the force of interaction between them can be calculated as a summation over all the induced multipoles, each having the form

$$(F_{\|})_{m,n} = (-1)^{n+1} \frac{(m+n+1)!\,p^{(m)}p^{(n)}}{4\pi\varepsilon_m r_{12}^{m+n+2}} \tag{1.71}$$

Jones [5] took F_p as positive for attraction and negative for repulsion. The terms $p^{(i)}$ are the mth and n^{th} multipolar moments. As the gap between the particles becomes large, the first term of Eq. 1.71, (m,n)=(1,1), is dominant.

$$F_{\mathrm{P}} = \frac{3!\,p^2}{4\pi\varepsilon_m r_{12}^4} \quad \text{for} \quad r_{12} \to \infty \tag{1.72}$$

The force is attractive and proportional to the inverse fourth power of the distance between particle centers. Equation 1.73 is a generalization of Eq. 1.72 for aligned dipoles whose line of centers makes an angle θ with the imposed electric field,

$$F(r_{12}, \theta) = \frac{(1 - 3\cos^2\theta)\, p_{eff}^2}{4\pi\varepsilon_m r_{12}^4}, \tag{1.73}$$

as used by Gong and Marr [24] (compare to Eq. 1.30) and as modified slightly by Zhou et al. [40], where p_{eff} is given by the real part of Eq. 1.58 and attractive force is taken as negative. The particles experience maximum attraction when a line joining their centers is aligned with the electric field and maximum repulsion when that line is perpendicular to the electric field.

The force dependence on particle separation changes as the gap between the particles decreases. In the limit of zero gap, the force approaches a constant asymptotic limit for any finite ratio of the permittivity of the particle to the permittivity of the medium [5]. The constant of proportionality depends on this permittivity ratio.

$$F_{\mathrm{P}} \propto \varepsilon_m \pi a^2 E_o^2 \quad \text{for} \quad r_{12} \to 2a \tag{1.74}$$

Equation 1.74 provides a basis for the chaining force cited by Velev when the particles are touching [7].

$$F_{\|} \sim \frac{p_{eff}^2}{4\pi\varepsilon_m a^4} \sim \frac{(4\pi\varepsilon_m a^3 K E_o)^2}{4\pi\varepsilon_m a^4} \sim \varepsilon_m \pi a^2 K^2 E_o^2 \tag{1.75}$$

1.8 Capabilities and Applications

In this final section we highlight a few demonstrated and prospective capabilities. There is a rich supply of both experimental and theoretical contributions for study. The interested reader is referred to the extensive descriptions of observations and demonstrated effects in the reviews cited in the beginning of this contribution [4, 5, 7, 8, 9, 10]. In addition to the work mentioned, but by no means an exhaustive list, we point out a few additional fundamental contributions. A series of thorough reports on *ac* driven electroosmotic flow described measurements [64], theory [65], and observations [66] on an in-plane geometry as in Fig. 1.3c. Castellanos et al. [67] surveyed scaling laws for a wide variety of phenomena. Bazant and Squires [68, 69] described the theory of induced charge electromosis. Duval et al. [70–72] investigated the electrified interface during passing of faradaic current. Prieve [73] obtained an asymptotic solution for one-dimensional, steady transport with of a symmetric binary electrolyte with a redox reaction for thin Debye lengths, which allowed theoretical exploration of the effect of faradaic current on zeta potential. Delgado et al. [74] discussed measurement and interpretation of electrokinetic phenomena. Squires [75] recently surveyed *ac* ICEO for use in on-chip technologies. Following the cue of Velev and Gupta, [9], we use dimensionality as a way to finish the discussion, stopping after 2-D ordering because 3-D assembly is beyond the scope of this contribution. Finally, the application of concepts described herein to what might be called zero dimension ordering, i.e. manipulation of single particles, is mentioned and a potential use of the principles discussed in this contribution is identified.

1.8.1 Directed Assembly of Particles in 1-D

Velev and co-workers [2, 76] investigated aggregation of 15–30 nm gold nanoparticles in one dimension as shown in Fig. 1.2 at the outset of this contribution. The deposition technique involved *ac* polarization of in-plane electrodes at 50–250 Hz. The aggregation produced wires that grew as fast as 50 μm per second over millimeter length scales. The geometry was in-plane as in Fig. 1.3c. Two types of growth were observed, one where the wires grew through the solution in a low arc and one where the wires grew along the surface. Nanoparticles concentrated in the area around the growing tip and there was electroosmotic flow in the vicinity. Substantial electric field gradients very near the growing tips indicated that dielectrophoresis was driving the growth locally, but the process seemed to be diffusion limited because the rate did not depend on the proximity of the growing conductor to the opposing electrode and did depend on the overall concentration of nanoparticles. Formation of the wires was irreversible and they behaved ohmically when connecting two conductors; the wires even were self-repairing; the large fields at the break brought particles back to make the connection. More recently Gupta et al. [77] demonstrated one dimensional chains of cells and particles. Particles moved to cells by dielectrophoresis and became inserted into the chains. Molecular binding ligands permanently tied cells together as shown in Fig. 1.29.

Hoggard et al. [48] reported 1-D reversible chaining of micron scale particles during *ac* polarization of densely packed 2D ensembles of particles. The geometry was parallel plates sandwiching electrolyte between them. Having previously found that particle pairs aggregate, for example, in bicarbonate solution and sepa-

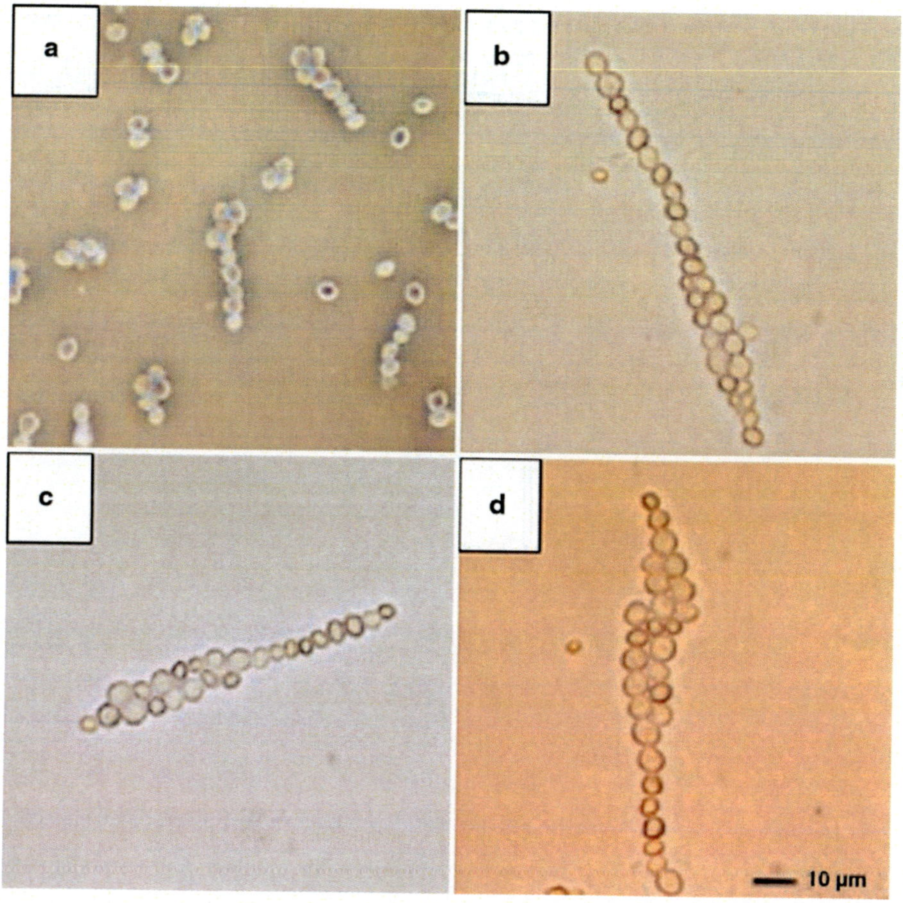

Fig. 1.29 Gupta et al. [77] demonstrated permanent binding of yeast cell chains after assembly using **a** poly(dimethyl aminoethyl methacrylate) (PDMAEMA), **b** concanavalin-A, **c** D-mannose, and **d** both concanavalin-A lectin and D-mannose. (Reprinted in part with permission from Gupta et al. [77], © 2010 American Chemical Society)

rate in KOH, Hoggard et al. [48] experimented with particle ensembles to determine what the conclusions reached for particle pairs meant for ensembles. Appearing in Fig. 1.30, particles aggregated in bicarbonate as expected and repelled each other in KOH, as found for particle pairs. The particles in bicarbonate formed an ordered array with spacing less than the particle diameter. The particles in KOH, however, had no vacant areas to occupy when repulsive forces were suddenly present; they responded by forming one dimensional "hairs" approximately 5–10 particles long extending along the normally directed field lines, another example of chaining. The particle hairs collapsed when the polarization was terminated.

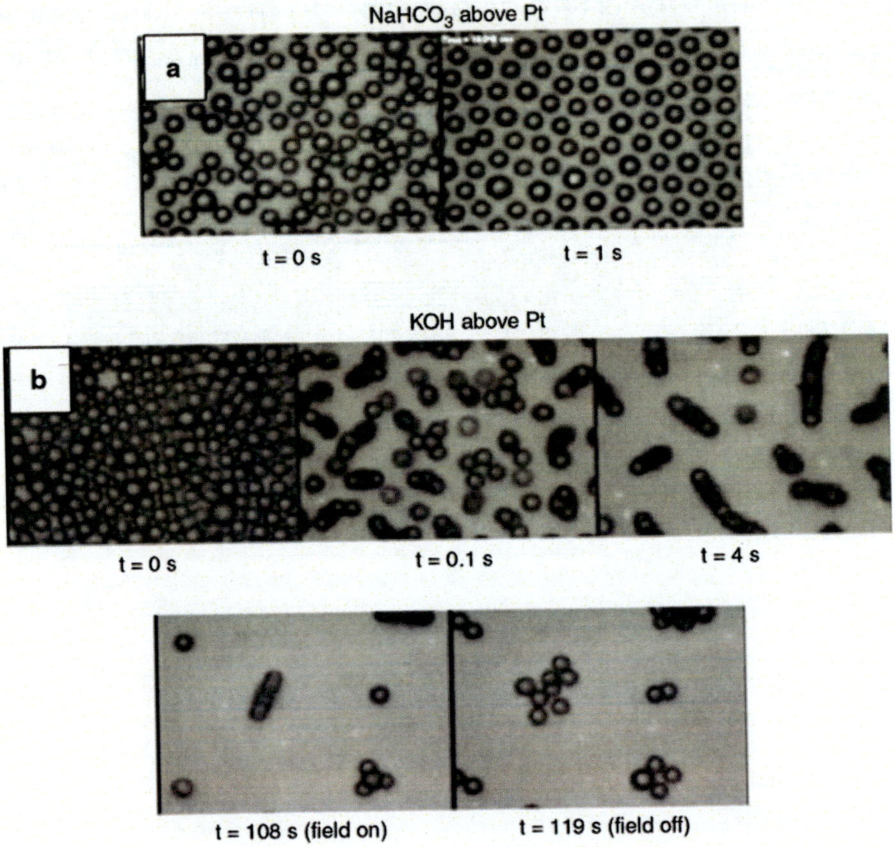

Fig. 1.30 Evidence of the contrasting behavior of particles in KOH resulting in particle based "hairs" [48]. A large spike of voltage was introduced at twice the frequency. The bicarbonate particles aggregated as usual, but the particles in KOH repelled each other so severely, they chained in the direction perpendicular to the electrode in the form of "hairs" that broke apart when the electric field was turned off. (Reprinted in part with permission from Hoggard et al. [48], © 2008 American Chemical Society)

1.8.2 Directed Assembly of Particles in 2-D

Ordering in two dimensions is well established; it is the main phenomenon driving this survey of mechanisms. Similar results can be reached from a variety of systems. Aggregates are formed by polarization between opposing planar electrodes [1, 14, 18, 19, 35] whether in *ac* or *dc* polarization. Particles can be assembled between electrodes of an in-plane geometry where the particles first form chains that are deposited and grow into ordered arrays as shown in Fig. 1.31 [78]. Particles in confinement form a variety of patterns [24, 25]. Patterned electrodes can draw particles in from neighboring areas; conductive silicon patches exposed in a field of silicon

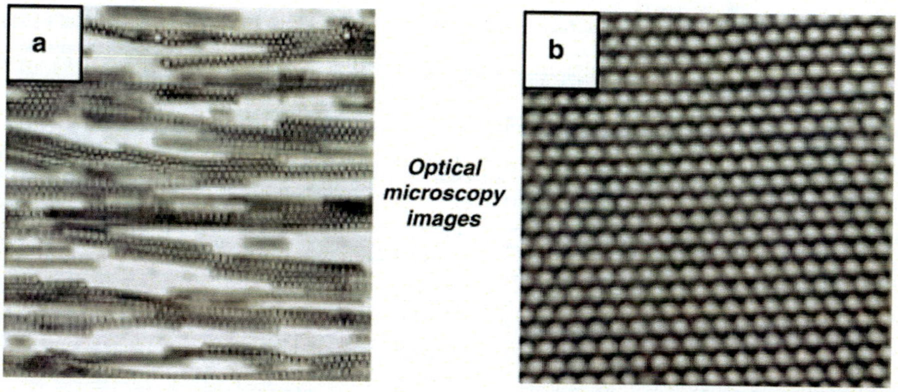

Fig. 1.31 Growth of ordered arrays in an in-plane geometry (such as Fig. 1.3c) by growing chains via field-induced dipoles that coalesce into *hexagonal arrays* [78]. The particles were 1.4 μm latex. (Reprinted with permission from Lumsdon et al. [78], © 2003, American Institute of Physics)

covered by photoresist formed conductive "corrals" for particles, [79] as shown in Fig. 1.32. Ristenpart et al. [80] demonstrated the formation of superlattices of colloidal particles. Two dimensional directed assembly can be applied to gathering of cells, as mentioned previously [33]. Zhou et al. [81] extended their observation of cell assembly to demonstrate the separation of particles from cells and different types of cells from each other by adjusting the mixture of dielectrophoresis and electroosmosis at the edges of patterned electrodes, as shown in Fig. 1.33. They also investigated processes by which commingled particles and cells were separated by exposure to *ac* electric fields [81]. They concluded that the interplay of electroosmotic and induced dipole effects governed the behavior. Their work highlighted the importance of differences in dielectric response between the cells and the polystyrene particles.

Fig. 1.32 Collection of 2 μm latex particles from 0.5 mM NaCl at 4 kV/m on exposed conductive "corrals" [79]. The length scale of the corral is the same as the photograph; the particles form the dark shape in the center, away from the edges. (Reprinted in part with permission from Bhatt et al. [79], © 2005 American Chemical Society)

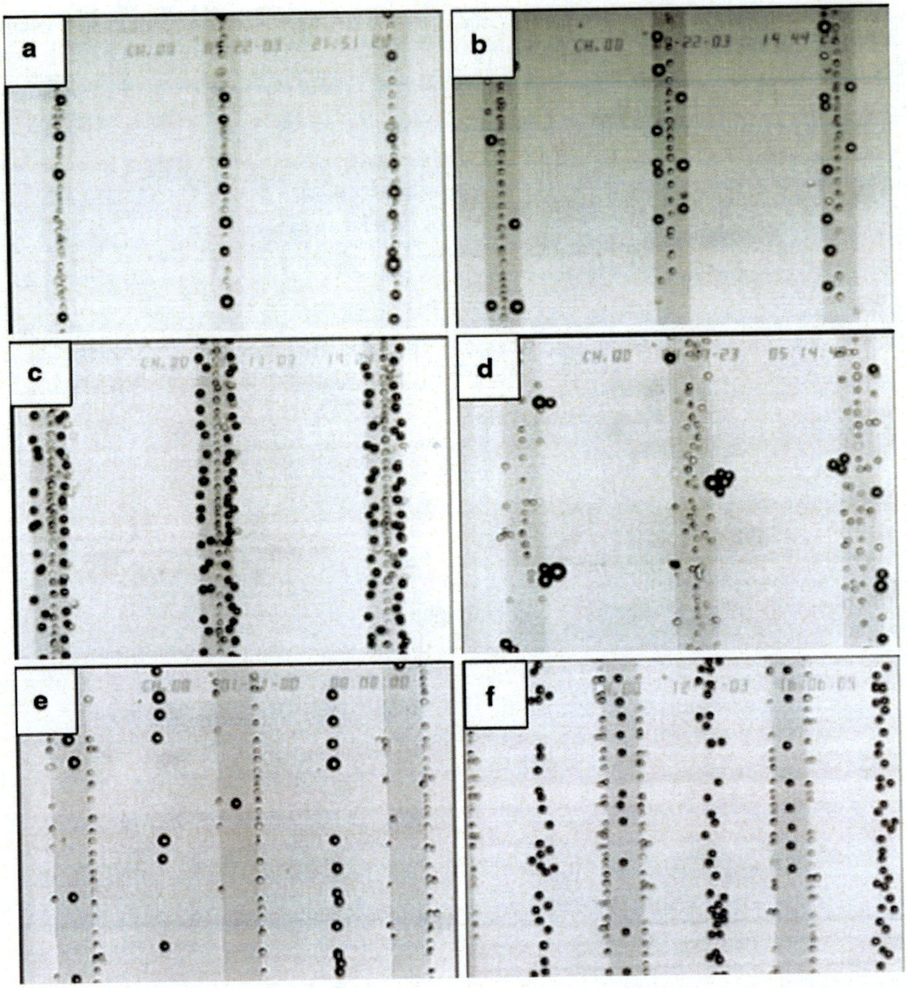

Fig. 1.33 Sorting of polystyrene-yeast mixtures by interactions at patterned electrodes [81]. The particles are the larger entities. The darker strips are ITO. **a** Particles and yeast cells mix on the centerline at 60 Hz and 0.17 V. **b** Yeast remains on the centerline but the particles move to the edges of the strips. **c** same as (b) but a higher concentration of particles. **d** separation and aggregation at 1 kHz. **e, f** complete separation at 1 MHz. (Reprinted from Zhou et al. [81], © 2005 with permission from Elsevier)

1.8.3 Directed Assembly of Particles in 0-D

Ordering in 0-D, though mildly oxymoronic, means control of a single particle by many of the mechanisms described. For example, Jones [5] describes levitation of particles by dielectrophoresis. One can write a force balance on a particle based on Eq. 1.69 and gravity.

$$\mathbf{F} = -2\pi a^3 \varepsilon_m \mathrm{Re}[\tilde{K}(\omega)] \nabla E_{rms}^2 + m\mathbf{g} \tag{1.76}$$

where $\mathbf{F}$ is the time-averaged net force on the particle. Clearly, the net force $\mathbf{F}$ on the particle only can be zero if the dipole coefficient is less than zero, but the question remains whether the particle is stable at that point. Consider a particle's electromechanical potential energy function ψ such that the force on a particle can be expressed as $-\nabla\psi$. A particle is stable at a point in space if $\nabla^2\psi > 0$ [5]. Applying this operator to a particle at a point in space, one obtains the relationship

$$\nabla^2\psi = -2\pi a^3 \varepsilon_m \mathrm{Re}[\tilde{K}(\omega)] \nabla^2 E_{rms}^2. \tag{1.77}$$

Using the properties of the electric field, one can show that the Laplacian of the electric field intensity is always greater than or equal to zero, $\nabla^2\mathbf{E}^2 \geq 0$, [5] away from distributions of unbalanced charge. Only minima in the electric field intensity can exist, which implies that the stability criterion also can be satisfied if the dipole coefficient is negative. The net force on a particle, therefore, can be zeroed in Eq. 1.76 and the particle rendered passively stable at that position if $\mathrm{Re}[\tilde{K}(\omega)] < 0$, which means only particles exhibiting negative dielectrophoresis. Jones [5] cites applications for such passive levitation, such as measurement of dipole coefficients and levitation of biological cells. Passive levitation requires working in frequency ranges where the dipole coefficient is negative; levitation of particles in positive dielectrophoresis is likewise possible if a feedback mechanism is provided [5].

Passive levitation is also possible by electroosmosis. If a particle sediments to a horizontal surface but does not adsorb, it samples a potential well created by the balance of double layer repulsion and gravity. Prieve and co-workers [31, 36, 37, 39] have shown that *ac* electric fields applied normal to the electrode move the particle toward the electrode or away, but it follows also that particles are pulled closer on average to the electrode or pushed away from it under *dc* polarization; the electric field contributes to the force balance. As a simple analysis, consider the balance of forces represented by $F_{dl} + F_g + F_{ep} = 0$ where the subscripts *dl, g,* and *ep* denote the double-layer, gravity, and electrophoresis, respectively. Using the expression for the electrophoretic force term, Eq. 1.5, one writes Eq. 1.78 for the forces on a particle near a charged surface when current is passing, including double layer repulsion, gravity, and the electrophoretic force, respectively.

$$\frac{B}{\lambda} e^{-h_m/\lambda} - \frac{4\pi a^3 g(\rho_p - \rho_e)}{3} + \frac{q(h_m) 6\pi a \varepsilon \zeta_p}{\sigma_e} i = 0 \tag{1.78}$$

(When the gap between the particle and the electrode is greater than approximately 100 nm, van der Waals forces can be assumed negligible and are neglected for simplicity.) In Eq. 1.78, λ is the Debye length, h_m is the gap between the particle and the electrode, a is the particle radius, ρ is density, ε is the permittivity of the solvent, ζ_p is zeta potential of the particle, and σ_e is the conductivity of the liquid. The third term of Eq. 1.78 expresses the effect of the imposed electric field as the electrophoretic "force" where use of Ohm's Law replaces the electric field by the quotient of the current density divided by the conductivity. The coefficient q in Eq. 1.78 expresses wall hindrance to motion normal to the electrode. A positive current and negative zeta potential on the particle pulls the particle closer to the electrode; thus, a particle can be raised or lowered, on average, relative to the electrode, an example of zero dimensional ordering in the sense of this section.

The balance expressed in Eq. 1.78 contains an interesting possibility for application of 0-D ordering as an imaging ammeter, [82] when viewed differently. The intensity of light scattered from a particle at height h_m above the electrode is expressed in Eq. 1.79,

$$I = I_o e^{-\beta h_m}, \tag{1.79}$$

where I is scattering intensity, I_o is the intensity of a particle in contact with the electrode, and β^{-1} is the decay length of the evanescent radiation used in TIRM [32]. Writing Eqs. 1.78 and 1.79 twice, once for the intensity at the equilibrium particle height in the absence of current, and once for the equilibrium height in the presence of current, one can solve for the current density as a function of the measured light intensity, Eq. 1.80.

$$i = \frac{2}{9}\frac{(\rho_p - \rho_w)g\sigma_e a^2}{q(h_m)\,\varepsilon\,|\zeta_p|}\left[\left(\frac{I}{I_{eq}}\right)^{\frac{1}{\beta\lambda}} - 1\right] \tag{1.80}$$

I_{eq} is the scattering intensity in the absence of current. Equation 1.80 is an equation of imaging amperometry based on scattered light. It makes possible the probing of local electrochemical current density on the length scale of a colloidal particle diameter without the need for isolated electrodes. Investigation of this potential application has begun in our laboratories.

1.9 Concluding Remarks

Investigators in laboratories around the globe have observed electrically driven motion and organization of colloidal particles and cells on electrodes in a variety of circumstances. Much work has been done and capabilities have been identified. The toolbox of electrically driven directed assembly, consisting of electrophoresis, dielectrophoresis, electroosmosis, and interparticle dipole interactions, provides the means for producing a variety of structures. The real issue henceforth is identifying valuable applications.

With regard to the ordering in the geometry of Fig. 1.3a, the evidence and foregoing discussion justify some assertions. For *dc* electric fields in the geometry of Fig. 1.3a, electroosmosis arising from the interaction of the primary electric field with the equilibrium charge on the particle accounts for particle behavior; this *dc* mechanism promotes fast and reversible assembly of colloidal particles on an electrode. *dc* polarization, however, continuously produces species that change the composition of the electrolyte in the vicinity of the electrode, which might limit its usefulness.

For *ac* electric fields in the geometry of Fig. 1.3a, however, the situation becomes complicated even though the ordering itself during *ac* polarization has been reproduced in multiple laboratories. It is tempting to accept the observed rectified flow predicted by the *ac* ICEO mechanism as the explanation. Two observations, however, need exploration. First, particles oscillate *at the frequency of the applied field*, as would be expected in *ac* ECEO, but the particle *does not* oscillate at the doubled frequency predicted by the *ac* ICEO mechanism. Second, the ICEO mechanism does not obviously predict both aggregation of particles in bicarbonate and chloride solution and the observed repulsion in potassium hydroxide and other solutions. The *ac* ECEO mechanism, in conjunction with the experimentally measured phase angle between the particle height and the electric field, predicts electrolyte dependent attraction or repulsion, and evidence exists that the electrolyte dependence holds both at low frequencies where faradaic reactions might occur and at high frequencies where the electrode behaves as ideally polarizable. Furthermore, the phase angle can change from greater than $\pi/2$ to less than $\pi/2$ in some cases, manifesting itself as an apparent "cutoff frequency". The complete explanation of the phase angle and its electrolyte dependence, while partially understood as arising from double layer repulsion, remains elusive. Nothing in the *ac* ECEO mechanism, however, generates the steady flow toward an individual particle observed in several laboratories.

A hypothesis might resolve the paradox described in the preceding paragraphs and serve as a basis for further inquiry. Since the *ac* ECEO mechanism and the *ac* ICEO mechanism are not mutually exclusive, both are active and superimposed in the geometry of Fig. 1.3a. If the *ac* ICEO mechanism governs the observed steady flow but not the aggregation, and if the *ac* ECEO mechanism governs the aggregation rate, the paradox is largely resolved.

Acknowledgement The authors acknowledge the support of the National Science Foundation for research on this topic through grants CTS0089875, CTS0338089, and CBET0730391.

References

1. Böhmer, M: In situ observation of 2-dimensional clustering during electrophoretic deposition. Langmuir **12**, 5747–5750 (1996)
2. Hermanson, K.D., Lumsdon, S.O., Williams, J.P., Kaler, E.W., Velev, O.D.: Dielectrophoretic assembly of electrically functional microwires from nanoparticle suspensions. Science **294**, 1082–1086 (2001)
3. Pohl, H.: The motion and precipitation of suspensoids in divergent electric fields. J. Appl. Phys. **22**, 871 (1951)
4. Pohl, H.A.: Dielectrophoresis. Cambridge University Press, Cambridge (1978)
5. Jones, T.B.: Electromechanics of Particles. Cambridge University Press, Cambridge (1995)

6. Green, N.G., Ramos, A., Morgan, H.: Ac electrokinetics: a survey of sub-micrometre particle dynamics. J. Phys. D: Appl. Phys. **33**, 632–641 (2000)
7. Velev, O.D.: Assembly of electrically functional microstructures from colloidal particles. In: Caruso F. (ed.) Colloids and Colloid Assemblies. Wiley, Weinheim (2004)
8. Velev, O.D., Bhatt, K.H.: On-chip micromanipulation and assembly of colloidal particles by electric fields. Soft Matter **2**, 738–750 (2006)
9. Velev, O.D., Gupta, S.: Materials fabricated by micro- and nanoparticle assembly—the challenging path from science to engineering. Adv. Mater. **21**, 1897–1905 (2009)
10. Velev, O.D., Gangwal, S., Petsev, D.N.: Particle-localized AC and DC manipulation and electrokinetics. Annu. Rep. Prog. Chem. Sect. C: Phys. Chem. **105**, 213–246 (2009)
11. Lyklema, J.: Fundamentals of Interface and Colloid Science, vol. 4. Elsevier, Amsterdam (2005)
12. Russel, W.B., Saville, D.A., Schowalter, W.R.: Colloidal Dispersions. Cambridge University Press, Cambridge (1989)
13. Prieve, D.C., Sides, P.J., Wirth, C.L.: Current Opinion in Colloid and Interface Science. doi:10.1016/j.cocis.2010.01.005 (2010)
14. Richetti, P., Prost, J.F., Barois, P.: Two-dimensional aggregation and crystallization of a colloidal suspension of latex spheres. J. Phys. Lett. **45**, 1137–1143 (1984)
15. Giersig, M., Mulvaney, P.: Formation of ordered two-dimensional gold colloid lattices by electrophoretic deposition. J. Phys. Chem. **97**, 6334–6336 (1993)
16. Giersig, M., Mulvaney, P.: Preparation of ordered colloid monolayers by electrophoretic deposition. Langmuir **9**, 3408–3413 (1993)
17. Van Der Biest, O.O., Vandeperre, L.J.: Electrophoretic deposition of material. Annu. Rev. Mater. Sci. **29**, 327–352 (1999)
18. Trau, M., Saville, D.A., Aksay, I.A.: Field-induced layering of colloidal crystals. Science **272**, 706–709 (1996)
19. Yeh, S.R., Seul, M., Shraiman, B.I.: Assembly of ordered colloidal aggregates by electric-field-induced fluid flow. Nature **386**, 57–59 (1997)
20. Solomentsev, Y., Böhmer, M., Anderson, J.L.: Particle clustering and pattern formation during electrophoretic deposition: a hydrodynamic model. Langmuir **13**, 6058–6068 (1997)
21. Trau, M., Saville, D.A., Aksay, I.A.: Assembly of colloidal crystals at electrode interfaces. Langmuir **13**, 6375–6381 (1997)
22. Solomentsev, Y., Guelcher, S.A., Bevan, M., Anderson, J.L.: Aggregation dynamics for two particles during electrophoretic deposition under steady fields. Langmuir **16**, 9208–9216 (2000)
23. Guelcher, S.A., Solomentsev, Y., Anderson, J.L.: Aggregation of pairs of particles on electrodes during electrophoretic deposition. Powder Technol. **110**, 90–97 (2000)
24. Gong, T.Y., Wu, D.T., Marr, D.W.M.: Electrically switchable colloidal ordering in confined geometries. Langmuir **17**, 2301–2304 (2001)
25. Gong, T.Y., Wu, D.T., Marr, D.W.M.: Two-dimensional electrohydrodynamically induced colloidal phases. Langmuir **18**, 10064–10067 (2002)
26. Sides, P.J.: Electrohydrodynamic particle aggregation on an electrode driven by an alternating electric field normal to it. Langmuir **17**, 5791–5800 (2001)
27. Sides, P.J.: Calculation of electrohydrodynamic flow around a single particle on an electrode. Langmuir **19**, 2745–2751 (2003)
28. Nadal, F., Argoul, F., Hanusse, P., Pouligny, B., Ajdari, A.: Electrically induced interactions between colloidal particles in the vicinity of a conducting plane. Phys. Rev. E. **65**, 061409-1-5 (2002)
29. Kim, J., Guelcher, S.A., Garoff, S., Anderson, J.L.: Two-particle dynamics on an electrode in AC electric fields. Adv. Coll. Interface Sci. **96**, 131–142 (2002)
30. Kim, J., Anderson, J.L., Garoff, S., Sides, P.J.: Effects of zeta potential and electrolyte on particle interactions on an electrode under AC polarization. Langmuir **18**, 5387–5391 (2002)
31. Fagan, J.A., Sides, P.J., Prieve, D.C.: Vertical oscillatory motion of a single colloidal particle adjacent to an electrode in an AC electric field. Langmuir **18**, 7810–7820 (2002)

32. Prieve, D.C.: Measurement of colloidal forces with TIRM. Adv. Coll. Interface Sci. **82**, 93–125 (1999)
33. Tilton, R.D., Brisson, V.: Self-assembly and two-dimensional patterning of cell arrays by electrophoretic deposition. Biotechnol. Bioeng. **77**, 290–295 (2002)
34. Ristenpart, W.D., Aksay, I.A., Saville, D.A.: Assembly of colloidal aggregates by electrohydrodynamic flow: kinetic experiments and scaling analysis. Phys. Rev. E. **69**, 021405 (2004)
35. Zhang, K.Q., Liu, X.Y.: In situ observation of colloidal monolayer nucleation driven by an alternating electric field. Nature **429**, 739–743 (2004)
36. Fagan, J.A., Sides, P.J., Prieve, D.C.: Vertical motion of a charged colloidal particle near an AC polarized electrode with a nonuniform potential distribution: theory and experimental evidence. Langmuir **20**, 4823–4834 (2004)
37. Fagan, J.A., Sides, P.J., Prieve, D.C.: Evidence of multiple electrohydrodynamic forces acting on a colloidal particle near an electrode due to an alternating current electric field. Langmuir **21**, 1784–1794 (2005)
38. Liu, Y., Narayanan, J., Liu, X.Y.: Colloidal phase transition driven by alternating electric field. J. Chem. Phys. **124**, 124906 (2006)
39. Fagan, J.A., Sides, P.J., Prieve, D.C.: Mechanism of rectified lateral motion of particles near electrodes in alternating electric fields below 1 kHz. Langmuir **22**, 9846–9852 (2006)
40. Zhou, H., White, L.R., Tilton, R.D.: Microphase separation during binary electrophoretic deposition of particles with dissimilar polarizabilities. Coll. Surf. A. **277**, 119–130 (2006)
41. Santana-Solano, J., Wu, D.T., Marr, D.W.M.: Direct measurement of colloidal particle rotation and field dependence in alternating current electrohydrodynamic flows. Langmuir **22**, 5932–5936 (2006)
42. Ristenpart, W.D., Aksay, I.A., Saville, D.A.: Electrohydrodynamic flow around a colloidal particle near an electrode with an oscillating potential. J. Fluid. Mech. **575**, 83–109 (2007)
43. Ristenpart, W.D., Aksay, I.A., Saville, D.A.: Electrically driven flow near a colloidal particle close to an electrode with a faradaic current. Langmuir **23**, 4071–4080 (2007)
44. Hoggard, J.D., Sides, P.J., Prieve, D.C.: Electrolyte-dependent pairwise particle motion near electrodes at frequencies below 1 kHz. Langmuir **23**, 6983–6990 (2007)
45. Liu, Y., Liu, X.Y., Narayanan, J.: Kinetics and equilibrium distribution of colloidal assembly under an alternating electric field and correlation to degree of perfection of colloidal crystals. J. Phys. Chem. C. **111**, 995–998 (2007)
46. Zhang, K.Q., Liu, X.Y.: Size-dependent planar colloidal crystals guided by alternating electric field. Appl. Phys. Lett. **90**, 111911(2007)
47. Liu, Y., Xie, R.G., Liu, X.Y.: Fine tuning of equilibrium distance of two-dimensional colloidal assembly under an alternating electric field. Appl. Phys. Lett. **91**, 063105 (2007)
48. Hoggard, J.D., Sides, P.J., Prieve, D.C.: Electrolyte-dependent multiparticle motion near electrodes in oscillating electric fields. Langmuir **24**, 2977–2982 (2008)
49. Xie, R.G., Liu, X.Y.: Epitaxial assembly and ordering of two-dimensional colloidal crystals. Appl. Phys. Lett. **92**, 083106 (2008)
50. Yariv, E.: Electro-hydrodynamic particle levitation on electrodes. J. Fluid. Mech. **645**, 187–210 (2010)
51. O'Brien, R.W., White, L.R.: Electrophoretic mobility of a spherical colloidal particle. J. Chem. Soc. Far. Trans. **2**(74), 1607–1626 (1978)
52. Khair, A.S., Squires, T.E.: The influence of hydrodynamic slip on the electrophoretic mobility of a spherical colloidal particle. Phys. Fluid. **21**, 042001 (2009)
53. Keh, H.J., Lien, L.C.: Electrophoresis of a dielectric sphere normal to a large conducting plane. J. Chin. Inst. Chem. Eng. **20**, 283–290 (1989)
54. Brenner, H.: The slow motion of a sphere through a viscous fluid towards a plane surface. Chem. Eng. Sci. **16**, 242–251 (1961)
55. Newman, J.: Electrochemical Systems, 3rd. edn. Wiley, Hoboken (2004)
56. Breiter, M., Kleinerman, M., Delahay, P.: Structure of the double layer and electrode processes. J. Am. Chem. Soc. **80**, 5111–5117 (1958)

57. Hu, K., Fan, F.-R.F., Bard, A.J., Hillier, A.C.: Direct measurement of diffuse double-layer forces at the semiconductor/electrolyte interface using an atomic force microscope. J. Phys. Chem. B. **101**, 8298–8303 (1997)
58. Hollingsworth, A.D., Saville, D.A.: A broad frequency range dielectric spectrometer for colloidal suspensions: cell design, calibration, and validation. J. Coll. Interface Sci. **257**, 65–76 (2003)
59. O'Konski, C.: Electric properties of macromolecules. V. Theory of ionic polarization in polyelectrolytes. J. Phys. Chem. **64**, 605–619 (1960)
60. Green, N.G., Morgan, H.: Dielectrophoresis of submicrometer latex spheres. 1. Experimental results. J. Phys. Chem. B. **103**, 41–50 (1999)
61. Saville, D.A., Bellini, T., Degiorgio, V., Mantegazza, F.: An extended Maxwell–Wagner theory for the electric birefringence of charged colloids. J. Chem. Phys. **113**, 6974–6983 (2000)
62. Zhou, H., Preston, M.A., Tilton, R.D., White, L.R.: Calculation of the electric polarizability of a charged spherical dielectric particle by the theory of colloidal electrokinetics J. Coll. Interface Sci. **285**, 845–856 (2005)
63. Bahaj, A.S., Bailey, A.G.: Dielectrophoresis of small particles. Proc. IEEE/IAS Annual Meeting Cleveland OH October:154–157 (1979)
64. Green, N.G., Ramos, A., Gonzalez, A., Morgan, H., Castellanos, A.: Fluid flow induced by nonuniform ac electric fields in electrolytes on microelectrodes. I. Experimental measurements. Phys. Rev. E. **61**, 4011–4018 (2000)
65. Gonzalez, A., Ramos, A., Green, N.G., Castellanos, A., Morgan, H.: Fluid flow induced by nonuniform ac electric fields in electrolytes on microelectrodes. II. A linear double-layer analysis. Phys. Rev. E. **61**, 4019–4028 (2000)
66. Green, N.G., Ramos, A., Gonzalez, A., Morgan, H., Castellanos, A.: Fluid flow induced by nonuniform ac electric fields in electrolytes on microelectrodes. III. Observation of streamlines and numerical simulation. Phys. Rev. E. **66**, 026305 (2002)
67. Castellanos, A., Ramos, A., Gonzalez, A., Green, N.G., Morgan, H.: Electrohydrodynamics and dielectrophoresis in microsystems: scaling laws. J. Phys. D: Appl. Phys. **36**, 2584–2597 (2003)
68. Bazant, M.Z., Squires, T.M.: Induced-charge electrokinetic phenomena: theory and microfluidic applications. Phys. Rev. Lett. **92**, 066101 (2004)
69. Squires, T.M., Bazant, M.Z.: Induced-charge electro-osmosis. J. Fluid. Mech. **509**, 217–252 (2004)
70. Duval, J., Lyklema, J., Kleijn, J.M., van Leeuwen, H.P.: Amphifunctionally electrified interfaces: coupling of electronic and ionic surface-charging processes. Langmuir **17**, 7573–7581 (2001)
71. Duval, J.F.L., Huijs, G.K., Threels, W.F., Lyklema, J., van Leeuwen, H.P.: Faradaic depolarization in the electrokinetics of the metal-electrolyte solution interface. J. Coll. Interface Sci. **260**(95–106), 1318 (2003)
72. Duval, J.F.L.: Electrokinetics of the amphifunctional metal/electrolyte solution interface in the presence of a redox couple. J. Coll. Interface Sci. **269**(211–223), 1321 (2004)
73. Prieve, D.C.: Changes in zeta potential caused by a DC electric current for thin double layers. Coll. Surf. A. **250**, 67–77 (2004)
74. Delgado, A.V., Gonzalez-Caballero, F., Hunter, R.J., Koopal, L.K., Lyklema, J.: Measurement and interpretation of electrokinetic phenomena. J. Coll. Interface Sci. **309**, 194–224 (2007)
75. Squires, T.M.: Induced-charge electrokinetics: fundamental challenges and opportunities. Lab. Chip. **9**, 2477–2483 (2009)
76. Bhatt, K.H., Velev, O.D.: Control and modeling of the dielectrophoretic assembly of on-chip nanoparticle wires. Langmuir **20**, 467–476 (2004)
77. Gupta, S., Alargova, R.G., Kilpatrick, P.K., Velev, O.D.: On-chip dielectrophoretic coassembly of live cells and particles into responsive biomaterials. Langmuir **26**, 3441–3452 (2010)

78. Lumsdon, S.O., Kaler, E.W., Williams, J.P., Velev, O.D.: Dielectrophoretic assembly of oriented and switchable two-dimensional photonic crystals. Appl. Phys. Lett. **82**, 949–951 (2003)
79. Bhatt, K.H., Sonia Grego, S., Velev, O.D.: An AC electrokinetic technique for collection and concentration of particles and cells on electrodes. Langmuir **21**, 6603–6612 (2005)
80. Ristenpart, W.D., Aksay, I.A., Saville, D.A.: Electrically guided assembly of planar superlattices in binary colloidal suspensions. Phys. Rev. Lett. **90**, 128303 (2003)
81. Zhou, H., White, L.R., Tilton, R.D.: Lateral separation of colloids or cells by dielectrophoresis augmented by AC electroosmosis. J. Coll. Interface Sci. **285**, 179–191 (2005)
82. Sides, P.J., Wirth, C.L., Prieve, D.C.: An imaging ammeter for electrochemical measurements. Patent Pending (2010)

Chapter 2
Nanoparticles Dispersion and the Effect of Related Parameters in the EPD Kinetics

Rodrigo Moreno and Begoña Ferrari

Nomenclature

a	Particle radius
A	Hamaker constant
B	London constant
C_d	Deposit concentration
C_s	Solid content of the suspension
$C_{s,0}$	Initial solid content of the suspension
D	Particle diameter
E	Electric field
E_{ef}	Effective electric field
f	Efficiency factor or sticking parameter
I	Electric current
I^+	Electric current transported by the cations
I^-	Electric current transported by the anions
I_p	Electric current transported by the particles
i_0	Initial current density
i	Current density
K	Kinetics parameter
L	Electrode distance
m	Deposited mass
m_0	Initial mass of powder in suspension
R	Roughness
S	Deposition surface area
S_{WE}	Conduction surface area
t	Deposition time
V	Volume of the suspension
V_d	Volume of the deposit

R. Moreno (✉)
Instituto de Cerámica y Vidrio, CSIC, Kelsen 5, 28049 Madrid, Spain
e-mail: rmoreno@icv.csic.es

J. H. Dickerson, A. R. Boccaccini (eds.), *Electrophoretic Deposition of Nanomaterials*, Nanostructure Science and Technology,
DOI 10.1007/978-1-4419-9730-2_2,

Greek Symbols

$1/\kappa$ Debye length
ε_d Dielectric constants of the deposit
ε_s Dielectric constants of the suspension
$\varepsilon_{r,l}$ Dielectric constant of the solvent
$\varepsilon_{r,p}$ Dielectric constant of the particles
ε_0 Vacuum dielectric constant
ϕ_d Volumetric fraction of the deposit
ϕ_s Volumetric fraction of the suspension
$\phi_{s,0}$ Initial volumetric fraction of the suspension
η Solvent viscosity
μ Electrophoretic mobility
v Electrophoresis rate
ρ_s Resistivity of the suspension
$\rho_{s,\infty}$ Resistivity of the suspension at infinite time
ρ_d Resistivity of the deposit
ρ_+ Resistivity associated to the cations
ρ_- Resistivity associated to the anions
ρ_p Resistivity associated to the particles
$\rho_{s,0}$ Initial resistivity of the suspension
δ Deposit thickness
σ_s Conductivity of the suspension
$\sigma_{S,\infty}$ Conductivity of the liquid medium
τ Characteristic deposition time
τ_0 Characteristic deposition time for initial conditions
τ_∞ The characteristic deposition time for final conditions
$\Delta\psi$ Potential drop and surface potential
$\Delta\psi_a$ Potential drop at the anode
$\Delta\psi_c$ Potential drop at the cathode
ζ Zeta potential

2.1 Nanoparticles and Nanomaterials

In recent years the emergence of a new generation of high technology materials involving nanoparticles and nanocomponents has increased exponentially [1, 2]. Nanomaterials are used in many different domains, such as chemistry, electronics, magnetic materials, biotechnology, etc. The great interest of the nanomaterials deals with the new properties associated to the small size that can strongly differ from those of the bulk conventional material.

Feynman pointed out that the designing of materials atom-by-atom was a real possibility, as it would not violate any physical laws. However this prediction could not be demonstrated until the arrival of very sophisticated instrumentation capable of viewing and manipulating materials at the nanoscale. The first use of the term "nanotechnology" was by Taniguchi in 1974 referring to "production technology to get extra high accuracy and ultra fine dimensions, i.e., the preciseness and fineness on the order of 1 nm (nanometer), 10^{-9} m in length" [3]. Although many definitions for nanotechnology have been suggested, NASA recently suggested the most thorough description: the creation of functional materials, devices and systems through control of matter on the nanometer length scale (1–100 nm), and exploitation of novel phenomena and properties (physical, chemical, biological) at that length scale [4].

A challenge of existing technology is to organize different functional components in two and three dimensions at the nanoscale. It is the cooperative interaction of different nanoscale components that allow the combination of desirable properties of single units into a larger integrated system [5, 6]. Nanoparticle assemblies have potential applications in photonic materials, advanced ceramics, and electronics, among other applications. Several approaches have been used to precisely position nanoparticles into two- and three-dimensional (2D/3D) structures, the most widely used being self-assembly processes, methods that utilize diblock copolymers, and nanoparticle infilling of colloidal crystals and other templates [7].

The progress in nanoscience has been associated with the elaboration of new methods for synthesizing, studying, and modifying nanoparticles and nanostructures [8]. In principle, there are two kinds of methods for synthesizing nanoparticles. The first group combines methods that allow preparation and studies of nanoparticles, and the second group includes methods that allow preparation of nanomaterials and nanocomposites, based on nanoparticles. This classification leads one to consider that particles can either be built from association of separate atoms through a chemical route (which constitutes the so-called *bottom-up* approach) or by dispersion or fracture of large units that are broken down through physical methods like milling (the so-called *top-down* approach). Figure 2.1 shows a schematic view of these two approaches to nanomaterials production.

For most ceramic processing techniques, the "bottom-up" approach is much more interesting, and this will be considered herein. The development of nanomaterials synthesis by this "bottom-up" approach has been determined by nanochemistry, which has two important aspects to be considered: on one hand, the different chemical properties and the reactivity of particles comprising a small number of atoms; on the other hand, its effect on the synthesis, modification, and stabilization of nanoparticles and the further assembly into more complex structures. Since the properties of synthesized structures directly depend on the size and shape of the original nanoparticles, the control of these parameters and their stability in a dispersing medium are critical for the manufacture of any nanostructured material. Nanochemistry studies the synthesis and modification of particles with sizes typically below 10 nm along one direction at least. Generally speaking, materials chemists efforts are devoted to the development of "bottom-up" techniques that afford the *self-assembly* of nanoscale species. There is also a parallel effort of nanophysics to allow nanoscale units by "top-down" routes.

There are two basic types of nanoscale building blocks that can be used for the fabrication of a device, zero-dimensional materials (0D, e.g. nanoparticles, nanoclusters, nanocrystals) and unidimensional ones (1D, e.g. nanowires, nanofibers,

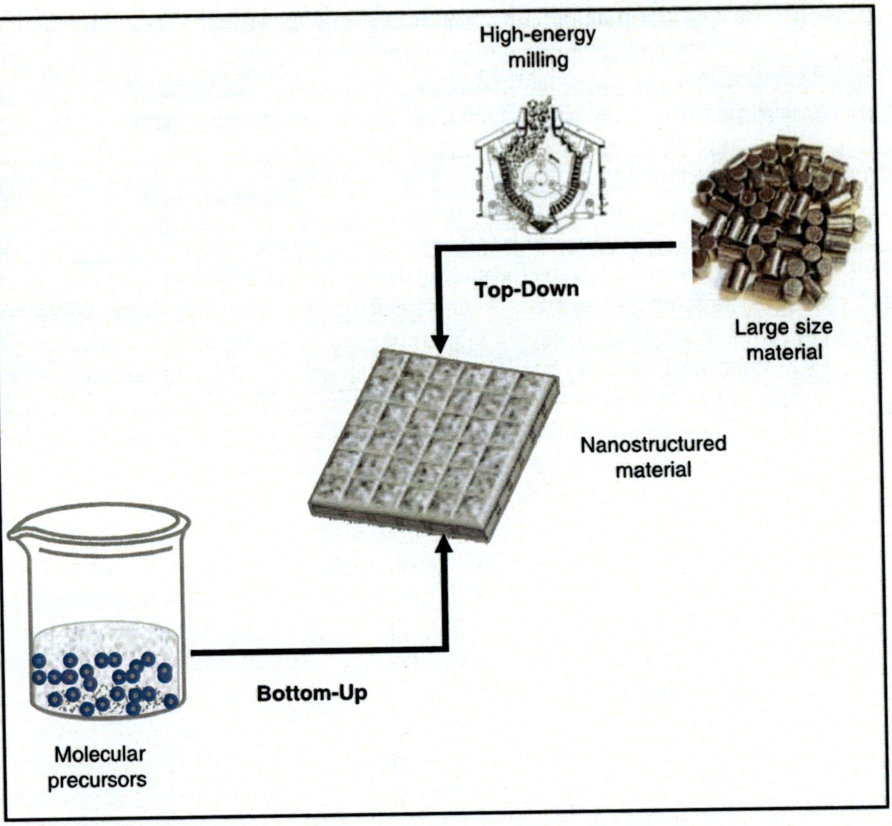

Fig. 2.1 The two approaches for the synthesis of nanomaterials

nanotubes). The assembly of these nanosized building blocks into 2D and 3D arrangements yields entirely new properties and functionalities. In this context it is important to make some basic definitions [8]. The term *nanoparticle* generally refers to 0D nanosized building blocks (regardless of size and morphology), or those that are amorphous and possess a relatively irregular shape; typical size in this case is > 10 nm. For amorphous/semicrystalline nanostructures smaller in size (i.e., 1–10 nm), with a narrow size distribution, the term *nanocluster* is considered more appropriate. The agglomeration of noncrystalline nanostructural subunits should best be termed a *nanopowder*. The term *nanocrystal* is reserved to nanomaterials that are single-crystalline. A special case of nanocrystal that is comprised of a semiconductor is known as a *quantum dot*. The nomenclature of 0D nanostructured materials related to the typical size intervals are shown in Table 2.1.

There are different classifications of nanoparticles, proposed by different authors based on the diameter of a particle expressed in nanometers and the number of atoms in a particle. These classifications also take into account the ratio of surface atoms to those in the bulk. Figure 2.2 shows a generally accepted classification of

Table 2.1 Nomenclature of 0D nanostructured materials

Structure	Size (nm)	Typical name
Nanostructures	<< 1	Clusters
	$1–10^2$	Nanoclusters and nanocrystals
	$10^2–10^3$	Nanoparticles and nanopowders
Microstructures	$10^3–10^5$	Sub-micron particles
Macrostructures	$>10^5$	Particulates (bulk powders)

N° atoms	N=10	$N=10^2$	$N=10^3$	$N=10^4$	$N=10^6$	Bulk
Diamteter (nm)	1	2	3 5	7	10	> 100
Chemistry	Nanoparticles					Solid state

Fig. 2.2 Classification of particles according to their sizes

particles according to their sizes [9]. According to this classification, nanoparticles can be considered as intermediate entities between individual atoms and the solid phase, measuring from 1 to 10 nm and built of atoms of one or several elements.

In order to tailor a device property to its function, a good knowledge of the correlation between property and microstructure is required. Because the microstructure depends on the processing route, processing methods are decisive factors in developing new materials. While applications are most of the time well identified, material scientists continue to develop new processing methods that can produce components of a complex shape and high reliability with a minimum of machining and cost. These fulfillments have been already reached for submicron sized powders by wet processing through a colloidal approach [10, 11]. However, when moving to the nanoscale, there is an important problem because interparticle interactions are in the same range of particle size and spontaneous agglomeration occurs.

It must be remarked that the fraction of atoms pertaining to the surface increases as particle size decreases and hence, the contribution of surface atoms to the system's energy increases. This has important thermodynamic consequences such as the fact that particle size is an active variable that determines the state of the system and its reactivity. That is, nanosized particles can enter into reactions untypical of coarser systems [12].

So far, we have considered the recent nomenclature related to nanomaterials and nanochemistry. However, it is also important to note the possible differences between nanoparticles and colloids, the last being known since the 1860's. Although both types lie in the same nanoregime, some authors have considered the existing differences between them, mainly related with the lower control over composition and morphology. In general nanoparticles are considered to lie in the range of 1–100 nm, whereas colloids are typically larger than 10 nm. Other important differences with respect to colloids are that nanoparticles have reproducible synthesis and physical properties and contain clean surfaces, while colloids contain surface-adsorbed species such as –OH, –X, $-OH_2$, etc. However, when introduced in a dispersing medium, as a liquid, the properties of the system become a function of the interactions among the different species in the system, including the interactions among particles and the interactions among particles and dispersing medium molecules. The dispersion in a liquid modifies completely the properties of the nanopowders that can be studied considering the laws of colloid chemistry.

2.2 Processing of Ceramics. The Colloidal Approach

Ceramic products have been used for more than 10,000 years. The term *ceramics* comes from the greek "κεραασ" (keras), which means *clay*. This word proceeds from a radical of Indo-European languages "κεραμ" (keram), which in turn means corresponds to the verb to burn. Thus, ancient ceramics were clay-based products subjected to a thermal treatment. Although there are different definitions to ceramics, it is generally accepted that ceramics are inorganic, non-metallic materials that are subjected to a thermal treatment to reach their final characteristics. However, these properties are strongly dependent not only on the composition, but also on the firing temperature. Due to the special characteristics of the ionic-covalent bonding, ceramics exhibit excellent properties like thermal and chemical stability, hardness and corrosion resistance, among others. However, ceramics have a great limitation, their inherent brittleness consequence of the high directionality of the bonds, so that catastrophic failure occurs when the fracture stress is exceeded.

The strength of a ceramic material can be described by Griffith's equation,

$$\sigma = YK_{IC}/\sqrt{C} \qquad (2.1)$$

where σ is the fracture stress, K_{IC} the fracture toughness, C the defect size, and Y a factor that depends on the position and shape of the defect. According to this law, there are two ways of increasing the strength of a ceramic material, by increasing the fracture toughness or by decreasing the flaw size. Toughness can be increased using composite materials, by the incorporation of secondary phases such as particulates, platelets and whiskers. Another approach to improve the toughness is through the design of materials with tailored microarchitectures, such as coatings, laminates and functionally graded materials, where the interfaces can arrest or even bifurcate the propagating crack. Toughness could be also enhanced by reducing the flaw size, C. This can be only achieved with a better processing control. Pores, inclusions, cracks and agglomerates can behave as flaws with a deleterious effect on the final properties. In all cases, the enhancement of toughness and the reduction of defects number and size are only possible if the reinforcing phase is well-dispersed in the matrix [13].

One major concern in the application of any material is the method of fabrication. Ashby et al. [14] have done a taxonomic classification of the types of materials and the types of processes for their manufacture. A summarized view of these

Table 2.2 Classes of manufacturing processes of materials

Classes of processes	Specific processes
Primary shaping	Deformation, casting, injection, powder processing, rapid prototyping, free-forming
Secondary processes	Machining, lamination and shaping, tempering, quenching
Joining & surface treatment	Adhesion, welding, fasteners Painting, printing, anodizing, plating

classes of processes is shown in Table 2.2. Polymers can be molded and ductile materials can be forged, rolled and drawn, but ceramics are brittle and must be shaped in other ways. Materials that melt at modest temperatures to low-viscosity liquids can be cast; those that do not have to be processed by other routes. Furthermore, shape influences the choice of a process, too. Slender shapes can be made easily by rolling or drawing but not by casting. Hollow shapes cannot be made by forging, but they can by casting or molding. The choice, for a given component, depends on the material, on its shape, dimensions and precision, and on the lot size [15–19].

Ceramics are mainly produced by powder processing techniques, which consist of four basic steps: (1) powder synthesis and/or preparation for further consolidation, (2) consolidation of powders into a self-supported shaped body, the so-called *green body*, (3) drying and burn out of organics, (4) sintering at high temperature to reach the final microstructure and properties, and (5) final machining and shaping, which is the most expensive step due to the hardness of ceramics. Maximum control at any stage is necessary since defects introduced in one step are very difficult to remove and will persist in the next steps.

A great effort has been devoted by ceramists to improve the quality of the starting powders [20], and to search novel forming processes capable to produce near-net shaped parts with high homogeneity and density and increased green strength in order to reduce as much as possible the machining step [21–23].

The exhaustive control of powders transformation (referred to as *beneficiation processes*, which include milling and mixing, washing, separation, filtration and granulation) has to be accomplished of a "clean" consolidation into bodies by non-conventional shaping procedures in which contamination by metal parts, high pressures or large concentration of additives is to be avoided. In this sense, colloidal processing methods have demonstrated their efficiency for producing complex shaped bodies with enhanced properties and increased reliability [10, 11, 24].

Forming methods can be classified according to three categories, depending on the relative content of liquid characteristic of the process: (1) dry pressing methods, where the liquid content is typically lower than 7%; (2) plastic forming, with typical liquid contents of 15–20 for extrusion or up to 30–40% in the case of injection molding, and (3) colloidal shaping methods that make use of suspensions, where the content of liquid is generally higher than 50%.

Suspensions have been used in ceramics from ancient civilizations, when it was observed that some water allowed clays to become plastic and to be easily handleable so that the first recipients could be created. During the decade of the 1980's the wet forming of ceramics was in focus, and a strong development was possible when ceramists learned the importance of colloid and surface science and its great influence in ceramic processing. This knowledge allowed the development of different near-net shaping techniques in the 1990's. The basis for such shaping techniques was the well-known slip casting process [25], which consists of pouring a suspension into the cavity of a permeable mould that allows the liquid to pass through the porosity channels of the walls. This is a solid-liquid separation process, used also in other technologies like filter-pressing. Consolidation of the particles occurs as the liquid flows through the porous medium under a pressure gradient. In slip casting, the driving force for the consolidation is just due to capillary forces. The low casting rates can be increased by using an external aid, like microwaves, centrifugation and, most common, pressure. The family of casting processes have led to other related processes like tape casting, coagulation casting, gelcasting, etc. These processes have different consolidation mechanisms, although all of them have a basic requirement: they require the preparation of a concentrated and well-dispersed suspension of particles in a liquid, preferably water. All these processes involve the preparation of a stable suspension of the powders in a liquid with the aid of deflocculants, which maintain particles apart each other by either the development of electrostatic charges or through the adsorption of polymers that promote steric hindrance, and other additives such as binders, plasticizers, anti-foaming agents, etc. [26–28]. Colloidal processing allows the manipulation and control of the interparticle forces operating not only in the suspension but also in the consolidation step, where repulsive forces maintain still active, so that the formation of agglomerates is dramatically reduced. As a consequence, if a proper dispersion state is reached, green bodies with high relative densities can be obtained, which result in dense materials with fine microstructures and controlled grain sizes [29, 30].

Table 2.3 summarizes some colloidal forming methods typically employed for the manufacture of ceramic bodies [31]. One possible classification of the methods can be done considering the consolidation mechanism. Although both concepts are intimately related, the differences between forming method and consolidation mechanism must be emphasized. The forming method is the technique employed to obtain the body, e.g. slip casting, injection molding, tape casting, etc. Consolidation refers to the mechanism through which the particles arrange into a consolidated body with the desired shape and size. The main consolidation mechanisms are the following: (1) *filtration*, which is the basis of slip casting, as described before; (2) *deposition-evaporation* methods, which are those methods based on the deposition of a thin layer of suspension that undergoes fast evaporation, like in tape casting, screen printing, etc. One particular case is that of electrophoretic deposition (EPD) where particles are forced to migrate to the electrode with opposite charge to that of the particles under the influence of an electric field, which can be considered the

Table 2.3 Conventional colloidal shaping techniques

Filtration	Deposition evaporation	Flocculation coagulation	Gelation
Slip casting	Screen printing	Short range forces	Gelcasting
– pressure	Tape casting	Temperature induced	Injection molding
– vaccuum	Electrophoresis	Coagulation casting	Thermogelation
– centrifugal	CVD, PVD...	Freeze casting	Protein casting
– microwaves	Dipping	Direct solidification	Starch consolidation
	Spin coating		

driving force for the deposition. Once the film is deposited onto the electrode material, coagulation occurs as a consequence of the rapid drying; (3) *Flocculation* and *coagulation* methods are based in the preparation of a stable suspension and its sudden destabilization after the formation of a physical gel formed by flocculation in a secondary minimum (when inorganic salts are added, for example) or coagulation in a primary minimum (when pH is moved towards the zero point of charge, and there are no repulsive forces that can overcome the attractive van der Waals forces); (4) *Gelation* methods are those based in the polymerization of small molecules (monomers or dimers) into long-chained macromolecules by either a chemical polymerization in the presence of a catalyst and an initiator or by thermogelation when the gel is formed as a consequence of a temperature change. In this case a chemical gel is formed that retains the ceramic particles in the developed network, while in the previous group the own particles are responsible for the development of a network structure that is considered as a physical gel.

Most of those shaping methods make use of moderately or highly concentrated suspensions, where sedimentation of large or dense particles is retarded while maintaining a flowing behavior. In general those suspensions are poured into moulds and the solids loading has to be the highest possible in order to minimize the drying shrinkage and the amount of liquid to be removed during drying. The main exception to this general rule is the case of EPD, where suspensions with low solids loading are normally used. It is mandatory, however, that colloidally stable suspensions have to be used and that the particles must carry a substantial charge. That is, EPD requires stable suspensions like any other wet forming procedure, although the stability cannot be evaluated by viscosity inspection. The stability has to be measured considering other parameters, the most important being the electrophoretic mobility or the zeta potential, and the conductivity, which in turn depend on the pH, the type and concentration of deflocculants, the temperature, etc.

If the suspension is completely stabilized gravity can be neglected, but if the particles tend to settle gravity becomes an influencing parameter, affecting mainly layer thickness and deposit quality. If the suspension is well stabilized, a vertical configuration will be used. The stability of the suspensions and their effects in the EPD process and the deposit characteristics will be studied in the following.

2.3 Colloidal Stability of Ceramic Suspensions

2.3.1 Colloidal Dispersions

The term 'colloid' comes from the Greek word '$\kappa o\lambda\lambda\alpha$' (kolla) for glue. It was originally used for gelatinous polymer colloids, which were identified by Thomas Graham in 1860 in experiments on osmosis and diffusion. The simplest definition of a colloidal dispersion is that it is a multi-phase system in which one phase (or more) is dispersed in a continuous one or medium. At least one dimension lies within the nanometre (10^{-9} m) to micrometre (10^{-6} m) range, so that colloidal dispersions are

Table 2.4 Types of colloidal dispersions

Dispersed phase	Dispersing medium		
	Gas	Liquid	Solid
Gas (bubbles)	–	Foam	Solid foam
Liquid (droplets)	Liquid aerosol	Liquid emulsion	Solid emulsion
Solid (particles)	Solid aerosol	Suspension/sol	Solid sol

mainly systems containing large molecules and/or small particles. The main factor determining the properties of a colloidal system are the particle size and shape, the surface properties, the interparticle interactions and the interactions between particles and dispersing medium. A fundamental issue of colloidal systems is that there is a well-defined separation surface between the dispersed phase and the dispersing medium. The interface plays an essential role in the surface properties, including adsorption, surface charge, electrical double layer, etc. There are many reference textbooks on colloid and surface science where the reader can find complementary information [31–39]. The colloidal dispersions may be classified considering the state in which either the dispersed phase or the dispersing medium are present (solid, liquid or gas). Table 2.4 shows the different types of colloidal dispersions.

To overcome the spontaneous tendency of particles to agglomerate by means of attractive forces (i.e. van der Waals), it is necessary to enhance the repulsive contribution to the interaction potential. This can be made by using dispersing aids, known as dispersants or deflocculants. Dispersants may be classified into several groups depending on the stabilising mechanism they promote, whose efficiency is a function also of the characteristics of the dispersing medium and the liquid/solid interface. In a polar medium, such as water, amphoteric ceramic particles develop an electrical double layer, whose thickness depends on the concentration of potential-determining ions (i.e. pH) that provides an electrostatic repulsion. The presence of a supporting electrolyte varies the surface charge and potential. Stability may arise also from the adsorption of surface active compounds or polymers, which provide a steric hindrance. They can be associated to electrical charge, thus providing an electrosteric stabilisation. Other possibilities are the use of alkoxydes that chemically react with the particle surface (coupling agents), but this is more properly a method for synthesising a coating material than a proper dispersing agent. Finally, non-adsorbing polymers could maintain the particles apart by means of a depletion mechanism. Table 2.5 summarizes the different kinds of dispersants used in ceramic processing.

Table 2.5 Kinds of dispersants used in ceramic processing

Substance	Mechanism
Potential-determining ions (pH)	Electrostatic repulsion
Electrolytes (inorganic salts)	Electrostatic repulsion
Surfactants (amphiphilic chains)	Adsorption + Electrostatic repulsion
Adsorbed polymers	Steric hindrance
Adsorbed charged polymers	Steric hindrance + Electrostatic repulsion
Coupling agents	Adsorption + Electrostatic repulsion
Non-adsorbing polymers	Depletion stabilisation

2.3.2 Intermolecular Potentials

At the most basic level, two approaching molecules have a natural tendency to interact, generating a pair potential. The interaction potential V(D) depends on the separation distance, and is related to the force (F) by:

$$F = -\frac{dV}{dD} \tag{2.2}$$

In colloidal suspensions, there are particles immersed in a surrounding medium. Thus, the interaction is not simply the result of a pair-wise addition, but a many-bodies effect, in which the total interaction is the result of particle-particle interactions, as well as medium-medium and particle-medium interactions, as illustrated in Fig. 2.3. The existence of a solvent produces different effects: (1) the displacement of solvent by two solute molecules, (2) reordering of solvent molecules by solute (solvation), (3) solute-solvent interactions can change the properties of dissolved molecules, (4) when new molecules are introduced the medium expends some energy to form a cavity to host the guest molecule. All these effects are referred to as solvent effects.

At a first level, three fundamental forces operate in a colloidal dispersion:

1. a gravitational force, tending to settle or raise particles depending on their density relative to the solvent:
2. a viscous drag force, which arises as a resistance to motion, since the fluid has to be forced apart as the particle moves through it;
3. the *natural* kinetic energy of particles and molecules, which causes Brownian motion.

The physical properties of a compound depend on the types of chemical bond between atoms. Ionic bonds result from electronic transfer because one atom attracts electrons more strongly than the other atom. In a covalent bond atoms attract electrons to a similar degree, thus resulting in a coordinate bond. Both kinds of bonds respond to the tendency of the atoms to reach the most stable electronic configuration. Neutral molecules can link together through *intermolecular forces*, such as electrostatic forces and dipole interactions.

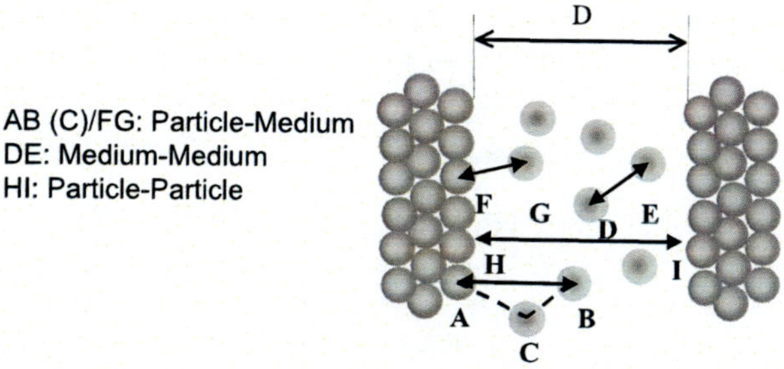

Fig. 2.3 Many-body interactions in suspensions

Intermolecular forces can be classified in three groups:

1. quantum mechanics forces, like covalent or chemical bonds.
2. electrostatic, arising from Coulomb force between charges,
3. polarization forces, produced by dipole moments induced by the electric fields of charges or permanent dipoles. All interactions in a solvent involve polarization effects.

When two or more atoms combine to form a molecule a covalent bond is formed. The same accounts for metal bonds. Both are based in sharing electrons between different atoms. Covalent forces are of short range, operating at distances of 0.1–0.2 nm. An important characteristic of chemical bonding is its directionality. Covalent bonds are typically in the range 100–300 kT (200–800 KJ/mol).

Coulomb forces (charge-charge interactions) are physical forces, but can be very strong. They lack the stoichiometry and directionality of covalent bonds. Therefore, these forces are essential for holding molecules together in liquids. The energy of the Coulomb interaction is given by:

$$V = \frac{Q_1 Q_2}{4\pi \varepsilon \varepsilon_0 d} = \frac{z_1 z_2 e^2}{4\pi \varepsilon \varepsilon_0 d} \tag{2.3}$$

Q_i being the charge of each molecule i, given by the elementary charge e ($1.602.10^{-19}$ C) multiplied by the valency z, d is the separation distance, and ε is the relative permitivity or dielectric constant of the medium. Coulombic forces are strong (~ 200 kT), long range forces; only at distances above 50 nm the Coulombic force falls below kT.

A molecule is polar when the centre of negative charge does not coincide with the centre of positive charge. The dipolar moment (μ) is then defined as the product of the charge density (e) and the separation distance (d) between the centres of charges with opposite sign,

$$\mu = e \cdot d \tag{2.4}$$

The unit for μ is the Debye (D). When two polar molecules approach a dipole-dipole interaction occurs in which the positive side of one molecule attracts the negative side of another molecule. One particularly strong dipole attraction is hydrogen bonding, in which a hydrogen atom links two electronegative atoms, one through a covalent bond, and the other by means of electrostatic forces.

A symmetric molecule (like methane) has no permanent dipole moment, but in the presence of an electric field an induced dipole can develop, whose moment (μ) is related to the electric field (E) by

$$\mu = \alpha \cdot E \tag{2.5}$$

α is referred to as the polarizability.

It is also possible to distinguish between short-range and long-range forces. The former appears when interacting species (atoms, molecules, particles, etc) approach very close together, resulting in chemical bonding or hydrogen bonding. These are,

Table 2.6 Energy (in kT units) for particles of given size

Type of force	Particle size range		
	0.1 μm	1 μm	10 μm
Van der Waals attraction	10	10^2	10^3
Electrostatic repulsion	$0–10^2$	$0–10^3$	$0–10^4$
Brownian motion	1	1	1
Kinetic energy of sedimentation	10^{-13}	10^{-6}	10
Kinetic energy of stirring	1	10^3	10^6

in fact, contact forces ranging up to 0.1–0.2 nm. Long-range forces operates at longer separation distances (several nanometers), as for van der Waals forces, solvation and hydration forces, capillary forces, all of which have a physical nature. However, some of these physical forces, solvation, capillarity, hydration, operate at short distances (typically below 10 nm), while van der Waals forces, hydrophobic forces and electrostatic repulsion operate to a much longer distances (even above 100 nm). Then, a further classification of these forces into short-range and long-range may be useful.

In principle, kinetic random motion should dominate the behavior of small particles, which will not settle but stay as a stable dispersion. However, these small particles can collide among them due to their kinetic energy and if there are attractive forces, the collisions might cause the growth of aggregates and the dispersion becomes unstable and settling occurs.

There is a ubiquitous force in nature, the so-called van der Waals force (vdW), which is one of the main forces acting between molecules and is responsible for holding together many condensed phases. Although electromagnetic in origin, vdW is much weaker than the Coulombic force acting between ions.

Table 2.6 shows the energies associated with some of these forces and the dependency with particle size for particles in the 0.1–10 μm size range. When particle size increases the tendency to sedimentation and the attractive vdW forces strongly increase and it becomes more and more difficult to maintain stability.

The first law for describing the intermolecular pair potential was that proposed by Mie (1903), who was first to introduce an attractive term and a repulsive one. A special case of the Mie potential is the Lennard-Jones potential, where the attractive term is the van der Waals interaction potential which varies with the inverse sixth power of the distance between molecules:

$$V(D) = -\frac{B}{D^6} + \frac{C}{D^{12}} \tag{2.6}$$

2.3.3 *The Attractive Potential*

The attractive term in the interaction pair potential is generally the result of London-van der Waals dispersion forces. Van der Waals forces arise from three different phenomena: (1) the interaction between a permanent dipole with other permanent dipole (Keesom interaction); (2) the interaction between a permanent dipole with a polarisable atom that produces an induced dipole (Debye interaction), and (3) when no permanent dipoles are present, instantaneous dipoles are formed due to fluctuations in the distribution of electronic charge (London dispersion interaction). If no permanent dipoles exist, the first two contributions do not occur but the last is always present and plays a key role in the stability of colloidal suspension. All these forces vary with separation distance as D^{-6}.

The attractive interaction between two similar spherical particles of radius, a, at a distance, D, and with the distance between the centres being R=2a+D (Fig. 2), as proposed by Hamaker, is:

$$V_A = -\frac{A}{6}\left(\frac{2a^2}{R^2-4a^2}+\frac{2a^2}{R^2}+\ln\frac{R^2-4a^2}{R^2}\right) \tag{2.7}$$

where A is the *Hamaker constant*, that depends on the properties of both the particle and the dispersing medium. In Hamaker's original treatment, known as the microscopic approach, the Hamaker constant is given by

$$A = \pi^2 \rho_1 \rho_2 B \tag{2.8}$$

ρ_i being the number of atoms (molecules) per cm^3 of each bulk phase, and B the London constant in Eq 2.4, where $B \sim 1/\alpha_0^2$, α_0 being the polarizability. This model considers the Coulomb's theory, valid for small separation distances. For longer distances (molecular range) there is a retardation effect associated to a characteristic time (λ_0) needed for the propagation, which makes the power law change to the inverse seventh power.

To transfer the van der Waals potential from atom interaction to body interaction, a pairwise additivity is considered, where the energies of all the atoms in each body must be integrated. In this way, the V_A term can be obtained for different geometries. Some non-retarded van der Waals interactions for different geometries are summarised in Fig. 2.4.

Another approach to calculate the Hamaker constant is due to Lifschitz, who proposed the modern theory for the dispersion force by considering each body as a continuum with certain dielectric properties. This model incorporates many-body effects, neglected in the microscopic theory. Hamaker constants of many ceramic materials have been calculated from the Lifschitz theory using optical data of the material and the medium, which can be measured by spectroscopic ellipsometry or refractometry. Table 2.7 shows the non-retarded Hamaker constants for ceramic materials interacting in a vacuum and in water. Values of A for most solvents are in the range $4–8.10^{-20}$ J, for polymers $6–10.10^{-20}$ J and for metals and ionic solids $10–30.10^{-20}$ J.

For two similar particles in a medium,

$$A = A_p + A_m - 2A_{pm} \tag{2.9}$$

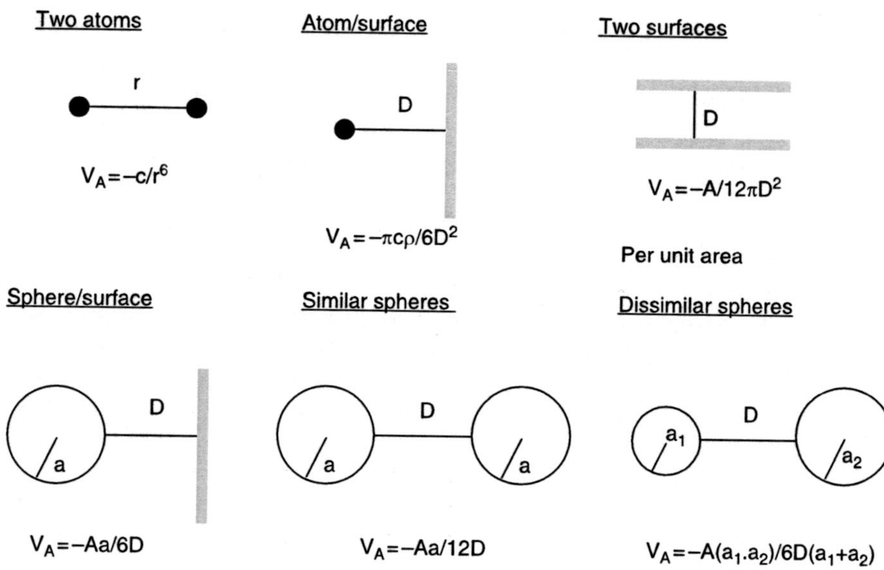

Fig. 2.4 Non-retarded van der Waals interactions for different geometries

Table 2.7 Non-retarded Hamaker constants between identical ceramics. [40]

Material	A (10^{-20} J)	
	Vacuum (air)	Water
α-Al_2O_3	15.2	3.67
$BaTiO_3$ (average)	18	8
BeO (average)	14.5	3.35
$CaCO_3$	10.1	1.44
CaF_2	6.96	0.49
CdS	11.4	3.40
MgO	12.1	2.21
Mica	9.86	1.34
PbS	8.17	4.98
6H-SiC	24.8	10.9
β-SiC	24.6	10.7
β-Si_3N_4	18.0	5.47
SiO_2 (quartz)	8.86	1.02
SiO_2 (amorphous)	6.50	0.46
$SrTiO_3$	14.8	4.77
TiO_2 (average)	15.3	5.35
Y_2O_3	13.3	3.03
ZnO	9.21	1.89
ZnS (cubic)	15.2	4.80
NS (hexagonal)	17.2	5.74
ZrO_2-3mol%Y_2O_3	20.3	7.23

where A_p and A_m are the constant for the particles and for the medium and A_{pm} is associated to the particle-medium interaction and is generally assumed to be $A_{pm}=(A_pA_m)^{1/2}$ and hence

$$A = (A_p^{1/2} - A_m^{1/2})^2 \tag{2.10}$$

where A is always positive. For dissimilar particles,

$$A = A_{12} + A_{33} - A_{13} - A_{23} \tag{2.11}$$

where subscripts 1 and 2 refer to the particles and 3 to the medium. If $A_{11}<A_{33}<A_{22}$, A becomes negative, that is, it is possible to develop a repulsive force. In the most general case of ceramic slip processing, the particles are immersed in a polar liquid, such as water. Thus, the presence of a electrolyte solution and hence, an adsorbed layer of this electrolyte develops onto the particle surface. This will be discussed in a later section that deals with stabilisation by adsorbing polymers and surfactants.

2.3.4 *The Repulsive Electrostatic Potential*

2.3.4.1 Charge Development

Hydrous oxide surfaces are amphoteric and can be modelled by the following acid-base reactions,

$$\begin{aligned} MOH + H^+ &\rightleftarrows MOH_2{}^+ \\ MOH + OH^- &\rightleftarrows MO^- + H_2O \end{aligned} \tag{2.12}$$

where protons and hydroxyl ions can specifically adsorb. For this reason, H^+ and OH^- are referred to as *potential determining ions*. The surface charge (σ_0) can be expressed by the difference in adsorption density between H^+ and OH^- adsorbed on the surface (Γ_{H^+} and Γ_{OH^-}, respectively),

$$\sigma_0 = F(\Gamma_{H^+} - \Gamma_{OH^-}) = F(\Gamma_{MOH_2^+} - \Gamma_{MO^-}) \tag{2.13}$$

F being the Faraday constant (96,485 C mol^{-1}).

The surface charge is negative at high pH values, and the protons are attracted. When the concentration of the potential determining ion is altered, the relative adsorption of ions on the surface varies. There is a defined concentration for which the activities of positive and negative species are equal, and the surface potential will be zero. This concentration is referred to as the *zero point of charge* (ZPC). The ZPC defines the maximum destabilisation, as there is not a net surface charge. Hence, suspensions are stable only if the pH is kept far enough from the ZPC.

When a particulate system is dispersed in a polar liquid (aqueous or non-aqueous) a double layer is developed around each particle. The separation of charge occurring at an interface between two phases is called an *electrical double layer* because it consists ideally of two regions of opposite charge.

Table 2.8 Typical values of ZPC of ceramic oxides

Material	PZC (pH)
α-Al_2O_3	8–9
FeOOH (goethite)	6,5–7,5
AlOOH (boehmite)	7,5–8,5
$3Al_2O_3$-$2SiO_2$ (mullite)	5–6
$BaTiO_3$	5–6
$CaCO_3$	7–9
MgO	>12
SiC	4–6
Si_3N_4	7–9
SiO_2 (quartz)	2–3
SiO_2 (amorphous)	2–3
TiO_2	5–6
Y_2O_3	8–9
ZnO	9
ZrO_2 (m)	4–5
ZrO_2-3 mol% Y_2O_3	5–6

Table 2.8 shows typical values of the ZPC of ceramic oxides [41]. In general, the ZPC decreases as the charge of the cation increases, and depends on the degree of hydration and the presence of impurities and dopants.

The first model of double layer is attributed to Helmholtz in which rigid charged layers are fixed in parallel planes to form a condenser. This model assumes that the capacity C of the condenser ($C=\sigma/\psi_r$) does not depend on the electrolyte concentration, which is not true. Gouy and Chapman proposed a diffuse double layer model, consisting in a monolayer of *counterions* (that is, inert electrolyte ions having the opposite charge of the surface charge) strongly adsorbed onto the surface (with an electrostatic potential ψ_0) and a diffuse double layer in which counterion concentration decreases as the distance from the surface increases (Fig. 2.5). When there is no added electrolyte, the concentration of each counterion in the diffuse layer can be obtained from the Boltzmann equation:

$$\rho = \rho_0 \cdot \exp\left(^{-ze\psi}/_{kT}\right) \tag{2.14}$$

where ρ and ρ_0 are the number of ions of valence z at a distance x and at the surface, k is the Boltzmann constant and T is the temperature. Considering also the net excess charge density at a distance x given by the Poisson equation:

$$ze\rho = -\varepsilon\varepsilon_0\left(^{d^2\psi}/_{dx^2}\right) \tag{2.15}$$

the Poisson-Boltzmann equation is obtained, where

$$\frac{d^2\psi}{dx^2} = -\frac{ze\rho}{\varepsilon\varepsilon_0} = -\left(\frac{ze\rho_0}{\varepsilon\varepsilon_0}\right) \cdot \exp\left(\frac{-ze\psi}{kT}\right) \tag{2.16}$$

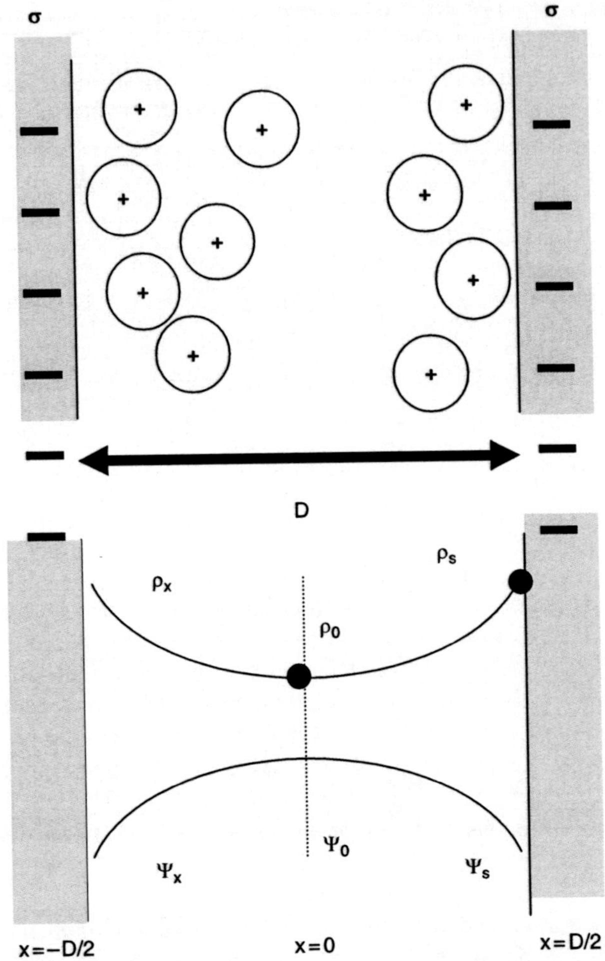

Fig. 2.5 Counterion density profile (ρ_x) and potential (Ψ_x) between charged surfaces in water (no added electrolyte). ρ_s and Ψ_s are the contact values

The solution of this equation gives the potential ψ, the electric field E=dψ/dx and the counterion density ρ at any point x between the two surfaces. To solve the equation we must impose that the field vanishes at the midplane E=dψ/dx=0 and that the total charge of counterions is equal and opposite to that on the surfaces to maintain the electroneutrality. For a surface charge density σ on each surface, the electroneutrality implies that:

$$\sigma = -\int_0^{D/2} ze\rho dx = \varepsilon\varepsilon_0 \int_0^{D/2} \left(\frac{d^2\psi}{dx^2}\right) dx = \varepsilon\varepsilon_0 E \qquad (2.17)$$

because the field at the surface (E) is equal to the field at the midplane ($E_{D/2}$). Thus, the relation between the concentration of counterions at the surface (ρ) and at the midplane ($\rho_{D/2}$) is obtained by differentiating Eq. 2.14 and using Eq. 2.16:

$$\rho - \rho_{D/2} = \int_0^{D/2} d\rho = \frac{\varepsilon\varepsilon_0}{2kT}\int_0^{D/2} d\left(\frac{d\psi}{dx}\right)^2 = \frac{\varepsilon\varepsilon_0}{2kT}\left(\frac{d\psi}{dx}\right)^2_{D/2} = \frac{\sigma^2}{2\varepsilon\varepsilon_0 kT} \tag{2.18}$$

That is, the concentration of counterions at the surface depends on the charge density and the concentration of counterions at the midplane. When different types of ions are present in the solution (e.g. when an electrolyte is added) the surface charge is given by the Grahame equation,

$$\sigma^2 = 2\varepsilon\varepsilon_0 kT\left[\sum_i \rho_{\infty i}\exp\left(\frac{-ze\psi_0}{kT}\right) - \sum_i \rho_{\infty i}\right] \tag{2.19}$$

where ψ_0 is the contact value of ψ_x (that at the surface) and ρ_∞ is the ionic concentration of ions i in the bulk, at x=∞, where $\psi_\infty=0$. For low potentials, below 25 mV, the Grahame equation simplifies to

$$\sigma = \varepsilon\varepsilon_0\kappa\psi_0 \tag{2.20}$$

$$\kappa = \left(\frac{\sum_i \rho_{\infty i} e^2 z_i^2}{\varepsilon\varepsilon_0 kT}\right)^{1/2} \tag{2.21}$$

defines a condenser whose two plates are separated by a distance $1/\kappa$, which is known as the *Debye length*, and gives the thickness of the double layer. Its magnitude depends only on the medium properties and not on the properties of the surface, such as the charge and the potential. At 25°C the Debye length of aqueous solutions is

$$^1/_\kappa = \left\{\begin{array}{ll} 0.304/\sqrt{(NaCl)}\ \text{nm} & \text{for electrolytes type } 1:1 \\ 0.176/\sqrt{(CaCl_2)}\ \text{nm} & \text{for electrolytes type } 1:2 \text{ or } 2:1 \\ 0.152/\sqrt{(\text{MgSO}_4)}\ \text{nm} & \text{for electrolytes type } 2:2 \end{array}\right\} \tag{2.22}$$

The potential gradient at any distance x from a surface in the presence of symmetrical electrolytes, Eq. 2.18 can be recast as:

$$\frac{d\psi}{dx} = -\left(\frac{8kT\rho_{\infty i}}{\varepsilon\varepsilon_0}\right)^{1/2}\sinh\left(\frac{ze\psi_x}{2kT}\right) = -\frac{2kT\kappa}{ze}\sinh\left(\frac{ze\psi_x}{2kT}\right) \tag{2.23}$$

and integrating, results in

$$\psi_x = \frac{2kT}{ze}\lg\left(\frac{1+\gamma\exp(-\kappa x)}{1-\gamma\exp(-\kappa x)}\right) \approx \frac{4kT}{ze}\gamma\exp(-\kappa x) \tag{2.24}$$

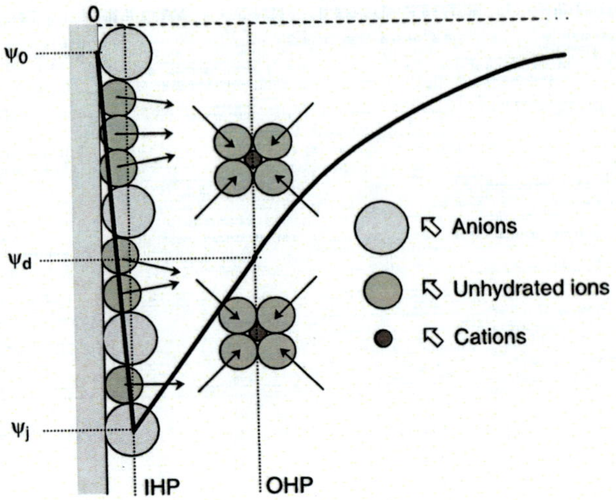

Fig. 2.6 Schematic representation of the Stern model of the double layer

which constitutes the Gouy-Chapman theory. For high potentials $\gamma \rightarrow 1$, while for low potentials the substitution $\mathbf{tanh}\,(x) \approx x$ reduces that equation to the so-called Debye-Hückel equation,

$$\psi_x = \psi_0 \exp\left(-\,\kappa x\right) \tag{2.25}$$

Stern proposed that the electrolyte ions could not be considered as point charges since the finite size of the ions, hydrated or not, limits the maximum concentration at the surface and their distance of closest approach to it. On the other hand, the high electric field near the surface must produce some ordering of solvent dipoles. Accordingly, the Stern modification of the Gouy-Chapman theory assumes the first layer of ions to adsorb specifically onto the surface, through a specific chemical adsorption potential $\boldsymbol{\theta}$. The distance of closest approach to the surface is x_1, and the diffuse part of the double layer starts at x_2, such that for $x > x_2$ specific interaction forces are negligible. The potential distribution at an interface is schematically represented in Fig. 2.6. If the surface is positive, the unhydrated anions are in contact with the surface and the plane formed by the centres of the anions is called the Inner Helmholtz Plane (IHP). The cations remain hydrated and the plane crossing their centres is the Outer Helmholtz Plane (OHP). When a particle is moving through the liquid, the Stern and part of the diffuse layer move with the particle. The potential at this plane of shear is referred to as the zeta potential ζ and it indicates the gradient of electrical potential when the surface potential is constant. The pH value at which the zeta potential is zero is called the isoelectric point (IEP). At the ZPC the surface charge is zero (σ_0=0), while at the IEP the sum of charge densities is zero ($\sigma_0+\sigma_r+\sigma_d$=0). Figure 2.7. illustrates the different planes and potentials at the double layer.

The interaction pressure between two similarly charged surfaces in electrolyte solution can be calculated considering that at any point x the pressure $P_x(D)$ is given by:

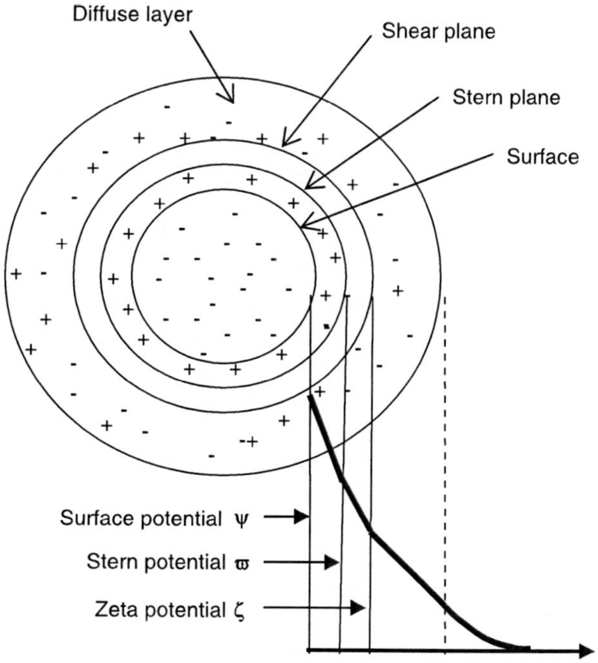

Fig. 2.7 The different planes and potentials at the double layer

$$P_x(D) = kT\left[\sum_i \rho_{mi}(D) - \sum_i \rho_{mi}(\infty)\right] \tag{2.26}$$

for $P_x(D=\infty)=0$, with ρ_{mi} as the total ionic concentration at the midplane. P is simply the excess of osmotic pressure of the ions in the midplane over the bulk pressure. For a 1:1 electrolyte. Assuming that ψ_m is the sum of the potential from each surface to x=D/2, the repulsive pressure between two planar surfaces is given by,

$$P = 64kT\rho_\infty\gamma^2 \exp(-\kappa D) \tag{2.27}$$

where $\gamma<1$. Integrating with respect to D yields the interaction potential,

$$V_R = \left(\frac{64kT\rho_\infty\gamma^2}{\kappa}\right)\exp(-\kappa D) \tag{2.28}$$

The interaction force (F) between two particles of radius a_1 and a_2 can be easily related to the interaction energy (V) between two planar surfaces by the Derjaguin approximation:

$$F(D) \approx \int_D^\infty 2\pi\left(\frac{a_1 a_2}{a_1+a_2}\right) f(z)dz = 2\pi\left(\frac{a_1 a_2}{a_1+a_2}\right) V(D) \tag{2.29}$$

F(z) being the normal force per unit area between two flat surfaces separated by a distance z. If $a_1 >> a_2$, then $F(D) \approx 2\pi aV(D)$, which is the case for a particle near a surface. If $a_1 = a_2 = a$, then $f(D) = \pi aV(D)$ and the interaction free energy is obtained by further integration,

$$V_R = \left(\frac{64\pi kTa\rho_\infty \gamma^2}{\kappa^2} \right) \exp(-\kappa D) \quad (2.30)$$

This indicates that the double layer interaction between surfaces or particles decays exponentially with the distance. The characteristic decay length is the Debye length.

2.3.4.2 The DLVO Theory

Derjaguin-Landau and Verwey-Overbeek proposed that the total interparticle potential in a polar liquid or electrolyte is obtained by addition of the electrostatic repulsion V_R due to overlap of the double layers and the van der Waals attraction V_A, arising from electromagnetic fluctuations. Substituting the respective forms of both potentials, V_R from Eq. 2.23 and the reduced form of V_A, the interparticle potential is given by:

$$V_T = V_R + V_A = \left(\frac{64\pi kTa\rho_\infty \gamma^2}{\kappa^2} \right) \exp(-\kappa D) - \frac{Aa}{12D} \quad (2.31)$$

Figure 2.8 shows the resulting interaction potentials as a function of the distance, in which the following three characteristics can be observed:

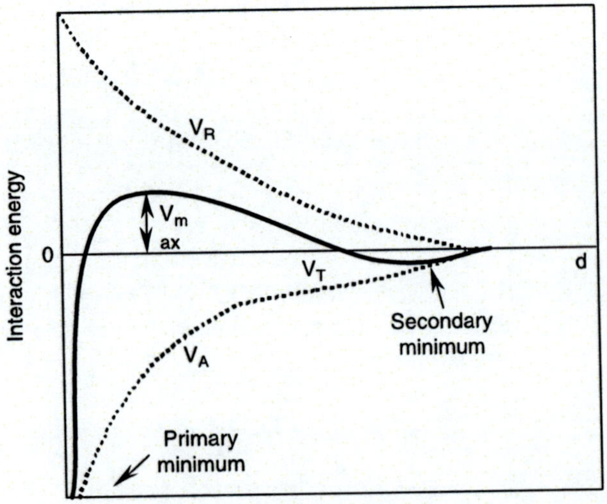

Fig. 2.8 Potential energy of interaction as a function of separation distance according to the DLVO theory

1. a potential barrier preventing contact between particles,
2. a primary minimum in which agglomerated particles are bound, and
3. a secondary minimum, not always present, that can provide some stabilisation by means of a liquid film between the particles.

If the height of the primary minimum is ≥25 kT, the system displays long-term stability. Coagulation occurs for higher depths in the primary minimum. This minimum is present at short separations between particles, in the range of molecular distances.

The secondary minimum can appear at higher separation between particles, in the range of the particle size (agglomeration effects). Flocculation takes place if the depth of this secondary minimum is large enough.

For highly charged surfaces in dilute electrolytes (long Debye length), a strong long-range repulsion exists that peaks at some distance. In more concentrated electrolyte solutions, there is a significant secondary minimum, usually beyond 3 nm before the energy barrier. For surfaces of low charge density or potential, the energy barrier will be much lower leading to coagulation or flocculation. Above a certain critical electrolyte concentration (known as *critical coagulation concentration*, ccc) the energy barrier falls below the $V=0$ axis, and particles coagulate rapidly. For increasing electrolyte concentrations, i.e., decreasing surface potentials, the total curve approaches to that of the van der Waals attraction, and particles attract each other strongly at any distance. The critical coagulation concentration occurs when $V=0$ and $dV/dD=0$. The so-called Schulze-Hardy rule states that the flocculation values for counterions with charges $1e^-$, $2e^-$, and $3e^-$ decrease in the ratio 100:1.6:0.13, respectively. This is why dispersants should contain monovalent cations, because a small concentration of a divalent one is enough to produce flocculation. The efficiency of an electrolyte to precipitate or sequestrate depends on the tendency of its ions to hydrate. Then, we can establish lyotropic series with decreasing precipitating power, in the form $Mg^{2+} > Ca^{2+} > Sr^{2+} > Ba^{2+} > Li^+ > Na^+ > K^+ > NH_4^+ > Rb^+ > Cs^+$. For anions, the series is Citrate $> SO_4^{2-} > Cl^- > NO_3^- > I^- > CNS^-$. These are referred to as Hofmeister series.

2.3.5 *Polymeric Stabilization*

The DLVO theory accurately quantifies the particle pair potential in most colloidal dispersions, either in water or in organic media. An exception is made of the non-DLVO interactions, like the oscillatory forces operating at separations near the contact distance. Another exception, one extremely important in the dispersion of ceramics, refers to those systems involving the presence of polymers providing an additional repulsive force related to the free energy change if there is adsorption or volume exclusion when polymers do not adsorb. There are two general ways in which polymers can impart colloidal stability: (1) through a *steric* stabilization mechanism, arising from the adsorption of the polymers to the particle surface, and

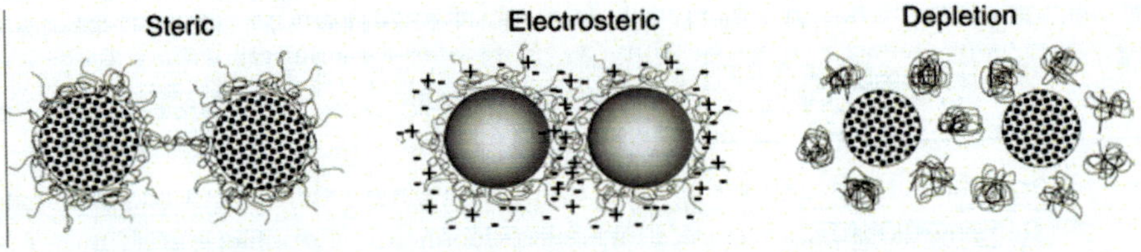

Fig. 2.9 Mechanisms of polymeric stabilization

(2) *depletion* stabilization, in which stability is provided not by attached polymers but by macromolecules that remain free in solution.

Moreover, steric stabilization can be divided into two groups, dealing with either uncharged or charged systems. The last mechanism is known as *electrosteric* mechanism and results from the combination of electrostatic and steric contributions to the total interaction pair potential. In electrosteric stabilization the electrostatic component may be due to a net charge on the particle surfaces and/or charges associated with the attached polymer. These ionic polymers are referred to as polyelectrolytes and are widely used in most ceramic processes. Combinations of depletion with steric and electrosteric mechanisms are also possible. Figure 2.9 illustrates the three general mechanisms of stability imparted by polymers.

Adsorbed polymer segments are classified into trains, loops and tails, depending on whether the segment is anchored at the interface, forms a loop out from the surface, or extends into the solution with only one end attached at the surface, as illustrated in Fig. 2.10. The relative contribution of trains, loops, and tails will depend on the interaction strength between the polymer monomers and the surface, the solvency of the polymer chain, and the polymer charge density in the case of polyelectrolytes. Increasing the molecular weight usually results in an increase of adsorption, thicker layers, and a larger relative fraction of loops at the expense of tails. Improved solvency reduces the overall adsorption, but frequently increases the size of loops and tails and their extension into solution. Finally, increasing the strength of the monomer surface interaction leads to an increase of adsorption and an increase of the loop and tail extensions into solution.

The best steric stabilizers are amphipathic block or graft copolymers, where one of the comonomers generates a homopolymer that is not soluble in the dispersion medium and the other polymerizes to give a polymer that is soluble in the dispersion medium [42]. The insoluble polymer attaches itself to the particle and is referred to

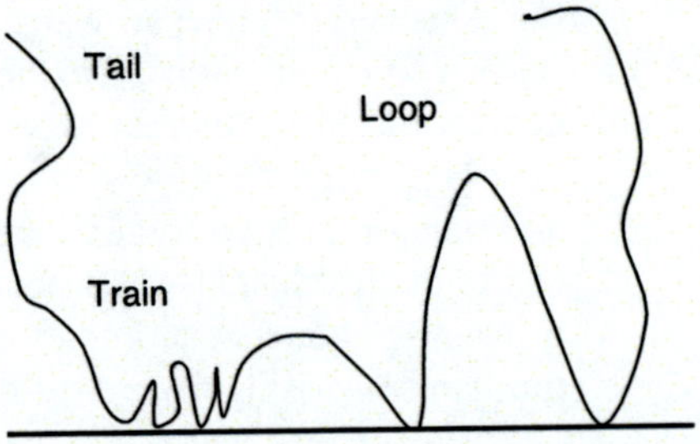

Fig. 2.10 Structure of adsorbed polymers with segments distributed into trains, loops and tails

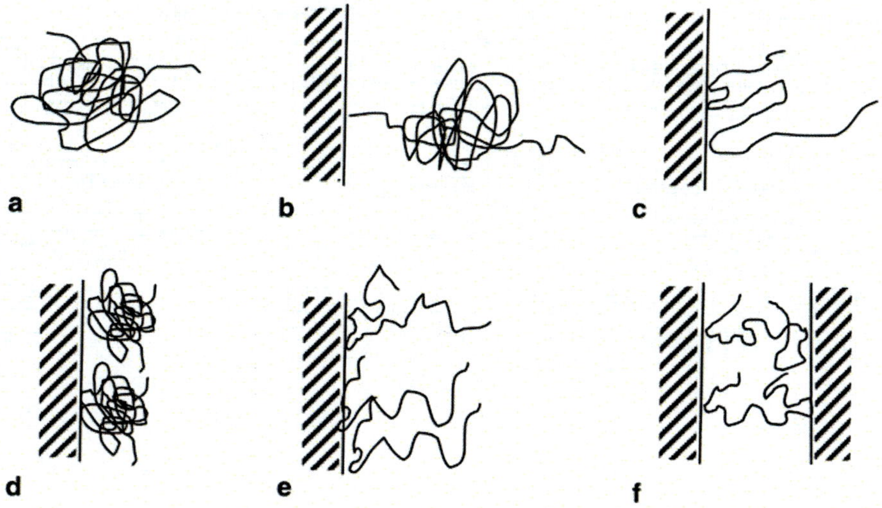

Fig. 2.11 Configurations adopted by polymer chains: free (**a**); grafted (**b**); adsorbed (**c, d**); brush (**e**); and bridging (**f**)

as the anchor polymer. The role of the soluble part is to provide steric stabilization. For this reason, such chains are known as the stabilizing moieties. This situation is illustrated in Fig. 2.11.

Steric stabilization occurs when particles having adsorbed layers approach each other. A useful model was proposed by Napper [43, 44] to illustrate steric stabilization that distinguish three domains of close approach determined by the relative spans of the attached polymer layers (L) and the separation distance between particles (d). The model first considers the particles as parallel flat plates. The three domains are (Fig. 2.12):

1. The non-interactional domain, $d > 2L$. The stabilizing moieties are far away from each other and there is no interaction, i.e. there is no free energy change.
2. The interpenetrational domain. $L \leq d \leq 2L$. There is interpenetration between the adsorbed polymer layers. In this zone, the polymer segment density increases leading to the exclusion of solvent molecules, hence the mixing of segments and solvents decreases. In good solvents the free energy of the system increases and there is repulsion. In worse than θ-solvents the free energy decreases, leading to attraction.
3. The interpenetrational-plus-compressional domain, $d < L$. As particles approach to near to contact distance, the separation can become lower than the layer thickness so that the moieties on a surface are compressed by the opposite surface. The free energy in this domain has two components, a solvent-segment mixing term, and an elastic term that arises from the compression of the moieties, which reduces the configurational entropy. Consequently, the elastic term is always repulsive.

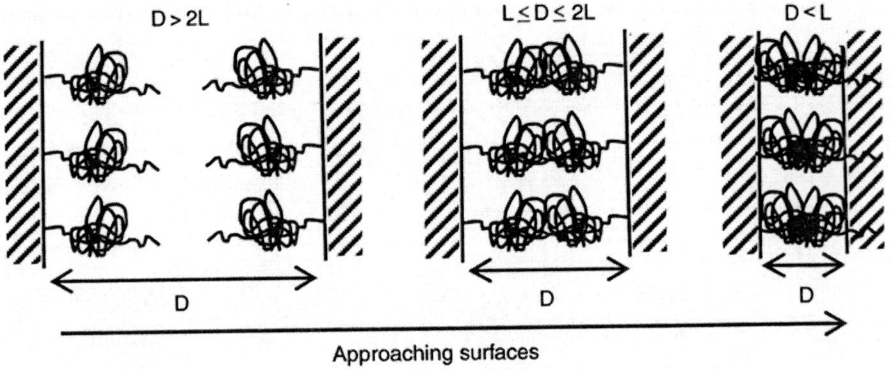

Fig. 2.12 The three domains of close approach of sterically stabilized surfaces

The model of the three domains of close approach of sterically stabilized particles considers two basic components, the mixing free energy and the elastic energy.

In the interpenetrational domain, the steric interaction is controlled by the increase in the segment density of the stabilizing moieties, which results in the exclusion of the solvent molecules into the bulk continuous phase. The free energy in the interpenetrational domain is called the *mixing free energy* (ΔG_{mix}), although the term mixing is used in most theories of polymer solutions to describe the reverse process. The mixing term has been also referred to as the osmotic pressure, because the steric layers generate a difference in chemical potential between the solvent molecules existing inside and outside the interpenetrational domain.

Where the operational distance between particles is lower than the span of the stabilizing moieties a compression effect takes place that reduces the configurational entropy. Thus, the elastic free energy is always repulsive, whereas the mixing free energy may be repulsive or attractive depending on the quality of the solvent. In the interpenetrational-plus-compressional domain both the mixing and the elastic contributions coexist.

The mixing free energy in the interpenetrational domain can be calculated as

$$\Delta G_{mix} = 4\Pi a \omega^2 N_A \left(\frac{\bar{v}_2^2}{\bar{V}_1} \right) kT \left(\frac{1}{2} - \chi_1 \right) \left(1 - \frac{D_0}{2L_s} \right)^2 \tag{2.32}$$

where Π is the osmotic pressure, ω is the weight of the adsorbed polymer layer, $\bar{v}_2$ is the specific partial volume, $\bar{V}_1$ is the molar volume of the dispersing medium, χ_1 is the Flory interaction parameter, and D_0 is the minimum distance between particle surfaces.

The interaction pair potential resulting from the adsorption of a charged polymer is essentially the same as for uncharged polymers, but modified with a term that accounts for the effects of the charge. For a polyelectrolyte having i charges per monomer unit, and considering an aqueous solvent containing c_s mol of 1:1 electrolyte per unit volume, the interaction energy may be expressed by

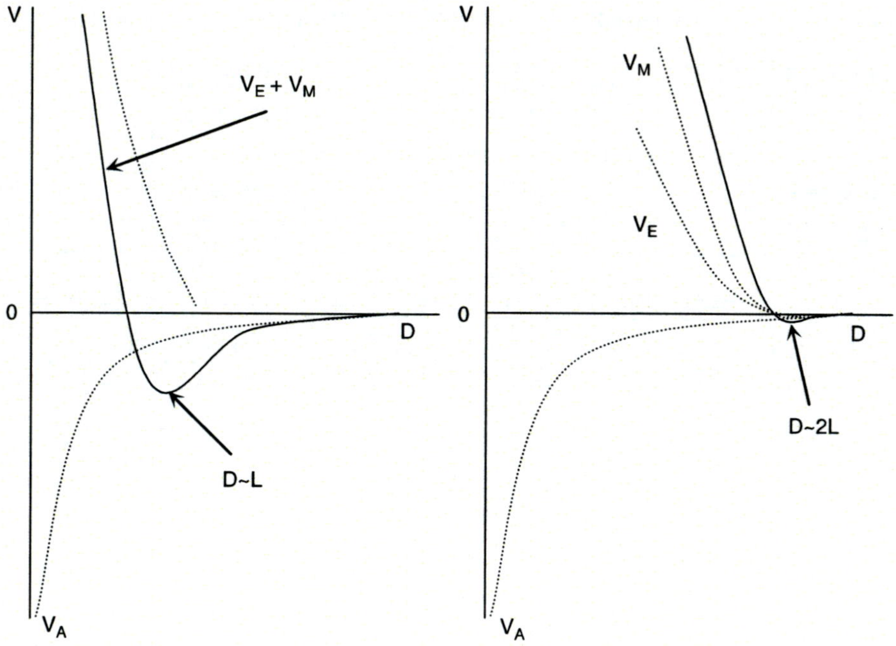

Fig. 2.13 Potential energy curves for sterically stabilized suspensions

$$\Delta G_M = 2\Pi akT\omega^2 N_A \left(\frac{\overline{v}_2^2}{\overline{V}_1}\right)\left(\frac{1}{2} - \chi_1 + \frac{i^2}{4\overline{V}_s c_s}\right) S_M \tag{2.33}$$

V_s is the molar volume of a monomer unit. At the temperature $-\theta$, i.e., for a specific concentration of salt and a certain temperature, $\Delta G_M = 0$.

The typical shape of the potential energy curve for a suspension stabilized by steric hindrance is that shown in Fig. 2.13. The same behavior is also shown by electrosterically stabilized suspensions, i.e., dispersed with charged polymers. The main difference with respect to the curve of electrostatically stabilized suspensions is that in the steric one there is no primary minimum, so that there is no coagulation at the primary minimum. This implies that steric stabilization provides thermodynamic stability, whereas electostatically stabilized suspensions are only metastable, since coagulation occurs when particles approach to near contact distances. Nevertheless, a secondary minimum can be present as in the electrostatic mechanism, and by analogy it is called pseudo-secondary minimum. At longer separation distances the stability is provided by the mixing term V_M.

The minimum of the curve shifts to longer distances (i.e. stability increases) when the polymer molecular weight or the thickness of the adsorbed layer increase.

Polyelectrolytes are the most widely used dispersants in ceramic technology, in particular anionic polyelectrolyte like polyacrylates, polycarboxylates, etc. The adsorption of an anionic polyelectrolyte on the particles surface provokes the IEP to

Table 2.9 Polymeric stabilizers commonly used in ceramic processing

Deflocculant	Formula	Stabilization mechanism
Dodecyl trimethyl ammonium bromide	$(CH_2)_{12}N(CH_3)_3Br$	Electrostatic, steric
Poly(acrylic acid)	$(CH_2\text{-}CH\text{-}COOH)_n$	Electrostatic, steric
Poly(ethylene oxide)	$(CH_2\text{-}CH_2\text{-}O)_n$	Steric
Polyacrylamide	$(CH_2\text{-}CH\text{-}CONH_2)_n$	Electrostatic, steric
Poly(vinyl alcohol)	$(CH_2\text{-}CHOH)_n$	Steric
Poly(methacrylic acid)	$(CH_2\text{-}CCH_3\text{-}COOH)_n$	Electrostatic, steric
Poly(vinyl amine)	$(CH_2\text{-}CHNH_3Cl)_n$	Electrostatic, steric

shift down to lower pHs. On the contrary, cationic polyelectrolytes like polyethyleneimine, shifts the IEP to higher pH values, although their dispersing efficiency is lower. In EPD technology, since the suspensions have usually low solids contents, both kinds of dispersants are able to provide the desired stability, but the use of cationic additives is preferred since cationic deposition is then allowed, so that the problem of electrode effects and the resulting contamination by corrosion disappear. Table 2.9 shows some common polymeric stabilizers.

2.4 State of the Art of the EPD Equation

Among colloidal routes applied to nanotechnology, EPD stands out as a powerful and versatile colloidal process to address the need for inexpensive and mass production using suspensions with relative low concentration (<20 vol.%). This drastically increases its range of applicability competing with other non-colloidal techniques, especially coating processes. In most of them, the transport of small particles, whereby colloidal forces overcome gravitational forces, is mainly governed by Brownian motion, resulting in completely random particle-particle and particle-substrate collisions [45]. In electrically driven processes, a directional force (external electric field) is applied, leading to 1D, 2D and 3D ordered structures at the nanoscale [46–50].

The EPD process is based on the migration and later deposition of charged particles in a stable suspension, induced by the application of an electric field between two electrodes. Particle packing by solvent evaporation is the common compaction mechanism in several shaping processes such as dipping, spin coating and tape casting. Solvent evaporation in EPD is preceded by the electrically-assisted formation of a cohesive layer of particles which reproduces the shape of the work electrode. Consequently, deposit yield and particle packing in electrophoresis depend not only on the stability and the solid content of the suspension, but also on the aggregation and the arrangement of the particles on the electrode surface. The performance of EPD in producing homogeneous and reliable films strongly depends on the particle surface chemistry, the behaviour of the surface-liquid interfaces under an electric field, and the development of particle-particle and particle-substrate networks during particle assembly.

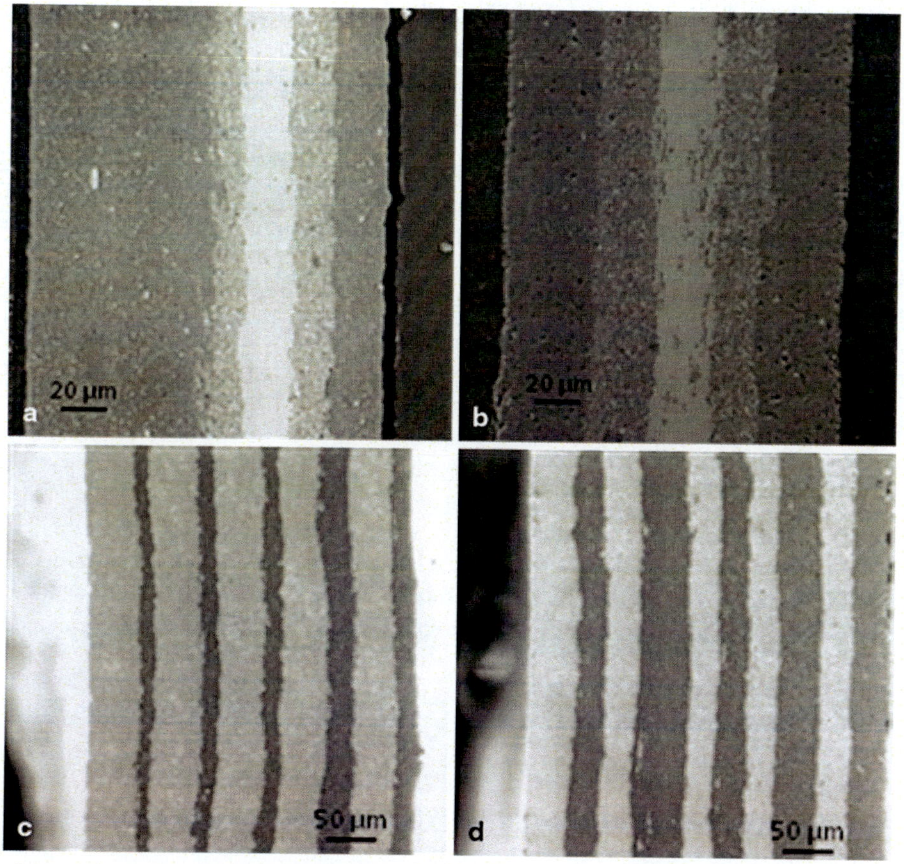

Fig. 2.14 Laminated materials shaped by sequential EPD controlling deposition time (**a** and **b**) and the stability of the suspension (**c** and **d**). [51, 52]

Figure 2.14 shows different materials obtained by EPD, acting over the deposition time and the suspension conductivity to design the layer thickness in a laminated structure. Differences in the dispersing conditions of the suspensions of dissimilar powders lead to a differentiate deposition. This fact can be used to build laminated structures (Fig. 2.14c, d) and in a similar way, time deposition can be used (Fig. 2.14a, b).

Suspension properties, such as conductivity, zeta potential, solid content, viscosity, etc., are key parameters of the EPD process. The colloidal systems used must be stable, meaning that the particles have to remain dispersed throughout the medium so they can move towards the electrode independently of each other. Then, particles can deposit separately, without agglomerates, keeping open the possibility of their rearrangement during packing through the action of an electric field, similarly to the deposit formation by gravitational forces (sedimentation) [53, 54]. To determine how suspension parameters affect to the EPD kinetics, criteria of different authors should be collected from the literature. In this sense, the nomenclature needs to be unified [54], as shown in the scheme appearing below.

The first model of EPD kinetics was proposed by Hamaker in 1940 (Eq. 2.34) for electrophoretic cells with a planar geometry [55]. It relates the deposited mass, m (g), with slurry properties, such as suspension concentration, C_s (g cm^{-3}), and electrophoretic mobility, μ ($cm^2s^{-1}V^{-1}$), with the physical and electrical conditions imposed on the system as electric field, E (V cm^{-1}), deposition area, S (cm^2), and deposition time, t (s),

$$m = C_s \mu S E t \tag{2.34}$$

A similar expression was deducted in 1991 by Hirata et al. [56] based on the application of the Faraday law to the deposition process, considering the particles as the unique charge carriers in the suspension. The linear variation of deposited mass with deposition time requires that the parameters of Eq. 2.34 remain unchanged with time. This fact limits the application of the Hamaker expression to short deposition times.

Sarkar and Nicholson [57] analyzed the dependence of kinetics with some of the experimental EPD conditions. They firstly introduce the efficiency factor or "sticking parameter", $f \leq 1$ (i.e. if all the particles reaching the electrode take part in the formation of the deposit $f = 1$) to quantify the effect of the undetermined factors affecting the process of deposition. Referring the Hamaker equation (Eq. 2.34), they quantified the effect of the variation of particle concentration on the EPD kinetics.

In the earlier stages of the process, the variation of bulk solid concentration in the suspension is negligible, since only a minor fraction of the powder is being deposited. Hence, for infinitesimal intervals of time, the Hamaker equation always holds, so that

$$\frac{dm}{dt} = f \mu S E C_s \tag{2.35}$$

The amount of powder excluded from the bulk suspension becomes significant for longer times, and consequently C_s decreases. In this case, the deposited mass and the solid content are expressed by:

$$m = V(C_{s,0} - C_s) = \frac{m_0}{C_{s,0}}(C_{s,0} - C_s) \tag{2.36}$$

where $C_{s,0}$ (g cm^{-3}) is the initial solid content of the suspension, V (cm^3) is the volume of the suspension considered constant, and m_0 (g) is the initial mass of powder in suspension. All of them are related by the expression,

$$C_{s,0} = \frac{m_0}{V} \tag{2.37}$$

Eq. 2.36 gives,

$$C_s = C_{s,0}\left(1 - \frac{m}{m_0}\right) \tag{2.38}$$

Combining Eqs. 2.35, 2.38

$$\frac{d}{dt}\left(\frac{m}{m_0}\right) = \frac{1}{\tau}\left(1 - \frac{m}{m_0}\right) \tag{2.39}$$

where τ defines a characteristic time scale given by

$$\tau = \frac{V}{f\mu SE} \tag{2.40}$$

The inverse value of the characteristic time was defined by Sarkar and Nicholson as a universal parameter, k, named the "kinetics parameter".

If no sedimentation occurs, and the only change of concentration is the mass of powder deposited by EPD for an initial time t=0, the deposited mass is m(0)=0. This leads to the solution of Eq. 2.39:

$$m(t) = m_0(1 - e^{-t/\tau}) \tag{2.41}$$

This solution for EPD kinetics has been widely applied by Sarkar and Nicholson [57], and completes the first description proposed by Zhang et al. [58] in 1994, concerning the incorporation of changes in particle concentration in the EPD kinetics.

Equation 2.41 can be reduced to the Hamaker model for short times, and it is widely accepted in the literature. In recent years several authors have proposed different mathematical models based on this equation to describe the deep electrophoretic penetration and deposition of ceramics in porous substrates [59–63] to determine deposit thickness [64], to control the homogeneity of the porous distribution in a ceramic membrane [65], or fabricate pieces with composition gradients combining EPD and sedimentation processes [66].

Furthermore, Sarkar and Nicholson model predicts deviations from linearity occurring when EPD is carried out under constant-voltage conditions, and the deposit resistivity is larger than that of the suspension. In this case, the electric field strength applied to the suspension can be considered to be:

$$E = \frac{\Delta\psi}{L + \delta\left(\frac{\rho_s}{\rho_d} - 1\right)} \tag{2.42}$$

where $\Delta\psi$ (V) is the potential drop between the electrodes, L (cm) the distance between electrodes, ρ_s and ρ_d (Ω cm) are the resistivities of the suspension and the deposit respectively, the deposit thickness, δ (cm), being defined as follows:

$$\delta = \frac{V_d}{S} = \frac{m/C_d}{S} \tag{2.43}$$

where, V_d (cm^3) is the deposit volume and C_d ($g\ cm^{-3}$) the deposit concentration.

Combining Eqs. 2.42, 2.43 yields:

$$E = \frac{\Delta\psi}{L + R'm} \tag{2.44}$$

where R′ is a constant, the value of which is given by

$$R' = \frac{\left(\frac{\rho_s}{\rho_d} - 1\right)}{C_d S} \tag{2.45}$$

Combining Eqs. 2.34, 2.36 and 2.42 yields:

$$\frac{d}{dt}\left(\frac{m}{m_0}\right) = k'\left(1 - \frac{m}{m_0}\right)\frac{\Delta\psi}{L + R'm} \tag{2.46}$$

where k′ is a redefined "kinetics parameter",

$$k' = \frac{fS\mu}{V} \tag{2.47}$$

Solving Eq. 2.46 with boundary conditions similar to those of Eq. 2.39 gives

$$R'm(t) + (R'm_0 + L)\ln\left(\frac{m_0 - m(t)}{m_0}\right) + k'\Delta\psi t = 0 \tag{2.48}$$

Equation 2.48 is a general expression to describe the deposition kinetics. It reduces to Eq. 2.41 when the resistivity of the deposit is similar to that of the suspension ($\rho_s = \rho_d$). Otherwise, the authors point out based on Eq. 2.48, the shaping of thick deposits involves the preparation of more concentrated suspensions than allowed for by the Hamaker model, but they did not quantify this effect. More recently, Biesheuvel et al. [67] have described a model for the growth of the deposit, based on the Kynch theory of sedimentation for planar and cylindrical geometries. This theory describes the bulk effect of the particle motion in the transport phenomena near the deposition electrode, based on the expression of the mass balance of the suspension-deposit boundary evolution, resulting in:

$$\frac{d\delta}{dt} = -v\frac{\phi_s}{\phi_d - \phi_s} \tag{2.49}$$

where, v (cm s^{-1}) is the electrophoretic rate of particles close to the electrode, ϕ_d is the volumetric fraction of the deposit and, ϕ_s is the volumetric fraction of the suspension.

This growth theory is limited to explain the evolution of the deposit-suspension boundary, and correspondingly the system studied was identified as a non-stirred and electrically neutral suspension. As a consequence, some phenomena occurring during the EPD process are neglected, including the decrease of particle concentration, the movement of particles by diffusion, and the local changes of the charge on the electrodes, both on the particle surface and its surrounding ionic clouds. Hence, as in the Sarkar and Nicholson model, the particle velocity depends only on the applied electric field, $v=E\mu$, and the effective electric field for a planar geometry was defined as $E=\Delta\psi/L$.

The deposited mass will be determined by:

$$m = SC_d\delta \tag{2.50}$$

Under these conditions, combining Eqs. 2.49 and 2.50 gives

$$m = \mu ESC_d \frac{\phi_d}{\phi_d - \phi_s} t \tag{2.51}$$

Comparing with the Hamaker expression (Eq. 2.34), where

$$C_s = \rho_p \phi_s \tag{2.52}$$

and the Biesheuvel and Verweij equation (Eq. 2.53), where

$$C_d = \rho_p \phi_d \tag{2.53}$$

this theory considers the deposition dependence of the deposit growth vs the starting solid content of the suspension, where a correction factor, X, should be incorporated to the kinetics expression for highly concentrated suspensions ($\phi_s > 0.2$ when $\phi_d \sim 0.6$):

$$X = \frac{\phi_d}{\phi_d - \phi_s} \tag{2.54}$$

Subsequently, the same researchers proposed a kinetics expression based on their previous model considering the effect of concentration decrease during the deposit formation [68]. Here a stirred and diluted suspension is the studied system, so in this case they correctly considered that there is no diffusion of particles because there is not a sufficient concentration gradient ($v=E\mu$). They also fix other system conditions such as a constant volume suspension, V, and the global electroneutrality ($E=\Delta\psi/L$).

Under these conditions the proposed mass balance of the particles in suspension was

$$V\frac{d\phi_s}{dt} = S(\phi_d - \phi_s)\frac{d\delta}{dt} = S(\phi_d - \phi_s)v\frac{\phi_s}{\phi_{d-}\phi_s} \tag{2.55}$$

Solving Eq. 2.55 by considering the characteristics of the described system, gives

$$\delta = \frac{V}{S}\frac{\phi_{s,0}}{\phi_d}(1 - e^{-t/\tau}) \tag{2.56}$$

where $\phi_{s,0}$ is the initial volumetric fraction of the suspension, and τ is the characteristic time defined by Eq. 2.40. Considering Eqs. 2.53, 2.54, 2.55 and 2.56 yields:

$$\delta = \frac{m_0}{SC_d}(1 - e^{-t/\tau}) \tag{2.57}$$

fitting the model proposed by Sarkar and Nicholson [57] for a similar system (Eq. 2.41).

In agreement with Sarkar and Nicholson, Biesheuvel and Verweij consider that the effective electric field under potentiostatic conditions depends on the electrical characteristics of the deposit. In this case, the relationship between the dielectric constants of the deposit, ε_d, and suspension, ε_s, was considered and the applied electric field was defined by

$$E = \frac{\Delta\psi}{L + \delta\left(\frac{\varepsilon_s}{\varepsilon_d} - 1\right)} \tag{2.58}$$

In addition to the Sarkar and Nicholson analyses (Eq. 2.50), the authors described how Eq. 2.58 predicts that no deposit growth will occur when $\varepsilon_s < \varepsilon_d$. In fact, when a low concentration is considered ($\phi_s << \phi_d$), the dielectric constant of the suspension can be approximated by the dielectric constant of the liquid medium ($\varepsilon_s = \varepsilon_{r,l}$), and similarly, the dielectric constant of the deposit can be approximated by the dielectric constant of the particles ($\varepsilon_d = \varepsilon_{r,p}$). Under these limits Eq. 2.58 indicates that homogenous layers can be only obtained *a priori* when the particle permitivity, $\varepsilon_{r,p}$, is lower than the liquid permitivity, $\varepsilon_{r,l}$.

To avoid the effect of the increase of the deposit resistivity in the effective electrical force acting on particles, Sarkar and Nicholson suggested working under galvanostatic or constant-current conditions [57]. In that way, the voltage drop across the two electrodes increases with time, but the voltage/unit length in the suspension remains constant. Working under galvanostatic conditions Ma and Cheng in 2002 [69] determined experimentally the relationship between the *kinetics parameter*, k, and the applied current intensity:

$$k = k_0(e^{i/i_0} - 1) \tag{2.59}$$

where i (mA cm^{-2}) is the current density, and i_0 (mA cm^{-2}) and k_0 (s^{-1}) are considered the reference conditions from which the expression predicts the kinetics constant from different values of the applied current, facilitating more effective modelling and controlling of the process.

However, most of the experimental work in the literature describes EPD processes performed under potenciostatic or constant-voltage conditions, and hence substantial efforts have been made to model the effective electric field applied for each system [70–77].

Several authors have proposed different secondary phenomena occurring during the EPD process that influence on the effectiveness of the applied electric field. In 1999 Van der Biest and Vanderperre [73], introduced an expression for the potential drop over an electrophoretic cell with flat, equal surface area electrodes which consists of four terms:

$$\Delta\psi = \Delta\psi_a + \frac{I}{S\rho_d}\delta + \frac{I}{S\rho_s}L + \Delta\psi_c \tag{2.60}$$

where I (A) is the current and $\Delta\psi_a$ and $\Delta\psi_c$ (V) are the potential drop at the anode and cathode respectively.

So, to consider the electric field strength in a suspension as: $E=\Delta\psi/L$, the resistivity of the deposit has to be of the same order as the resistivity of the suspension, and changes in the electrode polarization and in the suspension resistivity must be negligible. However, even when the electrode is selected with the aim of eliminating its boundary reactions, changes in suspension resistivity during the process must be considered.

Several authors [78, 79] have considered the effect of the suspension resistivity on electrophoretic feasibility. They have measured the suspension conductivity σ (S cm^{-1}) especially in suspensions stabilized by the addition of salts of different metals [80–82]. In 1996 Ferrari and Moreno studied the effect on EPD of different parameters of an aqueous suspension electrosterically stabilized [83–85]. These studies claimed a special role for the suspension conductivity in the effectiveness of the electrophoretic process. Other authors have then considered the influence of conductivity on the deposition in aqueous and non-aqueous systems [86–100]. Some of them have demonstrated that a high ionic concentration in the suspension could be deleterious for particle stability, inhibiting the movement of the ceramic particles, and thereby decreasing the deposit growth and sometimes its quality. In fact, these studies indicate that electrostatic and electrosteric stabilization in non-aqueous and aqueous media has to be adjusted in order to assure a low ionic concentration in stable suspensions. Then, the effectiveness of the electrophoretic process increases because the particles are the main current carriers.

Both surface-charged particles and ions contribute to the transport of the electric charge when an external electric field is applied to the system [100, 101]. The net current is

$$I = I^{+} + I^{-} + I_{p} \tag{2.61}$$

where I^+, I^- and I_p are the electric current (A) transported by the cations, anions and particles of the suspension respectively.

Assuming that free-ions and particles (with their ionic clouds) are homogeneously distributed in the bulk suspension, the total resistivity is

$$\rho_s = \left[\frac{1}{\rho_+} + \frac{1}{\rho_-} + \frac{1}{\rho_p}\right]^{-1} \tag{2.62}$$

where ρ_+, ρ_- and ρ_p represent the contribution of the cations, anions and particles respectively to the suspension resistivity.

The resistivity of the optimized suspensions mainly depends on the particle concentration. There are numerical models and theoretical expressions describing the relationship between conductivity (or resistivity) and solid content of concentrated suspensions with spherical, rigid [102, 103] or soft particles [104, 105], whose application is restricted to suspensions with uncharged particles. In practice, most experimental systems are formed by charged particles, where the conductivity (or resistivity) is affected by the thickness and properties of the electric double layer or the characteristics of the polyelectrolyte adsorbed at the particle surface. The effective conductivity of dilute suspensions of charged particles was first addressed by Saville in 1983 [105]. In this paper, the effect of counter-ions and non-specific adsorption are taken into account, and the expression proposed for the conductivity is:

$$\sigma_s = \sigma_{s,\infty}(1 + \alpha\phi_s) \tag{2.63}$$

where the suspension conductivity, σ_s (S cm^{-1}), depends on the conductivity of the liquid medium, $\sigma_{s,\infty}$ (S cm^{-1}), the volume fraction of particles in suspension, ϕ_s, and the zeta potential or double layer thickness, α. Today, different research papers have been published describing the properties and electrokinetics behavior of non-diluted colloidal suspensions, and presenting the suspension conductivity as a function of the particle volume fraction (Eq. 2.63) [106, 107].

The proportional dependence of the suspension conductivity on solid content has been determined experimentally by different authors working on the EPD process for different powders when the suspension vehicle and the amount of dispersant were fixed [78, 84, 99, 101].

Recently, G. Anné et al. [108] proposed a mathematical description of the kinetics based on the Hamaker model and the Biesheuvel correction for suspensions with a high solid loading considering that the suspension conductivity and the current density vary during EPD under constant voltage conditions. The expression of the electric field as a function of the conductivity of the suspension is

$$E = \frac{I}{S\sigma_s} \tag{2.64}$$

Considering Eqs. 2.35, 2.54:

$$\frac{dm}{dt} = f\mu\frac{I}{\sigma_s}C_s\frac{\phi_d}{\phi_d - \phi_s} \tag{2.65}$$

This model fits EPD results obtained from an alumina suspension prepared with ethanol and using different additives as dispersants. The proposed expression (Eq. 2.65) was verified using experimental data collected during the deposition.

In 2005 Ferrari et al. [109] proposed a resistivity model for the deposition kinetics considering the relationship between colloidal parameters such as suspension concentration and resistivity (Eq. 2.65) during the EPD process. This model describes experimental results obtained for longer deposition times and for suspensions in which resistivities change significantly during the deposition process.

Anné et al. assume in Eq. 2.65 that the conduction surface area was equal to the electrode deposition area and the cross-section of the EPD cell. Consequently, Eq. 2.65 as a function of the suspension resistivity is

$$\frac{dm}{dt} = f\mu I\rho_s C_s \tag{2.66}$$

The solution of this equation is similar to that of Eq. 2.41; the characteristic time is:

$$\tau = \frac{V}{f\mu I}\rho_s \tag{2.67}$$

which changes with the suspension resistivity. Moreover, a linear relation may be used to describe the dependence of resistivity on solid content, as follows:

$$\rho_s = \rho_{s,\infty} - (\rho_{s,\infty} - \rho_{s,0})\frac{C_s}{C_{s,0}} \tag{2.68}$$

where, $\rho_{s,0}$ is the initial resistivity of the suspension and $\rho_{s,\infty}$ is the resistivity at infinite time, when $C_s=0$. The substitution of Eqs. 2.38 and 2.68 into Eq. 2.66 yields a differential equation of the form:

$$\frac{d}{dt}\left(\frac{m}{m_0}\right) = \frac{1}{\tau_0}\left(1 - \frac{m}{m_0}\right)\left(1 + \left(\frac{\rho_{s,\infty}}{\rho_{s,0}} - 1\right)\frac{m}{m_0}\right) \tag{2.69}$$

where the characteristic deposition time, τ_0, is similar to that in Eq. 2.39, and is defined by

$$\tau_0 = \frac{V}{f\mu I \rho_{s,0}} \tag{2.70}$$

The solution of this new equation with the same initial condition as in Eq. 2.39, m(0)=0, is

$$m(t) = m_0\left(1 - \frac{1}{1 + \frac{\rho_{s,0}}{\rho_{s,\infty}}\left(e^{t/\tau_\infty} - 1\right)}\right) \tag{2.71}$$

where the characteristic deposition time, τ_∞, is defined in analogy to that in Eq. 2.70, with $\rho_{s,0}=\rho_{s,\infty}$.

According to the solution given in Eq. 2.71, it is evident that for long deposition times, $t\rightarrow\infty$, the deposited mass $m(t)\rightarrow m_0$. However, the qualitative behavior of the curve m(t) deserves some discussion. Eq. 2.69 shows that

$$\frac{d}{dt}\left(\frac{m}{m_0}\right) > 0 \tag{2.72}$$

regardless of the amount of deposited mass ($0 \leq m/m_0 < 1$). So the curve *m(t)* in Eq. 2.71 always increases. Differentiating Eq. 2.69 again, using the same equation to eliminate the derivative, leads to:

$$\frac{d^2}{dt^2}\left(\frac{m}{m_0}\right) = \frac{1}{{\tau_0}^2}\left(1 - \frac{m}{m_0}\right)\left(1 + \left(\frac{\rho_{s,\infty}}{\rho_{s,0}} - 1\right)\frac{m}{m_0}\right) \times \left(\frac{\rho_{s,\infty}}{\rho_{s,0}} - 2 - 2\left(\frac{\rho_{s,\infty}}{\rho_{s,0}} - 1\right)\frac{m}{m_0}\right) \tag{2.73}$$

The first three factors are positive because $\rho_{s,\infty} > \rho_{s,0}$. If $\rho_{s,\infty}/\rho_{s,0} \leq 2$, which includes the Sarkar & Nicholson model for which $\rho_{s,\infty}/\rho_{s,0} = 1$, the last factor will be negative and, hence, m(t) will rise with a continuously decreasing slope, as in a typical saturation curve. By contrast, if $\rho_{s,\infty}/\rho_{s,0} \geq 2$, the second derivative (Eq. 2.73) will be positive as long as:

$$0 \leq \frac{m}{m_0} < \frac{\rho_{s,\infty} - 2\rho_{s,0}}{2(\rho_{s,\infty} - \rho_{s,0})} \tag{2.74}$$

and will become negative above this value. Thus, initially the curve m(t) rises with an increasing slope, producing an S-shaped saturation curve. Converting this to deposition time using Eq. 2.71, the rising of the curve takes place within the range $0 \leq t < T$, where

$$T = \tau_\infty \log\left(\frac{\rho_{s,\infty}}{\rho_{s,0}} - 1\right) \tag{2.75}$$

The growth of the slope depends exclusively on the variation in resistivity, and increases as ρ_∞/ρ_0 increases. The shape of the m(t) curve is not affected by any other parameter. This S-shaped behavior has already been reported in the literature for long-lasting EPD tests, especially for aqueous suspensions, although it also appears in non-aqueous media when deposition time is long enough [109]

Finally, notice that if $\rho_{s,\infty} = \rho_{s,0}$ (and hence $\rho_s = \rho_{s,\infty}$ constant), the solution of this last kinetics model (Eq. 2.71) reduces to Sarkar & Nicholson model (Eq. 2.41). Also, if $t << \tau_\infty$, this solution becomes that of the linear model proposed by Hamaker (Eq. 2.34). So, Eq. 2.71 subsumes previous models of EPD kinetics.

Nowadays, alternatives to the standard electrophoresis-based process have been proposed to obtain free-standing laminate nanostructures [110] or design prototyping [111] and nanopatterning films [112], or avoid bubble formation at the electrodes during aqueous EPD, applying AC electric fields [113, 114] or a constant current pulse [115]. The kinetics of these processes could be also approximated by the described equations. Table 2.10 summarizes the most relevant equations, corrections and experimental expressions proposed by different authors up to date, highlighting the parameter on which each one is focused.

An example can illustrate the fitting of experimental data to the proposed equations. It starts with a stable aqueous suspension of yttria-doped tetragonal zirconia polycrystal (ZrO_2, 3 mol% Y_2O_3, Tosoh TZ3YS, Japan) powder, with a mean particle size of 0.35 μm, a density of 5.81 g cm^{-3} and a specific surface area of 6.7 m^2 g^{-1}. The suspension was prepared with a concentration of 0.052 g cm^{-3} (5 wt.%), adding 0.6 wt.% (on a dry solid basis) of a commercial carbonic acid-based polyelectrolyte (Dolapix CE64, Zschimmer-Schwarz, Germany). Powders were homogenized within the water using a high-shear mixer (Silverson L2R, UK) for 3 min. Electrophoretic mobility measurements were performed using the micro-electrophoresis technique (Zeta-meter 3.0+, Zetameter, USA) and the conductivity with a digital conductimeter (LF320 WTW, Germany). More extensive details of the experimental work can be found in reference 116.

Table 2.10 Summary of different equations, corrections and experimental expressions of the EPD kinetics

Kinetics equation	Characteristic time
Basic equation [55]	
$m = C_s \mu S E t$	0
Quantification of the deposition behavior: the sticking factor [57]	
$m = f \mu S E C_s t$	0
Considering the decrease of solid loading with time [57]	
$m(t) = m_0(1 - e^{-t/\tau})$	$\tau = \frac{V}{f \mu S E}$
Considering concentrated suspensions ($\phi_s > 0.2$) [67]	
$m = C_s \mu S E t \frac{\phi_d}{\phi_d - \phi_s}$	0
Considering solid loading and electric field variation with time [57]	
$R' m(t) + (R' m_0 + L) \ln\left(\frac{m_0 - m(t)}{m_0}\right) + k' \Delta\psi t = 0$	$k' = \frac{1}{\tau'} = \frac{f \mu S}{V}$
Variation of the kinetics parameter vs. applied current [69]	
$k = k_0(e^{i/i_0} - 1)$	$k = \frac{1}{\tau} = \frac{f \mu S E}{V}$
Considering suspension resistivity variation with time [108]	
$m(t) = m_0(1 - e^{-t/\tau})$	$\tau = \frac{V}{f \mu I} \rho_S(t)$
Considering the linear relationship of the suspension resistivity and solid loading, and their variation with time [109]	
$m(t) = m_0\left(1 - \frac{1}{1 + \frac{\rho_{S,0}}{\rho_{S,\infty}}(e^{t/\tau_\infty} - 1)}\right)$	$\tau_\infty = \frac{V}{f \mu I} \rho_{S,\infty}$

Suspensions had an electrical conductivity of 115±0.5 μS cm^{-1} and pH 7.9±0.1. After each experiment the conductivity and the pH were measured and the decrease of solid concentration was evaluated in relation to the amount of powder deposited. The deposition of charged particles increased the suspension resistivity. The relationship between the suspension resistivity and the variation of solid concentration (Eq. 2.63) is given by

$$\rho_s = 8806(1 + 38.12 C_s) \tag{2.76}$$

The electrophoresis cell had a planar geometry. The work electrodes were nickel foils with an average dipping area of 12.5 cm^2 and the counter electrode was a Pt foil. They were separated by a distance of 2 cm. EPD kinetic was determined applying a constant current density of 1.6 mA cm^{-2} for deposition times ranging from 60 s up to 5000 s. 200 g of fresh suspension was used for each test, resulting that a total powder mass of 10 g could be deposited by batch. This amount of suspension was considered enough to isolate and study the effects of any change of the suspension properties on the deposition kinetics in a reasonable time. During the deposition tests the suspensions were continuously stirred. The deposited mass was determined by weighting deposits and substrates with an analytical balance.

Table 2.11. summarizes the properties of as-prepared Y-TZP suspensions and the electrical conditions applied during EPD.

Table 2.11 Starting conditions

Initial mass of powder, m_0	10 g
Initial solid content, $C_{S,0}$	0.052 gcm^{-3} (=5 wt.%, =0.85 vol.%)
Volume of the suspension, V	192 cm^3
Suspension electrophoretic mobility, μ	3.35×10^{-4} cm^{2}s^{-1}V^{-1}
Initial suspension resistivity, ρ_0	8.7×10^3 Ω cm
Final suspension resistivity, ρ_∞	23.4×10^3 Ω cm
Current intensity, I	0.02 A
Deposition surface area, S	2×6.25 cm^2

Figure 2.15. shows the time evolution of the mass per unit area (mg cm^{-2}), for a deposition surface area, S, twice the geometric surface area of the substrate. The points correspond to the experimental values while the curves correspond to the fittings to different proposed equations. The slope of the experimental curve changes after the 600 s test, indicating that after this point deposition accelerates. Later, at longer times, the deposition rate starts to decrease. In this particular example, this happens after a deposition time of 3,000 s, when the deposit contained 73.8 wt.% of the starting mass of powder. The plot clearly reveals that the evolution of the weight per unit area with time follows an S-shaped trend. This tendency is confirmed for tests as long as 5,100 s, where 90 wt.% of the initial powder was deposited. Heavier

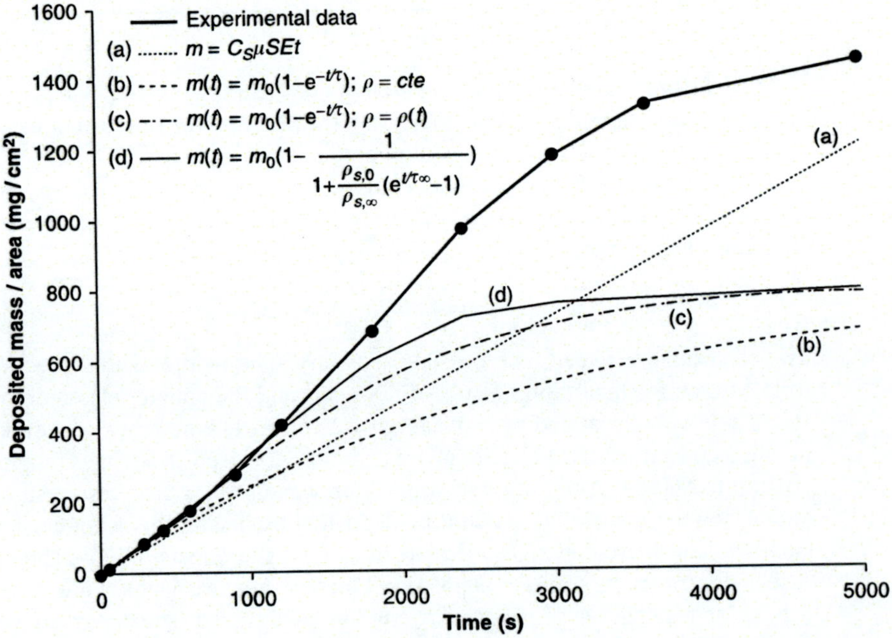

Fig. 2.15 Evolution of mass per unit area of Y-TZP deposits obtained on nickel substrates with a dispersed aqueous suspension. The *points* and the *line* connecting them correspond to experimental data, while *solid* and *dashed lines* correspond to the fitting for the different kinetic models

deposits could not be obtained because gravity made them to fall down from the substrate.

In this system, the volumetric fraction of powder in the suspension, ϕ_s, was lower than 0.2 ($C_{S,0}$<20 vol.%) hence, the effect of high concentration on the kinetics described in Eq. 2.51 could be neglected. The different kinetic curves are plotted using the data summarised in the Table 2.11, and following the models proposed by Hamaker[55], Sarkar & Nicholson [57], Anné et al. [108] and Ferrari et al. [109]. The Sarkar & Nicholson (b) and Anné et al. (c) models fit well to the experimental values at low deposition times; yet, they fail to fit the values registered above 600 s as the increment of the deposition rate is not predicted by those models. The solid line, (d) in Fig. 2.15., represents the EPD kinetics according to Eq. 2.71, taking into account that the ratio between resistivities of the final and the initial suspension, $\rho_{s,\infty}/\rho_{s,0}$, was higher than 2 ($\rho_{s,\infty}/\rho_{s,0}$=2.7). This is the expression which better fits the experimental results for short deposition times (<1200 s), as can be observed in the plot. However, none of the models fit correctly the deposited mass for longer electrophoresis times.

In fact, many parameters affecting the kinetics and deposit morphology have not been considered by the models proposed to date. Their effects can be only included within the sticking parameter, f (Eq. 2.35). Some of these effects have been described in the literature during the last decades, regarding phenomena that take place at the substrate surface, the interface substrate-suspension, the suspension itself and within the growing film. For example, surface charge coupling of the substrate and the particles during EPD has a relevant effect in the array ordering [117] and affects the movement of particles under the influence of the electric field [118]. In fact, the characteristics of the substrate surface, such as its nature [119], roughness, surface charge or electrical response [120], determine the morphology of the films and even the success of the assembly process.

Phenomena at the interface substrate-suspension can also disturb or promote deposition. The surrounding conditions of the working electrode change during the layer growth. Several authors have described the decrease of the H^+ concentration in the vicinity of the cathode surface, as a consequence of the reaction $H^+ + e^- \rightarrow 1/2\ H_2$. The reaction is controlled by the rate of H^+ diffusion from the suspension to the deposit/cathode interface [121–123]. However, the film itself can block the path of transportation of the electrolyte, reducing the number of active sites at the electrode surface. Then, the whole reaction should be governed by the charge transfer reaction rate. Accordingly, the amount of co-ions would increase at the substrate surrounding [124] promoting effects other than that firstly proposed.

Electrophoresis associated phenomena, such as solvent warming or electroosmosis, promote changes in both suspension and electric conditions when the process is running. Current passing through the system increases the suspension temperature, raising dielectric constants and lowering viscosities of the solvent, thus promoting a faster deposition rate. However, as the whole suspension (solvent and charged species) acts as a semiconductor, the increase of the conductivities with temperature oppositely leads to a fall of yield deposition. Other sources of divergences between experimentally obtained and theoretically expected results are particle aging and re-agglomeration, also as a consequence of the suspension warming. The later is a relevant problem regarding the electrophoretic assembly of nanoparticles, and its effect over the kinetics will be described later.

Also contradictory are the effect of parameters associated to the electric conditions. Generally, high voltage promotes disorder during particle compaction leading to low packing degrees in pieces or lacks of order in particle arrangement [125–127]. Oppositely, beneficial effects of the high voltage in EPD products, such as an increase of the green density [53, 128], have been also described.

Summarizing, the consideration of the time variation for the parameter involved in the EPD kinetics permits the standardization of the electrophoresis process. Then, whatever was the system under study (suspension/substrate), normalization of the EPD experiments leads to predictable and, more important, comparable results. Based on normalization of the EPD, process control and the application of new techniques for the experimental design [129] are affordable challenges to develop ordered structures within the nanosized range.

2.4.1 *Electrophoresis Standardization: Nanometric Approach*

Contrary to the Hamaker equation (Eq. 2.34), further electrophoretic models specifically consider changes of parameters during deposition time. Those parameters (today only solid loading and resistivity) are mainly involved in the electrophoretic behavior of the particles. Although theoretic and experimental studies lead to a clear dependence between suspension parameters and electrical conditions, only Eq. 2.71 considers the interdependence of both of them, the solid content of the suspension and its resistivity. Standardization of different variables contributes to establish relations among EPD parameters between them and with the final deposit characteristics. Approaches to normalized EPD parameters can be described from the point of view of the electrical conditions applied to a fixed suspension, or fixing electrical conditions and considering different suspensions.

2.4.1.1 Electrical Conditions

Particle movement in the liquid media depends on the geometry of the electrophoresis cell and on the value of current density passing through the suspension. The geometry of the cell defines the electric field lines, thus determining the regions where deposition takes place. Figure 2.16 shows four schemes of common electrophoresis cells of flat and wire coatings (Figs. 2.16a, 2.16b respectively), prototyping (Fig. 2.16c) and infiltration (Fig. 2.16d) performance.

The variation of the surrounding conditions at the electrodes alters the applied electric conditions of deposition. A loss of deposition capability due to the electrode overpotentials during the process has been assumed in the literature [73]. In fact, the real effect of overpotentials within the electrophoresis cell is a loss of particle

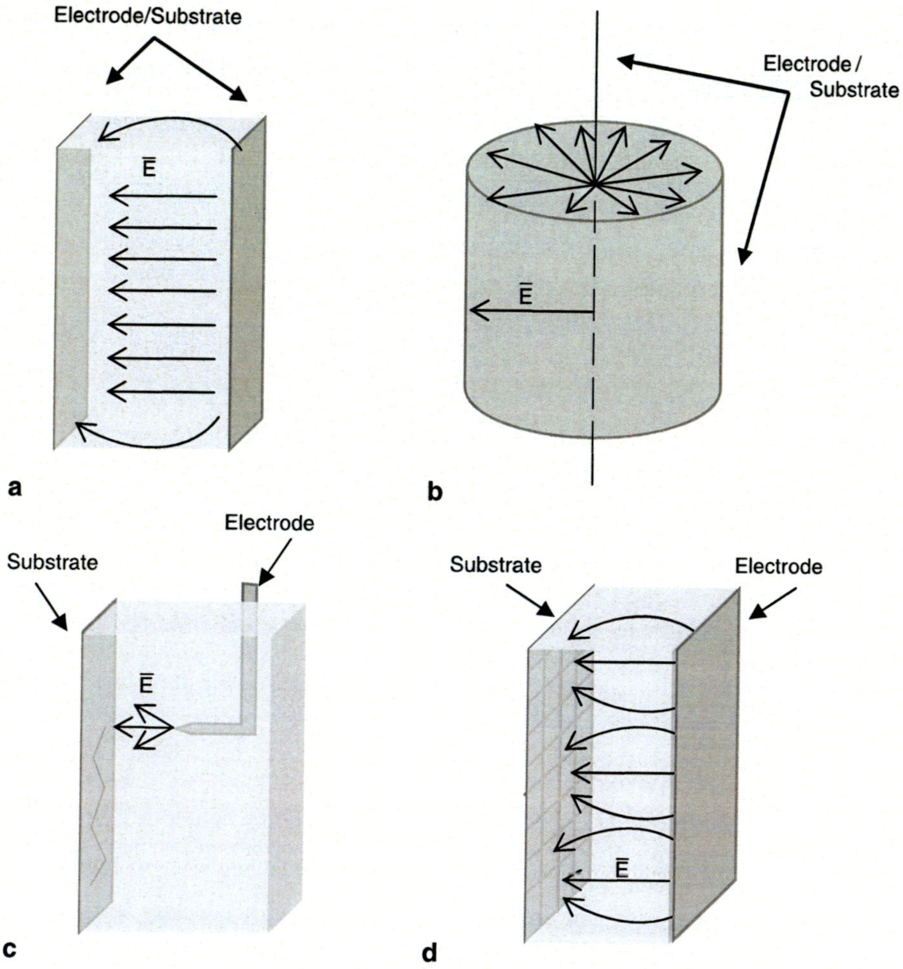

Fig. 2.16 Schemes of common geometries of the electrophoresis cell: planar geometry (**a**), axial geometry (**b**) point geometry (**c**) and irregular geometry (**d**)

movement through the liquid media. Whatever their nature, overpotentials have a large impact on the effective electric field applied to the suspension [73] (Eq. 2.79).

The effective electric field, E_{ef} (V/cm) is the active field promoting particle electrophoresis, and can be defined from Eq. 2.60, given

$$E_{ef} = \frac{I}{S} \times \rho_s \tag{2.77}$$

Then, electrophoresis rates (Eq. 49) in a stirred system, where concentration gradients of charged species are negligible, are provided by

$$\nu = E_{ef}\mu \tag{2.78}$$

Moreover, there is a cell parameter related to the effective electric field usually ignored in the literature: the conduction surface area of the work electrode. Considering Eqs. 2.77, 2.78 as a function of the suspension resistivity yields

$$\nu = \frac{I}{S_{WE}} \rho_s \mu \qquad (2.79)$$

where S_{WE} (cm^2) is the conduction surface area of the work electrode. In the electrophoresis cell, the movement of particles depends on the strength of the applied current density, i (A cm^{-2})

$$i = \frac{I}{S_{WE}} \qquad (2.80)$$

Figure 2.17. shows a simple equivalent circuit of an electrophoresis cell with a planar geometry. In this system, where the work electrode is the substrate, the conduction surface area, S_{WE}, has been usually indentified with the deposition or geometric surface area, S. However, the conduction surface area usually is higher than the geometric one, especially when the substrate/work electrode is a metal, as metals have often a certain surface roughness. So, both areas only match if each conduction point in the surface coincides with a deposited particle. In an electrophoresis cell, equivalence between both surfaces depends on the nature and surface roughness of the electrode, R (μm), and the particle diameter, D (μm).

Current density could be directly calculated from Eq. 2.80 if electrophoresis is performed under galvanostatic conditions. When EPD takes place under potentiostatic conditions, the effective electric field should be determined in a cell of three or four electrodes. But, in all cases, current density passing through the suspension (Eq. 2.80) depends on the geometry and roughness of the electrode surface.

In the case of a nanoparticle assembly on a polished electrode/substrate we can consider that surfaces have a null roughness (R=0). Also, surface roughness can be larger than particle diameter (R>>D); in these cases, the conduction surface area coincides with the deposition surface area, being $S = S_{WE}$. However, for most of the micronic or submicronic systems, surface roughness is of the same order, or lower, than the particle diameter (R≤D). Then, one particle covers more than one point of conduction in the electrode surface. Consequently, the conduction surface area, S_{WE}, and the geometric surface area, S, should be considered different.

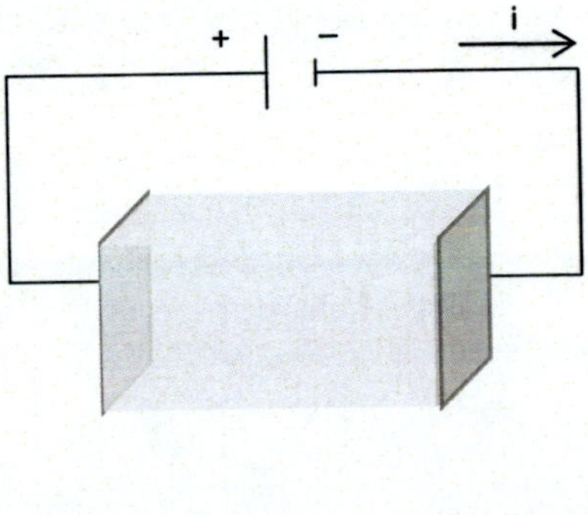

Fig. 2.17 Equivalent circuit of an electrophoresis cell with planar geometry

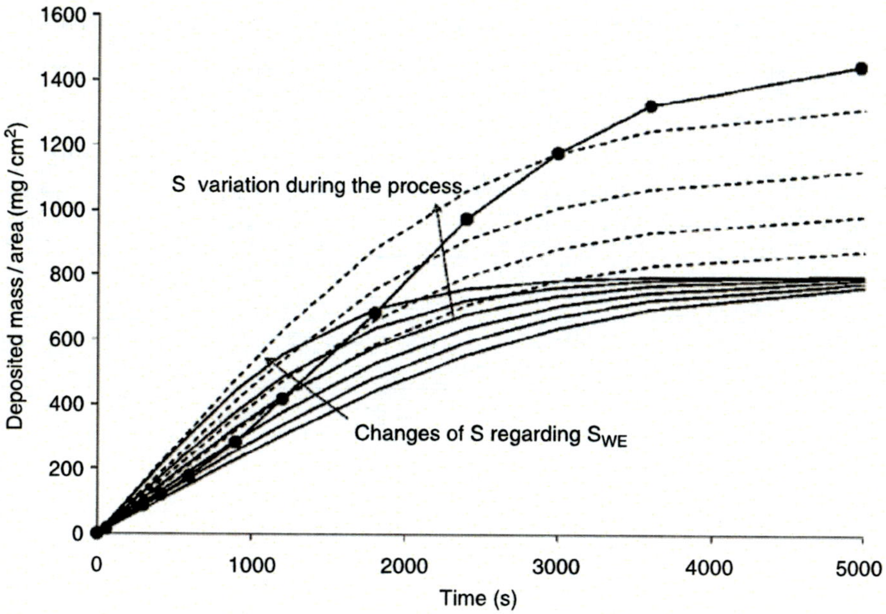

Fig. 2.18 Evolution of mass per unit area of Y-TZP deposits obtained in nickel substrates from a dispersed aqueous suspension. *Solid* and *dotted lines* correspond to simulations of the resistivity model (Eq. 2.71) considering different equivalences between conduction and deposition surface areas

Taking into account the Eqs. 2.77, 2.80 with Eq. 2.40, the characteristic time of the deposition kinetics gives

$$\tau = \frac{V S_{WE} \rho_S}{f \mu S I} \tag{2.80}$$

Similar expressions can be obtained for the characteristic times defined in other models in Eqs. 2.47, 2.67 and 2.70.

Figure 2.18 shows experimental data of the EPD kinetics of Y-TZP [116] and two different simulations of the theoretical approximation in Eq. 2.71. The first simulation, corresponding to solid lines in the plot, considers conduction surface areas up to twice the deposition surface area ($1 < S_{WE}/S < 2$ in Eq. 2.80). Moreover, simulation corresponding to dotted lines in the plot, considers film growth from one face to both faces of the substrate ($1 \times 6.25\ cm^2 < S < 2 \times 6.25\ cm^2$ in Table 2.11). Following the kinetic model in Eq. 2.71, the combination of both simulations fits the experimental curve of the Y-TZP deposition in aqueous media. So, the variation of the conduction area with the deposit area during the process should also be considered in the kinetics evolution.

Figure 2.19 shows the scheme of nanoparticle systems considering different degrees of equivalence between deposition and conduction surface areas.

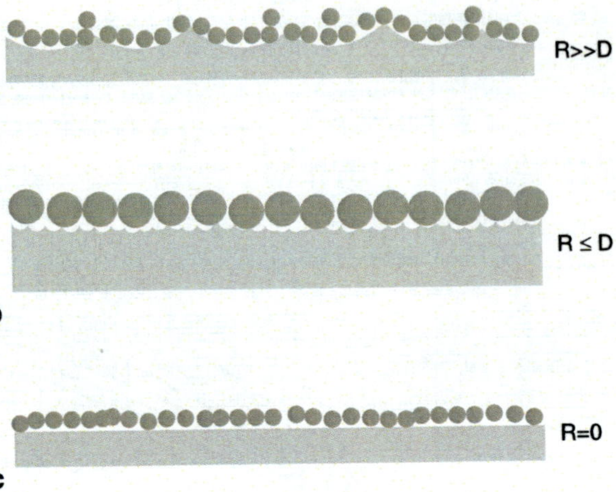

Fig. 2.19 Scheme of the effect of the ratio roughness/particle diameter in the equivalence of the surface areas considered in the EPD kinetics

Summarizing, the effective electric field, the current density passing through the suspension and the equivalence of the conduction and deposition surfaces, should be considered as electrical parameters of reference to compare results of the EPD processes in different suspensions.

2.4.1.2 Suspension Parameters

Relevant efforts during last decade have endeavored to study and to determine the stability of the suspensions as a necessary condition to shape/assemble particles by electrophoretic deposition. Particularly, mechanisms to stabilize particles in organic media have been widely studied [130–133]. The zeta potential quantifies electrostatic interactions, and is the most used parameter to determine particle stability. In fact, the zeta potential fits the potential developed in the shear plane of the double layer system [33–38]. Thus, the zeta potential is as a guideline of the magnitude of the electrostatic repulsive forces that keep particles "floating" within the solvent. The Henry equation relates the electrophoretic mobility to the zeta potential.

$$\mu = \frac{\zeta \varepsilon_0 \varepsilon_{r,l}}{\eta} f(1/\kappa, a) \tag{2.81}$$

where, ζ (mV), is the zeta potential, ε_0 (8.8544 10^{-12} A^2s^2/Nm^2), is the vacuum dielectric constant, ε_r, is the solvent dielectric constant, η (Pa s) is the solvent viscosity and f($1/\kappa$, a), is a function of the particle radius, a (nm), and the Debye length, $1/\kappa$ (nm). Depending on the dielectric constant and the viscosity of the solvent, suspensions with a similar zeta potential shows large differences in the electrophoretic mobility of the particles. Table 2.12. collects dielectric constants and viscosities of common solvents in EPD, and the corresponding scheme of double layer/particle systems. Considering the Smoluchouski approach for thin double layers and large particles (a>> $1/\kappa$) or the Hückel approach for thick double layers and small par-

Table 2.12 Dielectric constants and viscosity of solvents commonly used in EPD Also a double layer/particles scheme and the corresponding approximation to determine the electrophoretic mobility of the particles, based on zeta potential, is included

Solvent	*ε_r (25°C)*	*η (20°C)*	*System*	*Approach*
Water	80	1.00	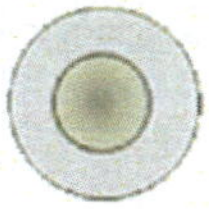	Hückel (a<<1/κ): $\mu = \frac{\zeta\varepsilon_o\varepsilon_r}{1.5\eta}$
Acetone	28.3	0.31	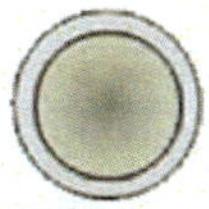	Smoluchouski (a>>1/κ): $\mu = \frac{\zeta\varepsilon_o\varepsilon_r}{\eta}$
Ethanol	20.7	1.20	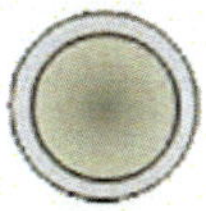	Smoluchouski (a>>1/κ): $\mu = \frac{\zeta\varepsilon_o\varepsilon_r}{\eta}$
Toluene	2.4	0.58	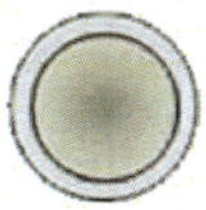	Smoluchouski (a>>1/κ): $\mu = \frac{\zeta\varepsilon_o\varepsilon_r}{\eta}$

ticles ($a \ll 1/\kappa$), powders with a similar zeta potential move faster in solvents with high dielectric constants and/or lower viscosities, such as water or acetone. Conversely, in solvents with lower dielectric constants and medium viscosities, such as ethanol and toluene, particles run slowly. In agreement with those equations, electrophoretic mobility should be considered as the stability parameter of reference to study EPD kinetics.

Besides electrophoretic mobility, resistivity of the suspension is a relevant parameter to be considered in the electrophoresis standardization. It plays an important role in the electrophoresis process, as its value depends on the charged species in the suspension, particles and ions, and the polar character of the solvent or of dispersant additives. During the stabilization of the suspension, the number of charge carriers should be diminished and restricted to the charged particles, in order to improve deposition yield. A high ionic concentration will diminish the deposition rate, if it does not contribute to the particle stabilization.

Complete studies have been published [134, 135] dealing with the electrophoretic deposition of commercial and synthetic nanoparticles stabilized in different alcohols (MeOH, EtOH, BuOH and IPA). In suspensions with similar zeta potentials, the electrophoretic mobility is higher in the suspension whose solvent has a lower viscosity and a higher dielectric constant: MeOH. However, suspensions in MeOH have also the lowest resistivity. Deposition rate in these studies depends on the applied electric conditions. For a similar current density passing through the suspension, the lowest electric field was applied to MeOH suspensions, resulting in the lowest deposition rate. Contrarily, if a constant voltage is applied the current density in MeOH suspension is higher than in others alcohols, and the coating process is significantly faster. Moreover, low film adhesion to the substrate is noted for butanol and propanol suspensions, leading to relevant effects of the solvent also at the process of deposit compaction. Consequently, differences between depositing behavior in selected solvents should be only determined if electric conditions are normalized, fixing in all cases a similar effective electric field during electrophoresis.

Then, to discuss standardized results, the suspension resistivity should be also considered. In this sense, particle size has also an effect over the suspension resistivity. Large specific surface area is associated with nanometric particles, determining the charge per unit area within the suspension. Moreover, the number of charged species in a nanoparticle suspension is higher than in submicronic systems for a previously fixed solid content. That means that suspensions of nanoparticles should be less resistive, leading to lower deposition yields. In fact, recent studies comparing the electrophoretic deposition among particles with similar zeta potential and different sizes, show that a lower deposited mass and lower green density is attained for particles in the nanometric range (<100 nm) [128].

Primary particle size has been often calculated by XRD pattern or TEM images. Yet in suspensions, the particle size is given by the degree of particle agglomeration. Mobility of particles in the liquid media favors the natural trend of nanoparticles to agglomerate. As has been discussed previously the interparticle potentials should be manipulated to avoid this agglomeration. Figure 2.20. shows (a) the evolution of the average particle size (D_V50) with ultrasound time during dispersion of three different suspensions, (b) and a HR-TEM micrograph of particles dispersed under optimized conditions (marked with a circle in the plot in (a)). Micrographs show agglomeration degree measured by Dynamic Light Scattering (DLS) [136]. Particles with a primary size lower than 10 nm are agglomerated forming clusters of approximately 25 nm. Then, nanoparticles will move by electrophoresis as 25 nm agglomerates, behaving as coarse particles.

As at the submicronic range [137], both electrostatic and steric interactions determine particle dispersion maintaining them isolated. In nanometric systems, the size of the suspended particles is as relevant as zeta potential and resistivity to complete the dispersion/stability optimization. In these studies, stability should be referred to the minimum particle size able to be measured in a suspension, and then not only volume concentration but also the number of isolated nanoparticles should be considered [138, 139].

Nanometric particles easily re-agglomerate because of a large specific surface area. Large efforts have been made in nanometric systems to determine stabilization mechanisms in aqueous and organic solvents. Dispersion in non-polar media [140, 141], mixtures of polar solvents [142] and additions of specific dispersants [137, 143] such as I_2 [131, 132, 144], phosphate ester (PE) [145], and polyelectrolytes such as polyacrylates (PAA) or polyethylene imine (PEI), have been chosen and their role investigated in depth. In fact, nanometric colloidal systems need the synergism of electrostatic and steric mechanisms to avoid aging. Also the addition

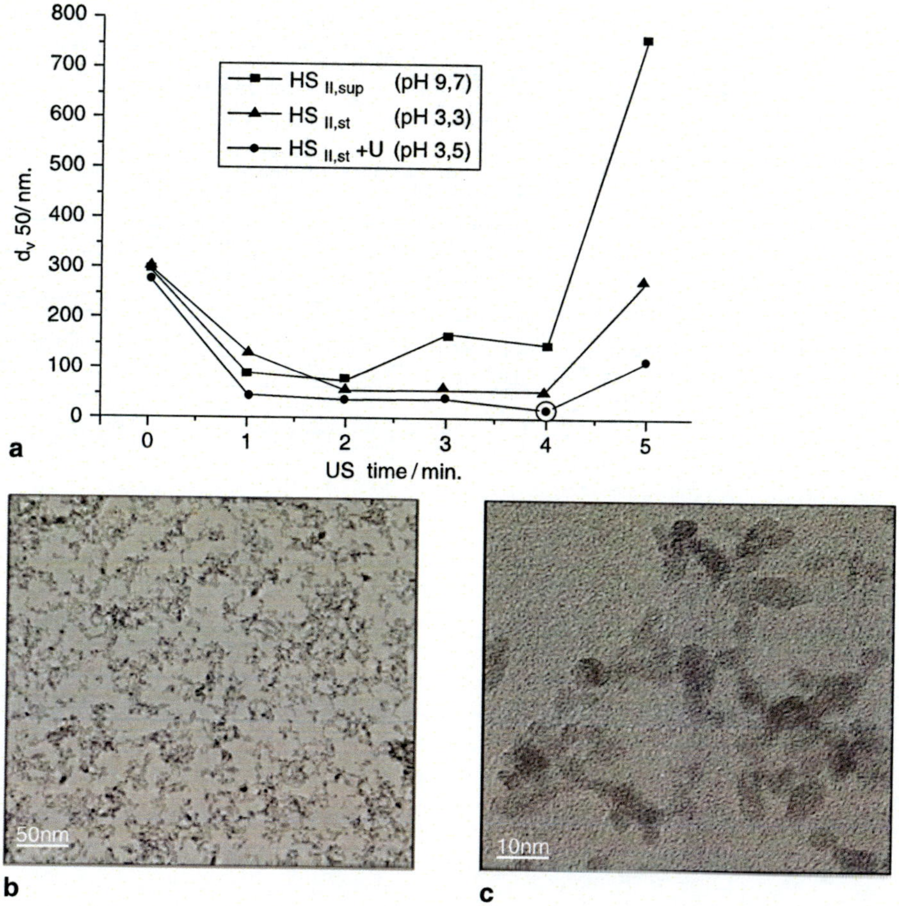

Fig. 2.20 **a** Evolution of the average particle size (D_V50) with ultrasound time during dispersion of three different suspensions, and **b** and **c** a HR-TEM micrograph of particles dispersed under optimized conditions. [136]

of binders or over-addition [145] amounts of polyelectrolytes has been studied to improve deposition rate and adjust film morphology.

As a final remark, we must note that optimization of parameters describing stability, such as electrophoretic mobility, resistivity and particle size distribution in a nanoparticle suspension are parameters that govern the deposition kinetics and deposit characteristics. The control over all these parameters allows the organization (high packing degree, orientation, etc.) of the particles at the nanometric range, leading to high structural integrity in materials, such as scaffolds, free-standing nanoparticle films, transparent films, etc, in functional applications [140, 141, 146–154]. As an example, Fig. 2.21 shows a micrograph of a deposit of flake-shaped nanoparticles of ZnO forming a film of particles oriented perpendicularly to the substrate [154].

Fig. 2.21 Micrograph of an oriented nanostructured film. [154]

Acknowledgements The authors greatly acknowledge the corrections and fruitful suggestions of Prof. J. E. Iglesias (ICV-CSIC). This work has been supported by MICINN (Spain, projects MAT2009-14448-C02-01 and MAT2009-14369-C02-01). B. Ferrari thanks to M. Verde and I. Gonzalo-Juan.

References

1. Pileni, M.P.: Nanosized particles made in colloidal assemblies. Langmuir **13**, 3266–3276 (1997)
2. Knauth, P., Schoonmann, J. (eds.): Nanostructured materials. Selected synthesis methods, properties and applications. Kluwer, New York (2004)
3. Taniguchi, N.: On the basic concept of nanotechnology. Proc. ICPE, Japan (1974)
4. http://www.ipt.arc.nasa.gov/nanotechnology.html. Accessed Jan 2010
5. Caruso, F.: Nanoengineering of particle surfaces. Adv. Mater. **13**, 11–22 (2001)
6. Zeng, H., Li, J., Liu, J.P., Huang, C.L., Sun, S.: Exchange-coupled nanocomposite magnets by nanoparticle selfassembly. Nature **420**, 395–398 (2002)
7. Caruso, F.: Colloids and colloid assemblies: synthesis, modification, organization and utilization of colloidal particles. Wiley, Germany (2003)
8. Fahlman, B.D.: Materials chemistry. Springer, The Netherlands (2007)
9. Sergeev, G.B.: Nanochemistry. Mos. Gos. Univ., Moscow. (2003)
10. Lange, F.F.: Powder processing science and technology for increased reliability. J. Am. Ceram. Soc. **72**, 3–15 (1989)
11. Sigmund, W.M., Bell, N.S., Bergström, L.: Novel powder-processing methods for advanced ceramics. J. Am. Ceram. Soc. **83**, 1557–1574 (2000)
12. Finke, R.G.: Transition metal nanoclusters in metal nanoparticles: synthesis, characterization, and applications. Dekker, New York (2002)
13. Richerson, D.W.: Modern ceramic engineering—properties, processing and use in design, 2nd edn. Dekker, New York (1992)
14. Ashby, M., Shercliff, H., Cebon, D. Materials. Engineering, science, processing and design. Elsevier, Oxford (2007)
15. Jaschinski, W., Nagel, A.: Possibilities and limits in the shaping of ceramic powdes. Interceram **42**, 135–139 (1993)

16. Onoda, G.Y., Hench, L.L. (eds.): Ceramic processing before firing. Wiley, New York (1978)
17. Reed, J.S.: Introduction to the principles of ceramic processing, 2nd edn. Wiley, New York (1995)
18. Buchkremer, H.P., Menzler, N.H.: Ceramic processing (Chap. 24). In: Groza, J.R., Shackelford, J.F., Lavernia, E.J., Powers, M.T. (eds.) Material processing handbook. CRC Press, Boca Raton (2007)
19. Wang, F.F.Y. (ed.): Treatise on materials science and technology, vol 9: ceramic fabrication processes. Academic Press, New York (1978)
20. Brinker, C.J., Clark, D.E., Ulrich, D.R. (eds.): Better ceramics through chemistry. Mat. Res. Soc. Proc., vol 32, North-Holland (1984)
21. Moreno, R.: Trends in slip forming of ceramics. Bol. Soc. Esp. Ceram. Vidr. **39**, 601–608 (2000)
22. Tari, G.: Gelcasting ceramics: a review. Am. Ceram. Soc. Bull. **82**, 43–47 (2003)
23. Wang, L.A., Aldinger, F.: Near-net shape forming of advanced ceramics. Adv. Eng. Mater. **3**, 110–113 (2000)
24. Lewis, J.A.: Colloidal processing of ceramics. J. Am. Ceram. Soc. **83**, 2341–2359 (2000)
25. Kingery, W.D. (ed.): Ceramic fabrication orocesses, pp. 5–51. MIT Press: Cambridge (1958)
26. Shanefield, D.J.: Organic additives and ceramic processing, 2nd edn. Kluwer, Norwell (1996)
27. Moreno, R.: The role of slip additives in tape casting technology. Part I: solvents and dispersants. Am. Ceram. Soc. Bull. **71**, 1521–1531 (1992)
28. Moreno, R.: The role of slip additives in tape casting technology. Part II: binders and plasticizers. Am. Ceram. Soc. Bull. **71**, 1647–1656 (1992)
29. Hench, L.L., Ulrich, D.R. (eds.): Ultrastructure processing of ceramics, glasses and composites. Wiley, New York (1984)
30. Aksay, I.A., Lange, F.F., Davis, B.I.: Uniformity of Al_2O_3/ZrO_2 composites by colloidal filtration. J. Am. Ceram. Soc. **60**, C190–C192 (1983)
31. Pashley, R.M., Karaman, M.E.: Applied colloid and surface chemistry. Wiley, West Sussex (2004)
32. Hiemenz, P.C.: Principles of colloid and surface chemistry, 3rd edn. Dekker, New York (1997)
33. Hunter, R.J.: Foundations of colloid science, vol. 1, Clarendon Press, Oxford (1987)
34. Israelachvili, J.N.: Intermolecular and surface forces. Academic Press, London (1985)
35. Shaw, D.J.: Introduction to colloid and surface chemistry, 4th edn. Butterworth Heinemann, Oxford (1992)
36. Cosgrove, T. (ed.): Colloid science. Principles, methods and applications. Blackwell, Oxford (2005)
37. Holmberg, K. (ed.): Handbook of advanced surface and colloid chemistry. Wiley, West Sussex (2002)
38. Everett, D.H.: Basic principles of colloid science. The Royal Society of Chemistry, London (1988)
39. Pugh, R.J., Bergström, L. (eds.): Surface and colloid chemistry in advanced ceramics processing. Dekker, New York (1994)
40. Bergstrom, L.: Hamaker constants of inorganic materials. Adv. Colloid Interface Sci. **70**, 125–169 (1997)
41. Parks, G.A., De Bruyn, P.L.: The zero point of charge of oxides. J. Phys. Chem. **66**, 967–972 (1962)
42. Tadros, T.F. (ed.): Surfactants. Academic Press, London (1984)
43. Napper, D.H.: Steric stabilization. J. Colloid Interface Sci., **58**, 390–407 (1977)
44. Napper, D.H.: Polymeric stabilization of colloidal dispersions. Academic Press, London (1983)
45. Ruda, H.E., Polanyi, J.C., Yang, S.Y.: Developing 1D nanostructure arrays for future nanophotonics. Nanoscale Res. Lett. **1**, 99–119 (2006)
46. Corni, I., Ryan, M.P., Boccaccini, A.R.: Electrophoretic deposition: from traditional ceramics to nanotechnology. J. Eur. Ceram. Soc. **28**, 1353–1367 (2008)

47. Cao, G., Liu, D.: Template-based synthesis of nanorod, nanowire, and nanotube arrays. Adv. Colloid Interface Sci. **136**, 45–64 (2008)
48. Besra, L., Liu, M.: A review on fundamentals and applications of electrophoretic deposition (EPD). Prog. Mater. Sci. **52**, 1–61 (2007)
49. Boccaccini, A.R., Cho, J., Subhani, T.: Electrophoretic deposition of carbon nanotube-ceramic nanocomposites. J. Eur. Ceram. Soc. **30**, 1115–1129 (2010)
50. Boccaccini, A.R., Roether, J.A., Thomas, B.J.C., Shapper, M.S.P., Chavez, E., Stoll, E., Minay, E.J.: The electrophoretic deposition of inorganic nanoscaled materials. J. Ceram. Soc. Jpn. **114**, 1–14 (2006)
51. Moreno, R., Ferrari, B.: Advanced ceramics via EPD of aqueous slurries. Am. Ceram. Soc. Bull. **79**, 44–48 (2000)
52. Ferrari, B., González, S., Moreno, R., Baudín, C.: Multilayer coatings with improved reliability produced by aqueous electrophoretic deposition. J. Eur. Ceram. Soc. **26**, 27–36 (2006)
53. Ji, C., Lan, W., Xiao, P.: Fabrication of yttria-stabilized zirconia coatings using electrophoretic deposition: packing mechanism during deposition. J. Am. Ceram. Soc. **91**, 1102–1109 (2008)
54. Ferrari, B., Moreno, R.: EPD kinetics: a review. J. Eur. Ceram. Soc. **30**, 1069–1078 (2010)
55. Hamaker, H.C.: Formation of a deposit by electrophoresis. Trans. Faraday Soc. **36**, 279–287 (1940)
56. Hirata, Y., Nishimoto, A., Ishisara, Y.: Forming of alumina powder by electrophoretic deposition. J. Ceram. Soc. Jpn. Int. Ed. **99**, 105–109 (1991)
57. Sarkar, P., Nicholson, P.S.: Electrophoretic deposition (EPD): mechanisms, kinetics, and application to ceramics. J. Am. Ceram. Soc. **79**, 1897–2002 (1996)
58. Zhang, Z., Huang, Y., Jiang, Z.: Electrophoretic deposition forming of SiC-TZP composites in a nonaqueous sol media. J. Am. Ceram. Soc. **77**, 1946–1949 (1994)
59. Patel, M.N., Williams, R.D., May, R.A.: Electrophoretic deposition of Au nanocrystals inside perpendicular mesochannels of TiO_2. Chem. Mater. **20**, 6029–6040 (2008)
60. Bao, Y., Nicholson, P.S.: Constant current electrophoretic infiltration deposition of fiber-reinforced ceramic composites. J. Am. Ceram. Soc. **90**, 1063–1070 (2007)
61. Wang, Y.C., Leu, I.C., Hon, M.H.: Size control of ZnO nano-fibril within template by electrophoretic deposition. Electrochem. Solid-State Lett. **7**, D15–D18 (2004)
62. Haber, S., Gal-Or, L.: Deep electrophoretic penetration and deposition of ceramic particles inside porous substrates I: analytical model. J. Electrochem. Soc. **139**, 1071–1077 (1992)
63. Haber, S., Liubovich, S., Gal-Or, L.: Deep electrophoretic penetration and deposition of ceramic particles inside porous substrates II: experimental model. J. Electrochem. Soc. **139**, 1078–1081 (1992)
64. Will, J., Hruschka, M.K.M., Gubler, L., Gauckler, L.J.: Electrophoretic deposition of zirconia on porous anodic substrates. J. Am. Ceram. Soc. **84**, 328–332 (2001)
65. Chen, C.Y., Chen, S.Y., Liu, D.M.: Electrophoretic deposition forming of porous Alumina membranes. Acta. Mater. **47**, 2717–2726 (1999)
66. Bonnas, S., Ritzhaupt-Kleissl, H.J., Haußelt, J.: Fabrication of particle and composition gradients by systematic interaction of sedimentation and electrical field in electrophoretic deposition. J. Eur. Ceram. Soc. **30**, 1177–1185 (2010)
67. Biesheuvel, P.M., Verweij, H.: Theory of cast formation in electrophoretic deposition. J. Am. Ceram. Soc. **82**, 1451–1455 (1999)
68. Gonzalez-Cuenca, M., Biesheuvel, P.M., Verweij, H.: Modeling constant voltage electrophoretic deposition from stirred suspension. Aiche. J. **46**, 626–631 (2000)
69. Ma, J., Cheng, W.: Electrophoretic deposition of lead zirconate titanate ceramics. J. Am. Ceram. Soc. **85**, 1735–1737 (2002)
70. Stappers, L., Zhang, L., Van Der Biest, O., Fransaer, J.: The effect of electrolyte conductivity on electrophoretic deposition. J. Colloid Interface Sci. **328**(2), 436–446 (2008)
71. Pignolet, C., Filiarte, C., Foissy, A.: Influence of surfactant counter-ions during electrophoretic particle deposition. Langmuir **24**, 10181–10186 (2008)

72. Anné, G., Vanmeensel, K., Vleugels, J., Van Der Biest, O.: Influence of the suspension composition on the electric field and deposition rate during electrophoretic deposition. Colloids Surf.: Phys. Eng. Aspects **245**, 35–39 (2004)
73. Van Der Biest, O., Vandeperre, L.J.: Electrophoretic deposition of materials. Annu. Rev. Mater. Sci. **29**, 327–352 (1999)
74. Ciou, S.J., Funga, K.Z., Chiang, K.W.: Behaviors and mechanism of electrolyte electrophoresis during electrophoretic deposition. J. Power Sources **175**, 33–39 (2008)
75. Ciou, S.J., Funga, K.Z., Chiang, K.W.: A comparison of the artificial neural network model and the theoretical model used for expressing the kinetics of electrophoretic deposition of YSZ on LSM. J. Power Sources **175**, 338–344 (2008)
76. Anné, G., Neirinck, B., Vanmeensel, K., Van Der Biest, O., Vleugels, J.: Origin of the potential drop over the deposit during electrophoretic deposition. J. Am. Ceram. Soc. **89**, 823–828 (2006)
77. Wang, Y.C., Leu, I.C., Hon, M.H.: Kinetics of electrophoretic deposition for nanocrystalline zinc oxide coatings. J. Am. Ceram. Soc. **87**, 84–88 (2004)
78. Vander Poorten, H.: Caracterisation de l'electrodeposition et des electrodepots de pates ceramiques. Silicates Ind. **41**, 159–172 (1981)
79. Choudhary, J.Y., Ray, H.S., Rai, K.N.: Electrophoretic depositon of Alumina from aqueous suspensions. Trans. J. Br. Ceram. Soc. **81**, 193–96 (1982)
80. Shane, M.J., Talbot, J.B., Schreiber, R.D.: Electrophoretic deposition of phosphors. I Conductivity and zeta potential measurements. J. Colloid Interface Sci. **165**, 325–333 (1994)
81. DeBeer, E., Duval, J., Meulenkamp, E.A.: Electrophoretic deposition: a quantitative model for particle deposition and binder formation from alcohol-based suspensions. J. Colloid Interface Sci. **222**, 117–124 (2000)
82. Bouyer, F., Foissy, A.: Electrophoretic deposition of silicon carbide. J. Am. Ceram. Soc. **82**, 2001–2010 (1999)
83. Ferrari, B., Moreno, R.: Electrophoretic deposition of aqueous alumina slips. J. Eur. Ceram. Soc. **17**, 549–556 (1997)
84. Moreno, R., Ferrari, B.: Effects of the slurry properties on the homogeneity of alumina deposits obtained by aqueous electrophoretic deposition. Mater. Res. Bull. **35**, 887–897 (2000)
85. Ferrari, B., Moreno, R.: The conductivity of aqueous Al_2O_3 slips for electrophoretic deposition. Mater. Lett. **28**, 353–355 (1996)
86. Popa, A.M., Vleugels, J., Vermant, J., Van Der Biest, O.: Influence of surfactant addition sequence on the suspension properties and electrophoretic deposition behavior of alumina and zirconia. J. Eur. Ceram. Soc. **26**, 933–939 (2006)
87. Uchikoshi, T., Sakka, Y.: Phosphate esters as dispersants for the cathodic electrophoretic deposition of alumina suspensions. J. Am. Ceram. Soc. **91**, 1923–1926 (2008)
88. Javidi, M., Javadpour, S., Bahrololoom, M.E., Ma, J., Electrophoretic deposition of natural hydroxyapatite on medical grade 316L stainless steel. Mater. Sci. Eng. C. **28**, 1509–1515 (2008)
89. Sun, H., Quan, X., Chen, S.: Preparation of well-adhered γ-Al_2O_3 washcoat on metallic wire mesh monoliths by electrophoretic deposition. Appl. Surface Sci. **253**, 3303–3310 (2007)
90. Plešingerová, B., Súèik, G., Maryška, M., Horkavcová, D.: Hydroxyapatite coatings deposited from alcohol suspensions by electrophoretic deposition on titanium substrate. Ceramics-Silikáty **51**, 15–23 (2007)
91. Tang, F., Uchikoshi, T., Ozawa, K., Sakka, Y.: Effect of polyethylenimine on the dispersion and electrophoretic deposition of nano-sized titania aqueous suspensions. J. Eur. Ceram. Soc. **26**, 1555–1560 (2006)
92. Lebrette, S., Pagnoux, C., Abeland, P.: Fabrication of titania dense layers by electrophoretic deposition in aqueous media. J. Eur. Ceram. Soc. **26**, 2727–2734 (2006)
93. Doungdaw, S., Uchikoshi, T., Noguchic, Y., Eamchotchawalit, C., Sakka, Y.: Deposition of lead zirconate titanate (PZT) powder from ethanol suspension prepared with phosphate ester. Sci. Tech. Adv. Mater. **6**, 927–932 (2005)

94. Wang, C., Ma, J., Cheng, W.: Formation of polyetherketone polymer coating by electrophoretic deposition method. Surf. Coat. Tech. **173**, 271–275 (2003)
95. Ma, J., Wang, C., Peng, K.W.: Electrophoretic deposition of porous hydroxyapatite scaffold. Biomater **24**, 3505–3510 (2003)
96. Wang, C., Ma, J., Cheng, W., Zhang, R.: Thick hydroxyapatite coatings by electrophoretic deposition. Mater. Lett. **57**, 99–105 (2002)
97. Yamada, N., Shoji, H., Kubo, Y., Katayama, S.: Preparation of inorganic-organic hybrid films containing particles using electrophoretic deposition method. J. Mater. Sci. **37**, 2071–2076 (2002)
98. Ogata, N., Van Tassel, J.J., Randall, C.A.: Electrode formation by electrophoretic deposition of nanopowders. Mater. Lett. **49**, 7–14 (2000)
99. Chen, F., Liu, M.: Preparation of YSZ films on LSM and LSM-YSZ substrates using an electrophoretic deposition process. J. Eur. Ceram. Soc. **21**, 127–134 (2001)
100. Tang, F.Q., Uchikoshi, T., Ozawa, K., Sakka, Y.: Electrophoretic deposition of aqueous nano-γ-Al2O3 suspensions. Mater. Res. Bull. **37**, 653–660 (2003)
101. Put, S., Veugels, J., Van Der Biest, O.: Gradient profile prediction in functionally graded materials processed by electrophoretic deposition. Acta. Mater. **51**, 6303–6317 (2003)
102. Carrique, F., Arroyo, F.J., Delgado, A.V.: Electrokinetics of concentrated suspensions of spherical colloidal particles: effect of a dynamic Stern layer on electrophoresis and DC conductivity. J. Colloid Interface Sci. **243**, 351–361 (2001)
103. Oshima, H.: Electrical conductivity of a concentrated suspension of soft particles. J. Colloid Interface Sci. **229**, 307–309 (2000)
104. Jonhson, T.J., Davis, E.J.: An analysis of electrophoresis of concentrated suspensions of colloidal particles. J. Colloid Interface Sci. **215**, 397–408 (1999)
105. Saville, D.A.: The electrical conductivity of suspensions of charged particles in ionic solutions: the role of added counter-ions and nonspecific adsorption. J. Colloid Interface Sci. **91**, 34–50 (1983)
106. Navaneetham, G., Posner, J.D.: Electrokinetic instabilities of non-dilute colloidal suspensions. Mech. Res. Commun. **36**, 22–32 (2009)
107. Posner, J.D.: Properties and electrokinetics behavior of non-dilute colloidal suspensions. J. Fluid Mech. **619**, 331–365 (2009)
108. Anné, G., Vanmeensel, K., Vleugels, J., Van Der Biest, O.: A mathematical description of the kinetics of the electrophoretic deposition process for Al_2O_3-based suspensions. J. Am. Ceram. Soc. **88**, 2036–2039 (2005)
109. Ferrari, B., Moreno, R., Cuesta, J.A.: A resistivity model of electrophoretic deposition. Key. Eng. Mater. **314**, 175–180 (2006)
110. Hasan, S.A., Kavich, D.W., Dickerson, J.H.: Sacrificial layer electrophoretic deposition of free-standing multilayered nanoparticle films. Chem. Commun. **25**, 3123–3725 (2009)
111. Nold, A., Zeiner, J., Assion, T., Clasen, R.: Electrophoretic deposition as rapid prototyping method. J. Eur. Ceram. Soc. **30**, 1163–1170 (2010)
112. Chang, Y., Huang, S., Chen, Y.: Biomolecular nanopatterning by electrophoretic printing lithography. Small **1**, 63–66 (2009)
113. Gardeshzadeh, A.R., Raissi, B., Marzbanrad, E., Mohebbi, H.: Fabrication of resistive CO gas sensor based on SnO_2 nanopowders via low frequency AC electrophoretic deposition. J. Mater. Sci: Mater. Electron. **20**, 127–131 (2009)
114. Neirinck, B., Fransaer, J., Van Der Biest, O., Vleugels, J.: Aqueous electrophoretic deposition in asymmetric AC electric fields (AC–EPD). Electrochem. Commun. **11**, 57–60 (2009)
115. Besra, L., Uchikoshi, T., Suzuki, T.S., Sakka, Y.: Application of constant current pulse to suppress bubble incorporation and control deposit morphology during aqueous electrophoretic deposition (EPD). J. Eur. Ceram. Soc. **29**, 1837–1845 (2009)
116. Ferrari, B., Moreno, R.: Zirconia thick films deposited on Nickel by aqueous EPD. J. Electrochem. Soc. **147**, 2987–2992 (2000)
117. Riahifar, R., Marzbanrad, E., Dehkordi, B.R., Zamani, C.: Role of substrate potential on filling the gap between two planar parallel electrodes in electrophoretic deposition. Mater. Lett. **64**, 559–561 (2010)

118. Lee, J., Leu, I., Chung, Y., Hon, M.: Fabrication of ordered ZnO hierarchical structures controlled via surface charge in the electrophoretic deposition process. Nanotech **17**, 4445–4450 (2006)
119. Rha, S., Chou, T.P., Cao, G.: Characteristics of silicon oxide thin films prepared by sol electrophoretic deposition method using tetraethylorthosilicate as the precursor. Curr. Appl. Phys. **9**, 551–555 (2009)
120. Van Tassel, J.J., Randall, C.A.: Ionic gradients at an electrode above the equilibrium limit current. 3. Stabilization of ion depleted conduction by a nanoporous alumina layer during electrophoretic deposition. J. Phys. Chem. C. **111**, 3358–3365 (2007)
121. Besra, L., Uchikoshi, T., Suzuki, T.S., Sakka, Y.: Experimental verification of pH localization mechanism of particle consolidation at the electrode/solution interface and its application to pulsed DC electrophoretic deposition (EPD). J. Eur. Ceram. Soc. **30**, 1187–1193 (2010)
122. De, D., Nicholson, P.S.: Role of the ionic depletion in deposition during the electrophoretic deposition. J. Am. Ceram. Soc. **81**, 3031–3036 (1999)
123. Zhitomirsky, I.: Cathodic electrodeposition of ceramic and organoceramic materials. Fundamental aspects. Adv. Colloid Interface Sci. **97**, 279–317 (2002)
124. Ciou, S., Fung, K., Chiang, K.: Behaviors and mechanism of electrolyte electrophoresis during electrophoretic deposition. J. Power Sour. **175**, 33–39 (2008)
125. Shan, M., Mao, X., Zhang, J., Wang, S.: Electrophoretic shaping of sub-micron alumina in ethanol. Ceram. Intern. **35**, 1855–1861 (2009)
126. Kinzl, M., Reichmann, K., Andrejs, L.: Electrophoretic deposition of silver from organic PDADMAC-stabilized suspensions. J. Mater. Sci. **44**, 3758–3763 (2009)
127. Simchi, S., Pishbin, F., Boccaccini, A.R.: Electrophoretic deposition of chitosan. Mater. Lett. **63**, 2253–2256 (2009)
128. Ji, C., Shapiro, I.P., Xiao, P.: Fabrication of yttria-stabilized-zirconia coatings using electrophoretic deposition: Effects of agglomerate size distribution on particle packing. J. Eur. Ceram. Soc. **29**, 3167–3175 (2010)
129. Corni, I., Cannio, M., Romagnoli, M., Boccaccini, A.R.: Application of a neural network approach to the electrophoretic deposition of PEEK–alumina composite coatings. Mater. Res. Bull. **44**, 1494–1501 (2009)
130. Kawakita, M., Uchikoshi, T., Kawakita, J., Sakka, Y.: Preparation of crystalline-oriented Titania photoelectrodes on ITO glasses from a 2-Propanol–2,4-Pentanedione solvent by electrophoretic deposition in a strong magnetic field. J. Am. Ceram. Soc. **92**, 984–989 (2009)
131. Kadam, M.B., Sinha, B.B., Kalubarme, R.S., Pawar, S.H.: Transformation of MgB2 powder into superconducting film via electrophoretic deposition technique. J. All. Comp. **478**, 467–473 (2009)
132. Santillán, M.J., Caneiro, A., Quaranta, N., Boccaccini, A.R.: Electrophoretic deposition of $La_{0.6}Sr_{0.4}Co_{0.8}Fe_{0.2}O_{3-\delta}$ cathodes on $Ce_{0.9}Gd_{0.1}O_{1.95}$ substrates for intermediate temperature solid oxide fuel cell (IT-SOFC). J. Eur. Ceram. Soc. **29**, 1125–1132 (2009)
133. Novak, S., König, K.: Fabrication of alumina parts by electrophoretic deposition from ethanol and aqueous suspensions. Ceram. Int. **35**, 2823–2829 (2009)
134. Dor, S., Rühle, S., Ofir, A., et al.: The influence of suspension composition and deposition mode on the electrophoretic deposition of TiO_2 nanoparticle agglomerates. Colloids Surf.: Phys. Eng. Aspects **342**, 70–75 (2009)
135. Plesingerova, B., Súèik, G., Maryska, M., Horkavcová, D.: Hydroxyapatite coatings deposited from alcohol suspensions by electrophoretic deposition on titanium substrate. Ceramics **51**, 15–23 (2007)
136. Gonzalo-Juan, I., Ferrari, B., Colomer, M.T.: Influence of the urea content on the YSZ hydrothermal synthesis under dilute conditions and its role as dispersant agent in the post-reaction medium. J. Eur. Ceram. Soc. **29**, 3185–3195 (2009)
137. Xu, H., Shapiro, I.P., Xiao, P.: The influence of pH on particle packing in YSZ coatings electrophoretically deposited from a non-aqueous suspension. J. Eur. Ceram. Soc. **30**, 1105–1114 (2010)

138. Radice, S., Bradbury, C.R., Michler, J., Mischler, S.: Critical particle concentration in electrophoretic deposition. J. Eur. Ceram. Soc. **30**, 1079–1088 (2010)
139. Uni, H.N., Yang, C.: Colloidal particle deposition from electrokinetics flow in a microfluidic cannel. Electrophoresis **30**, 732–741 (2009)
140. Jia, S., Banerjee, S., Herman, I.P.: Mechanism of the electrophoretic deposition of CdSe Nanocrystal films: Influence of the nanocrystal surface charge. J. Phys. Chem. C. **112**, 162–171 (2008)
141. Hasan, S., Kavich, D.W., Mahajan, S.V., Dickerson, J.H.: Electrophoretic deposition of CdSe nanocrystal films onto dielectric thin films. Thin Solid Films **517**, 2665–2669 (2009)
142. Ovtar, S., Lisjak, D., Drofenik, M.: Barium hexaferrite suspensions for electrophoretic deposition. J. Colloid Interface Sci. **337**, 456–463 (2009)
143. Wang, Y., Zhitomirsky, I.: Electrophoretic deposition of manganese dioxide-multiwalled carbon nanotube composites for electrochemical supercapacitors. Langmuir **25**, 9684–9689 (2009)
144. Bhosale, A.C., Kadam, M.B., Joshi, R., et al.: Studies on electrophoretic deposition of nanocrystalline SDC electrolyte films. J. Alloys Comp. **484**, 795–800 (2009)
145. Doungdaw, S., Uchikoshi, T., Noguchic, Y.: Electrophoretic deposition of lead zirconate titanate (PZT) powder from ethanol suspension prepared with phosphate ester. Sci. Tech. Adv. Mater. **6**, 927–932 (2005)
146. Chung, Y., Leu, I., Lee, J., Hon, M.: Electrophoretic deposition kinetics of ZnO nanoparticles into an opal template and fabrication of well-ordered macroporous structure. J. Electrochem. Soc. **156**, E91–E95 (2009)
147. Chung, Y., Leu, I., Lee, J., Hon, M.: Filling behavior of ZnO nanoparticles into opal template via electrophoretic deposition and the fabrication of inverse opal. Electrochem. Acta. **54**, 3677–3682 (2009)
148. Kurokawa, S., Kikuchi, T., Sakairi, M., Takahashi, H.: Fabrication of micro-dot arrays and micro-walls of acrylic acid/melamine resin on aluminum by AFM probe processing and electrophoretic coating. Electrochim. Acta. **53**, 8118–8127 (2008)
149. Xue, L., Kajiyoshi, K., Yan, Y.: Preparation of highly oriented titania nanosheet thin films by electrophoretic deposition. Thin Solid Films **518**, 10–15 (2009)
150. Lin, T., Huang, W., Jun, I., Jiang, P.: Electrophoretic co-deposition of biomimetic nanoplatelet-polyelectrolyte composites. Electrochem. Commun. **11**, 1635–1638 (2009)
151. Zhang, L., Vleugels, J., Van Der Biest, O.: Fraction of textural alumina by orienting template particles during electrophoretic deposition. J. Eur. Ceram. Soc. **30**, 1195–1202 (2010)
152. Zhao, L., Yu, J., Fan, J.: Dye-sensitized solar cells based on ordered titanate nanotube films fabricated by electrophoretic deposition method. Electrochem. Commun. **11**, 2052–2055 (2009)
153. Lin, T., Huang, W., Jun, I., Jiang, P.: Electrophoretic deposition of biomimetic nanocomposites. Electrochem. Commun. **11**, 14–17 (2009)
154. Verde, M., Caballero, A.C., Iglesias, Y., Villegas, M., Ferrari, B.: Electrophoretic deposition of flake-shaped ZnO nanoparticles. J. Electrochem. Soc. **157**, H55–H59 (2010)

Part II
Applications of Electrophoretic Deposition

Chapter 3
Electrophoretic Deposition of Nanocrystals in Non-polar Solvents

James H. Dickerson

3.1 Introduction

Nanoscience and nanotechnology are international household terms because of the substantial potential that exist in the physical, chemical, and biological phenomena discovered in these fields. From a physics perspective, a number of nanoscale systems show great promise to revolutionize many areas of society. The field of nanocrystalline materials exhibits the greatest potential to elucidate our understanding of fundamental physical phenomena (electricity, magnetism, etc.) and to be incorporated into devices that could improve our lives. For nanocrystalline materials to become widely employed and, hence, competitive with traditional macroscopic materials in electronic, optical, and magnetic device applications, more must be understood about their intrinsic properties and how to control, distribute, and integrate nanomaterials in an effective manner.

One approach that has received considerable attention is the electrophoretic deposition (EPD) of nanomaterials. The concept of electrophoresis has been used extensively in biology, chemistry, materials science, and bioengineering for the manipulation and processing of ceramics, biological materials, colloids, and other materials. In parallel, electric field mediated schemes, such as electrochemical deposition and electroplating, have been employed for several decades in industrial materials processing and thin film applications involving metal coatings, ceramic composites, bio-active materials, and paints and dyes.

Deposited nanocrystalline materials have as many potential applications as do individual nanocrystals (NCs). Thus, developing successful schemes to distribute NCs will play a substantive role in their implementation in industrial and commercial applications. Further, assembling NCs into tightly packed arrays requires an intimate knowledge of their dynamics in solution as well as the interactions that govern NC-substrate and NC-NC binding. State-of-the-art NC assembly schemes,

---

J. H. Dickerson (✉)
Department of Physics and Astronomy, Vanderbilt University, Station B #351807,
2301 Vanderbilt Place, Nashville, TN 37235-1807, USA
e-mail: james.h.dickerson@vanderbilt.edu

J. H. Dickerson, A. R. Boccaccini (eds.), *Electrophoretic Deposition of Nanomaterials*,
Nanostructure Science and Technology,
DOI 10.1007/978-1-4419-9730-2_3,

such as Langmuir-Blodgett (LB) and layer-by-layer (LbL), have recognized limitations, including the inability to achieve rapid and large-scale NC deposition, and the necessity (for LbL) for additives that can affect the physical characteristics of the constituent NCs [1–4]. Electrophoretic deposition is a technique that is superior (compared to LB and LbL) for the assembly of NC films due to its deposition rate, among other traits. EPD allows us to assemble NC solids without additives to explore basic science, such as fundamental NC-NC interactions, and to investigate actual device fabrication approaches.

EPD combines aspects of electrophoresis (the translation of charged particles, suspended in a solution, due to an ambient DC electric) and dielectrophoresis (the locomotion of dipolar, polarizable, or charged particles, also in solution, due to AC or gradient electric fields) to deposit nanostructures onto conducting electrodes. In typical EPD, a dc voltage is applied across the cell, thereby creating an electric field that transports charged particles to the electrodes where they deposit to form a cast film. The primary advantages of electrophoretic deposition as a technique to compose films of nanomaterials include site-selectivity, dense packing of the nanomaterials, size-scalability of the film, and control over the deposition thickness of the film. Thus, EPD can rapidly fabricate films comprising nanomaterial multilayers, with short-range van der Waals forces acting as a stabilizing influence.

Although other wet-casting processes are capable of producing micron-sized ordered superlattices of nanocrystals, the corresponding *fabrication rates* are at least one order of magnitude slower than EPD [1–4]. EPD also is *size-scalable*, as demonstrated by applications in the ceramics and coatings industries, where films are cast onto substrates as large as automotive bodies (e.g., primer coating) and as small as nanoscale electrodes. The advantages of electrophoretic deposition include site-selectivity, tight-packing of the NCs in the film, size-scalability, and marked control over the deposition thickness of the film [3]. This allows one to pattern the films as desired, using macroscopic electrodes (surface area $>1\ cm^2$) as well as nanoscale electrodes (surface area $\sim 100\ nm^2$) [5, 6]. The properties of the film can be tuned by manipulating process variables, such as applied dc and ac voltages, frequency, and nanocrystal surface chemistry, affirming the flexibility of the process. Finally, by preparing suitable templates as deposition electrodes, *patterned* films can be cast. These traits make EPD the ideal scheme to produce robust films of nanomaterials.

3.2 Background

First investigated by Hamaker [5], Koelmans and Overbeek [7], EPD has been applied to cast uniform layers of particles on various conducting and semiconducting surfaces from polar, aqueous, and non-polar suspensions [8]. From the early beginnings, the traditional approach toward EPD has been to use easily accessible industrial solvents, such as acetone and alcohols (methanol, isopropanol, etc.), to play the role of particle suspension medium. Along with water, these solvents have been the preferred fluids used to assist the electric field-assisted materials deposition because such polar solvents facilitate the suspension and the evolution of the charge stated on the deposited materials. However, a fundamental scientific argument does not exist explaining why notably dissimilar fluids, non-polar solvents, cannot be employed for EPD. Our philosophy toward EPD, originally developed and supported by others [5, 9, 10–16], visualizes the locomotion of individual, charged nanoscale objects in an applied electric field that emanates between two electrodes, typically in a parallel plate capacitor configuration. This approach, therefore, considers the non-polar solvent as an essentially stationary dielectric suspension medium through with the nanocrystals are transported.

By using non-polar solvents, we characterize the EPD scenario as the locomotion of the charged nanocrystals between the electric field-emanating substrates, suppressing electrochemical reactions at the electrodes, restricting hydrodynamics of the solvent to turbulent flows due to the motion of the nanocrystalline materials, and limiting the measured current between the electrodes to be comprised primarily of the charged nanocrystals' motion. Further, using non-polar solvents suppresses dramatic changes in the composition and conductivity of the medium due to the generation of charged species near the electrodes can be reduced. Arguably, non-polar EPD solutions yield greater control over the film thickness, film surface roughness, and other parameters, compared to other polar or aqueous solution techniques. Due to these advantages, research has increased associated with non-aqueous and non-polar media. Most of these studies have deposited semiconducting NCs on conducting electrodes using relatively strong electric fields (E≥100 V/mm), engendered through high voltages and low solution conductivities [5, 6, 10, 17–21, 120]. The use of these electrophoretic conditions is further motivated by observations [5, 6, 9–11, 20–25, 119, 120] that films produced by this technique can withstand degradation when exposed to various solvents.

A schematic of the EPD process, shown in Fig. 3.1, illustrates macroscopic NC EPD films on gold substrates. In our EPD scheme, a dc electric field is applied

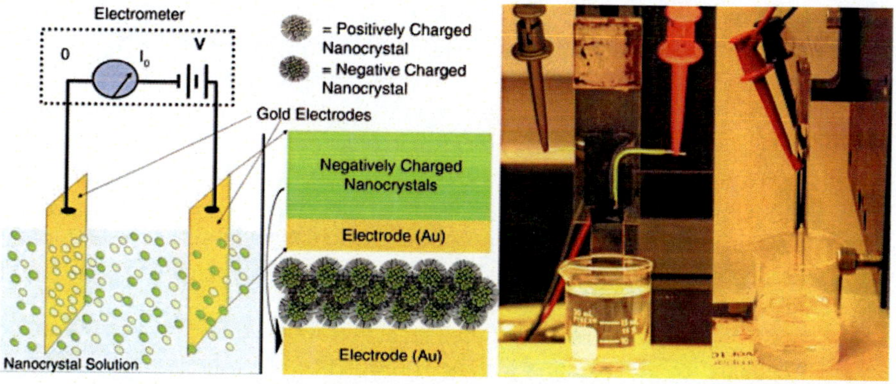

Fig. 3.1 *Left* scheme for NC EPD. *Middle & right* characteristic photographs of electrodes and a nanocrystal suspension for EPD

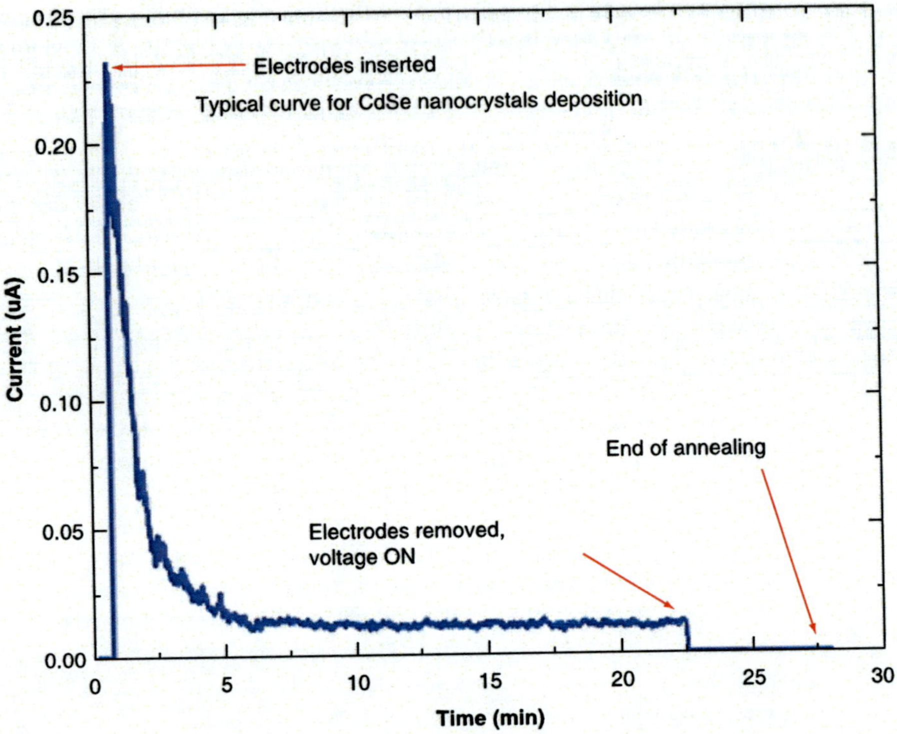

Fig. 3.2 Electrophoretic deposition current profile for deposition of CdSe nanocrystals. After the deposition, the film is annealed in air to improve its integrity. This profile is typical for non-polar solvent EPD of nanocrystals

between two planar electrodes that are inserted into a solution of charged, dipolar, or polarizable NCs in a non-polar, dielectric solvent. The typical electrophoretic current, measured between the electrodes (Fig. 3.2), is characterized as rapidly decaying (exponentially decaying), which can be explained by a combination of properties of the system: (a) materials deposition at the electrodes that creates both capacitive and resistive voltage drops across the deposit and, hence, effects on the current; (b) remnant charge accumulation at the electrodes due to incomplete charge transfer from the nanocrystals to the electrodes; (c) overall nanocrystal depletion from the suspension, which causes some change in the dielectric and resistive properties of the EPD suspension. Even uncharged NCs can be deposited by EPD due either to induced dipole moments or to the formation of aggregates in suspension (adhering to each over through van der Waals attraction). Indeed, a recent study by Islam et al. showed that **< 1%** of all the CdSe NCs, deposited by EPD from a hexane solution, indeed acquired any charge (positive or negative) [11].

3.3 Nanocrystals Employed for Non-polar EPD

3.3.1 Cadmium Selenide Nanocrystals

Cadmium selenide (CdSe) nanocrystals display archetypical size-dependent physical characteristics, the mot notable of which are quantum-confinement shifts in their absorption and luminescence energies. These features render the materials potentially useful for optical devices. CdSe, among the II-VI semiconductors, are one of the most widely studied nanomaterials, synthesized through conventional colloidal synthesis techniques. Because the radius of excitons in CdSe is approximately 5 nm, nanocrystals with a diameter smaller than this reside within the strong quantum confinement regime for their associated electrons and holes. Thus, the luminescence energies of the nanocrystals can be tuned in the visible spectrum by varying their size.

In addition to their strong optical properties, CdSe nanocrystals have been revealed to possess a strong, intrinsic electric dipole moment, which is a function of the size of the nanocrystal [26–31]. Dipole moments upwards of 40 Debye [32–36] in CdSe nanocrystals provide a physical parameter, in addition to the net surface charge, that facilitate both the electric field-mediated deposition of the nanocrystals as well as the tight packing and self-adhesion of the nanocrystals in thin film form. These properties, along with the organic surfactant molecule that usually coat the surface of the nanocrystals—the surfactant plays the dual role of terminating the growth of the NCs during synthesis *and* facilitating their suspension in a variety of solvents, such as hexane—allow CdSe NCs to be an excellent materials platform for non-polar solvent, nanocrystal EPD research.

3.3.2 Iron Oxide Nanocrystals

Several nanocrystals have been investigated to understand their size-dependent intrinsic physical properties and to explore their use in various applications. One family of materials, the iron oxides, has been produced for a number of years for their magnetic properties. Iron oxide NCs (Fig. 3.3) of different sizes exhibit ferrimagnetic or superparamagnetic properties, which have motivated the research on their application in devices. Magnetite (Fe_3O_4) NCs specifically have been of contemporary scientific interest as a result of their study in spin-valves [36] and magnetic tunnel junctions [37, 38] for possible use in magnetic read-head sensors. The synthesis of monodisperse, Fe_3O_4 NCs has involved a number of approaches, from colloidal chemistry to flame pyrolysis [39]. Since Fe_3O_4 NCs possess a strong magnetic dipole moment at room temperature ($\mu \approx 9000\ \mu_B$/NC for our materials [40–46]), this property will play a vital role in maintaining film integrity at room temperature ($U \cong k_B T$).

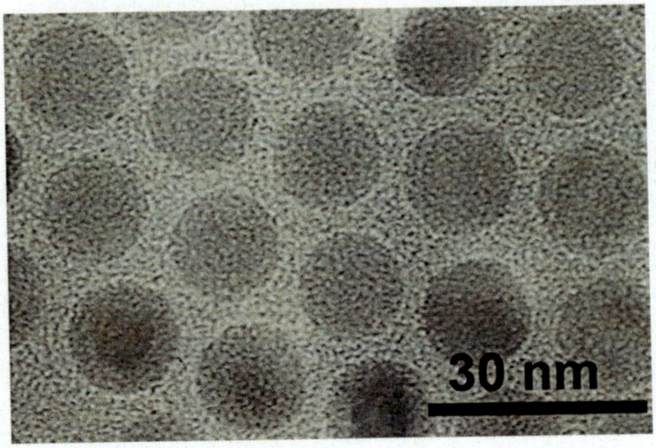

Fig. 3.3 Transmission electron microscopy (TEM) image of ~14 nm Fe_3O_4 nanocrystals

3.3.3 Rare Earth Sesquioxide Nanocrystals

Rare earth sesquioxide (RE_2O_3) NCs have attracted significant research attention recently for their potential use in industrial and commercial luminescent device applications, in next generation dielectric media, and for use as model systems to investigate size-dependent phenomena in zero-dimensional materials that may not be due to quantum confinement effects [47]. These interests are largely based on the use of bulk lanthanide oxide materials as visible light phosphors in cathode-ray tube monitors and field emission displays and their potential use as high-κ dielectric materials [6, 20, 21, 48–74]. Innovations, from biocompatible medical imaging reagents to energy efficient electroluminescent displays, provide motivation for the nanocrystal research [21, 56, 75–77]. RE_2O_3 nanomaterials are ***f***-block compounds that possess a large band gap and strongly bound ***4f*** electrons. These characteristics make them excellent dielectric and luminescent materials. Preliminary research suggests that the size and surface ligands of RE_2O_3 NCs can influence favorable their fluorescence intensity [55, 76]. These properties motivated our interest in RE_2O_3 NCs, particularly gadolinium oxide (Gd_2O_3, Fig. 3.4) and europium oxide (Eu_2O_3) nanocrystalline films [51].

3.4 Synthesis, Characterization, and Prepartion of Non-polar Solvent EPD

3.4.1 Cadmium Selenide Nanocrystals

Monodisperse CdSe nanocrystals with dodecylphosphonic acid capping ligands (length≈2 nm, Fig. 3.5) were synthesized according to the method of Peng and Peng [78]. After synthesis, the nanocrystals were precipitated from the reaction mixture with methanol (Fisher Scientific). Nanocrystal size and shape uniformity

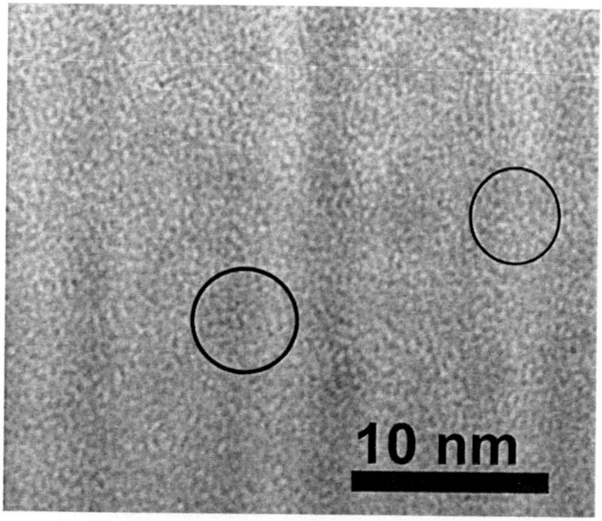

Fig. 3.4 TEM image of ~2.4 nm Gd_2O_3 nanocrystals. Circles are to guide the eye

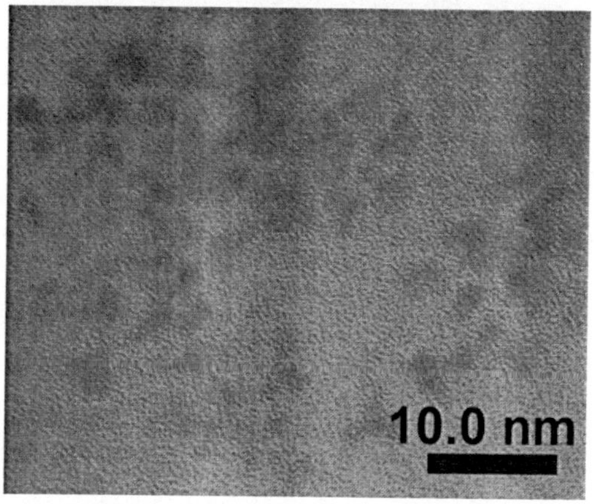

Fig. 3.5 TEM image of ~2 nm CdSe nanocrystals

for these and all other nanocrystals described in this chapter were verified by transmission electron microscopy (TEM) using a Philips CM-20 microscope.

Solutions of nanocrystals were prepared prior to deposition. Solution preparation involves cleaning the as-synthesized CdSe nanocrystals via the precipitation and re-suspension technique. Initially, a mixture of butanol and ethanol of volume ratio 3:2 was added to 50 ml of as-synthesized nanocrystal solution. This mixture was then centrifuged at 3500 rpm (2000 × **g**) to precipitate and separate the nanocrystals from the solution. Excess surfactant and other impurities remain dissolved or suspended in the butanol/ethanol solvent. The supernatant was discarded and the remaining nanocrystal sediment dried in air for several minutes. Additional butanol/ethanol solvent was added to the dried nanocrystal sediment and the centrifugation step repeated five additional times. After the last cleaning, the nanocrystals were dissolved in hexane at a ratio of 7.0 ml of hexane per 250 mg of nanocrystals. This typically yields a nanocrystal concentration of approximately 10^{17} nanocrystals/ml, which serves as the base suspension for preparing electrophoretic deposition dispersions. The EPD dispersion was prepared by diluting 100 μl of base nanocrystal suspension 15 ml of hexane, our non-polar solvent of choice, yielding a nanocrystal concentration of approximately 10^{15} nanocrystals/ml. Nanocrystalline CdSe films were prepared in the EPD cell, as depicted in Fig. 3.1. Voltages from 100 to 1100 V typically are applied across parallel conducting electrodes (3.0 cm^2 in surface area) that are spaced approximately 2.0 mm apart. Given the low conductivity of the nanocrystal solution, such voltages yield initial deposition currents of only 10–250 nA. (Fig. 3.2)

3.4.2 *Iron Oxide Nanocrystals*

Fe_3O_4 NCs were synthesized following the techniques of J. Park et al. The iron oleate precursor was prepared by reacting 2.2 g of iron chloride hexahydrate with 7.0 g of sodium oleate. During the reaction, sodium atoms are removed from the oleate; subsequently, three oleate molecules attach to individual, free iron ions. Next, 1.5 g of iron oleate precursor and 0.25 g of oleic acid (OA) were dissolved in 10 ml of 1-octadecene and heated to 320°C to yield a solution of Fe_3O_4 NCs in 1-octadecene. The mixture was held at 320°C for an additional 30 min before cooling to room temperature. Fe_3O_4 nanocrystals were precipitated by the subsequent addition of a butanol/ethanol mixture to the nanocrystal solution. The concentration was 10^{15} NCs/ml in the as-synthesized solution. We targeted ~14-nm NCs (Fig. 3.3) to maximize the monodispersity of the nanocrystals and to facilitate inter-nanocrystalline forces that promote film integrity. The procedure to produce EPD suspensions for these nanocrystals is identical to that used for CdSe NCs.

3.4.3 *Rare Earth Sesquioxide Nanocrystals*

Monodisperse, colloidally stable, RE_2O_3 NCs were synthesized following a novel approach that involves the production of an europium-oleate complex prior to the nanocrystal synthesis [79]. For example, the europium complexes were prepared from europium (III) chloride hexahydrate, oleic acid, sodium oleate, and tri-n-octylamine, based on the aforementioned Fe-oleate precursor preparation [51]. Europium (III) chloride hexahydrate, sodium oleate, ethanol, deionized (DI) water, and hexane were mixed and heated in a round-bottom flask to 70°C for four hours, while continuously stirring. A transparent paste of the europium-oleate was formed upon evaporation of the hexane and was used as a standard precursor solution for the synthesis of the NCs. To produce the NCs, 0.5 mM of the europium-oleate and 0.25 mM of oleic acid were mixed with 7 ml tri-n-octylamine in a three-neck, round-bottom flask. The solution was heated to ~350°C under constant argon flux and was cooled to room temperature. Then, nanocrystals were precipitated using an ethanol/butanol mixture and centrifugation. Through this technique, we can make Eu_2O_3 NCs, as well as Gd_2O_3 NCs, in a number of sizes. As mentioned previously, the Re_2O_3 NC suspensions for EPD were prepared in a fashion identical to that for CdSe NCs.

3.4.4 *Electrophoretic Mobility of Nanocrystals in Non-polar Solvents*

In non-aqueous media, charging of NCs can result from adsorption of uncharged ligands, ion exchange between the adsorbed ligands and the surface, and desorption of ionized ligands [79]. NCs may be electrically neutral or may adopt some effective charge ($\pm Ne^-$). Surface charge states may be altered through precipitation steps and by the addition of excess organic molecules. Identifying the magnitude and the mechanisms to alter surface charge states will provide systematic control over the EPD deposition process. To produce NC films, we must understand NC dynamics in non-polar solutions; identifying and tuning the NC mobility in solution, thus, are important.

Electric potentials develop when ionized ligands desorb into the medium, where they form the diffuse double layer [23, 80, 81]. The equilibrium charge state on the NCs can be determined through the electrophoretic mobility (μ_e) of the NCs in solution. Measured using a Zetasizer Nano ZS (Malvern), μ_e, was determined for a test case system of ~14 nm Fe_3O_4 NCs, synthesized as stated previously. Gaussian fits correspond to the ideal NC mobility, based on the Eq. 2.1:

$$\mu_e = \frac{Ze}{3\pi\eta a} \tag{3.1}$$

where η is the viscosity of the solvent, *a* is the hydrodynamic diameter of the NC, and Z is the charge of the nanocrystal (Fig. 3.6). When 0.4 mM of oleic acid is added to the suspension, the mobility shifted from negative to positive. This allows us to control the charge state on the NCs, which allows us to control the inter-nanocrystal forces that facilitate freestanding films. This suggests that NCs with targeted mobilities can be synthesized by varying the concentration of the ligand, which facilitates the production of freestanding films.

3.4.5 *Nanocrystal-Nanocrystal Interactions*

Forces between colloidal NCs are typically modeled by the Derjaguin-Landau-Verwey-Overbeek (DLVO) theory, which accounts for electrostatic and van der Waals forces between particles [80]. The DLVO approximation can be applied for spherical particles when the minimum separation and the range of the interaction for the NCs are small relative to the radii of curvature of the surfaces [82]. The validity of

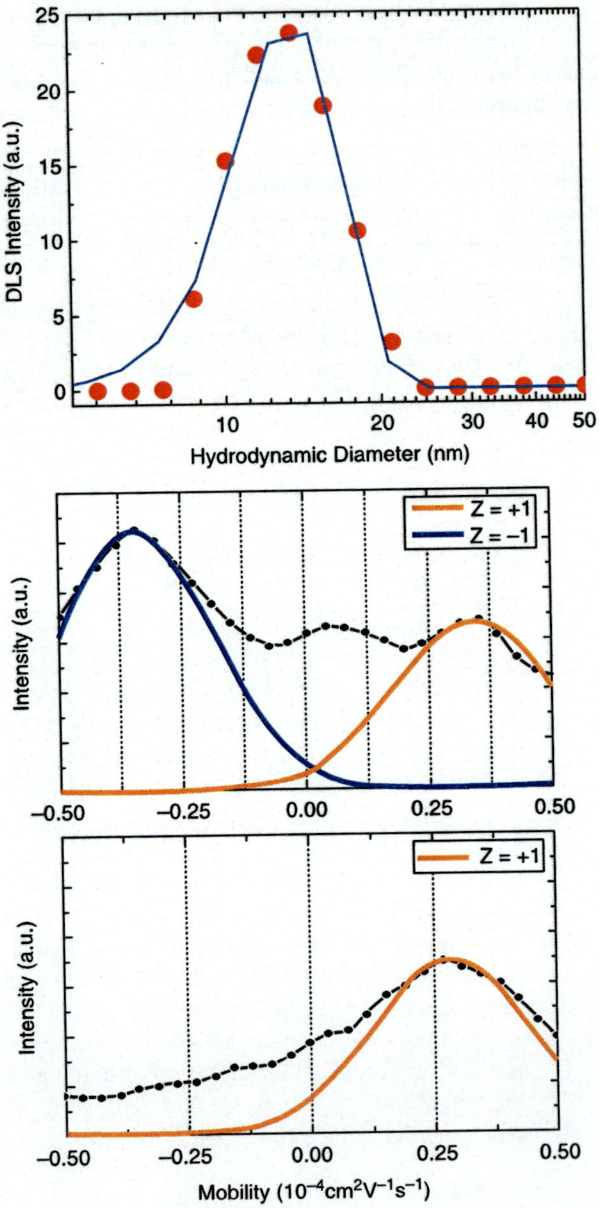

Fig. 3.6 *Top* dynamic light scattering data, from a Malvern Zetasizer, for Fe_3O_4 NCs to determine the hydrodynamic diameter of the nanocrystals in a hexane suspension. A Gaussian fit yields an average hydrodynamic diameter of ~ 13.7 nm. *Middle* electrophoretic mobility of the Fe_3O_4 NCs in hexane. Two Gaussian fits correspond to net charges of approximately ± 1e. *Bottom* the mobility distribution shifted toward positive charges after the titration of the suspension with oleic acid, the ligand molecule of the nanocrystals. Thus, the charge distribution in the suspension can be tuned, affecting the resulting location of the nanocrystal deposition

this approximation for NCs dispersed in non-aqueous media is arguable, since the electrical double layer thickness is comparable to the NC's size [83].

The stability of nanoparticulate films can be elucidated by considering the pairwise nanocrystal interactions summed over all said interactions in a superlattice. Stability may be determined by summing over all interactions and comparing the resulting potential energy well-depth to the thermal energy. Generally, stability results when the well depth is a few times greater than the thermal energy at room temperature. When considering monolayer assemblies, the binding energy is represented by:

$$
\begin{aligned}
U(r) = \sum_j \sum_k \Bigg[& \frac{-A_{\text{Hamaker}}}{6} \left\{ \frac{2R^2}{r_{jk}^2 - 4R^2} + \frac{2R^2}{r_{jk}^2} + \ln\left(\frac{r_{jk}^2 - 4R^2}{r_{jk}^2} \right) \right\} \\
& + \left\{ \frac{1}{4\pi\mu} \left(\frac{\vec{m}_j \cdot \vec{m}_k}{r_{jk}^3} - 3\frac{(\vec{m}_j \cdot \vec{r}_{jk})(\vec{m}_k \cdot \vec{r}_{jk})}{r_{jk}^5} \right) \right\} \\
& + \left(k_B T Z^2 \lambda_B \frac{e^{2R\kappa}}{(1+\kappa R)^2} \frac{e^{-\kappa r_{jk}}}{r_{jk}} \right) \Bigg]
\end{aligned}
\tag{3.2}
$$

The first two terms represent the summation over van der Waals [84] and dipole-dipole [85–89] interactions, while the last term represents the summation of electrostatic [86, 89] interactions through the distance r_{jk} between NCs within a wet film. R is the NC radius, λ_B is the Bjerrum length, r the center-to-center separation, and κ is the Debye screening length. The Bjerrum length, which relates the separation distance at which the Coulomb energy becomes comparable to the thermal energy, is given by

$$
\lambda_B = \frac{e^2}{4\pi \varepsilon \varepsilon_0 k_B T}
\tag{3.3}
$$

with e being the quantized electric charge, ε_0 the permittivity of free space, and ε the relative permittivity. The relative permittivity of hexane, our typical non-polar solvent for EPD, is $\varepsilon \approx 1.89$, yielding $\lambda_B \approx 30$ nm at room temperature. The Debye screening length is given by

$$
\kappa^{-1} = (8\pi \lambda_B c)^{-\frac{1}{2}}
\tag{3.4}
$$

with c being the number density of monovalent ions [90]. A graph of the potential energy (Eq. 3.2) is shown in Fig. 3.7. The ion density in colloidal solutions results from free carboxylic acids. These molecules are monovalent, since the COOH head group can dissociate into an H^+ cation and a COO^- anion. Since the degree of dissociation of inorganic salts in non-polar organic solvents is less than 1%, the degree of dissociation of oleic acid in hexane is should be low [91]. For nanocrystals in hexane without oleic acid titration, an ion concentration of 0.1–10 μM is assumed [92].

The electrostatic potential energy incorporates the dense charge screening engendered by the double layer that surrounds each nanocrystal, even after deposition into the film. In addition to van der Waals and electrostatic interactions, steric repulsion due to the adsorption of ligands at the surface of the nanocrystal impacts the stability of the film [93]. Steric interactions between two particles, when r_{jk} is

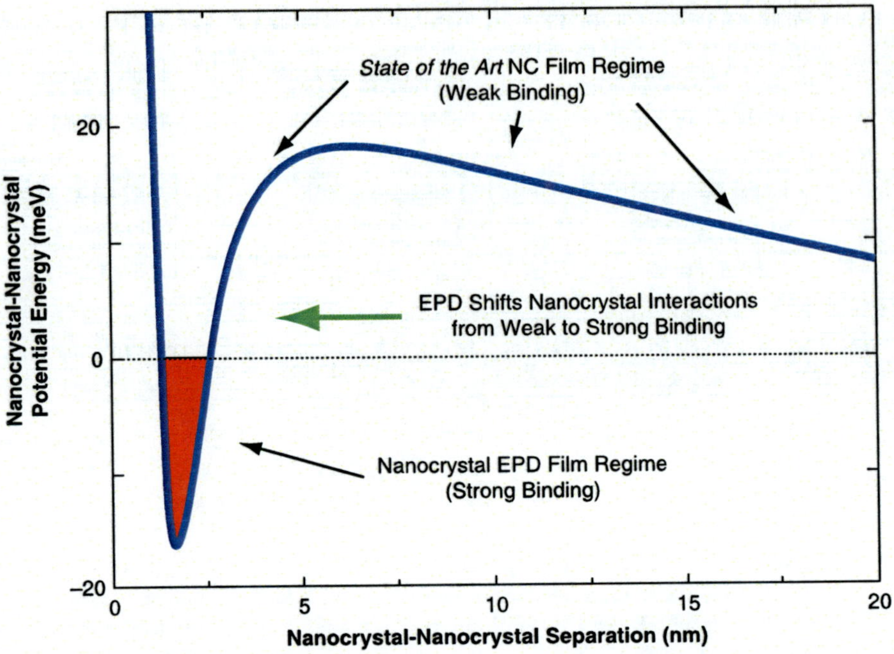

Fig. 3.7 Potential energy diagram of strong (EPD) and weak (state-of-the-art) nanocrystal-nanocrystal binding regimes

smaller than twice the thickness of the ligand layer at the surface (δ), can be expressed by the following [86, 89]:

$$U^{steric} = \pi d^2 N k_B T \sum_j \sum_k \left[1 - \frac{r_{jk} - d}{\delta} - \frac{r_{jk}}{\delta} \ln \left(\frac{d + \delta}{r_{jk}} \right) \right] \quad (3.5)$$

For example, a pair of 10 nm iron oxide nanocrystals with 11 nm of center-to-center separation has a binding energy of approximately −60 meV. The same system with a separation of 11.5 nm has a binding energy of approximately—4 meV. This calculation shows the sensitivity of the total binding energy to the thickness of the ligand layer. While the surface concentration of ligands for the nanocrystal-hexane-oleic acid system is not known, typical surface densities of highly passivated nanocrystals are approximately 10^{18} molecules/m^2. This is based on the findings of Taylor et al., who determined that the percent of surface ligand coverage of a nanocrystal can vary depending on the local stoichiometry [94]. The effects of ligand surface density on the stability of nanocrystal films cast by EPD are known qualitatively [95]. However, neither a quantitative analysis nor a theoretical survey of the relationship between the extant particle-particle forces, the magnitude of those forces, and the quality and stability of NC film has been reported. Optimal film stability and quality will be achieved at surface densities that are high enough to stabilize nanocrystals in the bulk, but not so high as to prevent dense packing.

3.5 Nanocrystalline Films

3.5.1 *Cadmium Selenide Nanocrystals*

In the majority of EPD experiments in which colloidally-synthesized nanocrystals are suspended in non-polar solvents, the nanocrystals deposited on both the positively biased and negatively biased electrodes over the entire inward-facing area dipped in solution (approximately 1 cm × 1 cm, Fig. 3.8) [11, 23, 120]. Notably, nanocrystals fail to deposit in a film covering the dipped area in the absence of applied electric field; this confirmed that the deposition of nanocrystals was due to the engendered electric field, not to the mere dipping of the electrodes into the nanocrystal suspension. The origin of the charge state of the nanocrystals helps to explain why they deposit on both electrodes. After synthesis and precipitation, the CdSe nanocrystals can be dispersed stably in hexane. In such non-aqueous solvents, nanocrystal charging—both positive and negative—has been suggested to result from adsorption or desorption of ligand molecules at the nanocrystal surface [5, 6], from thermal charging [6], and from ion exchange between nanocrystals and the suspension solvent [97].

We prepared homogeneous films of thickness up to tens of microns, depending on the initial EPD solution density, the applied voltage, the electrode surface area, and the deposition time. Figure 3.9 shows an optical micrograph, a SEM image, and an atomic force microscopy (AFM) image of the surface topology of a CdSe NC film. This film exhibits very close NC packing and relatively low roughness. The peak to valley roughness of the film is approximately the diameter of one NC, depicted in the color scale in the image. Although these images illustrate tightly packed films of the nanocrystals with volumetric packing fractions between 0.5 and 0.65 (constituting glassy packing of the particles), these films are not yet ordered. Recent results using one dimensional nanorods, however, have successfully produced ordered arrays of nanomaterials using electrophoretic deposition [98]. indeed, fabricating such films, *a crystal of nanocrystals*, could yield innumerable

Fig. 3.8 CdSe nanocrystal film, deposited onto gold substrates via EPD. Both (*left*) positively and (*right*) negatively based electrodes exhibit nanoparticle film

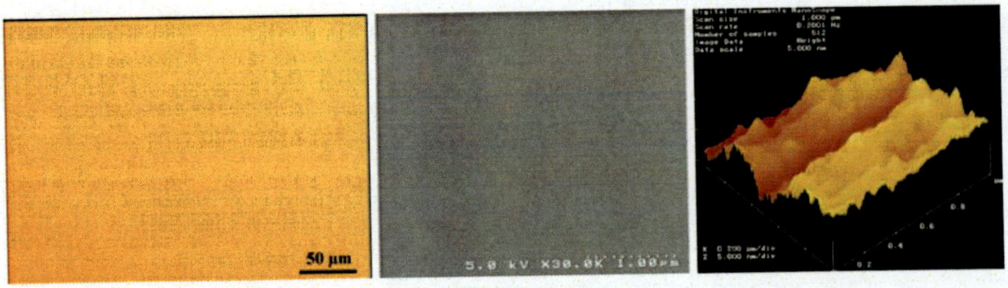

Fig. 3.9 *Left* optical micrograph, *middle* SEM image, and *right* atomic force microscope image of a CdSe nanocrystal film, cast via EPD. The AFM image illustrates the smoothness of the nanocrystal film

applications that may challenge conventional bulk materials in device applications [24, 99, 100].

3.5.2 *Iron Oxide Nanocrystals*

Instead of using a system of colloidally stable nanocrystals as a platform merely to study EPD, some have sought to use EPD to assemble NCs to explore magnetic NC-NC interactions in films for potential device applications. Indeed, magnetism at the nanoscale continues to be an active area of research as a consequence of its application to magnetic storage media, spin transport devices, and microwave emission in communication devices. Virtually all viable magnetic technologies depend upon two critical properties: (1) the stability of spin on time scales relevant to the particular application; and (2) the facile manipulation of dynamic or static spin orientations with magnetic fields, spin-polarized currents, or a combination of both. Continued miniaturization of nanomagnets presents new challenges in next generation devices because of the onset of superparamagnetism, which is the thermally activated spin relaxation that undermines the predictable control of spin. The superparamagnetic threshold with respect to magnetic nanocrystals depends upon size, shape, and material composition. Suppression of superparamagnetism was a key ingredient to successful device design and prompted Kavich et al. to explore the effect of NC-NC interactions on the relaxation of spins in iron oxide NCs [99].

The thermally activated spin relaxation of the nanocrystal film was characterized by measuring the magnetic moment vs. temperature via the technique of vibrating sample magnetometry. Iron oxide nanocrystal films, depicted in Fig. 3.10, exhibit packing that is similar to those previously observed in other non-polar solvent EPD films. The interactions among the nanocrystals, facilitated by this tight packing, had several profound effects on the thermally activated spin relaxation. The first effect was an increase in the superparamagnetic blocking temperature with respect to the value predicted by the Néel-Brown model. Further, a broad energy barrier distribution existed, which contrasted dramatically with non-interacting nanocrystals that

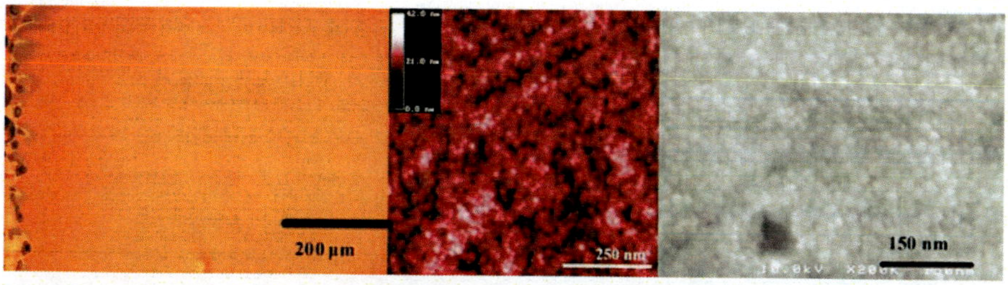

Fig. 3.10 *Left* optical micrograph, *middle* AFM image, and *right* SEM image of an EPD film of ~14 nm Fe_3O_4 NCs. Surface roughness ≈ one NC diameter. Individual NCs are evident. Packing fraction is 0.55

exhibit narrow energy barrier distributions for monodisperse samples. Finally, a monotonic decrease of the superparamagnetic blocking temperature with applied field was observed, which was in agreement with recent theoretical models of interacting nanocrystal assemblies. Thus, the tight packing of the NCs, achieved through EPD, in this uniform film allowed for the observation of magnetic phenomena only observable when the NCs are spaced closely enough to experience strong magnetic coupling [47, 25].

3.5.3 *Rare Earth Sesquioxide Nanocrystals*

Gd_2O_3 in its bulk crystalline and amorphous phases has been of research interest as a replacement material for silicon dioxide in transistors and other electronic devices because of its relatively high dielectric constant ($\kappa_{Gd_2O_3} = 14\ versus\ \kappa_{SiO_2} = 3.9$) [25]. Discovery of a high-κ dielectric material, nanocrystalline or otherwise, could make possible a number of technological advances regarding computing (energy efficiency, processing speed) telecommunications (bandwidth, speed), among others. Recently, dielectric studies of amorphous Gd_2O_3 films, embedded with Gd_2O_3 NCs, revealed intriguing charge storage characteristics of the NCs [101]. Similarly, metallic NCs (gold and cobalt [56]) and semiconducting NCs (silicon [102–104]) have exhibited charge-storage characteristics when they were embedded in the gate oxide of a metal–oxide–semiconductor (MOS) structure for non-volatile memory applications.

This motivated the exploration to determine whether films composed entirely of colloidal $Gd_2O_3Gd_2O_3$ nanocrystals could possess similar charge-storage capabilities. Charge storage in NC films arises from unpassivated surface states that can arise due to the detachment of some fraction of the surfactant molecules from the surface of the NCs during sample preparation [105]. Thus, EPD was used to produce films consisting only of colloidal Gd_2O_3 NCs to be used as the gate oxide layer in MOS architecture. Focus was placed on capacitance-voltage (*C-V*) measurements of these MOS structures to characterize both the charge-storage and the dielectric properties of these films [106, 119].

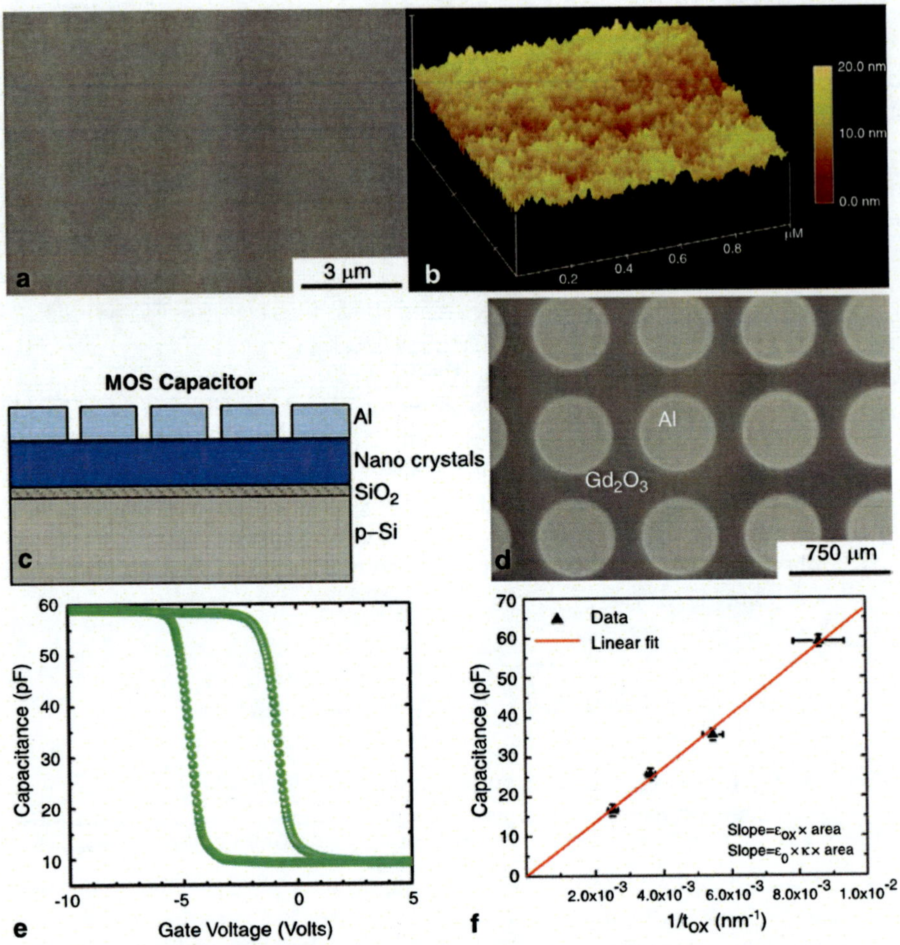

Fig. 3.11 **a** *Top* view SEM image of the Gd_2O_3 NC film prior to processing. **b** AFM image of NC film with RMS roughness of ~1.6 nm. **c** Schematic of the MOS capacitor structures. **d** *Top* view SEM image of the NC film, with aluminum contacts indicated, in the MOS capacitor configuration. **e** Capacitance *versus* voltage graph of a Gd2O3 NC film. **f** Graph of capacitance versus the inverse of the nanocrystal film thickness. Assessment of the linear fit of the data yields the effective dielectric constant

As shown in the schematic of Fig. 3.11, we fabricated MOS capacitor structures, composed of films of ultra-small colloidal Gd_2O_3 NCs employed as the dielectric oxide layer. The NCs were deposited onto p-(100) silicon substrates via EPD. To confirm the NC deposition, energy dispersive spectroscopy (EDS) was performed, which provide elemental analysis of a materials composition. Shown in Fig. 3.11c, the EDS spectrum of the film exhibits the gadolinium, oxygen, carbon, and silicon peaks, which are present due to the oleic acid-capped Gd_2O_3 NCs on the silicon substrate. This confirmed that no excess contamination was present in our films. To complete the capacitor fabrication, aluminum contacts were deposited on the Gd_2O_3 NC film. Figure 3.11d shows the SEM image of the capacitors.

In analyzing the films, unequivocal evidence was uncovered that because *C-V* measurements of the MOS capacitors exhibited hysteresis, charge can be stored across a purely nanocrystalline film, as seen in Fig. 3.11e. The hysteresis indicated storage of charge carriers. From the measurement of the *C-V* graphs of several MOS capacitors of different thicknesses, the effective dielectric constant for the films, $\kappa=3.90$ (Fig. 3.11f), was determined. This value is much lower than the dielectric constant of single crystalline Gd_2O_3 because the packing density of the NCs within the film was 0.66 ± 0.08. This meant that only 66% of the film consisted of material with $\kappa=14$. However, the net value was equivalent to the dielectric constant of silicon dioxide. This result provides substantial evidence that all-nanocrystalline assemblies could be employed as charge-storing, dielectric films in various architectures for electronics applications.

3.5.4 Freestanding Nanocrystal Films

One of the challenges to implement nanocrystals and nanocrystal films into device applications is incorporating the films into systems (circuitry architectures, substrate configurations, etc.) that exploit their properties. One approach to eliminate this challenge is to develop a versatile film casting system that creates freestanding, macroscopic films comprised exclusively of nanocrystals that are free of an underlying substrate. Nanocrystal films can be produced in a rapid, repeatable format for applications like chemical sensors and flexible video displays, such that rigid substrates are not needed to maintain the integrity of the film. Several methods exist that have yielded self-supporting architectures, not notably LbL and evaporation-assisted self assembly [90, 107–112]. However, many of these approaches require the addition of cross-linking molecules and other binders, which constitute an internal support scaffold for the film, or involve laminar, multilayered films of charged nanocrystals (such as LbL), both of which produce self-supporting, but very rigid and potentially brittle materials [2, 1, 113–116]. Further, potential applications that seek the production of nanostructured, nanocrystalline materials with minimal additives would not favor these methods. Thus, additional techniques should be explored to fabricate self-supporting, tightly packed NC films at high assembly rates and without the presence of additives or binders. Notably, prior to 2009, freestanding nanoparticle films had not been successfully fabricated by EPD [99, 119].

These macroscopic films, comprised exclusively of nanocrystals, are entirely independent of a supporting substrate, an external supportive matrix, or interstitial glue. In other words, these films are entirely self-supporting colloidal solids. However, what characteristics of the nanocrystals are crucial in making them freestand? Do steric and shape effects associated with the NC (spherical vs. cylindrical vs. planar; coated with surfactant molecule "hairs" vs. uncoated) play a substantial role? Does the presence of a dipole (magnetic or electric) significantly increase the particle-particle adhesion? Do the answers to these questions change when examining the films at the microscale versus the macroscale?

The scheme (Fig. 3.12) to produce freestanding nanocrystal films involves multiple elements: the electric field manipulation of nanocrystals in solution with dielectric layers, ideal densification and liberation of the films without degradation.

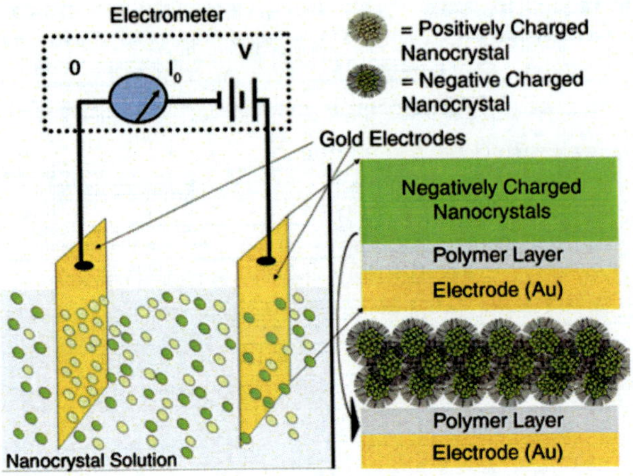

Fig. 3.12 *Left* schematic for nanocrystal EPD to fabricate freestanding films, using the sacrificial layer EPD technique. The sacrificial polymer layer is indicated. [119, 120]

Because of the small inter-NC separation engendered by the electrophoretic deposition, strong NC-NC interactions were relied upon to maintain the structural integrity of the films before, during, and after liberation from the underlying substrate. Freestanding NC films of CdSe [1, 101], Fe_3O_4 [119] NCs, (Fig. 3.13), as well as other nanomaterials (suspended in polar solvents [117, 18]) have been successfully fabricated. We have succeeded in depositing films onto sacrificial polymer layers, removing polymer layers, and transferring the NC films onto new substrates. This worked with films of area ~4.0 mm² for Fe_3O_4 NCs and ≥1 cm² for carbon nanotubes and exfoliated graphene oxide [117–119]. The extent of the mechanical properties of these films, elucidating the magnitude of the particle-particle interactions that exist within the EPD films remains a topic of future research activity.

Part of the success of previous results has been identifying possible candidates for the sacrificial polymer layer. The choice for the polymer had to satisfy three major criteria: (1) during EPD, the polymer must not suffer dielectric breakdown due

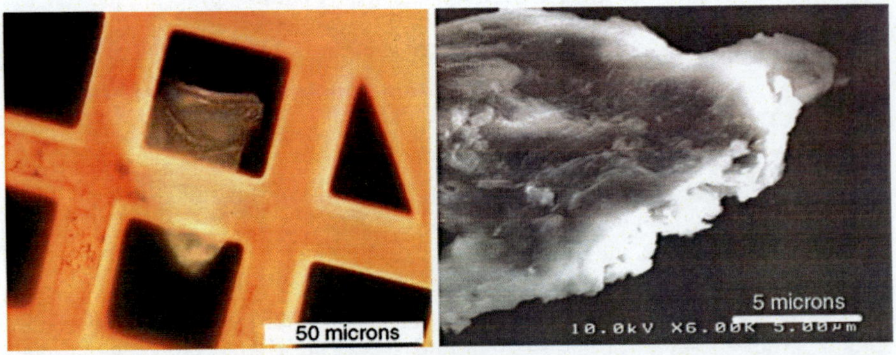

Fig. 3.13 *Left* optical micrograph of a freestanding iron oxide nanocrystal film resting atop a copper TEM grid. Scale bar is 50 μ. *Right* SEM image of the same film

to the applied voltage; (2) the polymer must not swell or dissolve in the nanocrystal solution during EPD; (3) the polymer must dissolve in a solvent that does not damage the nanocrystal film. Polystyrene was the initial choice since it has a dielectric constant of ~2.6 and experiences breakdown in fields stronger than 200 kV/cm, [119] a threshold not exceeded by our EPD protocol. Further, hexane, the nanocrystal solvent during EPD, is a non-solvent for polystyrene. Dichloromethane could be used to dissolve the polymer layer with minimal effect on the oleic acid-capped nanocrystals. Yet, its volatile nature while being dissolved during the NC film liberation process did not make polystyrene the preferred polymer for freestanding film production. Refinement of this technique, such that a wider variety of nanomaterials and solvents can be employed to fabricate freestanding films, remains an invigorating topic for continued exploration and development for the EPD and nanoscience communities.

3.6 Summary

This brief introduction to the use of non-polar solvents for the electrophoretic deposition of nanocrystals has reviewed the underlying science of the nanocrystal synthesis, the suspension preparation, the governing particle-particle interactions, and the some of the physical characteristics of the resulting films. Given the breadth of colloidal nanocrystals that have been synthesized and remain to be investigated in film form, the potential for substantial basic research discovery as well as the implementation of the nanocrystals and the discussed techniques into commercial and/or industrial fabrication protocols and systems will only increase over time. Recent developments for device fabrication, such as solar cells and charge storage dielectric layers, among others [24, 25, 47, 99, 100], have demonstrated that nanocrystalline films fabricated via EPD are on a positive trajectory.

References

1. Lee, D., Rubner, M.F., Cohen, R.E.: All-nanoparticle thin film coatings. Nano Lett. **6**(10), 2305–2312 (2006)
2. Shchukin, D.G., Zheludkevich, M., Yasakau, K., Lamaka, S., Ferreira, M.G.S., Mohwald, H.: Layer-by-layer assembled nanocontainers for self-healing corrosion protection. Adv. Mater. **18**(13), 1672–1678 (2006)
3. Maenosono, S., Okubo, T., Yamaguchi, Y.: Overview of nanoparticle array formation by wet coating. J. Nanopart. Res. **5**(1–2), 5–15 (2003)
4. Tian, Y.C., Fendler, J.H.: Langmuir-Blodgett film formation from fluorescence-activated, surfactant-capped, size-selected CdS nanoparticles spread on water surfaces. Chem. Mater. **8**(4), 969–974 (1996)
5. Islam, M.A., Herman, I.P.: Electrodeposition of patterned CdSe nanocrystal films using thermally charged nanocrystals. Appl. Phys. Lett. **80**(20), 3823–3825 (2002)

6. Mahajan, S.V., Kavich, D.W., Redigolo, M.L., Dickerson, J.H.: Structural properties of electrophoretically deposited europium oxide nanocrystalline thin films. J. Mater. Sci. **41**(24), 8160–8165 (2006)
7. Hamaker, H.C., Verwey, E.J.W.: Colloid stability. The role of the forces between the particles in electrodeposition and other phenomena. Trans. Faraday Soc. **36**, 180–185 (1940)
8. Koelmans, H.: Philips Res. Rep. **10**, 161 (1955)
9. Giersig, M., Mulvaney, P.: Formation of ordered 2-dimensional gold colloid lattices by electrophoretic deposition. J. Phys. Chem. **97**(24), 6334–6336 (1993)
10. Islam, M.A., Xia, Y.Q., Steigerwald, M.L., Yin, M., Liu, Z., O'Brien, S., Levicky, R., Herman, I.P.: Addition, suppression, and inhibition in the electrophoretic deposition of nanocrystal mixture films for CdSe nanocrystals with gamma-Fe_2O_3 and Au nanocrystals. Nano Lett. **3**(11), 1603–1606 (2003)
11. Islam, M.A., Xia, Y.Q., Telesca, D.A., Steigerwald, M.L., Herman, I.P.: Controlled electrophoretic deposition of smooth and robust films of CdSe nanocrystals. Chem. Mater. **16**(1), 49–54 (2004)
12. Teranishi, T., Hosoe, M., Tanaka, T., Miyake, M.: Size control of monodispersed Pt nanoparticles and their 2D organization by electrophoretic deposition. J. Phys. Chem. B. **103**(19), 3818–3827 (1999)
13. Wong, E.M., Searson, P.C.: Kinetics of electrophoretic deposition of zinc oxide quantum particle thin films. Chem. Mater. **11**(8), 1959–1961 (1999)
14. Wong, E.M., Searson, P.C.: ZnO quantum particle thin films fabricated by electrophoretic deposition. Appl. Phys. Lett. **74**(20), 2939–2941 (1999)
15. Zhao, S.Y., Lei, S.B., Chen, S.H., Ma, H.Y., Wang, S.Y.: Assembly of two-dimensional ordered monolayers of nanoparticles by electrophoretic deposition. Colloid Polym. Sci. **278**(7), 682–686 (2000)
16. Somarajan, S., Hasan, S.A., Adkins, C.T., Harth, E., Dickerson, J.H.: Controlled electrophoretic deposition of uniquely nanostructured star polymer films. J. Phys. Chem. B. **112**(1), 23–28 (2008)
17. Islam, M.A., Xia, S.G.: Electrostatic properties of maghemite (gamma-Fe_2O_3) nanocrystalline quantum dots determined by electrophoretic deposition. J. Phys.: Condens. Matter **21**(28) (2009). doi:28530110.1088/0953-8984/21/28/285301
18. Jia, S., Banerjee, S., Herman, I.P.: Mechanism of the electrophoretic deposition of CdSe nanocrystal films: influence of the nanocrystal surface and charge. J. Phys. Chem. C. **112**(1), 162–171 (2008). doi:10.1021/jp0733320
19. Jia, S.G., Banerjee, S., Lee, D., Bevk, J., Kysar, J.W., Herman, I.P.: Fracture in electrophoretically deposited CdSe nanocrystal films. J. Appl. Phys. **105**(10) (2009). doi:10351310.1063/1.3118630
20. Mahajan, S.V., Dickerson, J.H.: Understanding the growth of Eu_2O_3 nanocrystal films made via electrophoretic deposition. Nanotechnology **21**(14) (2010). doi:10.1088/0957-4484/21/14/145704
21. Mahajan, S.V., Dickerson, J.H.: Dielectric properties of colloidal Gd_2O_3 nanocrystal films fabricated via electrophoretic deposition. Appl. Phys. Lett. **96**(11) (2010). doi:10.1063/1.3359418
22. Giersig, M., Mulvaney, P.: Preparation of ordered colloid monolayers by electrophoretic deposition. Langmuir **9**(12), 3408–3413 (1993)
23. Mohammad, A.I.: Amphoteric CdSe nanocrystalline quantum dots. Nanotechnology **19**(25), 255708 (2008)
24. Smith, N.J., Emmett, K.J., Rosenthal, S.J.: Photovoltaic cells fabricated by electrophoretic deposition of CdSe nanocrystals. Appl. Phys. Lett. **93**(4) (2008). doi:04350410.1063/1.2965464
25. Kavich, D.W., Hasan, S.A., Mahajan, S.V., Park, J.H., Dickerson, J.H.: Field dependence of the spin relaxation within a film of iron oxide nanocrystals formed via electrophoretic deposition. Nanoscale Res. Lett. **5**(10), 1540–1545 (2010). doi:.1007/s11671-010-9674-2
26. Alivisatos, A.P.: Semiconductor clusters, nanocrystals, and quantum dots. Science **271**(5251), 933–937 (1996)

27. Bowers, M.J., McBride, J.R., Rosenthal, S.J.: White-light emission from magic-sized cadmium selenide nanocrystals. J. Am. Chem. Soc. **127**(44), 15378–15379 (2005). doi:10.1021/ja055470d
28. Murray, C.B., Norris, D.J., Bawendi, M.G.: Synthesis and characterization of nearly monodisperse CdE (E=S, Se, Te) semiconductor nanocrystallites. J. Am. Chem. Soc. **115**(19), 8706–8715 (1993)
29. Nirmal, M., Brus, L.: Luminescence photophysics in semiconductor nanocrystals. Acc. Chem. Res. **32**(5), 407–414 (1999)
30. Peng, X.G., Schlamp, M.C., Kadavanich, A.V., Alivisatos, A.P.: Epitaxial growth of highly luminescent CdSe/CdS core/shell nanocrystals with photostability and electronic accessibility. J. Am. Chem. Soc. **119**(30), 7019–7029 (1997)
31. Rogach, A.L., Kornowski, A., Gao, M.Y., Eychmuller, A., Weller, H.: Synthesis and characterization of a size series of extremely small thiol-stabilized CdSe nanocrystals. J. Phys. Chem. B. **103**(16), 3065–3069 (1999)
32. Blanton, S.A., Leheny, R.L., Hines, M.A., GuyotSionnest, P.: Dielectric dispersion measurements of CdSe nanocrystal colloids: observation of a permanent dipole moment. Phys. Rev. Lett. **79**(5), 865–868 (1997)
33. Dollefeld, H., Weller, H., Eychmuller, A.: Particle-particle interactions in semiconductor nanocrystal assemblies. Nano Lett. **1**(5), 267–269 (2001)
34. Nann, T., Schneider, J.: Origin of permanent electric dipole moments in wurtzite nanocrystals. Chem. Phys. Lett. **384**(1–3), 150–152 (2004)
35. Rabani, E.: Structure and electrostatic properties of passivated CdSe nanocrystals. J. Chem. Phys. **115**(3), 1493–1497 (2001)
36. Shim, M., Guyot-Sionnest, P.: Permanent dipole moment and charges in colloidal semiconductor quantum dots. J. Chem. Phys. **111**(15), 6955–6964 (1999)
37. van Dijken, S., Fain, X., Watts, S.M., Nakajima, K., Coey, J.M.D.: Magnetoresistance of Fe_3O_4/Au/Fe_3O_4 and Fe_3O_4/Au/Fe spin-valve structures. J. Magn. Magn. Mater. **280**(2–3), 322–326 (2004)
38. Gupta, A., Sun, J.Z.: Spin-polarized transport and magnetoresistance in magnetic oxides. J. Magn. Magn. Mater. **200**(1–3), 24–43 (1999)
39. Hu, G., Suzuki, Y.: Negative spin polarization of Fe_3O_4 in magnetite/manganite-based junctions. Phys. Rev. Lett. **89**(27), 276601 (2002)
40. Bate, G.: Ferromagnetic materials, recording materials, vol. 2. North Holland, Amsterdam (1980)
41. Ito, A., Shinkai, M., Honda, H., Kobayashi, T.: Medical application of functionalized magnetic nanoparticles. J. Biosci. Bioeng. **100**(1), 1–11 (2005)
42. Kang, Y.S., Risbud, S., Rabolt, J.F., Stroeve, P.: Synthesis and characterization of nanometer-size Fe_3O_4 and gamma-Fe_2O_3 particles. Chem. Mater. **8**(9), 2209 (1996)
43. Mornet, S., Vasseur, S., Grasset, F., Duguet, E.: Magnetic nanoparticle design for medical diagnosis and therapy. J. Mater. Chem. **14**(14), 2161–2175 (2004)
44. Raj, K., Moskowitz, R.: Commercial applications of ferrofluids. J. Magn. Magn. Mater. **85**(1–3), 233–245 (1990)
45. Vollath, D., Szabo, D.V., Taylor, R.D., Willis, J.O., Sickafus, K.E.: Synthesis and properties of nanocrystalline superparamagnetic gamma-Fe_2O_3. Nanostructured Mater. **6**(5–8), 941–944 (1995)
46. Ziolo, R.F., Giannelis, E.P., Weinstein, B.A., Ohoro, M.P., Ganguly, B.N., Mehrotra, V., Russell, M.W., Huffman, D.R.: Matrix-mediated synthesis of nanocrystalline gamma-Fe_2o_3—a new optically transparent magnetic material. Science **257**(5067), 219–223 (1992)
47. Kavich, D.W., Dickerson, J.H., Mahajan, S.V., Hasan, S.A., Park, J.H.: Exchange bias of singly inverted FeO/Fe_3O_4 core-shell nanocrystals. Phys. Rev. B. **78**(17), 6(2008). doi:17441410.1103/PhysRevB.78.174414
48. Bazzi, R., Flores, M.A., Louis, C., Lebbou, K., Zhang, W., Dujardin, C., Roux, S., Mercier, B., Ledoux, G., Bernstein, E.: Synthesis and properties of europium-based phosphors on the

nanometer scale: Eu_2O_3, Gd_2O_3:Eu, and Y_2O_3:Eu. J. Colloid Interface Sci. **273**(1), 191–197 (2004)
49. Eilers, H., Tissue, B.M.: Laser spectroscopy of nanocrystalline Eu_2O_3 and Eu_3^+:Y_2O_3. Chem. Phys. Lett. **251**(1–2), 74–78 (1996)
50. Mahajan, S.V., Dickerson, J.H.: Understanding the growth of Eu_2O_3 nanocrystal films made via electrophoretic deposition. Nanotechnology **21**(14). doi:10.1088/0957-4484/21/14/145704
51. Mahajan, S.V., Dickerson, J.H.: Synthesis of monodisperse sub-3 nm RE_2O_3 and Gd_2O_3:RE_3+nanocrystals. Nanotechnology **18**(32), 325605 (2007)
52. Wakefield, G., Keron, H.A., Dobson, P.J., Hutchison, J.L.: Synthesis and properties of sub-50-nm europium oxide nanoparticles. J. Colloid Interface Sci. **215**(1), 179–182 (1999)
53. Bazzi, R., Flores, M.A., Louis, C., Lebbou, K., Zhang, W., Dujardin, C., Roux, S., Mercier, B., Ledoux, G., Bernstein, E., Perriat, P., Tillement, O.: Synthesis and properties of europium-based phosphors on the nanometer scale: Eu_2O_3, Gd_2O_3:Eu, and Y_2O_3:Eu. J. Colloid Interface Sci. **273**, 191 (2004)
54. Feng, J., Shan, G.M., Maquieira, A., Koivunen, M.E., Guo, B., Hammock, B.D., Kennedy, I.M.: Functionalized europium oxide nanoparticles used as a fluorescent label in an immunoassay for atrazine. Anal. Chem. **75**(19), 5282–5286 (2003)
55. Shionoya, S., Yen, W.M.: Phosphor handbook. CRC Press, Boca Raton (1999)
56. Wang, J.C., Lai, C.S., Chen, Y.K., Lin, C.T., Liu, C.P., Huang, M.R.S., Fang, Y.C.: Characteristics of gadolinium oxide nanocrystal memory with optimized rapid thermal annealing. Electrochem. Solid State Lett. **12**(6), H202-H204 (2009). doi:10.1149/1.3109573
57. An, L.Q., Zhang, J., Liu, M., Wang, S.W.: Preparation and upconversion properties of Yb^{3+}, Ho^{3+}: LU_2O_3 nanocrystalline powders. J. Am. Ceram. Soc. **88**(4), 1010–1012 (2005)
58. An, L.Q., Zhang, J., Liu, M., Wang, S.W.: Synthesis and luminescence properties of Yb^{3+}/HO^{3+} co-doped Lu_2O_3 nanocrystalline powders. In: High-Performance Ceramics Iii, Pts 1 and 2, vol 280–283. Key engineering materials, pp. 521–524. Trans Tech Publications Ltd, Zurich-Uetikon (2005)
59. Azad, A.M., Matthews, T., Swary, J.: Processing and characterization of electrospun Y_2O_3-stabilized ZrO_2 (YSZ) and Gd_2O_3-doped CeO_2 (GDC) nanofibers. Mater. Sci. Eng. B-Solid State Mater. Adv. Tech. **123**(3), 252–258 (2005)
60. Camenzind, A., Strobel, R., Krumeich, F., Pratsinis, S.E.: Luminescence and crystallinity of flame-made Y_2O_3: Eu^{3+} nanoparticles. Adv. Powder Technol. **18**(1), 5–22 (2007)
61. Capobianco, J.A., Vetrone, F., Boyer, J.C., Speghini, A., Bettinelli, M.: Visible upconversion of Er^{3+} doped nanocrystalline and bulk Lu_2O_3. Opt. Mater. **19**(2), 259–268 (2002)
62. Chen, Q.W., Shi, Y., An, L.Q., Chen, J.Y., Shi, J.L.: Fabrication and photoluminescence characteristics of Eu^{3+}-Doped Lu_2O_3 transparent ceramics. J. Am. Ceram. Soc. **89**(6), 2038–2042 (2006)
63. Chen, Q.W., Shi, Y., Shi, J.L.: Preparation and characterization of a new phosphor Lu_2O_3: Eu_3^+. In: High-Performance Ceramics Iii, Pts 1 and 2, vol. 280–283. Key engineering materials, pp 525–528. Trans Tech Publications Ltd, Zurich-Uetikon (2005)
64. Dosev, D., Nichkova, M., Liu, M.Z., Guo, B., Liu, G.Y., Hammock, B.D., Kennedy, I.M.: Application of luminescent Eu: Gd_2O_3 nanoparticles to the visualization of protein micropatterns. J Biomed. Opt. **10**(6), 7(2005)
65. Flores-Gonzalez, M.A., Louis, C., Bazzi, R., Ledoux, G., Lebbou, K., Roux, S., Perriat, P., Tillement, O.: Elaboration of nanostructured Eu^{3+}-doped Gd_2O_3 phosphor fine spherical powders using polyol-mediated synthesis. Appl. Phys. A-Mater. Sci. Process **81**(7), 1385–1391 (2005).
66. Fu, X.Y., Niu, S.Y., Zhang, H.W.: Preparation and characterization of dysprosium oxide doped in ZrO_2 and yttrium-stabilized zirconium(YSZ) nanocrystalline. Abstr. Pap. Am. Chem. Soc. **230**, U2096-U2096 (2005)
67. Gedanken, A., Reisfeld, R., Sominski, L., Zhong, Z., Koltypin, Y., Panczer, G., Gaft, M., Minti, H.: Time-dependence of luminescence of nanoparticles of Eu_2O_3 and Tb_2O_3 deposited on and doped in alumina. Appl. Phys. Lett. **77**(7), 945–947 (2000)

68. Gogoi, P., Konwar, K., Baishya, B.: A study of Ge-thin film transistors with rare earth oxides as gate insulators. Indian J. Phys. Proc. Indian Assoc. Cultiv. Sci. **80**(10), 1021–1023 (2006)
69. Gordon, W.O., Carter, J.A., Tissue, B.M.: Long-lifetime luminescence of lanthanide-doped gadolinium oxide nanoparticles for immunoassays. J. Lumin. **108**(1–4), 339–342 (2004)
70. Hirai, T., Orikoshi, T.: Preparation of Gd_2O_3: Yb,Er and Gd_2O_2S: Yb, Er infrared-to-visible conversion phosphor ultrafine particles using an emulsion liquid membrane system. J. Colloid Interface Sci. **269**(1), 103–108 (2004)
71. Jenouvrier, P., Boccardi, G., Fick, J., Jurdyc, A.M., Langlet, M.: Up-conversion emission in rare earth-doped $Y_2Ti_2O_7$ sol-gel thin films. J. Lumin. **113**(3–4), 291–300 (2005)
72. Nichkova, M., Dosev, D., Perron, R., Gee, S.J., Hammock, B.D., Kennedy, I.M.: Eu^{3+}-doped Gd_2O_3 nanoparticles as reporters for optical detection and visualization of antibodies patterned by microcontact printing. Anal. Bioanal. Chem. **384**(3), 631–637 (2006)
73. Nyk, M., Hreniak, D., Strek, W., Misiewicz, J., Zych, E.: Photo- and cathodoluminescence properties of LU_2O_3: Tb^{3+} nanocrystallites embedded in TiO_2 films on silicon and quartz substrates. Opt. Mater. **26**(2), 129–132 (2004)
74. Pedersen, H., Ojamae, L.: Towards biocompatibility of RE_2O_3 nanocrystals—Water and organic molecules chemisorbed on Gd_2O_3 and Y_2O_3 nanocrystals studied by quantum-chemical computations. Nano Lett. **6**(9), 2004–2008 (2006)
75. Eilers, H., Tissue, B.M.: Synthesis of nanophase ZnO, Eu_2O_3, and ZrO_2 by gas-phase condensation with cw-CO_2 laser heating. Mater. Lett. **24**(4), 261–265 (1995)
76. Feng, J., Shan, G., Maquieira, A., Koivunen, M.E., Guo, B., Hammock, B.D., Kennedy, I.M.: Functionalized Europium oxide nanoparticles used as a fluorescent label in an immunoassay for atrazine. Anal. Chem. **75**(19), 5282–5286 (2003)
77. Wakefield, G., Keron, H.A., Dobson, P.J., Hutchinson, J.L.: Synthesis and properties of sub-50-nm europium oxide nanoparticles. J. Colloid Interface Sci. **215**, 179 (1999)
78. Peng, Z.A., Peng, X.: Mechanisms of the shape evolution of CdSe nanocrystals. J. Am. Chem. Soc. **123**(7), 1389–1395 (2001). doi:10.1021/ja0027766
79. Park, J., An, K.J., Hwang, Y.S., Park, J.G., Noh, H.J., Kim, J.Y., Park, J.H., Hwang, N.M., Hyeon, T.: Ultra-large-scale syntheses of monodisperse nanocrystals. Nat. Mater. **3**(12), 891–895 (2004)
80. Morrison, I.D., Ross, S.: Colloidal dispersions: suspension, emulsions, and foams. Wiley, New York (2002)
81. Pugh, R.J., Matsunaga, T., Fowkes, F.M.: The dispersibility and stability of carbon-black in media of low dielectric-constant.1. Electrostatic and steric contributions to colloidal stability. Colloids Surf. **7**(3), 183–207 (1983)
82. Israelachvili, J.N.: Intermolecular and surface forces, 2nd edn. Academic Press, New York (1992)
83. Russel, W.B., Saville, D.A., Schowalter, D.A.: Colloidal dispersions. Cambridge University Press, Cambridge (1989)
84. Kornbrekke, R.E., Morrison, I.D., Oja, T.: Electrophoretic mobility measurements in low conductivity media. Langmuir **8**, 1211–1217 (1992)
85. Tadmor, R.: The London-van der Waals interaction energy between objects of various geometries. J. Phys. Condens. Matter. **13**(9), L195–L202 (2001)
86. Morimoto, H., Maekawa, T.: Cluster structures and cluster-cluster aggregations in a two-dimensional ferromagnetic colloidal system. J. Phys. Math. Gen. **33**(2), 247–258 (2000)
87. Collier, C.P., Vossmeyer, T., Heath, J.R.: Nanocrystal superlattices. Annu. Rev. Phys. Chem. **49**, 371–404 (1998)
88. Ohara, P.C., Leff, D.V., Heath, J.R., Gelbart, W.M.: Crystallization of opals from polydisperse nanoparticles. Phys. Rev. Lett. **75**(19), 3466–3469 (1995)
89. Motte, L., Courty, A., Ngo, A.-T., Lisiecki, I., Pileni, M.-P.: Self-organization of inorganic nanocrystals. Nanocrystals forming mesoscopic structures. Wiley, Weinheim (2005)
90. Shevchenko, E.V., Talapin, D.V., Kotov, N.A., O'Brien, S., Murray, C.B.: Structural diversity in binary nanoparticle superlattices. Nature **439**(7072), 55–59 (2006)

91. Leunissen, M.E., Christova, C.G., Hynninen, A.P., Royall, C.P., Campbell, A.I., Imhof, A., Dijkstra, M., van Roij, R., van Blaaderen, A.: Ionic colloidal crystals of oppositely charged particles. Nature **437**(7056), 235–240 (2005)
92. Yethiraj, A., van Blaaderen, A.: A colloidal model system with an interaction tunable from hard sphere to soft and dipolar. Nature **421**(6922), 513–517 (2003)
93. van Roij, R., Hansen, J.P.: Van der Waals-like instability in suspensions of mutually repelling charged colloids. Phys. Rev. Lett. **79**(16), 3082–3085 (1997)
94. Rosensweig, R.E.: Ferrohydrodynamics. Dover, Mineola (1997)
95. Taylor, J., Kippeny, T.C., Rosenthal, S.J.: Surface stoichiometry of CdSe nanocrystals determined by Rutherford backscattering spectroscopy. J. Clust. Sci. **12**(4), 571–582 (2001)
96. Van Der Biest, O.O., Vandeperre, L.J.: Electrophoretic deposition of materials. Annu. Rev. Mate. Sci. **29**, 327–352 (1999)
97. Islam, M.A.: Amphoteric CdSe nanocrystalline quantum dots. Nanotechnology **19**(25), 255708 (2008)
98. Morrison, I.D.: Electrical charges in nonaqueous media. Colloids Surf. Physicochem. Eng. Aspects **71**(1), 1–37 (1993)
99. Ahmed, S., Ryan, K.M.: Centimetre scale assembly of vertically aligned and close packed semiconductor nanorods from solution. Chem. Commun. **14**(42), 6421–6423 (2009). doi:10.1039/b914478a
100. Salant, A., Shalom, M., Hod, I., Faust, A., Zaban, A., Banin, U.: Quantum dot sensitized solar cells with improved efficiency prepared using electrophoretic deposition. ACS Nano **4**(10), 5962–5968 (2010). doi:10.1021/nn1018208
101. Kwo, J., Hong, M., Kortan, A.R., Queeney, K.T., Chabal, Y.J., Mannaerts, J.P., Boone, T., Krajewski, J.J., Sergent, A.M., Rosamilia, J.M.: High epsilon gate dielectrics Gd_2O_3 and Y_2O_3 for silicon. Appl. Phys. Lett. **77**(1), 130–132 (2000)
102. Guan, W.H., Long, S.B., Liu, M., Li, Z.G., Hu, Y., Liu, Q.: Fabrication and charging characteristics of MOS capacitor structure with metal nanocrystals embedded in gate oxide. J. Phys. D-Appl. Phys. **40**(9), 2754–2758 (2007). doi:10.1088/0022-3727/40/9/012
103. Park, B., Cho, K., Kim, H., Kim, S.: Capacitance characteristics of MOS capacitors embedded with colloidally synthesized gold nanoparticles. Semicond. Sci. Technol. **21**(7), 975–978 (2006). doi:10.1088/0268-1242/21/7/025
104. Yim, S.-S., Lee, M.-S., Kim, K.-S., Kim, K.-B.: Formation of Ru nanocrystals by plasma enhanced atomic layer deposition for nonvolatile memory applications. Appl. Phys. Lett. **89**(9), 093115–093113 (2006)
105. Kanjilal, A., Hansen, J.L., Gaiduk, P., Larsen, A.N., Cherkashin, N., Claverie, A., Normand, P., Kapelanakis, E., Skarlatos, D., Tsoukalas, D.: Structural and electrical properties of silicon dioxide layers with embedded germanium nanocrystals grown by molecular beam epitaxy. Appl. Phys. Lett. **82**(8), 1212–1214 (2003). doi:10.1063/1.1555709
106. Islam, M.A., Herman, I.P.: Electrodeposition of patterned CdSe nanocrystal films using thermally charged nanocrystals. Appl. Phys. Lett. **80**(20), 3823–3825 (2002). doi:10.1063/1.1480878
107. Baker, J.L., Jimison, L.H., Mannsfeld, S., Volkman, S., Yin, S., Subramanian, V., Salleo, A., Alivisatos, A.P., Toney, M.F.: Quantification of thin film crystallographic orientation using X-ray diffraction with an area detector. Langmuir **26**(11), 9146–9151 (2010). doi:10.1021/la904840q
108. Baker, J.L., Widmer-Cooper, A., Toney, M.F., Geissler, P.L., Alivisatos, A.P.: Device-scale perpendicular alignment of colloidal nanorods. Nano Lett. **10**(1), 195–201 (2010). doi:10.1021/nl903187v
109. Redl, F.X., Cho, K.S., Murray, C.B., O'Brien, S.: Three-dimensional binary superlattices of magnetic nanocrystals and semiconductor quantum dots. Nature **423**(6943), 968–971 (2003). doi:10.1038/nature01702
110. Akey, A., Lu, C.G., Yang, L., Herman, I.P.: Formation of thick, large-area nanoparticle superlattices in lithographically defined geometries. Nano Lett. **10**(4), 1517–1521 (2010). doi:10.1021/nl1001291

111. Baranov, D., Fiore, A., van Huis, M., Giannini, C., Falqui, A., Lafont, U., Zandbergen, H., Zanella, M., Cingolani, R., Manna, L.: Assembly of colloidal semiconductor nanorods in solution by depletion attraction. Nano Lett. **10**(2), 743–749 (2010). doi:10.1021/nl903946n
112. Zanella, M., Bertoni, G., Franchini, I.R., Brescia, R., Baranov, D., Manna, L.: Assembly of shape-controlled nanocrystals by depletion attraction. Chem. Commun. **47**(1), 203–205 (2011). doi:10.1039/c0cc02477e
113. Bishop, K.J.M., Wilmer, C.E., Soh, S., Grzybowski, B.A.: Nanoscale forces and their uses in self-assembly. Small **5**(14), 1600–1630 (2009). doi:10.1002/smll.200900358
114. Podsiadlo, P., Sui, L., Elkasabi, Y., Burgardt, P., Lee, J., Miryala, A., Kusumaatmaja, W., Carman, M.R., Shtein, M., Kieffer, J., Lahann, J., Kotov, N.A.: Layer-by-layer assembled films of cellulose nanowires with antireflective properties. Langmuir **23**(15), 7901–7906 (2007). doi:10.1021/la700772a
115. Srivastava, S., Kotov, N.A.: Composite layer-by-layer (LBL) assembly with inorganic nanoparticles and nanowires. Acc. Chem. Res. **41**(12), 1831–1841 (2008). doi:10.1021/ar8001377
116. Wu, Z., Lee, D., Rubner, M.F., Cohen, R.E.: Structural color in porous, superhydrophilic, and self-cleaning SiO2/TiO2 Bragg stacks. Small **3**(8), 1445–1451 (2007). doi:10.1002/smll.200700084
117. Rigueur, J.L., Hasan, S.A., Mahajan, S.V., Dickerson, J.H.: Buckypaper fabrication by liberation of electrophoretically deposited carbon nanotubes. Carbon **48**(14), 4090–4099 (2010). doi:10.1016/j.carbon.2010.07.016
118. Hasan, S.A., Rigueur, J.L., Harl, R.R., Krejci, A.J., Gonzalo-Juan, I., Rogers, B.R., Dickerson, J.H.: Transferable graphene oxide films with tunable microstructures. ACS Nano **4**(12), 7367–7372 (2010). doi:10.1021/nn102152x
119. Hasan, S.A., Kavich, D.W., Dickerson, J.H.: Sacrificial layer electrophoretic deposition of free-standing multilayered nanoparticle films. Chem. Commun. **7**(25), 3723–3725 (2009). doi:10.1039/b902622c
120. Hasan, S.A., Kavich, D.W., Mahajan, S.V., Dickerson, J.H.: Electrophoretic deposition of CdSe nanocrystal films onto dielectric polymer thin films. Thin Solid Films **517**(8), 2665–2669 (2009). doi:10.1016/j.tsf.2008.10.122

Chapter 4
Electrophoretic Deposition of Carbon Nanotubes (CNTs) and CNT/Nanoparticle Composites

Aldo R. Boccaccini, Cengiz Kaya and Milo S. P. Shaffer

4.1 Introduction

Electrophoretic deposition (EPD) is a well-known colloidal processing method which is gaining increasing interest as a simple and versatile technique for the manipulation and deposition of nanoparticles and carbon nanotubes (CNTs) [1–3]. The technique allows the fabrication of coatings, thin and thick films, the shaping of bulk and free-standing objects and the infiltration of porous substrates with ceramic (nano)particles [1–5]. EPD is achieved via the motion of charged particles, dispersed in a suitable solvent or aqueous solution, towards an electrode under an applied electric field. Electrophoretic motion of charged particles during EPD results in the accumulation of particles and the formation of a relatively homogeneous deposit on the electrode. The success of EPD is based not only on its wide applicability to numerous materials and combinations of materials, but also to the fact that it is a fast, cost-effective technique requiring simple equipment. Comprehensive reviews on the applications of EPD in conventional materials processing are available [3–5]. In addition, EPD is being increasingly considered as one of the processing methods of choice in the field of nanomaterials [1, 2].

CNTs, often used in combination with other nanoparticles, are materials of choice for the synthesis of a variety of advanced (nano)structures, including functional thick and thin films and coatings, nanocomposites, field emission devices, catalyst supports, bioactive materials, as well as heterostructures and components for a variety of electrochemical and electromechanical devices [6–8]. Due to their extraordinary properties, which are the result of their unique structure, aspect ratio and size, CNTs attract enormous attention in the field of nanotechnology [9]. The remarkable properties of CNTs are well established [8–11] and there are increasing efforts devoted to the exploitation of these properties in specific applications [9–11]. One key challenge is to tackle the strong tendency of CNTs to agglomerate,

A. R. Boccaccini (✉)
Department of Materials Science and Engineering, University of Erlangen-Nuremberg, 91058 Erlangen, Germany
e-mail: aldo.boccaccini@ww.uni-erlangen.de

J. H. Dickerson, A. R. Boccaccini (eds.), *Electrophoretic Deposition of Nanomaterials*, Nanostructure Science and Technology,
DOI 10.1007/978-1-4419-9730-2_4,

which frequently limits the ability to manipulate CNTs into the particular arrangement required for a given application. For example, in the case of CNT composites destined for structural applications, it is essential to develop processing methods that produce a homogeneous dispersion of CNT in the appropriate (polymeric, ceramic or metallic) matrix or alternatively, an ordered array of CNTs on a surface of interest. The intrinsic challenge of maintaining CNT dispersions both during processing and in the resulting microstructure is a significant area of research in the materials science community [12].

EPD represents a convenient technique for manipulating individual CNTs in liquid suspensions with the aim of producing ordered thin layers, films and coatings [13], as discussed in the present Chapter. Increasingly, research efforts are being devoted to the exploitation of CNT EPD for a variety of specific applications such as field emission devices, supercapacitors, bioactive materials and photocatalytic coatings; these studies confirm the excellent capability of EPD to manipulate, arrange and orientate both multi-walled (MWCNTs) and single-walled (MWCNTs) nanotubes [14–18]. In this context, arrangements of CNTs, exhibiting high aspect ratio and surface charge, are also suitable substrates for inclusion or entrapment of other (functional) nanoparticles to form CNT nanocomposites [1].

This Chapter comprehensively describes the field of CNT and CNT-inorganic nanocomposite EPD, discussing in detail some of the typical systems developed recently. The relevant experimental characteristics, such as CNT functionalisation, the design of (multi-component) suspensions for EPD, as well as the specific EPD parameters, such as deposition time, electric field and electrode materials, are discussed, addressing also the effects of processing conditions on the final quality of CNT-nanoparticle composites.

4.2 CNT and Ceramic Particle Suspensions for EPD

A detailed discussion of the preparation and characterisation of CNT suspensions for use in EPD has been presented in a previous review paper [13]. In this section, we include a brief discussion of the key aspects involved in the development of stable CNT suspensions using suitable solvents which can be combined with ceramic nanoparticle suspensions for EPD of CNT-ceramic nanocomposites. A summary of suspensions developed for EPD of multi-walled and single-walled CNT is presented in Table 4.1 [19–35].

It is evident that several types of solvents have been employed, including distilled water, acetone and ethanol mixtures and several other organic solvents such as isopropyl alcohol, n-pentanol, ethyl alcohol, tetrahydrofuran (THF) and dimethlyformamide (DMF) [13].

It is well known that as-produced CNT are intrinsically inert, often aggregated or entangled, and may contain impurities (such as amorphous carbon or catalytic metal particles) [36]. A post-synthesis treatment is usually required to purify and

Table 4.1 Examples of suspensions used for CNT EPD

CNT type	Suspension Properties	Ref.
MWCNT	Aqueous solutions of 0.25 and 0.55 mg/mL (CNT/H_2O)	[19]
SWCNT	10 mg of SWNTs mixed with 0.13 g of tetraoctylammonium bromide (TOAB) were dried and resuspended in 25 ml of Tetrahydrofuran (THF) + quaternary ammonium salt	[20]
SWCNT	10 mg of SWNTs with 100 mg of tetraoctylammonium bromide (TOAB) or Nafion in 25 ml of Tetrahydrofuran (THF)	[21]
MWCNT	Ethanol with EPI-Rez resin and EPI-CURE curing agent (aliphatic amine)	[22]
MWCNT	Mixtures of acetone and ethanol in different volume ratios	[23]
MWCNT	0.125 mg/ml (CNT/isopropyl alcohol) and small amount of bezalkonium chloride	[24]
SWCNT	Isopropyl alcohol with $NiCl_2$	[25]
MWNTs	Isopropyl alcohol solution containing $Mg(NO_3)_2$; ethyl cellulose was also added to improve dispersion	[26]
SWCNT	De-ionised water mixed with pyrrole and lithium perchlorate	[27]
SWCNT	Ethyl alcohol with $MgCl_2$	[28]
SWCNT	Tetraoctylammonium broamide (TOAB) in tetrahydrofuran (THF)	[29]
SWCNT	Tetraoctylammonium broamide (TOAB) in tetrahydrofuran (THF)	[30]
SWCNT	Methanol or dimethylformamide (DMF) with < 1 wt% concentrated sodium hydroxide	[31]
SWCNT	Distilled water with $Mg(NO_2)_2{\cdot}6H_2O$	[32]
MWCNT	Isopropyl alcohol (IPA)	[33]
MWCNT	2 g of CNTs mixed with 500 ml n-pentanol with a small amount of anhydrous $Mg(NO_3)_2$	[34]
MWCNT	Isopropyl alcohol (IPA)	[18]
SWCNT	Deionised water or butyl alcohol and surfactants (sodium dodecyl sulphate, hexadecyl trimethyl ammonium bromide or trioctylphosphine oxide	[35]

disperse the CNT in a suitable solvent. Typical treatments involve various combinations of thermal gas-phase oxidations, liquid phase oxidations with nitric acid or other reagents, treatment with HCl to remove catalytic metal contaminants, and (vacuum) thermal annealing. In addition, surfactants can be used to disperse CNT using sonication [37] either as part of the purification process, or as a means to produce a stable dispersion. One particularly common approach uses reflux in a mixture of concentrated nitric and sulphuric acids simultaneously to purify, shorten and functionalise CNTs [20, 38]. This treatment reacts primarily at defect sites in the CNT atomic framework, producing shortened CNTs decorated with carboxylic acid and other oxygen-containing groups. These acidic groups electrostatically stabilize CNTs in water or in other polar liquids by developing a negative surface charge, which leads to a significant increase in the stability of the suspension, making it suitable for EPD [39]. More recent data highlights the presence of adsorbed oxidation debris on the oxidised CNT surface, which is readily removed by treatment with aqueous base, at the cost of reducing the effective concentration of surface functional groups [40].

As mentioned previously, dispersing CNT homogeneously in a suitable solvent is the first step leading to controlled manipulation of CNT by electrophoresis. The main aim of using EPD is thus to avoid CNT agglomeration and entanglement to produce dense, locally ordered CNT films [13, 20, 41]. The stability of CNT suspensions, determined by measuring ζ-potential values, has been considered mainly in aqueous and ethanol-based suspensions, as reviewed in the literature [13]. As-produced CNT have usually low and positive ζ-potential in the acidic region with an isoelectric point in the pH range 5–8 [19]. After oxidation, the presence of surface acid groups decreases the isoelectric point to values often <pH 2. The adhesion of CNT to solid substrates, the suspension stability, and the deposition rate during EPD can be improved by including charger salts in the suspension [24, 25, 28, 37]. It is also evident that the direction of CNT movement during EPD is determined by the surface charge: oxidised nanotubes are typically negatively-charged and they migrate to the positive electrode in the EPD cell [19].

4.3 EPD of CNT

Since electric fields can align individual CNT, it is possible to obtain oriented arrays of CNTs by EPD and, indeed, this effect was one of the original motivations for applying electric fields to CNT suspensions [35, 42, 43]. Early studies by Yamamoto et al. [43] investigated AC electrophoresis to orient MWCNTs of 1–5 μm length and 5–20 nm in diameter suspended in isopropyl alcohol (IPA). MWCNTs in IPA are positively charged and covered with negative dipoles of the IPA carboxyl groups. At low frequency MWCNTs were observed to move towards the negative electrode, however at high frequencies negative ions surrounding MWCNTs were not able to move in response to the change of the polarity of the electric field. Under these conditions an electrical dipole forms in CNTs. Since the Coulomb force attracting the dipole is stronger for the nearer electrode than for the other electrode, the dipole moves towards the nearer electrode [43]. The degree of orientation of CNTs in such conditions depends on CNT length and on the frequency of the electric field, increasing with increasing frequency (up to 10 MHz) [42]. Most research conducted on oriented CNTs by electrophoresis has been focused on developing CNT arrangements for field emission applications [18, 31, 32, 35, 43]. In this context, the combination of EPD and patterned substrates has led to the best results in terms of the deposition of well-oriented CNT emitters for field emission display devices [32].

The application of EPD to the fabrication of CNT films or coatings on both conductive and insulating substrates (in this case applied as an infiltration technique) has focussed mainly on developing relatively thick (>5 μm) porous layers of non aligned CNT. Figures 4.1a, b show the upper surface and the cross section of a typical MWCNT film produced by EPD from aqueous suspensions [41], where the isotropic appearance of the porous deposit can be observed exhibiting a degree of two-dimensional orientation of the CNTs parallel to the electrode surface. Experiments evidence suggests [19] that the final morphology of the EPD CNT layers is

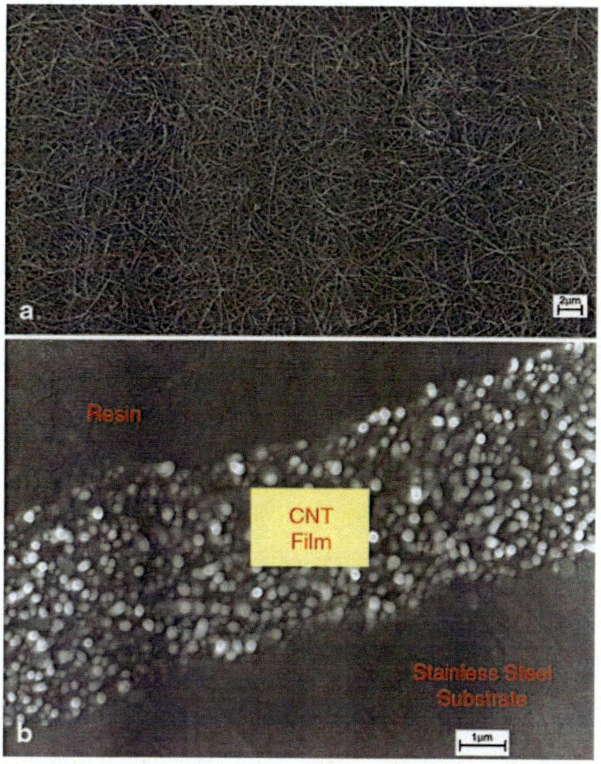

Fig. 4.1 SEM images of CNT films obtained by EPD on stainless steel substrates: **a** surface view and **b** cross sectional view. The CNT film was produced at 20 V/cm for 5 min. ([41] Reproduced with permission of Elsevier)

generated during drying of the gel-like initial deposit, leading to a large out-plane contraction.

Figure 4.2 shows a schematic diagram of the typical arrangement for EPD of CNT under DC conditions, featuring an EPD cell with vertical parallel electrodes. In DC EPD, the voltage is usually kept constant for the duration of the experiment. Once EPD has been completed, the coated electrodes are carefully removed from the EPD cell to avoid any influence of a drag force between the wet coating and the remaining suspension and are usually dried horizontally in air at room temperature. As mentioned above, CNTs that have been treated by acid oxidation and dispersed in water acquire a negative surface charge in neutral solution [20] and deposit on the anode during EPD.

For EPD of MWCNTs from aqueous suspensions, deposits of adequate quality in terms of uniform thickness and homogeneous covering of planar metallic electrodes have been obtained by applying electric fields >20 V/cm [41]. The minimum threshold for a coherent, robust film may be due to the larger number of contacts generated between more densely-packed CNTs, as the individual CNTs are drawn closer to each other at higher fields [18]. Aqueous suspensions have the advantages of low-cost processing, lower electric potential requirements and better environmental credentials than organic suspensions. However, the use of aqueous

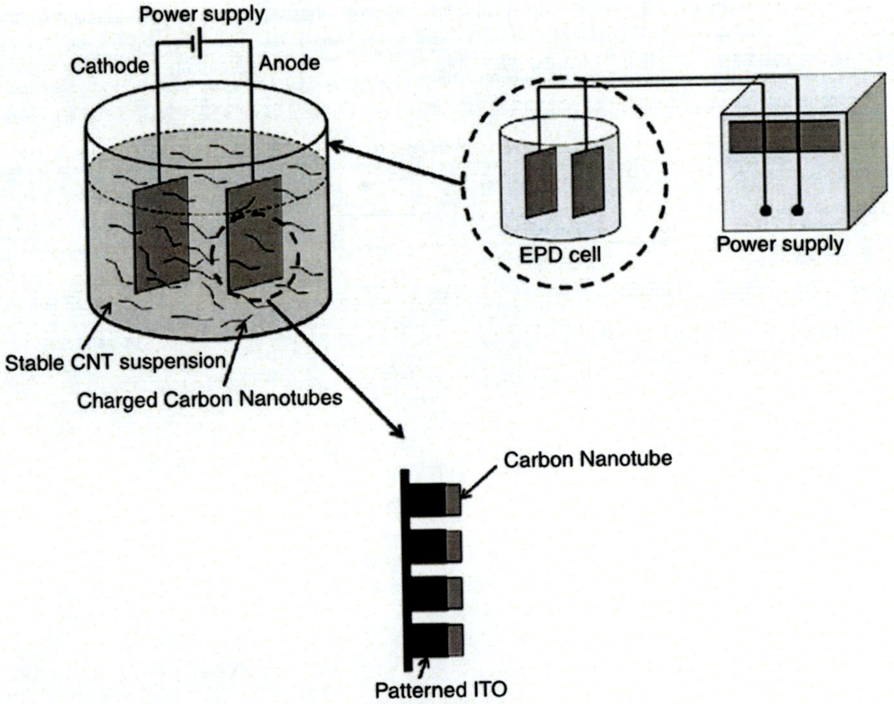

Fig. 4.2 Schematic diagram of the typical and simple EPD cell for processing CNT, showing also the option of depositing CNT onto a patterned ITO substrate, according to ref. [32]

suspensions can suffer the potential problem of gas evolution by water electrolysis (1.23 V). Hydrogen gas evolution occurs at the cathode while oxygen gas is generated at the anode. These effects can be apparent at higher electric fields (>35 V/cm), at lower pHs, and for long deposition times: excessive hydrogen gas evolution has been seen to occur at the cathode [41]. Similar results were reported by Thomas et al. [19], who showed that at relatively long EPD times (>5 min) MWCNT aggregation occurred, resulting in a large scatter of the yield and poor homogeneity of the deposits. Single, shorter EPD runs led to CNT films of ~3 μm [19]; several successive depositions on the same electrode resulted in homogeneous films of up to 100 μm in thickness, however the reason why short depositions avoid the problems of one long deposition have not been discussed [44].

The EPD of MWCNTs from ethanol/acetone suspensions on metallic substrates was first reported by Du et al. [23]. It was shown that the volume ratio of acetone-to-ethanol affected the microstructure of the electrophoretic CNT film. The EPD fabrication of CNT-reinforced resin composites was reported by the same group [22]. An ethanol solution containing both the resin and the curing agent (aliphatic amine) was sonicated for 30 min to form a stable suspension. A constant deposition voltage of 45 V was used for EPD, with aluminium electrodes placed 50 mm apart. This early work demonstrated the successful use of EPD for developing polymer/

CNT composite coatings and, in principle, the method could be applied to a range of polymer matrices. Recent investigations have shown, for example, the development of MWCNT/polyimide composite films by EPD of MWCNT-polyamic acid colloidal suspension which was derived from carboxylated MWCNTs and poly(pyromellitic dianhydride-*co*-4,4′-oxydianiline) (PMDA-ODA). Under an electric field, both negatively charged MWCNT and PMDA-ODA colloid particles migrate to the anode simultaneously, and are converted to a coherent MWCNT/polyimide composite film [45].

Research work has been also carried out on EPD of oxidized or surfactant-modified SWCNTs. Jin et al. [27] co-deposited pyrrole and SWCNTs on indium-tin-oxide (ITO) from an aqueous suspension, fabricating a 1 μm thick composite film where the CNTs were perpendicularly oriented to the substrate; in this case, the pyrrole polymerises at the surface producing a composite film. In-situ deposition and electropolymerisation processes have been also reported for MWCNTs [46]. Chemically-oxidized SWCNTs have been used to fabricate CNT patterns exhibiting 20 μm features on ITO-coated and plain glass [28]. Functionalised SWCNTs were stabilised in $MgCl_2$/ethanol and deposited on ITO-coated glass. The nanotubes were seen to strongly adhere to the ITO coating, which was attributed to two factors: (1) interaction between the hydrophilic CNT and the ITO surface [28, 47] and (2) the presence of the charger salt, $MgCl_2$, since Mg^{2+} ions form hydroxides at the surface of the negative electrodes which assist interfacial bonding [47]. Girishkumar et al. [48] have used EPD to deposit SWCNT thin films modified with tetraoctylammonium bromide (TOAB) in tetrahydrofuran (THF) on aminopropyltriethoxysilane (APS) coated optically transparent electrodes (OTE) made of conductive glass. The intended application of these SWCNT layers is for electrodes in portable fuel cells, in particular for methanol oxidation and oxygen reduction reactions. The same researchers [21] have reported the fabrication of a SWCNT-based membrane electrode assembly for hydrogen fuel cells by EPD.

As indicated above, beyond the fabrication of planar CNT-based coatings and films, EPD can be applied to deposit CNT onto curved or complex structures, including microwires, fibres, porous substrates and textile structures [13, 49]. The fabrication of more complex patterns of CNT deposits by EPD can be realised by using masks or by designing combinations of conductive and non-conductive surfaces as deposition electrodes. Thick electrophoretic CNT films, displaying relatively high packing density, also exhibit some degree of flexibility [50].

In recent experiments, functional interphases based on MWCNTs in glass fibre/epoxy composites have been developed by EPD [51]. The deposition process of MWCNTs onto glass fibre surfaces by EPD leads to a more homogeneous and continuous CNT distribution on the glass fibre surface than a conventional dip-coating method. It was shown that interphases formed by MWCNT, mimicking a biological fibrous nanostructure, can greatly improve the interfacial shear strength of the composites. In addition, it was found that the semiconductive interphase results in a high sensitivity of the electrical resistance of single glass fibre model composites to strain. These novel composites provide a possible in-situ mechanical load sensor to achieve early warning of fibre composite damage [51]. In the field of ceramic matrix composites

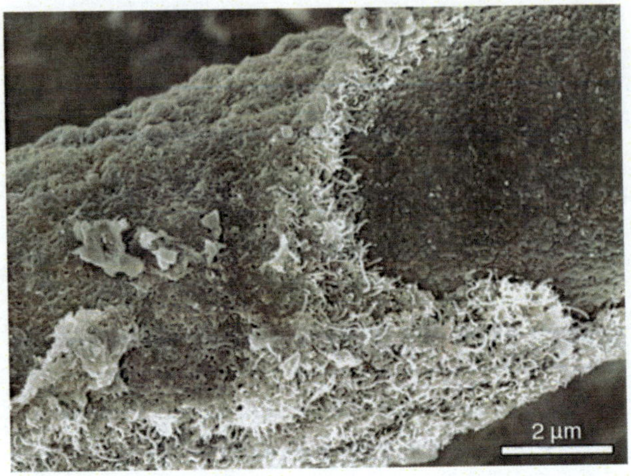

Fig. 4.3 SEM micrograph of the fracture surface of a CNT-coated SiC fibre after SiC deposition by electrophoresis. ([49] Reproduced with permission of Elsevier)

based on SiC fibres, the homogeneous CNT coating of SiC fibres provided by EPD may improve the interfacial strength and enhance the load transfer capability of the composites. Figure 4.3 demonstrates the electrophoretic CNT coating on a SiC fibre obtained in a recent investigation [49]. With increasing interest in the use of CNT coated reinforcing fibres in composites [52], it is highly likely that developments of the EPD technique in this application area will continue to flourish.

The results discussed in this Section demonstrate that manipulation of CNT by EPD is a very attractive approach for a wide range of applications, which is expected to attract further research efforts in the near future. In this context, EPD represents a powerful method to assist the fabrication of CNT-based devices and composite materials, particularly because there are few alternatives for depositing (and aligning) CNT on the planar and curved substrates required. Similarly, EPD of CNT represents a very effective process to create CNT membranes and nanofilters which are otherwise fabricated by slow and tedious filtration of CNT suspensions.

4.4 EPD of CNT-Nanoparticle Composites

EPD can be employed to deposit ceramic or metallic nanoparticles onto or into pre-existing CNT films [1]. In this approach, EPD can be used either to infiltrate the porous CNT structure with nanoparticles or to fabricate layered composite heterostructures. Composite CNT/nanoparticulate coatings can be obtained also by co-deposition from stable suspensions containing CNTs and one or more other components (microsized or nanoscale particles). The various components in suspension may be separately dispersed, coming together only during EPD, or may be agglomerated forming composite building blocks in suspension (Fig. 4.4) [1]. In this last case, nanoparticles can be made to uniformly coat individual CNTs by tai-

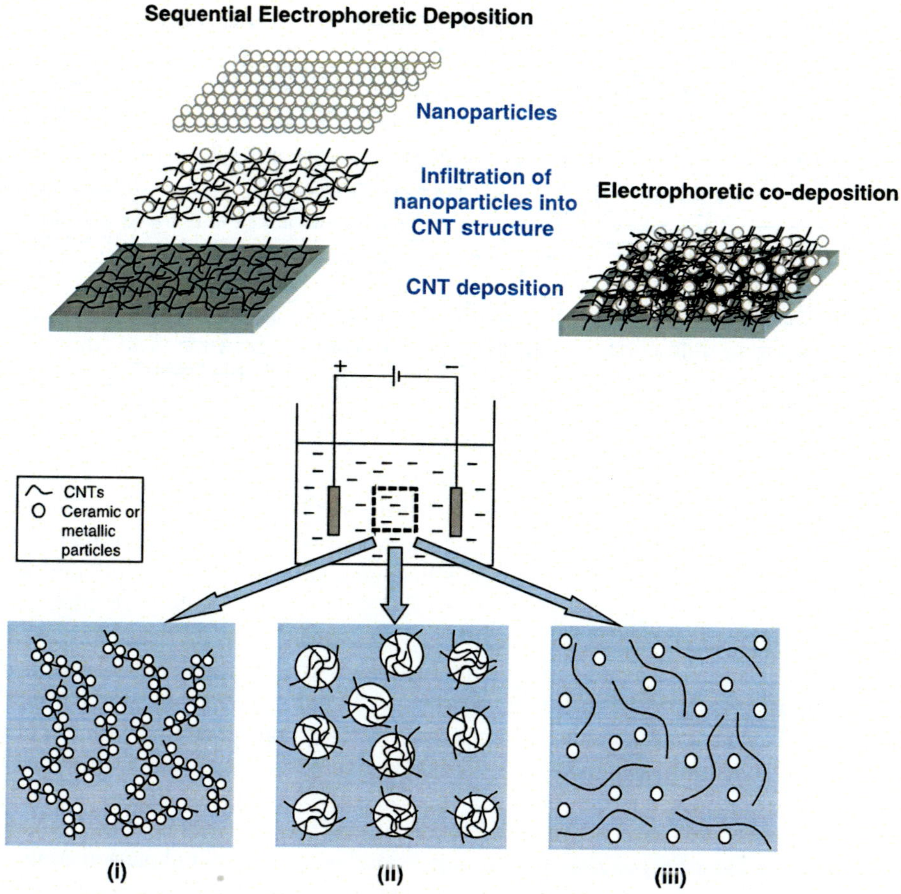

Fig. 4.4 Fabrication routes for CNT/nanoparticle composites by EPD: **a** sequential deposition of CNTs and nanoparticles to form layered heterostructures and electrophoretic co-deposition, **b** different alternatives to produce CNT/particulate composites by electrophoretic co-deposition: (*i*) self-assembly of nanoparticles coating individual CNTs, (*ii*) heterocoagulation of CNTs onto larger particles, (*iii*) simultaneous deposition of CNTs and ceramic (or metallic) particles exhibiting the same charge polarity in suspension. (According to ref. [1] Published with permission of Elsevier)

loring their respective zeta-potential values, thus promoting heterocoagulation and self-assembly of the different species in suspension. Alternatively, if large (micron-sized) particles are used, CNTs can coat individual particles forming charged CNT-coated composite particulates in suspension, which will migrate to the deposition electrode upon application of the electric field. Schematic diagrams showing the different EPD approaches are shown in Fig. 4.4 [1]. The following sections include specific examples of these different EPD routes to fabricate CNT-containing nanocomposites.

4.4.1 SiO_2/CNT Composites

The combination of silica nanoparticles with MWCNTs in EPD was investigated by Chicatun et al. [53], who prepared SiO_2/CNT composite films on metallic substrates. Commercially-available MWCNTs were used without any post-synthesis treatment. Aqueous dispersions of hydrophilic fumed silica were used as the source of SiO_2. Several surfactants were investigated for their suitability to produce stable CNT suspensions, including octylphenolpoly(ethyleneglycolether)$_x$ (Triton X-100), polyacrylic acid (PAA), cetyltrimethyl ammonium bromide (C16TMAB) and sodium dodecyl sulfate (SDS). Iodine (99,999%) was used as a charge promoter. The suspension for electrophoretic co-deposition of CNTs and SiO_2 nanoparticles was prepared by mixing an aqueous MWCNT suspension with different concentrations of silica nanoparticles. EPD involved different applied voltages in the range 20–55 V and constant deposition time (in the range 1–5 min). Layered CNT/SiO_2 porous composites were obtained by sequential EPD experiments alternating the deposition of CNTs and SiO_2 nanoparticles [53]. After deposition, the first CNT film was dried in normal air for 24 h and the CNT-coated electrode was used to deposit the silica nanoparticle layer. As in all colloidal processing methods, in order to reduce crack formation in the films, controlled drying methods are required. SEM results showed that CNT were efficiently mixed with SiO_2 nanoparticles to form a composite network structure [53]. Possible applications of MWCNT/SiO_2 films are as porous coatings in the biomedical field, thermal management devices, biomedical sensors and other functional applications [54]. CNT/silica nanocomposites can be also attractive for field emission devices. For this application, novel Ag/SiO_2/CNT composite films have been produced by pulsed voltage electrophoretic co-deposition [55]. MWCNTs suspended in a mixture of isopropyl alcohol were used combined with ethanol suspensions containing SiO_2 particles (1.55 μm) and Ag particles (0.9 μm). Two types of surfactants, polyethyleneimine and Disperbyk-184 (BYK Chemie, Germany), were added to the suspension as dispersants and surface charger for the CNTs and the suspended particles, respectively. SiO_2 and Ag particles were shown to enhance the bonding between CNTs and ITO glass (used as substrate) and the conductivity of the films, respectively. A pulsed voltage method, instead of conventional constant electric field, was used, as pulsed plating can result in coatings with improved surface appearance and properties [55]. MWCNTs deposited by the pulsed voltage method tended to align uniaxially in the direction of the applied electric field although the reasons for this behaviour have not been discussed in detail. The study of Wang et al. [55] represents thus a novel approach showing the electrophoretic co-deposition of CNTs and two particulate species to produce complex CNT-containing nanocomposites. Future applications of EPD in this field will, no doubt, consider other combinations of ceramic and metallic particles with SWCNT and MWCNT for specific applications.

4.4.2 CNT/TiO_2 Composites

Porous TiO_2 nanostructured coatings are of interest for photocatalytic applications. Titania is also a biocompatible material and titania coatings on stainless steel or titanium alloys can find applications in orthopaedic implants. The fabrication of CNT–TiO_2 composites offers the potential benefit of combining these two materials to increase the photocatalytic effectiveness of TiO_2 [56, 57] and to improve the mechanical properties of TiO_2 films [58]. Early investigations have demonstrated the EPD-based fabrication of homogeneous and thick deposits of CNTs either coated or infiltrated with TiO_2 nanoparticles (P25, Degussa, Frankfurt, Germany) [59]. The coating and infiltration of porous CNT assemblies with TiO_2 nanoparticles by EPD leads to homogeneous incorporation of CNTs in structural and functional titania matrices. CNTs were refluxed in mixed (1:3) concentrated nitric and sulphuric acids at 130°C for 30 min followed by washing in water to pH 7. This acid oxidation treatment produced well-dispersed aqueous CNT suspensions which were then centrifuged to remove remaining agglomerates. For use in EPD, suspensions were further diluted. For the co-deposition of MWCNTs and TiO_2 nanoparticles on stainless steel substrates, Singh et al. [59] prepared suspensions by mixing TiO_2 nanopowder with CNT aqueous solutions of concentration 0.6 mg/ml. Either sodium hydroxide (NaOH) or hydrochloric acid (HCl) was added to influence the suspension pH. At pH=5, it was found that co-deposition occurred due to the opposite signs of the surface charges (zeta potential values) of CNTs and TiO_2 nanoparticles. Thus the EPD mechanism is likely based on the electrostatic attraction between CNTs and TiO_2 nanoparticles which leads to the creation of composite particles in suspension (see Fig. 4.4) consisting of TiO_2 nanoparticles homogenously attached onto the surface of individual CNTs [59]. Under the applied electric field, these net negatively-charged "composite TiO_2/CNT" particles should migrate to and deposit on the anode. The CNT/titania nanostructure is potentially useful for nanoelectronic devices and photocatalytic substrates [60]. After sintering, dense CNT-reinforced TiO_2 structural coatings can be obtained.

TiO_2/CNT composite layers have been also made by sequential deposition of alternating layers of CNT and TiO_2 nanoparticles and by co-deposition [61]. Electrophoretic co-deposition of equally charged CNTs and nanoparticles can be explained by considering the different trajectories of nanoparticles in suspension [61]. The first region occurs in the suspension at a given distance from the working electrode, where both CNT and titania particles are moving towards the deposition electrode under the influence of the externally applied electric field.

The second region occurs close to the deposition electrode where the charge of the CNT deposit influences the motion of incoming charged particles. When CNTs and TiO_2 particles possess the same charge, repulsive forces act between them and particles are repelled before they can reach the surfaces of the charged CNTs. As suggested in the literature [61], under the effect of these repulsive forces the particles will follow the path with the fewest possible obstacles until reaching the next interstice in the CNT deposit. This mechanism of infiltration could be equally applied to the sequential EPD process, where titania nanoparticles initially infiltrate the porous CNT layer and then build up a deposit to form a second layer. Using sequential EPD, four-layer TiO_2/CNT laminate composite coatings have been produced [61]. The CNT layer should act as reinforcement by providing a crack deflection and delamination path. The effect of MWCNT in reducing microcrack formation in EPD TiO_2 films deposited on F-SnO_2 conducting substrates has been investigated [58]. Suspensions of TiO_2 nanoparticles in acetylacetone with addition of iodine were used. The presence of oxidised CNTs was confirmed to reduce the problem of microcracking in TiO_2 films due to bonding with the TiO_2 nanoparticles through the interaction of hydroxyl and carboxyl groups [58].

4.4.3 Fe_3O_4/CNT and Eu_2O_3/CNT Layered Composites

Functional CNT nanocomposites are being developed to integrate CNTs and functional nanoparticles into single structures for a range of different devices for potential applications in next-generation luminescent, magnetic and energy-storage devices [62, 63]. There are three major research topics in this field: functionalization of CNT surfaces with nanoparticles, EPD of alternating CNT and nanoparticle layers and co-deposition of CNTs and nanoparticles to form multilayered films. These heterostructures are relevant for several existing and proposed magnetic, optical and energy-storage devices [64]. The fabrication of functional CNT-nanoparticle layered composites is based on the ability of EPD to deposit homogeneous layers of CNT and nanoparticles on top of one another. As mentioned above, EPD provides substantial control over film thickness and deposition rate enabling a degree of site-selectivity when depositing both nanoparticles and CNTs. In an early published paper [65], iron oxide (Fe_3O_4) nanocrystal/MWCNT composite films were fabricated by EPD on stainless steel and gold substrates. Low field-high current and high field-low current EPD conditions were integrated to produce the composite films. The low field-high current EPD approach produced porous mats from aqueous CNT suspensions, while the high field-low current EPD approach produced tightly-packed nanoparticle films from a dispersion of Fe_3O_4 nanoparticles in hexane. The surface coverage and homogeneity of the iron oxide films improved with repeated deposition on the same CNT layer. Large electric fields applied during EPD of Fe_3O_4 and strong van der Waals interactions among the nanoparticles led to tightly-packed Fe_3O_4 films on the MWCNT porous layer to create CNT/Fe_3O_4 composite films [65]. The thicknesses of the first CNT layer, the Fe_3O_4 film, and the second CNT layer were approximately 6 μm, 150 nm and 1 μm, respectively. Recently, this work has been expanded to fabricate by EPD Eu_2O_3/CNT composites for optical and energy-storage device applications [66]. Figure 4.5a, b shows the top view and cross section of a three-layered heterostructure sandwich (CNTs:Eu_2O_3 nanocrystal:CNTs).

To fabricate this structure, the oxide nanocrystal film coated CNT mat was employed as the anode to deposit a second CNT mat by EPD (Fig. 4.5a). A significant difference in the thickness of the two CNT mats was observed for the same EPD

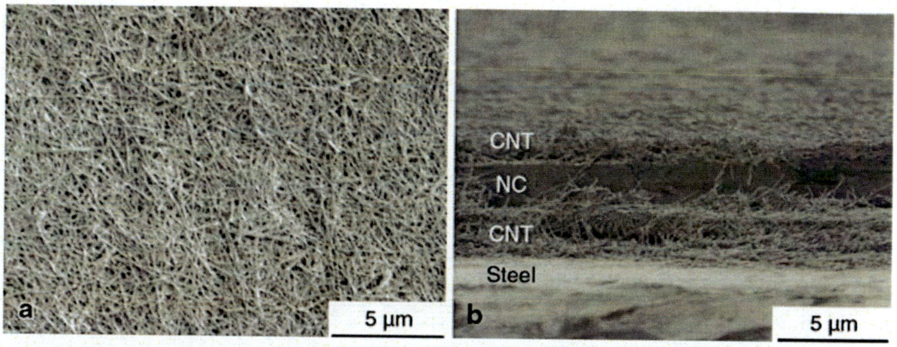

Fig. 4.5 **a** Top view and **b** side view SEM images of a composite tri-layered heterostructure: CNT mat-europium oxide nanocrystal film-CNT mat, produced by EPD. ([66] Reproduced with permission of Elsevier)

conditions. The first CNT mat was ~2 µm thick, while the second mat was ~500 nm thick, as shown in Fig. 4.5b. During the second CNT deposition, the applied voltage is reduced across the low-conductivity CNT mat-oxide film layer. Hence, the effective EPD electric field is reduced, leading to the thinner CNT mat. The second CNT mat thickness can thus be tuned by varying the EPD parameters.

4.4.4 Manganese Dioxide/CNT Composites

EPD has been applied for fabrication of composite films containing manganese dioxide nanofibers and MWCNTs for application in electrochemical supercapacitors (ES) [67]. Manganese dioxide nanofibers with length in the range 0.1–1 µm and diameter of 2–4 nm exhibiting low crystallinity were combined with commercially-available MWCNTs with average diameter ~15 nm and length ~0.5 µm. EPD was performed from aqueous suspensions containing 0–5 g/L manganese dioxide, 0.1–0.5 g/L sodium alginate and 0–0.5 g/L CNTs. Sodium alginate polyelectrolyte was used as dispersant, charging additive, and binder for EPD of both manganese dioxide nanofibers and MWCNTs. Figure 4.6 shows the nanocomposite produced which exhibited a fibrous, crack free and porous microstructure with pore size in the range 10–100 nm [67].

Composite films containing manganese dioxide and CNTs exhibit higher specific capacitance (SC) compared to pure manganese dioxide films and can be interesting electrode materials for electrochemical supercapacitors [68]. Recently, the synthesis of a new composite electrode based on nanosized-manganese oxide and CNTs by EPD followed by direct spontaneous reduction of MnO^{4-} ions to MnO_2 to form the multi-scaled CNT/MnO_2 electrode has been reported [69].

EPD can thus help to develop MnO_2-based electrochemical capacitors with the highest possible capacitance. More generally, EPD can be applied in the development of high performance nanocomposite electrodes based on CNT and metal oxide nanoparticles for the future generation of electrochemical power sources [69].

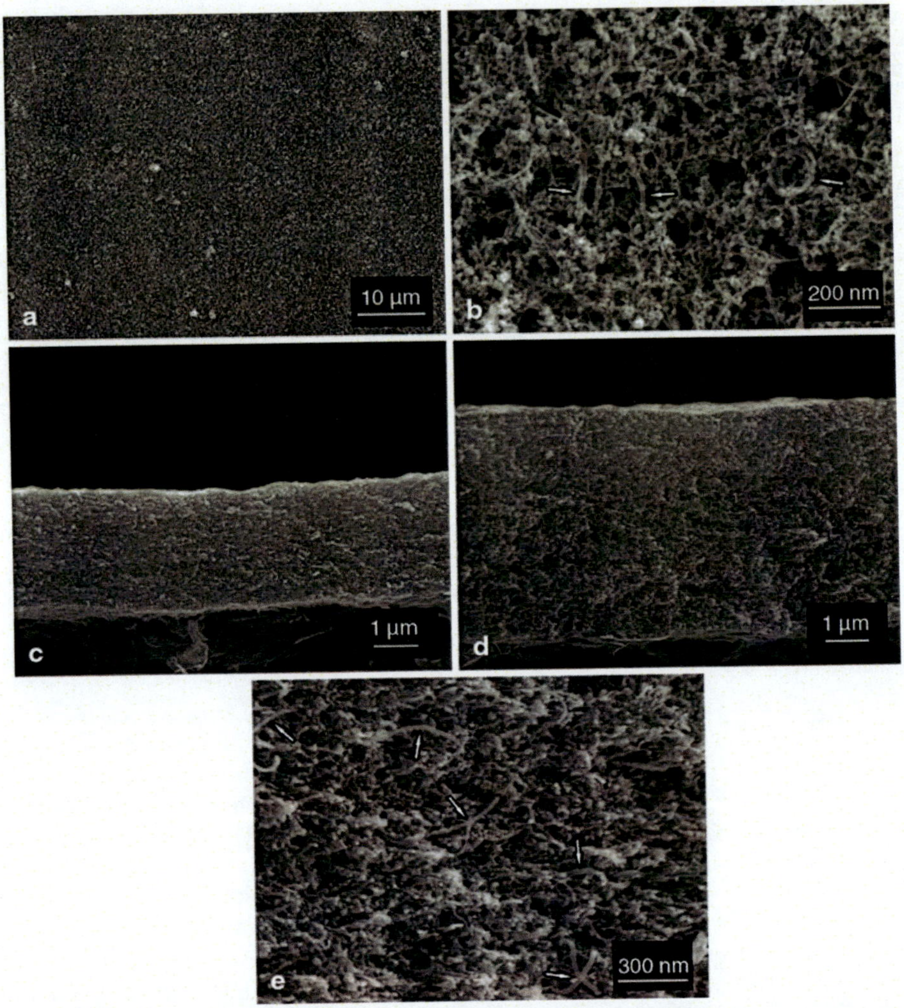

Fig. 4.6 SEM micrographs showing the microstructure of MnO_2/CNT deposits obtained by electrophoretic co-deposition (deposition voltage 15 V) from a sodium alginate solution: **a** and **b** surface of the deposits on a stainless steel substrate at different magnifications, **c** and **d** cross-sections of the deposits of different thickness on graphite substrates and **e** high magnification image of a cross-section (*arrows* show carbon nanotubes). ([67] Images published with permission of Elsevier)

4.4.5 CNT/Inorganic Matrix Bioactive Composites

The development of hydroxyapatite, $Ca_{10}(PO_4)_6(OH)_2$, (HA) and bioactive glass (BG) coatings by EPD is an area of increasing interest [70–72]. The remarkably high mechanical strength and nanoscaled morphology of CNTs make them attractive for biomedical applications, particularly for developing nanofibrous bioactive surfaces in combination with hydroxyapatite or BG [73–75]. Combined with hydroxyapatite and BG, CNTs also promote the formation of bone-like nanostructured calcium phosphate crystals (biomineralisation) when the coatings are in contact with biological fluids [76]. Additionally, CNTs represent optimal reinforcing elements for BG or HA matrices for orthopaedic applications, whereby electrical conductive CNTs can enhance also the function of biomedical coatings, for example for tracking of cells, sensing of microenvironments in addition to enabling nanostructured surfaces for optimal integration with bone tissue [77]. The combination of CNTs with hydroxyapatite or bioactive glass particles to form novel nanostructured inorganic bioactive coatings by using EPD is discussed in the following sections.

4.4.5.1 Hydroxyapatite (HA)/CNT Composites

Synthetic hydroxyapatite (HA) is a biologically-active calcium phosphate commonly used to coat orthopaedic metallic implants or directly as bone replacement material [78]. Bioactive HA promotes bone growth along its surface but it exhibits poor mechanical properties. Composites of HA reinforced by CNTs are being fabricated by EPD [59, 73, 75]. CNT functionalization is required to achieve the desired mechanical reinforcement, based on a good dispersion of CNTs in the HA matrix and on an improved CNT/HA interfacial bond strength. It has been shown that CNT content affects the hardness and elastic modulus of HA/CNT composites, with low CNT concentrations (0.1 wt.%) providing the maximum hardness of 3.46 GPa and elastic modulus of 69.5 GPa [79]. Both plasma spraying [80] and laser surface alloying [75] have been applied to produce HA/CNT composite layers on Ti–6Al–4V alloys. These two techniques, however, are cost-intensive technologies, and it is often difficult to control the coating microstructure and thickness during processing. Research efforts are, therefore, increasingly exploring the use of EPD to develop HA/CNT coating layers [59, 79, 81]. For example, homogeneous HA/CNT composite coatings have been obtained using different applied voltages and EPD times [79]. Moreover, Lin et al. [81] used HA nanoparticles and MWCNTs dispersed in ethanol to coat Ti alloys by EPD, employing an applied voltage of 30 V and a deposition time of 50 s, to obtain a deposit thickness of 10 μm. Densification of the layer was achieved by sintering at 700°C for 2 h in argon. The bonding strength between the substrate and the HA/CNT composite layer improved in comparison to that of the pure HA coating [81]. This result indicates that the presence of CNTs improves the adhesion of the HA coating to the substrate. An alternative method to functionalise CNTs is hydrothermal treatment (HT), as recently suggested for boehmite/CNT mixtures [82]. This method does not give rise to a high concentration of defects on CNT surfaces. HT involves chemical reactions at temperatures greater than room temperature and pressures above 1 atm in a closed system in a liquid medium [82]. Under hydrothermal conditions surface groups, e.g. (–COOH) and (–OH), are generated on the surfaces of CNTs, leading to a good dispersion and formation of stable suspensions for EPD. For the synthesis of HA/CNT mixtures, MWCNTs can be mixed with $Ca(OH)_2$ and H_3PO_4 and then hydrothermally processed at 200°C for 2 h. Subsequently, HA solutions are mixed with MWCNTs and ultrasonicated. Since pristine CNTs have poor wetting properties, carboxylic acid and other oxygen-containing groups on CNT surfaces must be introduced as discussed above. Hydrothermally processed HA/CNT mixtures containing 1 wt.% CNTs have been used for EPD conducted at constant DC voltage for different deposition times. The propagation of cracks induced by indentation of monolithic HA and an HA/CNT composite (2 wt.% CNTs) has shown that toughening mechanisms such as nanotube bridging and debonding appear to be active in the composite. No delamination or detachment of the coating layer was observed, which is ascribed to the presence of the CNT network acting as reinforcement, as has been also observed in TiO_2/CNT composite layers [58]. More recently, EPD of CNTs and HA on a titanium substrate was investigated by Bai et al. [83] showing the effect of applied voltage and deposition time on coating morphology, thickness, and roughness. The CNT–HA coatings were seen to provide efficient protection of the titanium substrate in SBF, promoting also good biocompatibility with osteoblast cells.

4.4.5.2 Bioactive Glass/CNT Composites

Bioactive silicate glasses, for example 45S5 Bioglass® of composition (in weight%): 45% SiO_2, 24.5% Na_2O, 24.5% CaO and 6% P_2O_5 [84] have been used in a range of biomedical applications such as non-load-bearing implants, bioactive coatings, bone cements, bone fillers and tissue engineering scaffolds, due to their excellent bioactivity and biocompatibility [85–87]. In order to resolve the problem of the brittleness and low mechanical strength of bioactive glasses, they are usually applied as coatings on metallic substrates, e.g. Ti alloys or stainless steel, in some cases replacing HA coatings [71, 86]. These coatings improve both the bioactivity and the biocompatibility of implants, as well as protecting the metallic implants and orthopaedic/dental devices from corrosion when in contact with body fluids. EPD has been used to deposit bioactive glass coatings on metallic substrates [71] and EPD was further developed for production of CNT/Bioglass® layered composite films [88, 89]. CNTs of diameter ~20 nm were used. After acid treatment, a well-dispersed aqueous CNT suspension with a concentration of 0.45 mg/ml was prepared. A sequential deposition method was applied to produce a coating of CNTs on bioactive glass layers [88]. Electric field strengths in the ranges 10–40 V/cm and 2–10 V/cm were used for CNT and Bioglass® particle suspensions, respectively, together with deposition times of 1–4 min and 3–6 min, respectively [88]. EPD has been also shown to be an innovative technique to deposit CNTs onto 3D Bioglass® scaffolds; a suitable EPD cell is shown schematically in Fig. 4.7 [76, 90]. SEM images showing the morphology of the deposited CNT layer on Bioglass® scaffolds are included in Fig. 4.8 [90], indicating that CNTs form a compact nanotopography on the scaffold strut surface. Several applications of these novel nanocomposites are possible. For example, the porous CNT network on the Bioglass® surface represents a suitable bioactive nanofibrous and porous substrate to study cell attachment and growth [91, 92]. Moreover incorporation of CNTs on the bioactive glass surface

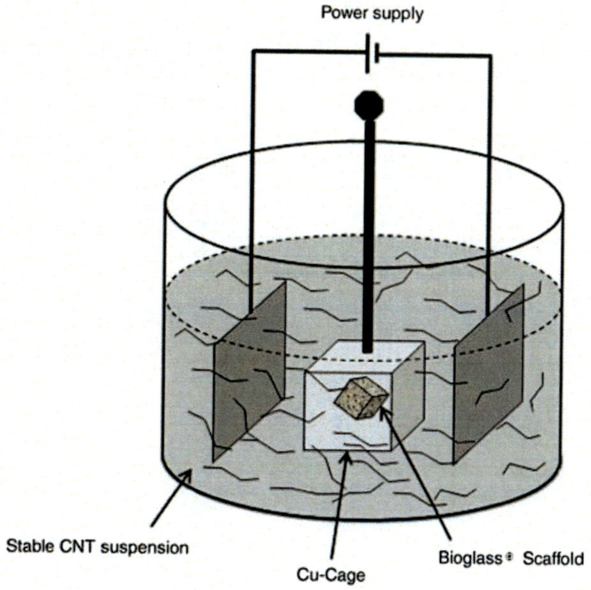

Fig. 4.7 Schematic diagram showing the EPD cell developed to infiltrate porous bodies, e.g. Bioglass® scaffolds, with CNT. (according to ref. [76])

provides improved bioactive function for applications as coatings on orthopaedic implants for strong bonding with bone, similar to those discussed above for HA/CNT composites. Considering biomimetic materials for bone tissue engineering, CNT coated Bioglass® scaffolds produced by EPD represent suitable substrates for rapid growth of nanoscale hydroxyapatite crystals, when the composites are immersed in relevant biological fluids, such as simulated body fluid (SBF) [93]. In this context, the CNT mesh can induce the ordered growth of a nanoscale HA layer leading to a nanostructured topography resembling the structure and composition of natural bone [88, 89]. Thus the presence of CNTs can enhance the bioactive behaviour of the Bioglass® surface, indicating the promise of these scaffolds (as fabricated by EPD) for bone tissue engineering. Clearly, the cellular biocompatibility and in-vivo behaviour of the CNT/Bioglass® composites remain to be investigated, including the assessment of possible toxicity effects of CNTs [77].

4.5 Conclusions

This Chapter has summarised the application of EPD for the ordered deposition of CNTs and CNT-containing nanocomposites for a variety of applications. Successful examples involving combinations of CNTs with SiO_2, TiO_2, MnO_2, Fe_3O_4, HA and bioactive glass, were discussed. EPD has great potential for manipulation of CNTs and for their assembly into ordered layers including thin films and coatings. Moreover, there have been recent reports [94, 95] of anodic EPD for fabrication of

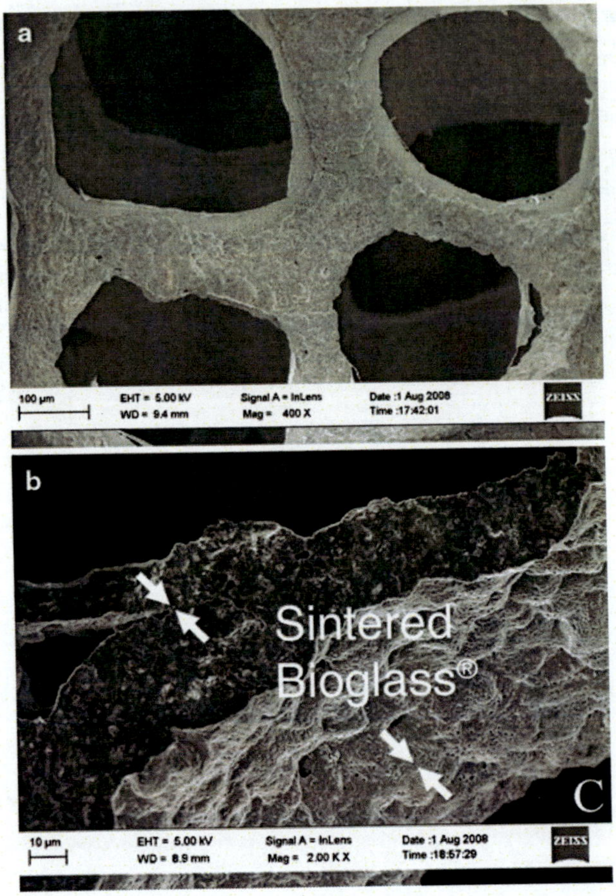

Fig. 4.8 SEM images showing the typical microstructure of a CNT-coated scaffold obtained by EPD (2.8 V, 10 min) at (**a**) low and (**b**) high magnifications. The CNT coating is indicated by the *arrows* in (**b**). ([90] Reproduced with kind permission from Springer Science+Business Media)

composite films containing HA nanoparticles and CNT in biodegradable polymer matrices, which were not discussed here. The current literature confirms that the key step for successful EPD of CNTs is the design of suitable stable suspensions using either aqueous or organic solvents. Based on the ability of EPD to manipulate CNTs, further applications of CNT-containing composite structures are likely to emerge, including, for example,functional CNT-nanoparticle heterostructures for use in field emission displays, nanoelectronics, electrochemical supercapacitors, gas sensors, photocatalytic devices, biomedical platforms, drug delivery systems, antibacterial films, and biosensors. EPD of CNTs is also now considered to be a promising means to fabricate nanoporous functional layers of other materials, for example for solid oxide fuel cells (IT-SOFC). [96]. In this case, CNTs were used as a sacrificial phase that was subsequently removed by a thermal process, leading to a nanoporous active structure.

EPD will play a significant role in the development of advanced CNT/nanoparticle composite nanostructures. Further improvements of the EPD process and better understanding of the kinetics of CNT EPD will help to link EPD parameters to the physical characteristics of the resulting CNT deposits, offering improved control of three dimensional nanostructures containing CNTs, either in the form of dense materials or with a predetermined pore structure. In the future, graded, aligned and patterned features based on predetermined arrays of CNTs may also be obtained by EPD and thus the technique has future potential in a range of functional applications requiring the local manipulation and arrangement of CNTs.

Acknowledgements The authors acknowledge the support of Ms S. Keim (University of Erlangen-Nuremberg, Germany) with some of the illustrations. CK acknowledge financial support from TUBITAK (The Scientific and Technological Research Council of Turkey) under the contract number 108T651.

References

1. Boccaccini, A.R., Cho, J., Subhani, T., Kaya, C., Kaya, F.: Electrophoretic deposition of carbon nanotube–ceramic nanocomposites. J. Europ. Ceram. Soc. **30**, 1115–1129 (2010)
2. Corni, I., Ryan, M.P., Boccaccini, A.R.: Electrophoretic deposition: from traditional ceramics to nanotechnology. J. Europ. Ceram. Soc. **28**, 1353–1367 (2008)
3. Besra, L., Liu, M.: A review on fundamentals and applications of electrophoretic deposition. Prog. Mater. Sci. **52**, 1–61 (2007)
4. Boccaccini, A.R., Zhitomirsky, I.: Application of electrophoretic and electrolytic deposition techniques in ceramics processing. Curr. Opin. Solid. St. M. **6**, 251–260 (2002)
5. Van Der Biest, O., Vandeperre, L.J.: Electrophoretic deposition of materials. Ann. Rev. Mater. Sci. **29**, 327–354 (1999)
6. Eder D.: Carbon nanotube-inorganic hybrids. Chem. Rev. **110**, 1348–1385 (2010)
7. Meyyappan, M.: Carbon nanotubes: science and applications. CRC Press, London (2004)
8. Lau, K. T., Hui, D.: The revolutionary creation of new advanced materials – carbon nanotube composites. Compos. Part B-Eng. **33**, 263–277 (2002)
9. Endo, M., Hayashi, T., Kim, Y. A., Terrones, M., Dresselhaus, M.S.: Applications of carbon nanotubes in the twenty-first century. Phil. Trans. R. Soc. Lond. A. **362**, 2223–2238 (2004)
10. Andrews, R., Jacques, D., Qian, D., Rantell, T.: Multiwall carbon nanotubes: synthesis and application. Acc. Chem. Res. **35**, 1008–1017 (2002)
11. Vairavapandian, D., Vichchulada, P., Lay, M.D.: Preparation and modification of carbon nanotubes: review of recent advances and applications in catalysis and sensing. Anal. Chim. Acta. **626**, 119–129 (2008)
12. Xie, X.-L., Mai, Y.-W., Zhou, X.-P.: Dispersion and alignment of carbon nanotubes in polymer matrix: a review. Mater. Sci. Eng. R. **49**, 89–112 (2005)
13. Boccaccini, A.R., Cho, J., Roether, J.A., Thomas, B.J.C., Minay, E.J., Shaffer, M.S.P.: Electrophoretic deposition of carbon nanotubes. Carbon **44**, 3149–3160 (2006)
14. Kim, S.K., Lee, H.W., Tanaka, H., Weiss P.S.: Vertical alignment of single-walled carbon nanotube films formed by electrophoretic deposition. Langmuir **24**, 12936–12942 (2008)
15. Minnikanti, S., Skeath, P., Peixoto, N.: Electrochemical characterisation of multi-walled carbon nanotube coated electrodes for biological applications. Carbon **47**, 884–893 (2008)
16. Kim, S.-K., Lee, H.: Fabrication of patterned single-walled carbon nanotube film using electrophoretic deposition. Ultramicroscopy **108**, 1005–1008 (2008)
17. Pimanpang, S., Maiaugree, W., Jarernboon, W., Maensiri, S., Amornkitbamrung, V.: Influences of magnesium particles incorporated on electrophoretically multiwall carbon nanotube film on dye-sensitized solar cell performance. Synth. Met. **159**, 1996–2000 (2009)

18. Qin, Y., Hu, M.: Field emission properties of electrophoretic deposition carbon nanotubes film. Appl. Surf. Sci. **255**, 7618–7622 (2009)
19. Thomas, B.J.C., Boccaccini, A.R., Shaffer, M.S.P.: Multi-walled carbon nanotube coatings using electrophoretic deposition (EPD). J. Am. Ceram. Soc. **88**, 980–982 (2005)
20. Shaffer, M.S., Fan, X., Windle, A.H.: Dispersion and packing of carbon nanotubes. Carbon **36**(11), 1603–1612 (1988)
21. Girishkumar, G., Rettker, M., Underhile, R., Binz, D., Vinodgopal, K., McGinn, P., Kamat, P.: Single-wall carbon nanotube-based proton exchange membrane assembly for hydrogen fuel cells. Langmuir. **21**, 8487–8494 (2005)
22. Du, C.S., Heldebrant, D., Pan, N.: Preparation of carbon nanotubes composite sheet using electrophoretic deposition process. J. Mater. Sci. Lett. **21**, 565–568 (2002)
23. Du, C.S., Heldbrant, D., Pan, N. Preparation and preliminary property study of carbon nanotubes films by electrophoretic deposition. Mater. Lett. **57**, 434–438 (2002)
24. Bae, J., Yoon, Y., Lee, S., Baik, H.: Field emission properties of carbon nanotubes deposited by electrophoresis. Physica B. **323**, 169–170 (2002)
25. Kurnosov, D., Bugaev, A.S., Nikolski, K.N., Tchesov, R., Sheshin, E.: Influence of the interelectrode distance in electrophoretic cold cathode fabrication on the emission uniformity. Appl. Surf. Sci. **215**, 232–236 (2003)
26. Zhao, H., Song, H., Li, Z., Yuan, G., Jin, Y.: Electrophoretic deposition and field emission properties of patterned carbon nanotubes. Appl. Surf. Sci. **251**, 242–244 (2005)
27. Jin, Y.W., Jung, J.E., Park, Y.J., Choi, J.H., Jung, D.S., Lee, H.W., et al.: Triode-type field emission array using carbon nanotubes and a conducting polymer composite prepared by electrochemical polymerization. J. Appl. Phys. **92**, 1065–1068 (2002)
28. Oh, S., Zhang, J., Cheng, Y., Shimoda, H., Zhou, O.: Liquid-phase fabrication of patterned carbon nanotube field emission cathodes. Appl. Phys. Lett. **84**, 3738–3740 (2004)
29. Barazzouk, S., Hotchandani, S., Vinodgopal, K., Kamat, P.: Single-wall carbon nanotube films for photocurrent generation. A prompt response to visible light irradiation. J. Phys. Chem. B. **108**, 17015–17018 (2004)
30. Kamat, P., Thomas, K., Barazzouk, S., Girishkumar, G., Vinodgopal. K., Meisel, D.: Self-assembled linear bundles of single wall carbon nanotubes and their alignment and deposition as a film in a dc field. J. Am. Chem. Soc. **126**, 10757–10762 (2004)
31. Gao, B., Yue, G.Z., Qiu, Q., Cheng, Y., Shimoda, H., Fleming, L., et al.: Fabrication and electron field emission properties of carbon nanotube films by electrophoretic deposition. Adv. Mater. **13**, 1770–1773 (2001)
32. Choi, W.B., Jin, Y.W., Kim, H.Y., Lee, S.J., Yun, M.J., Kang J.H., et al.: Electrophoresis deposition of carbon nanotubes for triode-type field emission display. Appl. Phys. Lett. **78**, 1547–1549 (2001)
33. Nakayama, Y., Akita, S.: Field-emission device with carbon nanotubes for a flat panel display. Synth. Met. **117**, 207–210 (2001)
34. Yu, K., Zhu, Z., Li, Q., Lu, W.: Electronic properties and field emission of carbon nanotube films treated by hydrogen plasma. Appl. Phys. A. **77**, 811–817 (2003)
35. Lin, C., Chen, Y.-C., Wang, T., Kuo, C.-T.: Feasibility study of high performance field emitter pattern with the horizontally oriented carbon nanotubes by electrophoresis. Diamond Rel. Mater. **18**, 520–523 (2009)
36. Singh, C., Shaffer, M.S., Windle, A.H.: Production of controlled architectures of aligned carbon nanotubes by an injection chemical vapour deposition method. Carbon **41**, 359–368 (2003)
37. Moon, J.M., An, K.H., Lee, Y.H., Park, Y.S., Bae, D.J., Park, G.S.: High-yield purification process of single walled carbon nanotubes. J. Phys. Chem. B. **105**, 5677–5681 (2001)
38. Esumi, K., Ishigami, M., Nakajima, A., Sawada, K., Honda, H.: Chemical treatment of carbon nanotubes. Carbon **34**(2), 279–281 (1996)
39. Du, C., Yeh, J. Pan, N.: Carbon nanotube thin films with ordered structures. J. Mater. Chem. **15**, 548–550 (2005)

40. Fogdena, S., Verdejo, R., Cottam, B., Shaffer, M.S.: Purification of single walled carbon nanotubes: The problem with oxidation debris. Chem. Phys. Lett. **460**, 162–167 (2008)
41. Cho, J., Konopka, K., Rożniatowski, K., Eva García-Lecina, E., Shaffer, M.S., Boccaccini, A.R. Characterisation of carbon nanotube films deposited by electrophoretic deposition. Carbon **47**, 58–67 (2009)
42. Quale, S.L., Talbot, J.B. Electrophoretic deposition of substrate-normal-oriented single-walled carbon nanotube structures. J. Electrochem. Soc. **154**, 25–28 (2007)
43. Yamamoto, K., Akita, S., Nakayama, Y.: Orientation and purification of carbon nanotubes using AC electrophoresis. J. Phys. D: Appl. Phys. **31**, L34–L36 (1998)
44. Thomas, B.J.C., Shaffer, M.S.P., Freeman, S., Koopman, M., Chawla, K.K., Boccaccini, A.R.: Electrophoretic deposition of carbon nanotubes on metallic surfaces. Proc. 2nd Int. Conference on Electrophoretic Deposition. Key Eng. Mater. **314**, 141–146 (2006)
45. De Cheng, W., Lu, S., Jia, E.L., Siew, Y.W., Xu, L., Wuiwui, C.T., Ye, L., Chao, B.H.: Multi-walled carbon nanotube/polyimide composite film fabricated through electrophoretic deposition. Polymer **51**, 2155–2160 (2010)
46. Chen, G.Z., Shaffer, M.S.P., Coleby, D., Dixon, G., Zhou, W., Fray, DJ., et al.: Carbon nanotube and polypyrrole composites: coating and doping. Adv. Mater. **12**, 522–526 (2000)
47. Shimoda, H., Oh, S.J., Geng, H.Z., Walker, R.J., Zhang, X.B., McNeil, L.E. et al.: Self-assembly of carbon nanotubes. Adv. Mater. **14**, 899 (2002)
48. Girishkumar, G., Vinodgopal, K., Kamat, PV.: Carbon nanostructures in portable fuel cells: single-walled carbon nanotube electrodes for methanol oxidation and oxygen reduction. J. Phys. Chem. B. **108**, 19960–19966 (2004)
49. König, K., Novak, S., Ivekovič, A., Rade, K., Meng, D., Boccaccini, A.R., Kobe, S.: Fabrication of CNT-SiC/SiC composites by electrophoretic deposition. J. Eur. Ceram. Soc. **30**, 1131–1137 (2010)
50. Dickerson, J.H.: Personal communication (2010)
51. Zhang, J., Zhuang, R., Liu, J., Mäder, E., Heinrich, G., Gao, S.: Functional interphases with multiwalled carbon nanotubes in glass fibre/epoxy composites. Carbon **48**, 2273–2281 (2010)
52. Qian, H., Greenhalgh, E.S., Shaffer, M.S.P., Bismarck, A.: Carbon nanotube-based hierarchical composites: a review. J. Mater. Chem. **20**, 4751–4762 (2010)
53. Chicatun, F., Cho, J., Schaab, S., Brusatin, G., Colombo, P., Roether, J.A., Boccaccini, A.R.: Carbon nanotube deposits and CNT/SiO2 composite coatings by electrophoretic deposition. Adv. Appl. Ceram. **106**, 186–195 (2007)
54. Colorado, R., Barron, A.R.: Silica-coated single-walled nanotubes: nanostructure formation. Chem. Mater. **16**, 2691–2693 (2004)
55. Wang, S.-C., Huang, B.-C.: Field emission properties of Ag/SiO2/carbon nanotube films by pulsed voltage co-electrophoretic deposition. Thin Solid Films **517**, 1245–1250 (2008)
56. Lee, S., Sigmund, W.M.: Formation of anatase TiO2 nanoparticles on carbon nanotubes. Chem. Commun. **6**, 780–781 (2003)
57. Jitianu, A., Cacciaguerra, T., Benoit, S., Delpeux, S., Béguin, F., Bonnamy, S.: Synthesis and characterization of carbon nanotubes-TiO2 nanocomposites. Carbon **42**, 1147–1151 (2004)
58. Jarernboon, W., Pimanpang, S., Maensiri, S., Swatsitang, E., Amornkitbamrung, V.: Effects of multiwall carbon nanotubes in reducing microcrack formation on electrophoretically deposited TiO2 film. J. Alloys. Compd. **476**, 840–846 (2009)
59. Singh, I., Kaya, C., Shaffer, M.S.P., Thomas, B.C., Boccaccini, A.R.: Bioactive ceramic coatings containing carbon nanotubes on metallic substrates by electrophoretic deposition. J. Mater. Sci. **41**, 8144–8151 (2006)
60. Yu, Y., Yu, J.Y., Yu, J.-G., Kwok, Y.-C., Che, Y.-K., Zhao, J.-C., Ding, L., Ge, W.-K.: Enhancement of photocatalytic activity of mesoporous TiO2 by using carbon nanotubes. Appl. Catal. A. **289**, 186–96 (2005)
61. Cho, J., Schaab, S., Roether, J.A., Boccaccini, A.R.: Nanostructured carbon nanotube/TiO2 composite coatings using electrophoretic deposition (EPD). J. Nanopart. Res. **10**, 99–105 (2008)

62. Cava, C.E., Possagno, R., Schnitzler, M.C., Roman, P.C., Oliveira, M.M., Lepiensky, C.M., Zarbin, A.J.G., Roman, L.S.: Iron- and iron oxide-filled multi-walled carbon nanotubes: electrical properties and memory devices. Chem. Phys. Lett. **444**, 304–308 (2007)
63. Georgakilas, V., Gournis, D., Tzitzios, V., Pasquato, L., Guldi, D. M., Prato, M.: Decorating carbon nanotubes with metal or semiconductor nanoparticles. J. Mater. Chem. **17**, 2679–2694 (2007)
64. Du, C.S., Pan, N.: High power density supercapacitor electrodes of carbon nanotube films by electrophoretic deposition. Nanotechnology **17**, 5314–5318 (2006)
65. Mahajan, S.V., Hasan, S.A., Cho, J., Shaffer, M.S.P., Boccaccini, A.R., Dickerson, J.H.: Carbon nanotube–nanocrystal heterostructures fabricated by electrophoretic deposition. Nanotechnology **19**, 195301 (8pp) (2008)
66. Mahajan, S.V., Cho, J., Shaffer, M.S., Boccaccini, A.R., Dickerson, J.H.: Electrophoretic deposition and characterization of Eu2O3 nanocrystal – carbon nanotube heterostructures. J. Europ. Ceram. Soc. **30**, 1145–1150 (2010)
67. Li, J., Zhitomirsky, I.: Electrophoretic deposition of manganese dioxide-carbon nanotube composites. J. Mater. Proc. Technol. **209**, 3452–3459 (2009)
68. Lee, C.Y., Tsai, H.M., Chuang, H.J., Li, S.Y., Lin, P., Tseng, T.Y.: Characteristics and electrochemical performance of supercapacitors with manganese oxide-carbon nanotube nanocomposite electrodes. J. Electrochem. Soc. **152**, A716–A720 (2005)
69. Bordjiba, T., Bélanger, D.: Development of new nanocomposite based on nanosized-manganese oxide and carbon nanotubes for high performance electrochemical capacitors. Electrochimica Acta. **55**, 3428–3433 (2010)
70. Ma, J., Wang, C., Peng, K. W.: Electrophoretic deposition of porous hydroxyapatite scaffolds. Biomaterials **24**, 3505–3510 (2003)
71. Krause, D., Thomas, B., Leinenbach, C., Eifler, D., Minay, E.J., Boccaccini, A.R.: The electrophoretic deposition of Bioglass® particles on stainless steel and Nitinol substrates. Surf. Coat. Technol. 200, 4835–4845 (2006)
72. Zhitomirsky, I.: Electrophoretic hydroxyapatite coatings and fibres. Mater. Lett. **42**, 262–271 (2000)
73. White, A.A., Best, S.M., Kinloch, I.A.: Hydroxyapatite-carbon nanotube composites for biomedical applications: A Review. Int. J. Appl. Ceram. Tech. **4**, 1–13 (2007)
74. Singh, K.M., Shokuffar, T., Gracio, J.J.A., Sousa, A.C.M., Fereira, J.M.D.F., Germestani, H., Ahzi, S.: Hydroxyapatite modified with carbon-nanotube-reinforced poly(methyl methacrylate): A nanocomposite material for biomedical applications. Adv. Functional Materials **18**, 694–700 (2008)
75. Chen, Y., Zhang, Y.Q., Zhang, T.H., Gan, C.H., Zheng, C.H., Yu, G.: Carbon nanotube reinforced hydroxyapatite composite coatings produced through laser surface alloying. Carbon **44**, 37–45 (2006)
76. Boccaccini, A.R., Chicatun, F., Cho, J., Bretcanu, O., Roether, J.A., Novak, S., Chen, Q.Z.: Carbon nanotube coatings on Bioglass-Based tissue engineering scaffolds. Adv. Funct. Mater. **17**, 2815–2822 (2007)
77. Harrison, B.S., Attala, A.: Carbon nanotube applications for tissue engineering. Biomaterials **28**, 344–353 (2007)
78. Lange, G.L., Donath, K.: Interface between bone tissue and implants of solid hydroxyapatite or hydroxyapatite-coated titanium implants. Biomaterials **10**, 121–125 (1989)
79. Kaya, C., Singh, I., Boccaccini, A.R.: Multi-walled carbon nanotube-reinforced hydroxyapatite layers on Ti6Al4V medical implants by electrophoretic deposition (EPD). Adv. Eng. Mater. **10**, 1–8 (2008)
80. Balani, K., Anderson, R., Laha, T., Andara, M., Tercero, J., Crumpler, E., Agarwal, A.: Plasma-sprayed carbon nanotube reinforced hydroxyapatite coatings and their interaction with human osteoblasts in vitro. Biomaterials **28**, 618–624 (2007)
81. Lin, C., Han, H., Zhang, F.: Electrophoretic deposition of HA/MWNTs composite coating for biomedical applications. J. Mater. Sci: Mater. Med. **19**, 2569–2574 (2008)

82. Zaman, A. C., Üstündağ, C. B., Kuşkonmaz, N., Kaya, F., Kaya, C.: 3-D micro-ceramic components from hydrothermally processed carbon nanotube–boehmite powders by electrophoretic deposition. Ceram. Int. **36**, 1703–1710 (2010)
83. Bai Y., Neupane M.P., Park, I.S., Lee, M.H., Bae, T.S., Watari F., Uo, M.: Electrophoretic deposition of carbon nanotubes–hydroxyapatite nanocomposites on titanium substrate. Mater. Sci. Eng. C. **30**, 1043–1049 (2010)
84. Hench, L.L., Splinter, R.J., Allen, W.C., Greenlee, T.K.: Bonding mechanisms at the interface of ceramic prosthetic materials. J. Biomed. Mater. Res. **2**, 117–141 (1971)
85. Hench, L.L.: Bioceramics. J. Am. Ceram. Soc. **81**, 1705–1728 (1998)
86. Hench, L.L., Andersson, Ö.: Bioactive glass coatings. An Introduction to Bioceramics 239–259 (1993)
87. Chen, Q.Z., Thompson, I.D. Boccaccini, A.R.: 45S5 Bioglass®-derived glass-ceramic scaffolds for bone tissue engineering. Biomaterials **27**, 2414–2425 (2006)
88. Cho, J., Cannio, M., Boccaccini, A.R.: The electrophoretic deposition of Bioglass®/carbon nanotube composite layers for bioactive coatings. Int. J. Mater. Product Technol. **35**, 260–270 (2009)
89. Schausten, M.C., Meng, D., Telle, R., Boccaccini, A.R.: Electrophoretic deposition of carbon nanotubes and bioactive glass particles for bioactive composite coatings. Ceram. Int. **36**, 307–312 (2010)
90. Meng, D., Ioannou, J., Boccaccini A.R.: Bioglass-based scaffolds with carbon nanotube coating for bone tissue engineering. J. Mater. Sci. Mater. Med. **20**, 2139–2144 (2009)
91. MacDonald, R.A., Laurenzi, B.F., Viswanathan, G., Ajayan, P.M., Stegemann, J.P.: Collagen-carbon nanotube composite materials as scaffolds in tissue engineering. J. Biomed. Mater. Res. **74**, 489–495 (2005)
92. Zanello, L.P., Zhao, B., Hu, H., Haddon, R.C.: Bone cell proliferation on carbon nanotubes. Nano Lett. **6**, 562–567 (2006)
93. Aryal, S., Bhattari, S.R., Bahadur, R., Khil, M.S., Lee, D.R., Kim, H.Y.: Carbon nanotubes assisted biomimetic synthesis of hydroxyapatite from simulated body fluid. Mater. Sci. Eng. A. **426**, 202–207 (2006)
94. Sun, F., Zhitomirsky, I.: Electrodeposition of hyaluronic acid and composite films. Surface Engineering **25**, 621–627 (2009)
95. Grandfield, K., Sun, F., Fitzpatrick, M., Cheong, M., Zhitomirsky, I.: Electrophoretic deposition of polymer-carbon nanotube-hydroxyapatite composites. Surf. Coatings Technol. **203**, 1481–1487 (2009)
96. Santillán, M.J., Caneiro, A., Lovey, F.C., Quaranta, N., Boccaccini, A.R.: Electrophoretic codeposition of La0.6Sr0.4Co0.8Fe0.2O3 and carbon nanotubes for developing composite cathodes for intermediate temperature solid oxide fuel cells. Int. J. Appl. Ceram. Technol. **7**, 30–40 (2010)

Chapter 5
Electrophoretic Deposition (EPD): Fundamentals and Novel Applications in Fabrication of Advanced Ceramic Microstructures

Partho Sarkar, Debnath De, Tetsuo Uchikochi and Laxmidhar Besra

5.1 Introduction

Electrophoretic Deposition (EPD) is a colloidal forming technique where charged, colloidal particles from a stable suspension are deposited onto an oppositely charged substrate by the application of a dc electric field. In recent years, the ceramic community has come to understand that good forming techniques are the key to achieving reliable product properties and performance. Good forming techniques should have three major capabilities: ability (1) to produce a dense and homogeneous green body, (2) to produce complicated shapes effectively and easily, and (3) to allow flexibility in microstructural manipulation i.e., ability to fabricate a wide spectrum of composite microstructures ranging from dispersed, laminated, fiber-composites to functionally graded materials etc. Illustrations in this chapter will demonstrate that EPD has all of these capabilities. Most significant in the EPD of laminates is the high-perfection of the microstructures produced, particularly the interfaces [1]. This chapter will address the fundamental mechanisms and kinetics of the EPD process, critical issues and myths of the EPD process, and why EPD is a potential processing technique for fabrication of high-performance and high-perfection ceramics and coatings that can have a wide spectrum of applications.

In a homogeneous green body i.e., unsintered powder compact, the spatial distribution of all of its constituent phases, including voids, should be uniform and narrow. Lange et al. [2] have shown that voids with high particle coordination number are difficult to remove during sintering and will exist in the sintered product as flaws. These flaws or defects in the final products will degrade their properties, rendering them unreliable and weak. Thus, it is prudent to choose an effective forming or consolidation technique that can produce a dense and homogeneous green body where particles are tightly, closely and uniformly packed and distributed homogeneously. This will, in turn, reduce the size and distribution of voids, thus minimiz-

---

P. Sarkar (✉)
Clean Energy, Environment & Carbon Management Division, Alberta Innovates-Technology Futures, 250 Karl Clark Road, Edmonton, AB, Canada T6N 1E4
e-mail: Partha.Sarkar@albertainnovates.ca

J. H. Dickerson, A. R. Boccaccini (eds.), *Electrophoretic Deposition of Nanomaterials*, Nanostructure Science and Technology,
DOI 10.1007/978-1-4419-9730-2_5,

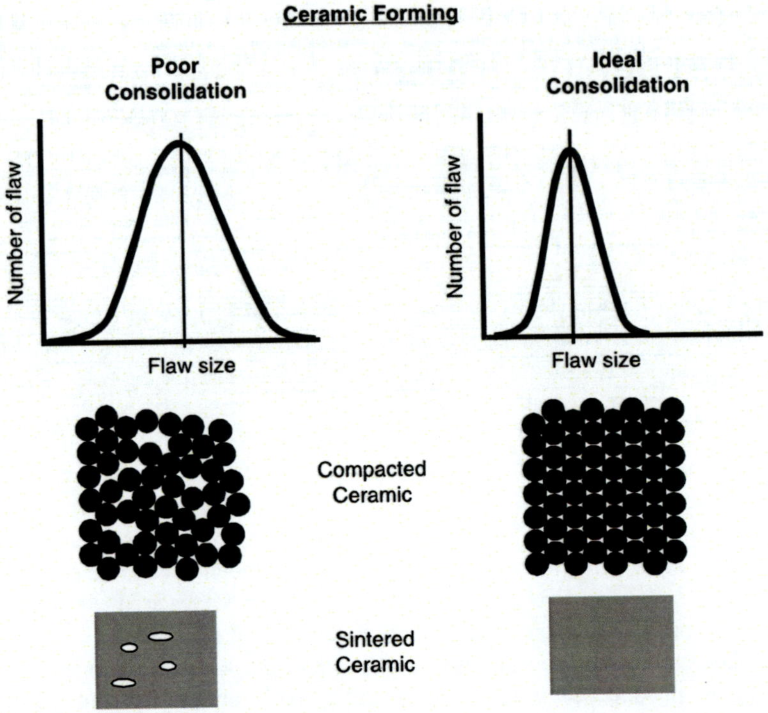

Fig. 5.1 Schematic illustration of a poor and ideal consolidation of ceramics [3]

ing defects in the sintered products. Figure 5.1 is a schematic illustration of a poor and ideal consolidation process [3]. In contrast to poor consolidation, in an ideal consolidation the product will be less prone to cracking and warpage during drying and sintering. Furthermore, the probability of incorporation of flaws or defects in the sintered product from ideal consolidation will be significantly low, thus lowering sintering temperatures and/or shorter soaking times, and resulting in more reproducible final densities ensuring better dimensional control of finished products.

5.2 Fundamentals of the Process of Electrophoretic Deposition (EPD)

EPD is a facile colloidal ceramic forming technique where one can control "the state of dispersion of particles in suspension" or "suspension structure" (via careful manipulation of interparticle colloidal forces) and its evolution during consolidation (via proper choice of processing parameters) in order to produce near ideal dense and homogeneous green bodies and microstructures.

Sarkar et al. [4] have provided examples of near ideal dense and homogeneous green bodies and microstructures made from mono-size silica spheres (dia. ≈ 0.5 μm) by the EPD process. The microstructure [4] shows the close packing of the silica spheres and a very narrow distribution of voids that can be easily removed during sintering. Sarkar et al. [4] have clearly demonstrated the capability of EPD forming technique to deliver a dense and homogeneous green body. The EPD process can also deposit powder uniformly on electrodes having complicated shapes and, as a result, can produce geometrically complicated shapes. However, in the case of bulk ceramics, after shape forming, the substrate (depositing electrode) needs to be separated from the deposit. Commercial viability of EPD depends on the effective separation of the substrate from the deposit. In the case of simple geometry, separation can be done by careful and controlled drying of the deposit. For complicated shapes, a combustible substrate that can be removed during the sintering process could be used. In the case of coatings, the sample often develops cracking during drying and sintering, and the success of EPD to this area depends on overcoming this problem. In general, commercial success of EPD is limited by a better understanding of how to control the EPD process precisely and consequently better trained operators as compared to other common forming methods such as slip casting or injection molding.

5.2.1 The Process of Deposition/Coagulation During EPD

In other casting (deposition) processes such as in slip-, centrifugal-, or pressure-casting, the liquid is induced to move and carries the particles with it. The casting process involves the liquid moving, carrying the ions through the building cast. The flight of ions reduces the zeta-potential and the particles coagulate. In contrast, in EPD the particles move while the liquid remains approximately motionless. Deposition during EPD involves a different mechanism. The liquid in EPD experiences no bulk force; the ions and the particles do. Several explanations of particle coagulation during EPD have been advanced. Koelmans and Overbeek [5, 6] proposed that the increase of electrolyte concentration caused by electrochemical reactions at one of the electrodes induces particle coagulation. Shimbo et al. [7] suggested that polymerization of adsorbed hydroxides produced by secondary processes at the depositing electrode causes the particles to adhere. Mizuguchi et al. [8] hold that charge neutralization of particles upon touching the electrode brings them together. Brown and Salt [9] calculated the minimum deposition field strength (25–500 V cm^{-1}) required for oxide particles in organic media but found actual experimental values are 5–10 times less. Hamaker and Verwey [10, 11] explained this discrepancy in terms of a "pressure" generated by accumulated particles causing deposition. This explanation is doubtful since Giersig and Mulvaney [12] recently deposited a monolayer of gold particles by EPD at the relatively low field strength of 0.5–5 mV cm^{-1}. Since particles can deposit on either the cathode or anode depending on the sign of their surface charge and even noble metals and carbon can also be deposited, explanations of hydroxide formation and polymerization are not plausible [7]. Recently, De and Nicholson [13] proposed a model for the process of deposition during the electrophoretic deposition process. Their model has demonstrated that the mandatory pre-step to deposition, i.e., particle coagulation, is the local pH change of the suspension due to ionic discharge. The model suggested that ions that move with the charged particles in suspension are depleted at the depositing electrode, locally changing the pH towards the isoelectric point ($pH_{i.e.p.}$) to give coagulation. They modeled the variation of zeta potential via chemical-equilibrium and surface-adsorption isotherms. Their model predictions successfully fit the experimental data for Al_2O_3 particles in ethanol when the Freundlich surface-adsorption isotherm is assumed. The model also predicts the co-ion concentration gradient as a function of location within the suspension, and the deposition time and its role in the coagulation process during EPD.

Further experimental evidence of the pH localization model of De and Nicholson [13] for particle consolidation at the electrode during EPD is reported recently by Besra et al. [14]. Using a specially designed ion sensitive field effect transistor (ISFET) pH meter that compactly houses the electrodes at its tip (Fig. 5.2), they successfully measured the local pH variations near the electrodes during electrolysis of water by both continuous as well as pulsed DC as a function of time and observed a significant change in pH near the electrodes compared to the bulk initial pH. They withdrew about 50 μl liquid from the electrode surface during electrolysis at varying time intervals for pH measurement. They observed that the initial (at time t=0) pH at cathode/solution interface was at bulk water pH of 4.5 but increased signifi-

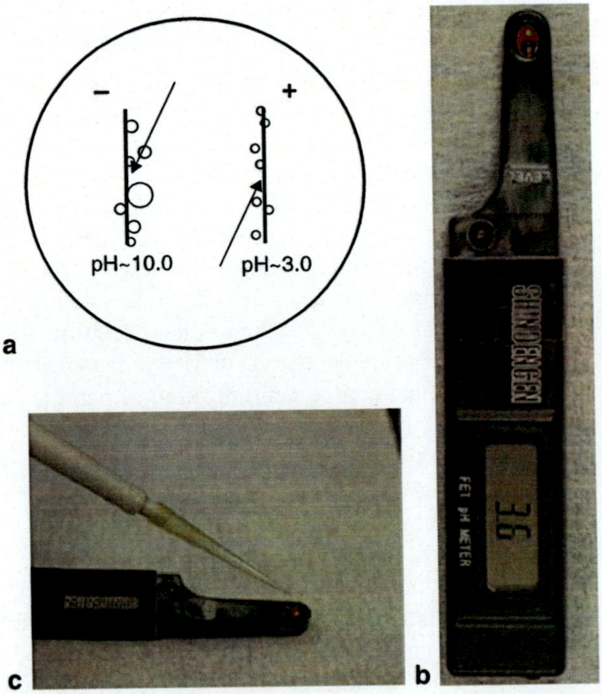

Fig. 5.2 Method of pH measurement near the electrodes during electrolysis. **a** Electrolysis bath showing bubble generation at the electrodes. Samples were drawn from the electrode surface from near the arrow head. **b** Special type of pH meter with micro-electrodes with which 50 μl sample is enough to give a reproducible measure of pH. **c** Electrolyzed water sample being dropped onto the microelectrode for pH measurement. [14]

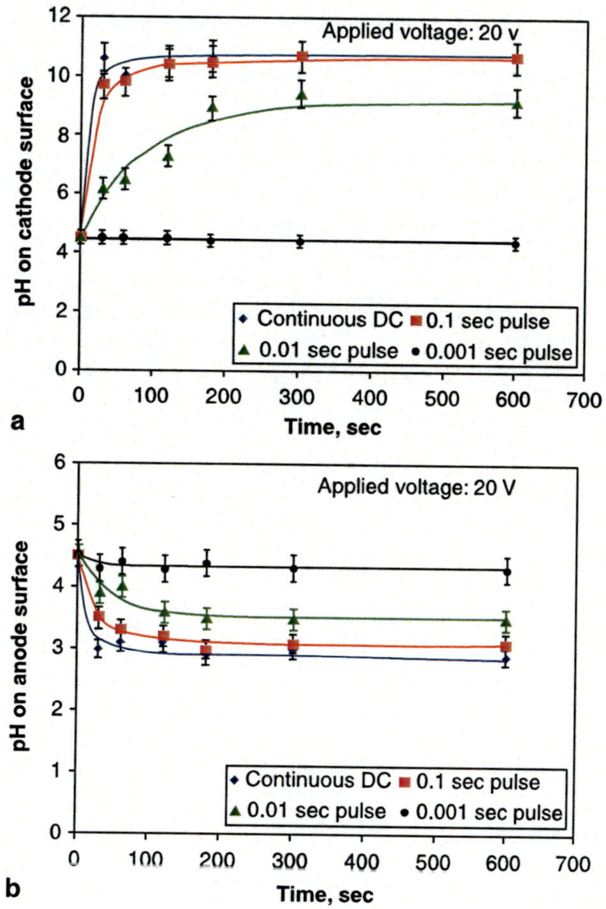

Fig. 5.3 pH localization during constant voltage electrolysis of water at **a** cathode/solution interface, and **b** anode/solution interface (initial bulk pH 4.5, conductivity = 110 μS/cm; applied voltage = 20 V). [14]

cantly to about 10.7 within 30 s of application of a continuous DC voltage of 20 V (Fig. 5.3a) and remained constant with time thereafter even though electrolysis continued. They also observed that the rise in pH was gradual on application of a pulse voltage, the plateau value of pH decreased with decreasing pulse size and longer electrolysis time was needed to attain the plateau pH for smaller pulse size. Furthermore, the rise in pH was negligible for a pulse size of 0.001 s for 20 V applied potential. Besra et al. [14] also observed that the local pH at the anode/electrolyte interface (Fig. 5.3b) dropped from bulk pH 4.5 to as low as pH 2.9 within 30 s of electrolysis on application of continuous voltage of 20 V and remained unchanged thereafter in spite of the fact that electrolysis continued. Also, they noticed that the decrease in pH was less on application of pulse voltage and lower the pulse size, less was the decrease in pH at anode/solution interface. Table 5.1 [14] describes the localized pH values during an actual EPD process for alumina suspension and clearly demonstrates that the local pH of the deposited mass at cathode is indeed higher than the bulk initial pH 4.5 of the suspension. Also, data in Tab. 5.1 clearly

Table 5.1 pH of alumina deposited at cathode on application of 20 V for 3 min [Suspension: 5 vol%, bulk pH 4.5]

Sl No	Pulse size	pH of deposited mass at cathode	Deposit yield, mg/cm^2	pH at anode
1	Continuous DC	8.4	180.77	4.5
2	0.1 s pulse	6.9	112.55	4.5
3	0.01 s pulse	5.4	46.50	4.6
4	0.001 s pulse	–	0.00	4.6

establish a direct correlation among the deposit yield, the closeness of shifted pH to the isoelectric point (i.e.p.) of alumina ($pH_{i.e.p} \approx 7.9$) and the rate of the shift of the local pH towards the $pH_{i.e.p}$ of alumina. Deposit yield in Tab. 5.1 was also observed to be the highest when a continuous DC voltage was applied for which the local pH shifted the fastest and closest to the isoelectric point of alumina, $pH_{i.e.p} \approx 7.9$. For the case of pulse DC application, the deposit yield decreased in tandem with decreasing pulse size since the pH shifted less and less towards $pH_{i.e.p} \approx 7.9$ with decreasing pulse size.

5.2.2 *Kinetics of Electrophoretic Deposition Process*

To make EPD commercially more viable, one needs to understand the kinetics of the EPD process in order to (a) control and manipulate deposition rate and (b) achieve flexibility in microstructural manipulation. In 1940, during a study of the phenomena of the EPD, Hamaker [10] observed that the deposited weight or yield of the EPD varies linearly with the amount of the charge passed. Hamaker proposed that the amount deposited or yield is proportional to the concentration of the suspension, time of deposition, surface area of the deposit, and the electric field.

It is important to mention that EPD is a non-Faradic process [1]. It follows laws different from those governing electroplating. Electrophoretic deposition can be conducted under either constant voltage or constant current conditions. Deposition can be performed under either of these conditions keeping the suspension concentration either constant or changing (concentration of suspension decreasing) with deposition time. Figure 5.4 demonstrates the schematic plots of weight of deposit as a function of time of deposition for these four deposition conditions: curve A (constant-current and constant-suspension concentration), curve B (constant-current but decreasing suspension concentration), curve C (constant-voltage and constant- suspension concentration) and curve D (constant-voltage but decreasing suspension concentration). Except in Curve A where the rate of deposition is constant with time, the rate of deposition decreases asymptotically with deposition time in either curve B, or C, or D. The final yield (after sufficient deposition time allowed) and rate of deposition are the highest in Curve A, followed by curves B, C, and D respectively. The effect of decreasing suspension concentration on the reduction of the final yield and rate of deposition is obvious during either constant-current (curves A and B) or constant-voltage (curves C and D) EPD. Comparison of curves A (constant-current) and C (constant-voltage)

Fig. 5.4 Schematic of Kinetics of EPD process [1]

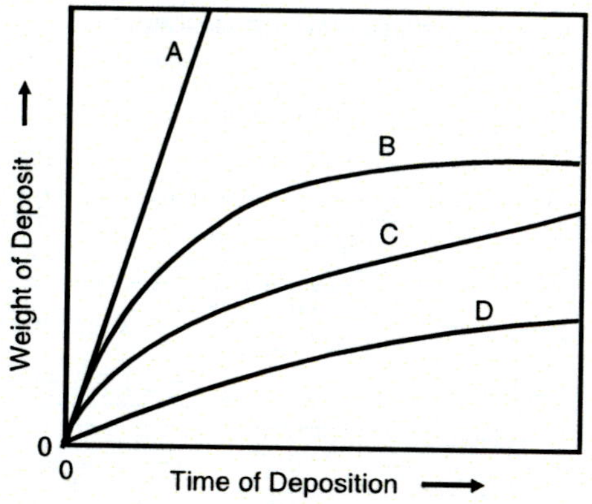

clearly reveals that even if the suspension concentration is kept constant during deposition in both of them, (a) the rate of deposition was constant in curve A while it decreased asymptotically with time in curve C and (b) final yield was considerably higher in curve A than that in curve C. Thus, the deviation of curve A from curve C is not due to a decreasing suspension concentration but is due to a decrease of particle velocity as function of deposition time. Why is the particle velocity decreasing during constant-voltage (Curve C) deposition? Typically, the EPD deposit has higher electrical resistance than the suspension from which deposition takes place. Consequently, as the deposit grows with deposition time, the available electrical driving force or voltage per unit length of suspension decreases with time. This results in decreasing velocity of the particles, and thus the EPD yield and rate of deposition with time. EPD of the alumina-ethanol system that is stabilized by hydrochloric acid serves a good demonstration of this phenomenon. Sarkar et al. [3] reported that under constant-voltage/constant-suspension concentration conditions (curve C), it is nearly impossible to fabricate millimeters thick samples from the hydrochloric acid-stabilized-alumina-ethanol suspension because of high resistivity of deposit. In contrary, when the alumina-ethanol system is stabilized by acetic acid, the resistivity of the deposit is only marginally higher than the suspension and thick deposits can be produced without any problem under constant-voltage/constant-suspension concentration conditions. During EPD, passage of current causes electrode reactions, producing electrolytes. Probably, the extent of electrolytes produced and any polarization in the deposit resulted in the difference in deposit resistivity in alumina-ethanol system that was stabilized either with strong acid: HCl or weak acid: acetic acid. This needs further investigation. Thus, under constant-voltage/ constant-suspension concentration condition (curve C), deposit resistivity plays a significant role on determining the EPD yield and rate of deposition. If the deposit resistivity equals that of the suspension during deposition, Curve A and C will be identical; i.e., there will be no difference between constant-current/ and constant-voltage/constant suspension concentration deposition conditions. Curve D deviates from Curve A due to (a) the decrease of suspension concentration and (b) increase of deposit resistance with deposition. Thus, Fig. 5.4 clearly demonstrates that constant-current depositions are efficient and can provide better and easier control over the EPD deposition process.

5.2.3 Factors Influencing Electrophoretic Deposition

The EPD process involves deposition of charged particles in a suspension onto an electrode under the influence of an applied electric field. Recently, Besra and Liu [15] summarized the influence of different parameters influencing the EPD process. Two groups of parameters have been identified to determine the characteristics of this process; (1) suspension parameters, and (2) process parameters including the physical parameters such as the electrical nature of the electrodes, the electrical conditions (voltage/intensity relationship, deposition time, etc).

Ideally in an EPD process, all of the applied electric field should be utilized to effect particle electrophoresis, the driving force for EPD. However, a real EPD suspension is far from ideal because of the presence of free ions in addition to the particles in suspension. A portion of the current is carried not only by the charged particles but by the free ions co-existing in the suspension. Therefore, unlike Hamaker's suggestion, the deposit yield in reality is not directly related to the current. The current carried by free ions can be ignored when their presence, as in the organic suspensions, is negligible. On the other hand, it is also believed that the accumulation of anionic and/or cationic charge at the electrodes during electrophoresis suppresses the subsequent deposition rate. However, the effect of accumulated ions is negligible in the initial period.

The kinetics of electrophoretic deposition and the quality of deposits formed is dependent on a large number of parameters. It is required to have careful control of these individual parameters during electrophoretic deposition, although many of the parameters are inter-related to one another. The first attempt to correlate the amount of particles deposited during EPD with different influencing parameters was described by Hamaker [10] and later by Avgustnik et al. [16]. The representation of Hamaker's law has changed over the years, but it relates the deposit yield (w) to the electric field strength (E), electrophoretic mobility (μ), surface area of the electrode (A), and the particle mass concentration in the suspension (C) through the following equation:

$$w = \int_{t_1}^{t_2} \mu \cdot E \cdot A \cdot C \cdot dt$$

Avgustinik et al. (1962) modified the Hamaker's law for deposition on cylindrical electrode with a coaxial counter-electrode to arrive at the following equation for deposit yield (w) in terms of permittivity (ε), the zeta potential (ξ), and the viscosity of the suspension (η):

$$w = \frac{l \cdot E \cdot \varepsilon \cdot \xi \cdot C \cdot t}{3 \ln (a/b) \cdot \eta}$$

where, l and a are the length and radius of the deposition electrode, respectively, b is the radius of the coaxial counter electrode (b>a).

Biesheuval and Verweij [17] improved upon these classical equations and developed more complex model of the deposition process by considering the presence of three distinct phases namely (1) a solid phase (the deposit), (2) a suspension phase, and (3) a phase containing little or no solid particles. The deposit phase and the particle-free liquid phase both grow at the expense of the suspension phase. By considering the movement of the boundary between the deposit (solid phase) and the suspension phase with time along with the continuity equation and expression for velocity of particles in the suspension and the movement of the cast-suspension boundary, Biesheuval and Verweij [17] derived the following equation based on that of Avgustinik [16].

$$w = \frac{2 \cdot \pi \cdot \mu \cdot l \cdot E \cdot C_d}{\ln\,(a/b)} \cdot \frac{\phi_s}{\phi_d - \phi_s} \cdot t$$

ϕ_s and ϕ_d in the above equation are the volumetric concentration of particles in suspension and deposit, respectively, C_d is the mass concentration of particles in the deposit, μ is the electrophoretic mobility ($=\varepsilon\xi/6\pi\eta$). Using this equation, it is possible to understand better the more-than-linear increase of cast formation with increasing suspension concentration as reported by several authors.

Ishihara et al. [18] and Chen and Liu [19] used the following equation to determine the weight (w) of charged particles deposited per unit area of electrode in the initial period, ignoring the charge carried by the free ions:

$$w = \frac{2}{3} C \cdot \varepsilon_o \cdot \varepsilon_r \cdot \xi \cdot \left(\frac{1}{\eta}\right) \cdot \left(\frac{E}{L}\right) \cdot t$$

where, C is the concentration of the particle, ε_o is the permittivity of vacuum, ε_r is the relative permittivity of the solvent, ξ is the zeta potential of the particles, η is the viscosity of the solvent, E is the applied potential, L is the distance between the electrodes, and t is the deposition time. The above equation suggests that the deposition weight of the charged particles under ideal electrophoretic deposition depends on the above parameters. For a given solvent, particles, and EPD apparatus, the factors ξ, ε_r, η and L in the above equation are constant. Hence, the weight of deposited particles (w) in the EPD method is a function of E, t and C. Therefore, the mass of the deposited particles, and the thickness of the films formed during EPD can be readily controlled by controlling the concentration of suspension, applied potential, and deposition time.

5.2.3.1 Suspension Parameters

The suspension parameters include several aspects such as the physicochemical nature of both suspended particle and the liquid medium, surface properties of the powder, and the influence of type and concentration of additives, mainly the dispersant.

Particle Size

Although there is no general rule of thumb to specify particle sizes suitable for electrophoretic deposition, good deposition for a variety of ceramic and clay systems have been reported to occur in the range of 1–20 μm [20]. But this does not necessarily mean that deposition of particles outside this size range is not feasible. Recently, with increasing thrust on nanostructured materials, the EPD technique is being viewed with more interest for assembly of nanoparticles as well. It is important that the particles remain completely dispersed and stable for homogeneous and smooth deposition. For larger particles, the main problem is that they tend to settle due to gravity. Ideally, the mobility of particles due to electrophoresis must be higher than that due to gravity. It is difficult to get uniform deposition from sedimenting suspension of large particles. Electrophoretic deposition from settling suspension will lead to gradient in deposition, i.e. thinner at the top and thicker deposit at the bottom when the deposition electrode is placed vertical. In addition, for electrophoretic deposition to occur with larger particles, either a very strong surface charge must be obtained, or the electrical double layer region must increase in size. Particle size has also been found to have a prominent influence on controlling the cracking of the deposit during drying. Sato et al. [21] observed that during EPD of $YBa_2Cu3O_{7-\delta}$ (YBCO), cracks formed in films deposited from a suspension consisting of relatively smaller particle (0.06 micron) was much less than that in films deposited from the suspension containing larger particles (3 micron). Hence, reduction in particle size improved the morphology of the YBCO superconducting film fabricated by electrophoretic deposition suggesting that it is a useful technique to minimize cracking of deposits.

Dielectric Constant of Liquid

Powers [22] studied the deposition of beta-alumina suspended in numerous organic media as a function of the dielectric constant of the liquids and the conductivity of the suspension. For liquids in their pure state, a sharp increase in conductivity with dielectric constant was noted. Impurities, in particular water, affected the conductivity of the suspension. The conductivity of milled suspension was found to be very different from that of pure liquid, as a consequence of dissociative or adsorptive charging modes. Powers [22] obtained deposits only with liquid for which the dielectric constant was in the range of 12–25. With too low a dielectric constant, deposition fails because of insufficient dissociative power; whilst with a high dielectric constant, the high ionic concentration in the liquid reduces the size of the double layer region and consequently the electrophoretic mobility. The ionic concentration in the liquid must remain low, a condition favored in liquids of low dielectric constant. Following Table 5.2 shows physical properties such as viscosity and relative dielectric constant of some solvents used in EPD [23].

Table 5.2 Physical properties of some solvents used in EPD. [23]

Solvents	Viscosity (cP) $= 10^{-3}$ (N s m^{-2})	Relative dielectric constant
Methanol	0.557	32.63
Ethanol	1.0885	24.55
n-Propanol	1.9365	20.33
Iso-propanol	2.0439	19.92
n-Butanol	2.5875	17.51
Ethylene glycol	16.265	37.7
Acetone	0.3087	20.7
Acetyl acetone	1.09	25.7

Conductivity of Suspension

Ferrari and Moreno [24], after a careful study, proposed that the conductivity of the suspension is a key factor and needs to be taken into account in EPD experiments. It has been pointed out that if the suspension is too conductive, particle motion is very low, and if the suspension is too resistive, the particles charge electronically and the stability is lost. They observed increase in conductivity of the suspension with both temperature and with polyelectrolyte (dispersant) concentration; but not all conductivity values were found useful for electrophoretic deposition. They found the existence of a narrow band of conductivity range at varying dispersant dosage and temperature, in which the deposit is formed. Conductivity out of this region is not suitable for EPD, limiting the forming possibilities. This suitable region of conductivity is however expected to be different for different systems. The margin of conductivity region suitable for EPD, however, can be increased by the applied current assuring the success of the EPD process [25].

Viscosity of Suspension

In casting processes, the main controlling parameter is the viscosity. Rheological measurements on concentrated slips give us a good idea about the optimum dispersing state when adding dispersants. In EPD process, the solid loading is very low and the viscosity cannot be used to evaluate the dispersion state [24, 25]. But the desired properties in the suspension vehicle are low viscosity, high dielectric constant and low conductivity.

Zeta Potential

The zeta potential of particles in suspension is a key factor in the electrophoretic deposition process. It plays an important role in: (1) stabilization of the suspension by determining the intensity of repulsive interaction between particles, (2) determining the direction and migration velocity of particle during EPD, (3) determining the green density of the deposit. The overall stability of a system depends on the interaction between individual particles in the suspension. Two mechanisms affect this interaction, which are due to electrostatic and van der Waals forces. The probability of coagulation of a disperse system depends on the interaction energy resulting from these forces. A high electrostatic repulsion due to high particle charge is required to avoid particle agglomeration. The particle charge also affects the green density of the deposit. During formation of the deposit, the particles become closer to each other and with increasing attraction force. When the particle charge is low, the particles would coagulate even for relative large inter-particle distances, leading to porous, sponge-like deposits. On the contrary, if the particles have a high surface charge during deposition they will repulse each other, occupying positions which will lead to high particle packing density [26]. Therefore, control of the solids loading and concentration of solvents and additives in the EPD suspension is very important to reach the highest possible green density of the deposit. The zeta potential can be controlled by addition of a variety of charging agents such as acids, bases and specifically adsorbed ions or polyelectrolytes, to the suspension [27]. While the deposition rate is directly dependent on the zeta potential, the influence of the additives is exerted by its effect on the ionic conductivity of the suspension. The ionic conductivity determines the potential drop in the bulk suspension, which constitute the driving force for the transfer of particles to the electrode.

Stability of Suspension

Electrophoresis is the phenomenon of motion of particles in a liquid medium under the influence of an electric field, and occurs when the distance over which the double layer charge falls to zero large is compared to the particle size. The particles generally move relative to the liquid phase when the electric field is applied. Colloidal particles which are 1 μm or less in diameter tend to remain in suspension for long periods due to Brownian motion. Particles larger than 1 μm require continuous hydrodynamic agitation to remain in suspension. The suspension stability is characterized by settling rate and tendency to undergo or avoid flocculation. Stable suspensions show no tendency to flocculate, settle slowly and form dense and strongly adhering deposits at the bottom of the container. Flocculating suspensions settle rapidly and form low density, weakly adhering deposits. When the suspension is too stable, the repulsive forces between the particles will only be overcome by high electric field, and typical EPD condition deposition may not occur. The requirement that suspension should be unstable in the vicinity of the electrodes for occurrence of deposition has already been discussed earlier. This local instability could be caused by the formation of ions from electrolysis or discharge of the particles; these ions then cause flocculation close to the electrode surface. It is desirable to find suitable physical/chemical parameters that characterize a suspension sufficiently in order that its ability to deposit can be predicted. Most investigators use zeta potential or electrophoretic mobility, but these do not uniquely determine the ability of a suspension to deposit. For example, Brown et al. [28] observed that addition of electrolyte to a suspension of aluminium in alcohol caused no significant change to the zeta potential, but deposits were obtained only in the presence of the electrolyte. The stability of the suspension is evidently its most significant property, but this is a somewhat empirical property not closely related to fundamental parameters.

5.2.3.2 Process Parameters

Effect of Applied Electric Field

The deposit yield is related directly to the applied electric field strength and increases with increase in applied potential. Although powders can be deposited more quickly if greater applied fields are used, the quality of the deposit can suffer. Basu et al. [29] observed that more uniform films are formed at moderate applied fields (25–100 V/cm), whereas the film quality deteriorates if relatively higher applied fields (>100 V/cm) are used. Because the formation of particulate film on the electrode is a kinetic phenomenon, the accumulation rate of the particles influences their packing behavior in coating. For a higher applied field, which may cause turbulence in the suspension, the coating could be disturbed by the flows in the surrounding medium, even during its deposition. In addition, particles can move so fast that they cannot find enough time to sit in their best positions to form a close-packed structure. Finally, high field situations restricts the lateral motion of particles on the surface of the already deposited layer, because higher applied potential exerts more pressure on particle flux and movement affecting the deposition rate and structure of the deposit.

Negishi et al. [23] observed flat and smooth morphology of YSZ deposited from its suspension in n-propanol; the morphology became rougher with increasing applied voltage. Such changes in morphology could be directly correlated to the stability of the solvent in absence of any powder, under the influence of an applied electric field. Figure 5.5 depicts the change in stability of n-propanol in terms of variation in current density profile with increasing applied voltages. Such stability data serves as a good guideline for deciding the deposition parameters and consequently the quality of deposit formed by EPD. It is considered that the unstable current density adversely influences the quality of deposition morphology. From the current density profile in Fig. 5.5, it is reasonable to suggest that the applied voltage should be less than 100 V to obtain smooth deposits in the case of n-propanol.

Effect of Deposition Time

In a typical constant voltage EPD process, many researchers observed high deposition rates during the initial period of deposition, which then decreased and attained plateau at very high deposition times [19, 29, 30]. In a constant voltage EPD,

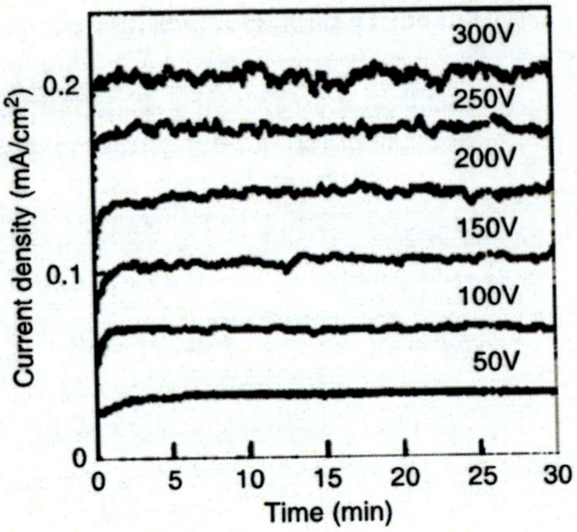

Fig. 5.5 Stability of solvent n-propanol at varying applied potential. [23]

this is expected because while the potential difference between the electrodes is maintained constant, the electric field influencing electrophoresis decreases with increasing deposition time because of the formation of an insulating layer of deposited particles on the electrode surface [31]. But during the initial period of EPD, there is generally a linear relationship between deposition mass and time.

Concentration of Solids in Suspension

The volume fraction of solid in the suspension plays an important role, particularly for multi-component EPD. In some cases, although each of the particle species has same sign of surface charge, they could deposit at different rates depending on the volume fraction of solids in the suspension. When the volume fraction of solids is high, the powders deposit at an equal rate. However, if the volume fraction of solids is low, the particles can deposit at rates proportional to their individual electrophoretic mobility [32].

Conductivity of Substrate

The uniformity and conductivity of substrate electrode is an important parameter critical to the quality of the green deposited film by EPD. Peng and Liu [33] observed that low conductivity of the $La_{0.9}Sr_{0.1}MnO_3$ (LSM) substrate leads to non-uniform green film and slow deposition. Chen and Liu [19] noticed that when as pressed LSM or LSM-YSZ composite pellets were used as substrate for EPD, the deposition rate of YSZ was slow and obtained film was non uniform. This was attributed to be due to the high resistance of the substrates resulting from the added binder. When the pellets were fired at 700°C for 0.5 h to remove the binder, the conductivity of the substrates increased substantially and a high quality green YSZ film was obtained.

It may be noted that many of the parameters discussed above are related to one another. The ionic concentration, for example, not only affects the zeta potential, but is also closely related to the suspension electrical conductivity. Generally, the electrical conductivity is low when the ionic concentration in the suspension is low. However, as the ionic concentration in the suspension increases, the conductivity of the suspension also increases. At high ionic concentration, larger agglomerates are formed, and the large amount of free ions in the suspension becomes the main current carrier, resulting in reduced electrophoretic mobility of the particles. The conductivity of suspension is also directly related to dielectric constant of the suspending medium and it increases with increase in dielectric constant [22]. Hence the choice of the suspension parameters needs to be made judiciously for preparation of a suitable EPD suspension. Once the parameters related to the suspension are fixed, the process parameters can be altered conveniently for attaining desired deposition. Obviously, the most dominant parameters influencing the electrophoretic deposition are the process parameters such as applied voltage, deposition time and particle concentration in the suspension. Invariably, high applied potential leads to higher deposition rate but care has to be taken to ensure stable current density to obtain uniform deposit.

5.2.4 Aqueous Electrophoretic Deposition

In general, organic liquids are used as the suspending medium in EPD because the use of water results in electrolysis of water and evolution of hydrogen and oxygen gases at the electrodes on application of electric field. Incorporation of these gas bubbles in the deposit leads to low deposit quality. But the organic liquids are hazardous, expensive and not environmental friendly. The use of aqueous system has important advantages such as requirement of much lower voltage/current than that for organic solvents, higher temperature control during the process, faster deposition kinetics, in addition to important health benefits, benign environment and low cost. These advantages have promoted interest in many research groups to develop water based EPD to process particulate materials and technical ceramics. Several different approaches including use of porous mold/membrane [34–36], oxidizable anode [37, 38], addition of chemical compounds that consumes hydrogen for its reduction [39], or oxygen for oxidation [40], use of hydrogen absorbing electrode such as palladium (Pd) [41–43], and use of two chamber cell with electrolytes separated by membranes for preventing gas transport [44] etc. have been reported in the literature. Each of the above approaches suffers from some limitation or the other. In the quest for development of a generic and better aqueous EPD system applicable to all types of substrates, Besra et al. [45, 46] reported the use of pulse DC for suppressing and controlling the extent of bubble incorporation during EPD from aqueous suspension. Pulse EPD was conducted at constant voltage or current mode by application of a series of direct voltage or current pulse of equal amplitude separated by periods of zero current, using a source meter. Simple square-wave

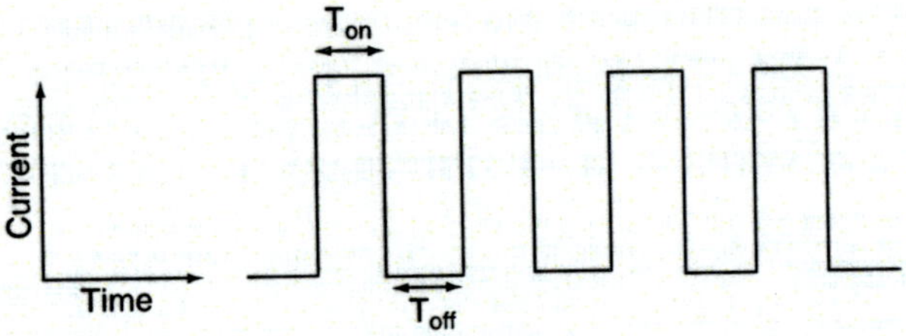

Fig. 5.6 Schematic of constant current pulse of 50% duty cycle. [45]

pulses of desired duty cycles were used. The duty cycle (d.c.) of the pulse [i.e. d.c. = $T_{on}/(T_{on}+T_{off})$] were varied by keeping the pulse ON time (T_{on}) constant and varying the pulse OFF time (T_{off}). Figure 5.6 shows a typical current pulse wave of 50% duty cycle.

As an example, Fig. 5.7 presents the surface morphology of α-alumina (Al_2O_3) deposited on stainless steel (316 L) at different pulse widths of 50% duty cycle for an applied current of 4 mA. Application of continuous DC resulted in maximum bubble incorporation in the deposit. The bubble incorporation in deposits invariably decreased on application of pulse current instead of continuous current. In general, more bubbles were found in deposits formed with larger pulse widths than those formed with smaller pulse widths. Below a suitable pulse width (0.005 s in this case) the deposits were devoid of any incorporated bubbles.

Figure 5.8 shows the deposit yield as a function of pulse width at varying applied currents. It must be noted that a current pulse of 180 s is equivalent to continuous DC EPD for 3 min. The deposit yield is maximum at continuous DC for each of the applied current. Analogous to bubble incorporation, the application of pulse current invariably decreased the deposit yield. The yield is higher for high applied currents at a given pulse widths. For a given applied current the yield decreases progressively with decrease in pulse width. The decrease in deposit yield with decrease in pulse width is attributed to lower pH localization effect at smaller pulse compared to larger pulse currents. The pH shifted faster and closer to the isoelectric point (i.e.p.) for larger pulse compared to the smaller ones.

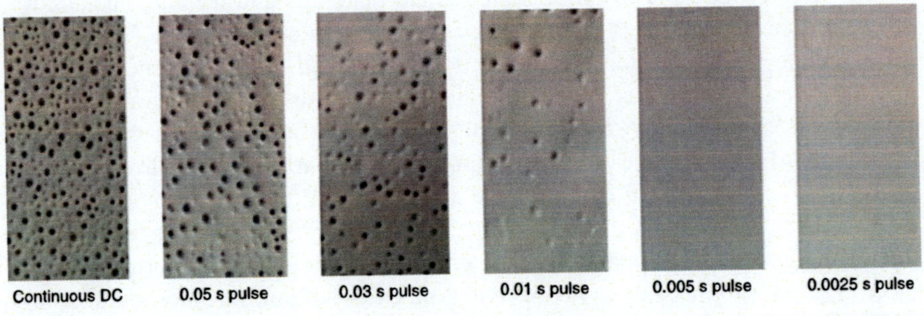

Fig. 5.7 Surface morphology of deposits obtained by pulsed DC EPD in constant current mode at an applied current of 4.0 mA. (Suspension: 5 vol%, pH 4.5, Substrate: stainless steel (316 L), Pulse ON time: 3 min pulse duty cycle: 50%, Inter-electrode distance: 20 mm). [46]

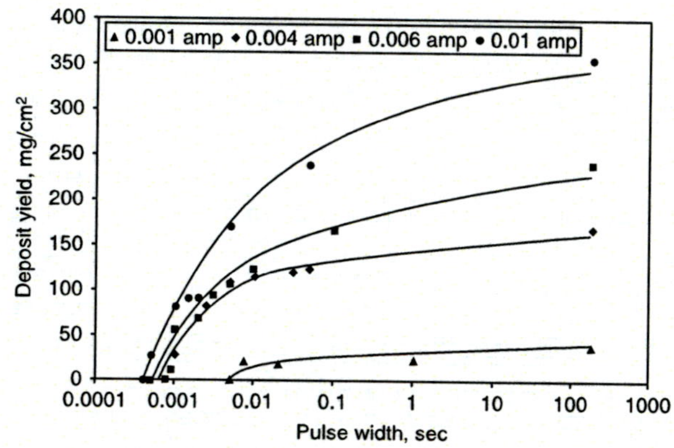

Fig. 5.8 Deposit yield as a function of pulse width during pulsed DC EPD at constant current mode (Suspension: 5 vol%, pH 4.5, Substrate: stainless steel (316 L), Pulse ON time: 3 min pulse duty cycle: 50%, Inter-electrode distance: 20 mm, pulse width of 180 s is equivalent to continuous DC EPD of 3 min). [46]

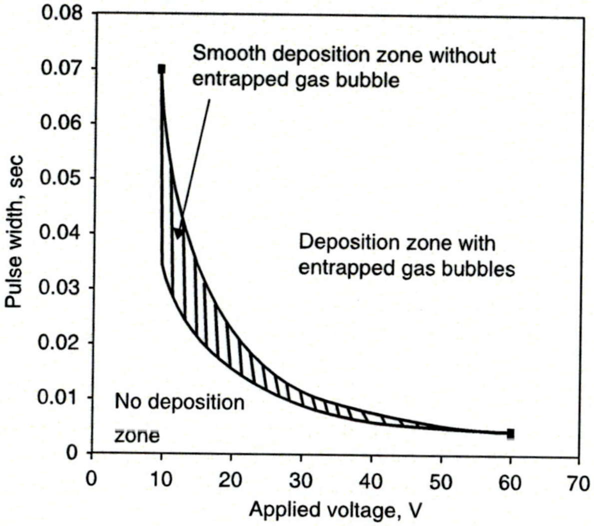

Fig. 5.9 Pulse width vs. applied potential diagram showing regions of bubble-free deposition from aqueous suspension for alumina suspension onto stainless steel (316 L) (powder concentration: 5.0 vol%, pH. 4.5, pulse ON time (T_{on})=3 min). [45]

Besra et al. [46] conducted a series of experiments for each applied current or voltage and observed that there exists a window in the plot of pulse width against pulse current or voltage within which bubble-free deposits are obtained (Fig. 5.9). At pulse widths above the window, the deposits invariably contained incorporated bubbles. At pulse widths below the window, no deposition was observed.

5.2.5 *Submonolayer Formation During Electrophoretic Deposition: Similarity with Atomic-Scale Molecular Beam Epitaxial (MBE) Growth Process*

Sarkar et al. [4] have followed the process of submonolayer formation as a function of deposition time during colloidal film growth of silica particles on a silicon wafer substrate by EPD. They compared the process of nucleation (or deposition of single particle), growth (or formation of a cluster of particles) and aggregation (or merging together of several particle clusters) of silica monolayers by the EPD technique with that for atomic thin-film growth (10,000 times smaller scale) process via molecular-beam epitaxy (MBE). They have shown a striking similarity between the two growth processes. Like the atomic thin-film growth process (MBE), the entire nucleation, growth and aggregation process during EPD of silica particles can be broadly classified into two regions. Diffusion-limited aggregation (DLA) is the mechanism at low surface coverage when silica particles are deposited outside of clusters, diffuse randomly and stick to a cluster on touching them and the fractal dimension of the two-dimensional clusters is ~ 1.65 [4]. As deposition proceeds, the clusters grow in size, deposition of particles inside the clusters become more and more important, and the cluster become increasingly more compact, producing a dense and closed-packed monolayer. This is termed as consolidation region where the fractal dimension of clusters rapidly changes from ~ 1.65 towards 2 as the surface coverage increases [4].

5.3 Ceramic Coating by EPD: How to Make It Survive During Drying and Sintering

Depositing ceramic or other powder on a metal (or a conductive surface) by EPD is relatively simple, but the most important factor is how to avoid cracking in the ceramic coating during drying and sintering. During drying and sintering, the coating densifies and shrinks, but the substrate typically does not change dimension. As a result, tensile stresses are developed in the coating and are relieved by the formation and propagation of cracks that originate from flaws or defects in the coating.

There are several approaches that can be adopted to avoid cracking during drying. When a wet coating or green body undergoes drying, the origin of stresses in it is due to capillary forces. During drying, cracking can be avoided by minimizing capillary stresses either by using a low surface tension solvent like ethanol or avoiding fine pore structure in the coating. In freeze/supercritical drying, the solvent phase is sublimed instead of vaporization and capillary stresses can be removed altogether. Furthermore, controlled slow drying via adjustment of vapor pressure of solvent in the drying medium will help control the capillary stresses and as a result, formation of drying cracks may be avoided.

During sintering, ceramic coatings typically suffer 10–15% linear shrinkage and the critical issue is how to avoid cracking. Traditionally, cracking during sintering is avoided by using a liquid phase during sintering; a good example is sintering of glass enamel on a metal substrate. The enamel composition is adjusted in such a way that its thermal expansion is closely matched with the substrate. As a result, it does not form crack during cooling from the sintering temperature. Sarkar et al. [47, 48] have used liquid phase sintering to avoid cracking in a superconductor coating on a silver substrate. The approach of liquid phase sintering is also equally effective in avoiding cracking in fiber composites. To avoid cracking during sintering, one can also use a substrate that also shrinks during sintering. Recently this approach has been used, particularly in SOFC fabrication where typically an YSZ electrolyte is added to a partially sintered or an unsintered anode substrate. During sintering, both the substrate and coating shrink, thereby avoiding cracks. Furthermore, the compositions of the substrate and coating are chosen such a way that their thermal expansion coefficients match to each other, minimizing any thermal stress and resultant cracking during cooling from the sintering temperature.

The problem of cracking in electrolytic zirconia deposits which usually occurs upon drying has recently been addressed by the use of polymer additives [49, 50]. It was shown [49] that poly(diallyldimethylammonium chloride) (PDDA) acts as a binder, providing better adhesion of zirconia deposits and preventing cracking. An important finding was that the amount of organic phase in the deposits could be changed by variation of PDDA concentration in solutions. These results pave the way for EPD of thick films.

Another unique approach to avoid cracking during sintering is by using a reaction-bonding technique. Sarkar et al. [51] have found the development of tunnel and radial cracks during sintering of a polycrystalline alumina coating on a sapphire filament substrate. This is an interesting example where the substrate and the coating are from the same material. The substrate is a single crystal alumina filament that suffers no change in dimension during sintering, whereas the coating is a polycrystalline alumina made from alumina powder and has considerable shrinkage during sintering. To overcome these cracking, the RBAO (Reaction Bonded Aluminum Oxide) process [51] is used. In this process, the starting materials for the coating are alumina and aluminum powder. The sapphire filament is initially coated with a thin carbon or gold layer to make it electrically conductive. On this filament substrate, alumina and aluminum powder are co-deposited from an ethanol based Al_2O_3–Al suspension of predetermined composition by EPD. During sintering, Al oxidizes to alumina and, as a result, expands and counter balances the sintering densification shrinkage of Al_2O_3, thereby avoiding cracking [51].

5.4 Application: Flexibility of EPD Towards Synthesis of High-Performance Microstructures and Their Manipulation in Various Areas

5.4.1 Structural and Functional Composites

Sarkar et al. [52–56] have demonstrated the versatility of EPD process towards synthesizing a wide spectrum of microstructures having different morphology and composition. They used EPD efficiently and with imagination to fabricate ZrO_2/Al_2O_3 continuous functionally graded material (FGM) [53, 54] and planar and non-planar laminates of ZrO_2/Al_2O_3 having as many as 80 interlayers [52, 54, 55]. They also demonstrated that the interface of layers in the planar and non-planar laminates are smooth and of high-perfection. They created the wavy layers in non-planar laminates by depositing on line electrodes on a printed circuit [54]. EPD has also been efficiently used to fabricate ceramic-fiber composites [54–56]. Using EPD, Sarkar et al. [56] fabricated composites where a graphite cloth was impregnated with zirconia powder and a Nicalon fiber mat was impregnated with ZrO_2/Al_2O_3 mixture and consolidated by hot pressing [54]. Recently, Illston et al. [57] and Boccaccini et al. [58] have fabricated ceramic-fiber composites where SiC (Nicalon) fiber mats were impregnated with SiO_2 and Al_2O_3 sol from their aqueous suspensions by a combination of EPD and dip coating process.

5.4.2 *Solid Oxide Fuel Cell*

The EPD technique has been increasingly considered for the fabrication of SOFCs due to the simplicity of the experimental setup, possibility of mass production, high deposition rate, short formation times, and the ease of thickness control. A single SOFC cell consists of a solid electrolyte sandwiched between a cathode on one side and an anode on the other side. Both electrodes are required to be porous for transport of reactants and products through them, while the electrolyte must be dense to prevent leakage of gas and shorting between the two electrodes. The electrolyte layer, which is basically an ion conducting oxide such as YSZ, is required to be as thin and dense as possible to minimize polarization losses of SOFC during its operation at high temperatures. Several researchers have demonstrated the application of EPD as a low-cost technique for fabrication of thin and dense electrolyte on porous anode or cathode for SOFC application [18, 59, 60]. Earlier literatures reported deposition of electrolyte materials such as YSZ, GDC or LSGM on electrically conductive cathode layer (e.g. $La_{0.8}\ Sr_{0.2}\ MnO_3$ (LSM) or LSM-YSZ) understandably because of the pre-requisite that the electrodes for EPD process are required to be electrically conductive. But the cathode supported SOFC have their own limitations and not very popular because of the difficulties in fabricating a dense thin electrolyte on a porous cathode by conventional methods. For instance, densification of thin electrolyte in the well-known wet-ceramic co-firing technique is generally achieved by co-sintering the laminated electrolyte with the support electrode at a relatively high temperature, i.e., 1,400°C for a typical Ni–cermet anode-supported cell. This represents a high risk of introducing several severe issues for the cathode-supported cell that includes: (1) chemical reactions between cathode and electrolyte, (2) excessive coarsening of cathode particles with loss of triple phase boundaries (TPB) for oxygen reduction reaction, and (3) dense electrode structure with little porosity, which significantly raises the impedance to gas transport and ultimately decreases the SOFC performance.

Recently, Matsuda et al. [61, 62] and Besra et al. [63] showed that deposition of YSZ on non-conducting NiO-YSZ substrate by EPD is possible by using an

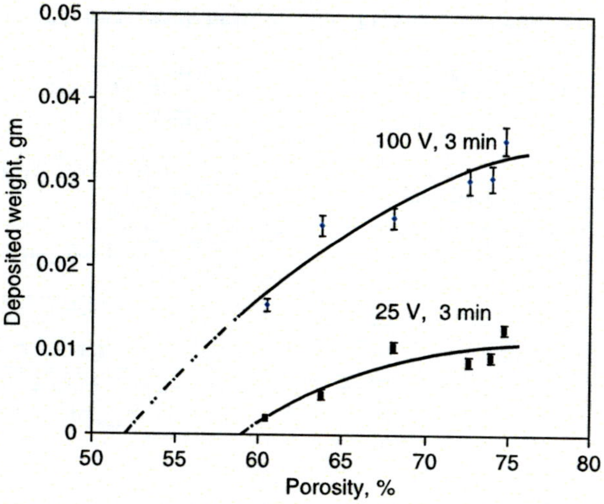

Fig. 5.10 Influence of substrate porosity on electrophoretic deposition of YSZ on non-conducting NiO–YSZ substrate from its suspension in acetylacetone (deposition area: 1.08 cm^2). [63]

electrically conductive contact on the back side of a porous substrate. Deposition of YSZ occurred on the front face of the substrate in contact with the suspension. Besra et al. [63] further showed that the porosity of substrate is the key parameter that decides the success or failure of deposition using this approach. They suggested that deposition is possible by using adequately porous substrate, which on saturation with the liquid medium establishes a "conductive path" through itself to ensure the presence of electric field on its front face where deposition occurs. It has also been shown that a threshold porosity exist for each applied voltage above which deposition by EPD is feasible. No deposition occurred for denser substrates that had porosity below the threshold porosity (Fig. 5.10).

EPD can also be very well-suited for fabrication of small complex shapes. One area where the authors are actively involved and foresee considerable potential of EPD is the fabrication of Micro-Solid Oxide Fuel Cell (μSOFC). Small diameter SOFC has two main potential advantages, substantial increase in the electrolyte surface area per unit volume of a stack and also quick start up. Since fuel cell power is directly proportional to the electrolyte surface area, a μSOFC stack has high potential to substantially increase the power per unit volume. Sarkar et al. [3, 64] have demonstrated that power output of a cell synthesized by EPD is comparable to any standard tubular SOFC. A SEM micrograph of cross-sectional fracture surface of a single cell produced by EPD is shown in Fig. 5.11. The electrolyte layer of the cell is <10 μm and anode functional layer is ~5 μm. Figure 5.12 shows the Current-Voltage (IV) and Current-Power (IP) plots of the single cell at three different temperatures. A mixture of 30% hydrogen and 70% helium was used as a fuel gas. The fuel gas contained 3% moisture as it was bubbled through water at 25°C. The fuel gas flow rate was 30 SCCPM. This cell has produced theoretical open circuit voltage indicating that the membrane does *not* have pinholes. Figure 5.12 demonstrates that at all three temperatures, activation polarization are absent. Concentration polarization was not observed in the measurement range. Approxi-

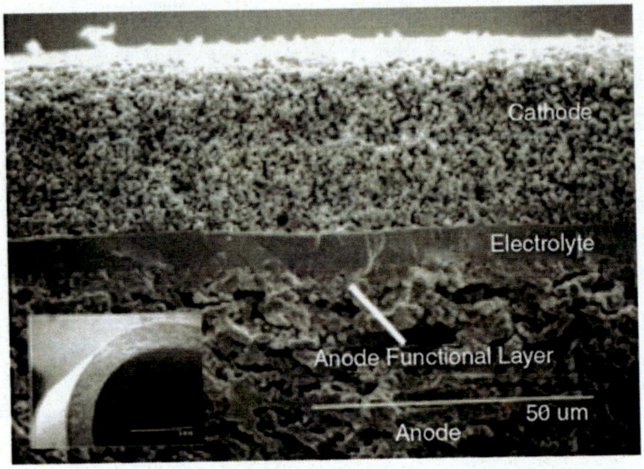

Fig. 5.11 SEM micrograph of cross-sectional fracture surface of a single cell [3]

mately a 2 cm length of the cell was coated with cathode; this is the active length of the cell during measurement. At 800°C, peak power output is over 300 mW with a corresponding voltage and current at ~0.5 V and ~600 mA respectively. At 750°C, peak power is ~255 mW and at 700°C, peak power is ~215 mW. Thus, Sarkar et al. [3, 64] have clearly demonstrated that EPD can simplify the forming process and reduce the production cost considerably, which is one of the major barriers for commercialization of fuel cells.

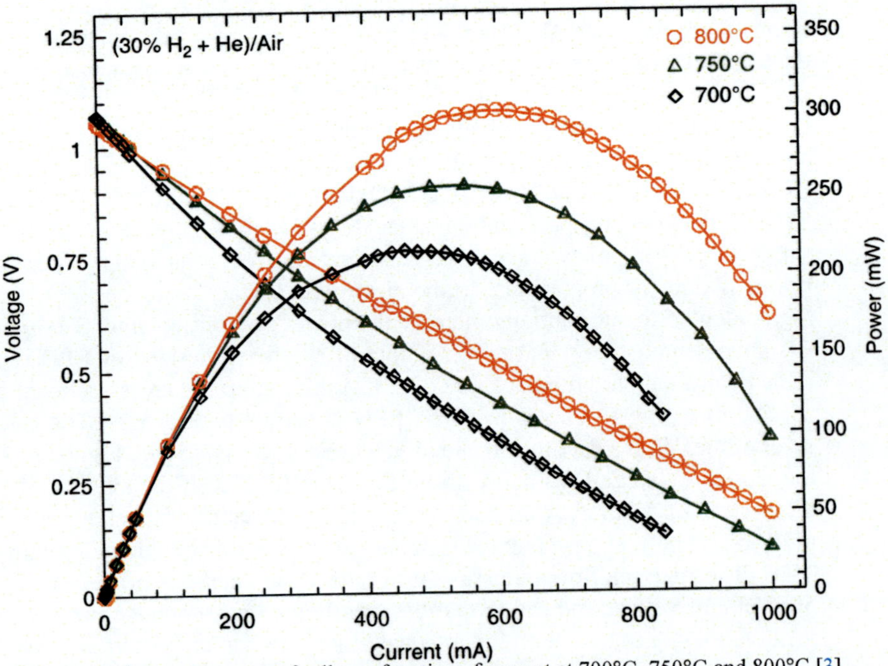

Fig. 5.12 Voltage and power of cell as a function of current at 700°C, 750°C and 800°C [3]

5.4.3 Textured Ceramics with Crystalline Oriented Microstructure

Extended and oriented micro and nanostructures are desirable for many applications and the controlled development of texture in polycrystalline materials is a topic of recent interest in ceramic processing, since it allows improvement of physical properties [65]. Physical properties of ceramics that could be tailored by texture control include electrical and thermal conductivity, superconductivity, and magnetic, dielectric, and piezoelectric properties [47, 66, 67]. Sarkar and Nicholson [48] deposited $YBa_2Cu_3O_7$ superconductor precursor powder on silver foil by EPD and then texture was developed by performing reactive sintering under magnetic field. Crystallographic texture reflects the preferred orientation of crystal lattice in the material, while morphological texture is a measure of the preferred orientation of particle morphology. But direct fabrication of complex micro and nanostructures with controlled crystalline morphology, orientation and surface architectures remains a significant challenge.

Textured monolithic and laminate of ceramics were fabricated by EPD in strong magnetic field by Uchikoshi et al. [68–70]. They showed that it is possible to control the crystalline orientation of ceramic film laminates by conducting the EPD in a strong magnetic field (10 T) and varying the angle between the applied electric field ***E*** relative to the magnetic field ***B***. This approach exploits the anisotropic magnetic susceptibilities associated with materials with asymmetric (non-cubic) crystalline structures. When a single crystal of these materials is placed in a magnetic field, the crystal is rotated and the crystallographic axis of high susceptibility (χ) is aligned in the direction of the magnetic field when the energy of anisotropy is higher than the energy of thermal motion (Fig. 5.13). When the electric field is then applied to the oriented particles in the solvent, they move along with the electric field while retaining their orientation relative to the magnetic field, and then deposit on the substrate.

Figure 5.14 shows variation in XRD pattern of the top plane of α-alumina electrophoretically deposited in 10T magnetic field with varying angle of 0, 45 and

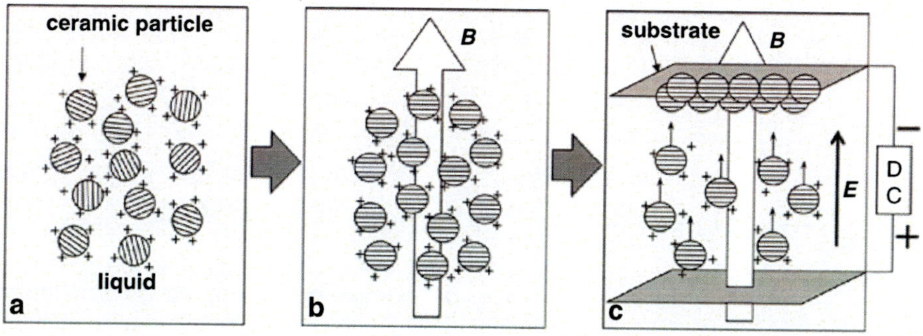

Fig. 5.13 Schematic diagram of the concept of the electrophoretic deposition in a high magnetic field; **a** deflocculated suspension, **b** orientation of the particles, **c** deposition of the particles. [69]

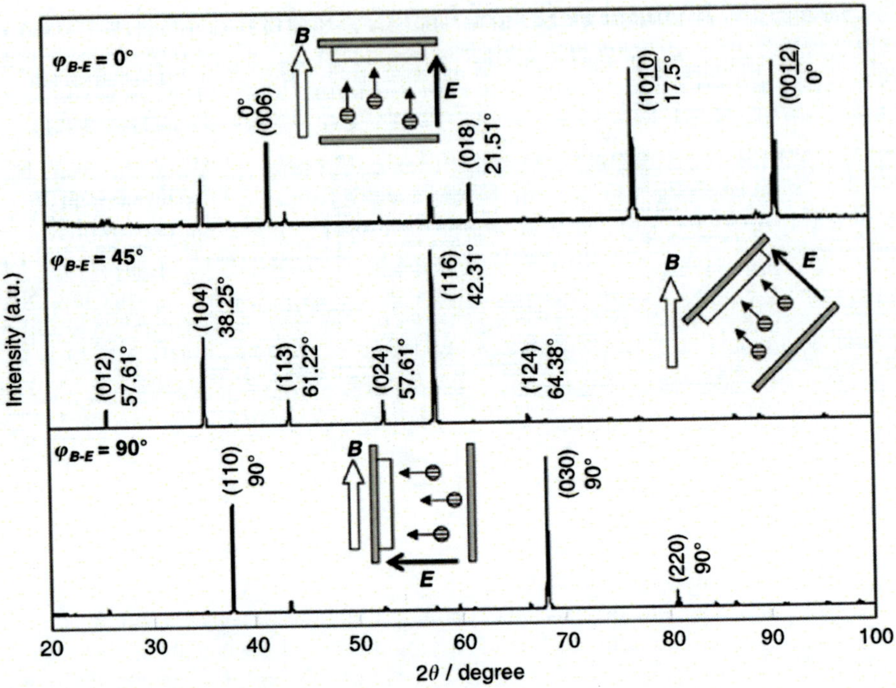

Fig. 5.14 Changes in the XRD pattern of the top plane with change in angle between *B* and *E*. [71]

90° between **B** and **E** after sintering at 1873 K for 2 h [71]. When the **E** is parallel to **B** (φB–E=0°), the diffraction peaks of the planes at low interplanar angles ($\Phi_{h\,k\,l}$ is close to 0°) are predominant. When the $\Phi_{h\,k\,l}$ is changed to 45°, the dominant diffraction peaks change to the planes of the intermediate interplanar angles (φB–E is close to 45°). When **E** is perpendicular to **B** (φB–E=90°), the dominant diffraction peaks changed to the planes of high-interplanar angles ($\Phi_{h\,k\,l}$ is close to 90°). These results suggest that the dominant crystal faces can be controlled by varying the angle between **B** and **E**.

An example of textured laminated alumina prepared by EPD is shown in Fig. 5.15. Here the cross-sectional microstructure was prepared by alternately changing φB–E=0° and 90° layer by layer. The alignment of grains in the microstructure compare well with the alteration in φB–E during EPD.

5.4.4 Bio-compatible Coatings on Metal Implants

It is now well established to have bio-compatible coating on metallic implants for improving the life of the implants in the body. Uncoated implants are susceptible to reaction with body fluids leading to their premature degradation. Because of the osteoconductive properties of calcium phosphate, hydroxyapatite $Ca_{10}(PO_4)_6\,(OH)_2$

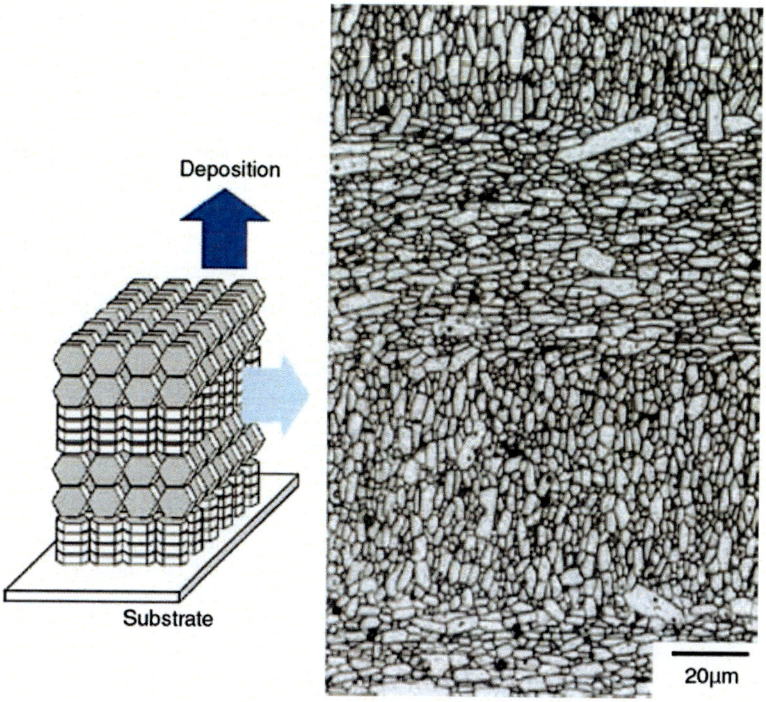

Fig. 5.15 Cross-sectional image of laminar composite prepared by EPD at changing relative angle of 0 and 90° between *B* and *E*. [71]

(HAp) has long been recognized as the suitable candidate as bio-compatible coating material. Conventionally, the coating of hdroxyapatite on metal implants is done using expensive plasma spray technique, sputter coating, biomimetic coating, doctor blade, etc. However, the plasma spray method has some important drawbacks; non-uniformity in coating density resulting in inability to cover porous implants and to incorporate biologically active agents, delamination and particle release [72]. Secondly, since the plasma spraying is done at high temperature, there is a tendency of HAp getting decomposed into undesirable soluble phases namely tricalcium phosphate (TCP). Ideally the HAp coating must be sintered at temperatures below 1,000 °C to prevent decomposition of HAp. For this, the HAp nanoparticles must be well packed in the deposit. The coating should be of good adhesive strength to avoid spalling, and be crack-free for its application as bio-compatible coating. As a result, it is necessary to develop inexpensive deposition methods for achieving all the above requirements, which is a major challenge.

EPD represents an important technological alternative due to rapid coating production, high reproducibility and low cost of the process. Deposition of HAp has been successfully carried out on Ti, Ti_6Al_4V and 316 L stainless steel metal substrates [31, 73, 74] by EPD to form uniform and crack-free film of HAp with good adhesion from stable suspensions in organic solvents like ethanol, and isopropyl alcohol, as well as from aqueous suspension [75]. Stoch et al. [76] pre-covered the

Ti surface with silica or calcium silica sol-gel film prior to EPD coating of HAp to improve its adhesion and performance.

Hamagami et al. [77] prepared HAp coating with highly ordered macropores on Titanium sheet by EPD from a mixture of nanodisperse polystyrene (PS) sphere (3 μm) and HAP (200–300 nm) in ethanol to prepare organic-inorganic composite film. Then, it was annealed at 900°C to remove organic PS and sinter the inorganic HAP. Ti plate, with such ordered porous HAp macropores, possessed good biocompatibility as confirmed by in-vitro testing using simulated body fluid. Sridhar et al. [78] demonstrated through measurement of corrosion current (E_{corr}), breakdown potential (E_b) and re-passivation potential (E_p) in a simulated body fluid that all the HAp coated samples were nobler and exhibited improved corrosion resistance compared to uncoated 316 L steel.

5.4.5 *EPD for Pseudo Light Emitting Devices*

Recently, the rare earth doped nitrides and oxynitrides phosphor materials have attracted significant attention because of their interesting photoluminiscent properties, non-toxicity and thermal stability. Among such nitrides/oxynitrides, the yellow Eu^{2+} doped Ca-α-SiAlON have been shown to exhibit photoluminiscent properties commensurate to pseudo-white light emitting diodes (LEDs). Hence, the SiAlON phosphors are seen as materials for application to general illumination and liquid crystal display (LCD) in near future. It is essential to establish particle packaging technologies for its application. Kitabatake et al. [79] demonstrated deposition of Eu^{2+} doped Ca-α-SiAlON phosphor films on ITO glasses by electrophoretic deposition (EPD) for its application in pseudo-white light emitting diode. The SiAlON phosphor powder contained large secondary aggregates of 1–20 μm size and hence required stirring to keep the powder in suspension. Deposition was conducted under two conditions. In condition 1, the EPD was conducted immediately after stirring was stopped. In condition 2, EPD was conducted 1 min after stirring was stopped. The thickness of the SiAlON phosphor films was controlled by altering the deposition time. The photoluminescent (PL) properties of the SiAlON phosphor films were characterized by blue-light ($\lambda=450$ nm) irradiation through the ITO glass substrates as shown in Fig. 5.16.

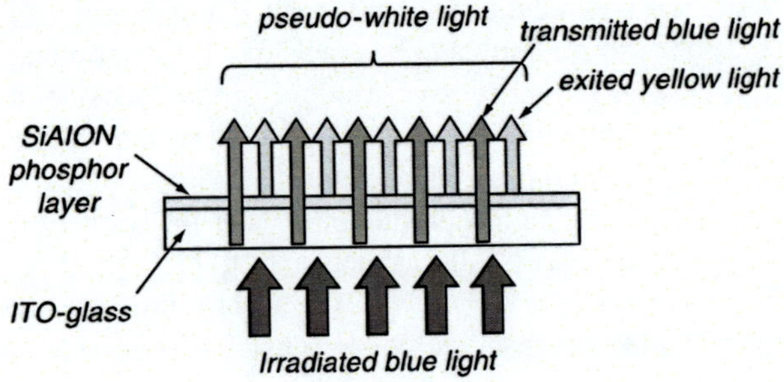

Fig. 5.16 Schematic illustration of a flat pseudo-white light emitting device. [79]

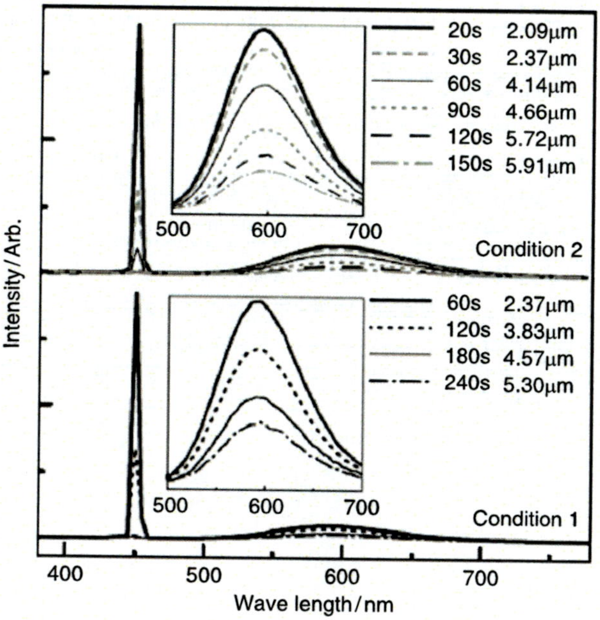

Fig. 5.17 Photoluminiscent spectra of Eu^{2+} doped Ca-α-SiAlON phosphor films. [79]

The light emission from the deposit surfaces was characterized by comparing the intensities of the excited yellow light and the transmitted blue light through the deposit. The chromaticity coordinates were calculated from the PL spectra and compared with actual colors perceived by human eyes. Upon irradiation by blue LED, from behind the substrate, the films showed visible colours from blue-white to yellow-white. Depending on the particle size and their packing obtained in condition 1 and condition 2, the visible light colour in the two films were different because of the different absorption and scattering of light by the films. The intensity of transmitted blue light at 450 nm as well as the excited yellow light at 585 nm decreased with increasing film thickness Fig. 5.17.

5.4.6 Ordered Assembly of Micro/Nanoscale Particulate Thin Films

Recently, there has been a rising interest in nanostructured materials because they often exhibit new properties that are remarkably different from those of bulk materials, allowing the manipulation of mechanical, electrical, magnetic, optical, catalytic and electronic functions. The ability to fabricate materials and structures with submicrometer and nanoscale features reliably and economically is of interest in many areas of science and technology such as photonic materials, high density magnetic data storage devices, microchip reactors, and biosensors.

The creation of nanophase or cluster-assembled materials is usually based upon the creation of separated small clusters as building blocks which are then fused into a bulk-like materials or as a thin film or even by embedding into a solid matrix, polymers or glass either in a disordered or ordered array. One specific goal in these investigations has been to tailor the size, shape, and geometrical arrangement of nanoscopic particles, clusters, and aggregates in an effort to produce desired properties. In order to make full use of this potential for materials engineering, it is important to identify and gain control over the relevant growth and ordering parameters.

Conventionally, micropatterned assembly of colloidal or nanoparticles have been produced with lithographically patterned electrodes, or micro moulds. The majority of these methods suffer from drawbacks such as: (1) slow kinetics of assembly (hours to days), (2) surface particle density is sometimes difficult to regulate, (3) difficulty in creating patterned assembly over large area. These drawbacks are largely a consequence of dealing with very small particles for which the colloidal forces i.e. buoyancy, frictional, and coulombic forces outweigh the gravitational force and act to keep the particles in suspension. Consequently, the particle transport is governed solely by Brownian motion, resulting in completely random particle-particle and particle-substrate collision.

Since the colloidal particles are usually charged, the concept of electrophoretic deposition utilizing a directional force (external electric field) has generated significant interest recently for patterned assembly of nanoparticles [80]. EPD of nanaoparticles was first used by Giersig and Mulvaney [81, 82] to prepare ordered monolayers of gold nano particles. Prior to that this method had only been used to produce 2D and 3D ordered structures of latex particles, such as polystyrene and silica [83, 84]. It has now been put forward by some researchers that the EPD method can be developed into a general method to prepare 2D and 3D ordered structures of nanoparticles by means of using sols that maintain colloidal stability on the electrode/sol interface; however, this conclusion is based on the work done on micrometer-sized polystyrene, silica particles, and nanometer-sized gold particles and it requires further evidence. Tabellion and Clasen [85] investigated the deposition structure of gold, silver and rare-earth carbonate nanoparticles by EPD and found that the composition and surface condition as well as the size distribution of the particles can affect the ordering of the particles in the monolayer. It was also observed that particles having similar diameter tended to pack together to form more orderly packed area compared to particles with wide size distribution. The ordering of Au nanoparticles thus obtained for different deposition conditions are shown in Fig. 5.18.

Sarkar et al. [4] observed interactions between particles over distances of several particle diameters, especially for interactions between single particle and clusters. Even they reported cluster–cluster aggregation phenomena. Once the clusters are formed, they rapidly grow. The distance over which they attract newly-arrived particles on the surface increases. They also reported that a particle arriving at the top of an existing cluster diffuse until it falls off the edge of cluster and the particles at the cluster surface have surface mobility that drive them to attain their most energetically favourable position, i.e. the equilibrium position where they may be

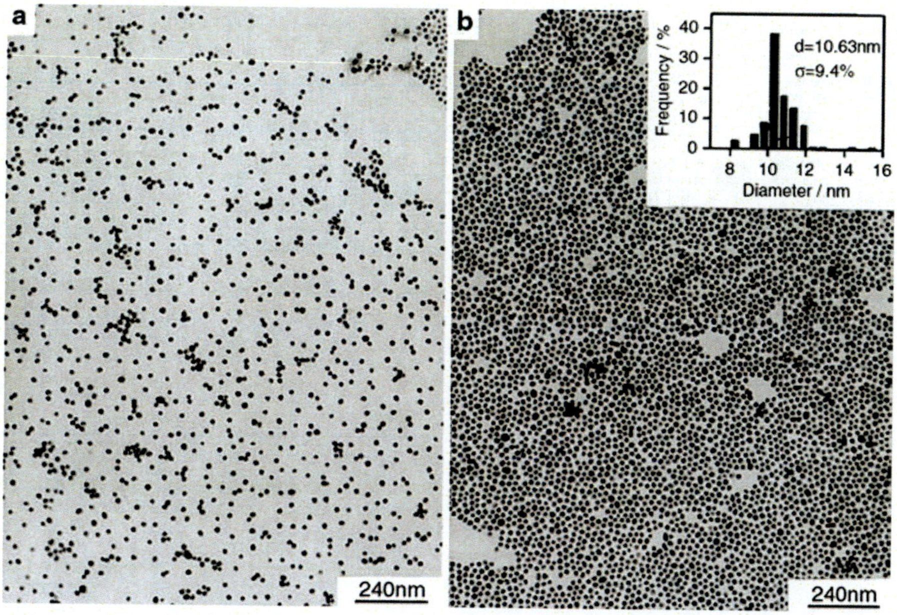

Fig. 5.18 Transmission electron microscope (TEM) images of Au nanoparticles on carbon coated copper grid formed by EPD at 20°C **a** at 100 mV cm^{-1} for 5 min, **b** at 500 mV cm^{-1} for 5 min. [85]

incorporated into the already existing lattice producing a closed-packed structure. It was also reported by Boehmer [86] that with increasing salt concentration, the deposition rate decreases because the conductivity of the suspension increases and, consequently, the potential drop across the suspension becomes smaller. The decrease in deposition/ aggregation of the particles at the electrode with increasing ionic strength does not agree with the DLVO theory. So the clustering is not a consequence of DLVO type interaction as proposed by Giersig and Mulvaney [81, 82] for gold nanoparticles on graphite. Here the interactions were long range, extending over one or more particle diameter, which is much larger than the Debye length. Boehmer [86] suggested that over this long range the distortion of the electric field by the particles and hydrodynamic interaction is important and has to be considered to formulate possible interpretation of such observation. Hayward et al. [87] proposed that once the particles are close to the surface, where they remain mobile, electro-hydrodynamic or electro-osmotic effects assemble them into arrays. Particles can be permanently attached to the surface by increasing the attractive forces between the particle and the electrode. When the attractive forces exceed that due to steric repulsion, entry into the primary minimum creates a permanent bond.

Kumacheva et al. [88] showed that transition from a disordered state to strongly ordered state of colloidal microsphere (PMMA particles of 0.58 μm size) occurs following their confinement to thinner gaps (Fig. 5.19). They have also observed that at the beginning of the electrophoretic deposition process, the structures of colloidal array in the groove are essentially random (Fig. 5.19b). However, as more particles

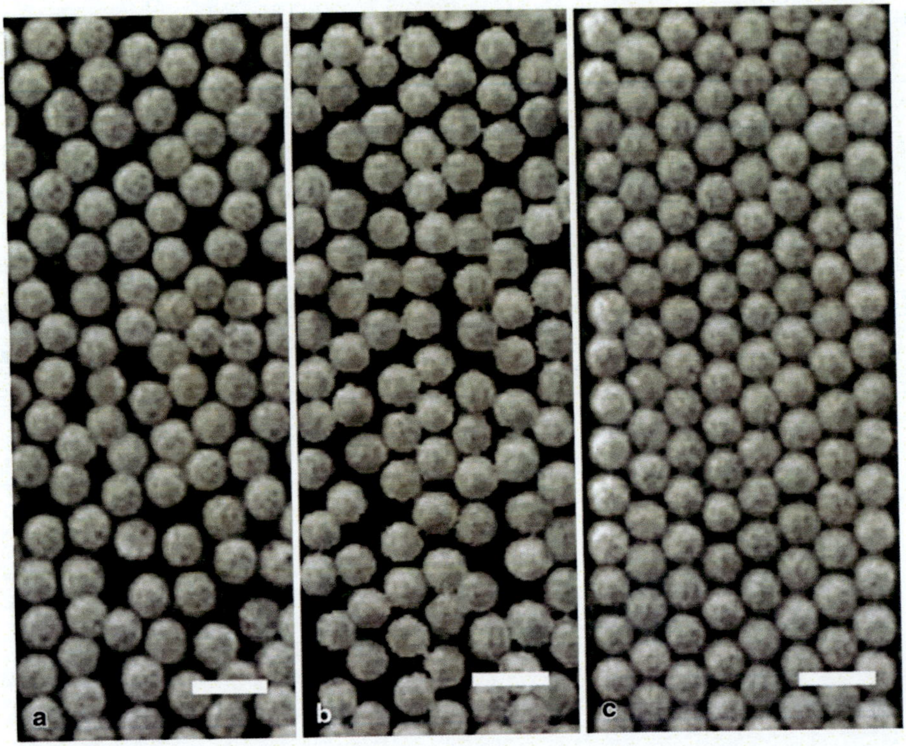

Fig. 5.19 SEM images showing PMMA particle arrays electrophoretically deposited on **a** non-patterned, and **b**, **c** patterned indium tin oxide (ITO) surface. The size of PMMA particles are 0.58 μm and the width of the grooves are 5.5 μm for (**a**) and 4.2 μm for (**b**, **c**). [88]

reach the electrode, reorganization of the particles happen in the groove by squeezing of the newly arriving particles between the already deposited ones and follows a synergistic particle rearrangement. This resulted in large scale particle ordering for assembly of the colloidal spheres (Fig. 5.19c).

5.5 Critical Issues, Myths and Future Applications of EPD

To produce a dense, homogeneous and close-packed green body, EPD must be performed from a dilute, stable colloidal suspension where interparticle forces keep the particles well dispersed. During EPD from a dilute and stable suspension, particles are expected to move and deposit individually. In unstable suspensions, loose flocs or an ensemble of particles are expected to move and deposit together. Unlike in stable suspension, the growth and consolidation of these loose flocs or ensemble of particles during deposit formation may be different and needs further research.

Furthermore, the electrical resistivity of these flocs (larger volume fraction of them is occupied with continuous phase) may be substantially different from that of a compact deposit (achieved in stable suspension) and the kinetics of deposition may be significantly different. Other areas of potential research are the dynamics of deposit formation from stable and unstable suspensions with polydisperse particles. It will be interesting to investigate whether the particles are segregated in suspension (different electrophoretic mobility) or they move as an ensemble of particles and how does it affect the homogeneity and density of the green deposits. Also, more imaginative research is needed to identify similarities and dissimilarities of the process of EPD with that of other external-field induced consolidation techniques such as sedimentation, slip-, pressure-, and centrifugal casting.

5.6 Summary and Closing Comments

Discussions in this chapter clearly establish that EPD is a powerful and versatile forming or consolidation technique. Like an ideal consolidation process, it can produce homogeneous, dense green bodies and complicated shapes effectively and easily. Furthermore, if practiced right and with imagination, a facile technique like EPD allows flexibility in microstructural manipulation i.e., a wide montage of advanced micro- and nano-microstructures ranging from dispersed, laminated, fiber-composites to functionally graded materials can be fabricated by EPD. There are still critical fundamental issues that need to be understood to understand the EPD process better.

Acknowledgements The authors are thankful to several coworkers for useful discussions during the course of the research discussed in this manuscript.

References

1. Sarkar, P., Nicholson, P.S.: Electrophoretic deposition (EPD): Mechanisms, Kinetics and Applications to Ceramics. J. Am. Ceram. Soc. **79**, 1987–2002 (1996)
2. Kellet, B.J., Lange, F.F.: Thermodynamics of densification: I, sintering of simple particle arrays, equilibruim configurations, pore stability, and shrinkage. J. Am. Ceram. Soc. **72**, 725–734 (1989)
3. Sarkar, P., De, D., Rho, H.: Synthesis and microstructural manipulation of ceramics by electrophoretic deposition. J. Mater. Sci. **39**, 819–823 (2004)
4. Sarkar, P., De, D., Yamashita, K., Nicholson, P.S., Umegaki, T.: Mimicking nanometer atomic processes on a micrometer scale via electrophoretic deposition. J. Am. Ceram. Soc. **83**, 1399–1401 (2000)
5. Frens, G., Overbeek, JT.: Repeptization and theory of electrocratic colloids. J. Colloid Interface Sci. **38**, 376–387 (1972)
6. Koelmans, H., Van Boxte, A.M.: Electrohydrodynamic flow in nematic liquid crystals. Mol. Cryst. Liq. Cryst. **12**, 185–191 (1971)

7. Shimbo, M., Tanzawa, K., Miyakawa, M., Emoto, T.: Electrophoretic deposition of glass powder for passivation of high-voltage transistors. J. Electrochem. Soc. **132**, 393–398 (1985)
8. Mizuguchi, J., Sumi, K., Muchi, T.: A Highly Stable Non-aqueous Suspension for the Electrophoretic Deposition of Powdered Substances. J. Electrochem. Soc. **130**, 1819–1825 (1983)
9. Brown, D.R., Salt, F.W.: The mechanism of electrophoretic deposition. J. Appl. Chem. **15**, 40–48 (1965)
10. Hamaker, H.C.: Formation of deposition by electrophoresis. Trans. Faraday Soc. **36**, 279 –287 (1940)
11. Hamaker, H.C., Verwey, E.J.W.: Colloid stability: the role of the forces between the particles in the electrodeposition and other phenomena. Trans. Faraday Soc. **36**, 180–185 (1940)
12. Giersig, M., Mulvaney, P.: Formation of ordered two-dimensional gold colloid lattices by electrophoretic deposition. J. Phys. Chem. **97**, 6334–6336 (1993)
13. De, D., Nicholson, P.S.: Role of ionic depletion in deposition during electrophoretic deposition. J. Am. Ceram. Soc. **81** (11), 3031–3036 (1999)
14. Besra, L., Uchikoshi, T., Suzuki, T.S., Sakka, Y.: Experimental verification of pH localization mechanism of particle consolidation at the electrode/solution interface and its application to pulsed DC Electrophoretic deposition (EPD). J. Eur. Ceram. Soc. **30**, 1187–1193 (2010)
15. Besra, L., Liu, M.: A review on fundamental and applications of Electrophoretic deposition. Prog. Mater. Sci. **52**(1), 1–61 (2007)
16. Avgustinik, A.I., Vigdergauz, V.S., Zharavlev, G.I.: Electrophoretic deposition of ceramic masses from suspension and calculation of deposit yields. J. Appl. Chem. USSR (English Translation). **35**(10), 2175–2180 (1962)
17. Biesheuvel, P.M., Verweij, H.: Theory of cast formation in electrophoretic deposition. J. Am. Ceram. Soc. **82**(6), 1451–1455 (1999)
18. Ishihara, T., Shimise, K., Kudo, T., Nishiguchi, H., Akbay, T., Takita, Y.: Preparation of Yttria-stabilised zirconia thin-films on strontium doped $LaMnO_3$ cathode substrate via Electrophoretic deposition for solid oxide fuel cells. J. Am. Ceram. Soc. **83**(8), 1921–1927 (2000)
19. Chen, F., Liu, M.: Preparation of yttria-stabilised zirconia (YSZ) films on $La_{0.85}Sr_{0.15}MnO_3$ (LSM) and LSM-YSZ substrate using an electrophoretic deposition (EPD) process. J. Eur. Ceram. Soc. **21**, 127–134 (2001)
20. Heavens, N.: Electrophoretic deposition as a processing route for ceramics. In: Binner, G.P. (ed.) Advanced ceramic processing and Technology, vol. 1. Noyes Publications, Park Ridge (1990)
21. Sato, N., Kawachi, M., Noto, K., Yoshimoto, N., Yoshizawa, M.: Effect of particle size reduction on crack formation in electrophoretically deposited YBCO films. Physica C. **357–360**, 1019–1022 (2001)
22. Powers, RW. The electrophoretic forming of beta-alumina ceramic. J Electrochem Soc. **122**, 482–486 (1975)
23. Negishi, H., Yanagishita, H., Yokokawa, H.: Electrophoretic deposition of solid oxide fuel cell material powders. In: Proc. Electrochemical society on Electrophoretic deposition: Fundamentals and Applications, vol. 2002–21. The electrochemical society Inc, Pennington, USA (2002)
24. Ferrari, B., Moreno, R.: The conductivity of aqueous Al_2O_3 slips for electrophoretic deposition. Mater. Lett. **28**, 353–355 (1996)
25. Ferrari, B., Moreno, R.: Electrophoretic deposition of aqueous alumina slip. J. Eur. Ceram. Soc. **17**, 549–556 (1997)
26. Krueger, H.G., Knote, A., Schindler, U., Kern, H., Boccaccini, A.: Composite ceramic metal coatings by means of combined electrophoretic deposition. J. Mater. Sci. **39**, 839–844 (2004)
27. Zarbov, M., Schuster, I., Gal-Or, L.: Methodology for selection of charging agents for electrophoretic deposition of ceramic particles. In: Proc. of the International symposium on Electrophoretic deposition: Fundamentals and applications, vol. 2002–21. The electrochemical society Inc, Pennington, USA (2002)
28. Brown, D.R., Salt, F.W.: The mechanism of electrophoretic deposition. J. Appl. Chem. **15**, 40–48 (1965)

29. Basu, R.N., Randall, C.A., Mayo, M.J.: Fabrication of dense zirconia electrolyte films for tubular solid oxide fuel cells by electrophoretic deposition. J. Am. Ceram. Soc. **84**(1), 33–40 (2001)
30. Wang, Y.C., Leu, I.C., Hon, M.H.: Kinetics of electrophoretic deposition for nanocryatalline zinc oxide coatings. J. Am. Ceram. Soc. **87**(1) 84–88 (2004)
31. Zhitomirsky, I., Gal-or, L.: Electrophoretic deposition of hydroxyapatite. J. Mater. Sci: Mater. Med. **8**, 213–219 (1997)
32. Vandeperre, L., Van Der Biest, O., Clegg, W.J.: Silicon carbide laminates with carbon interlayers by electrophoretic deposition. Key Eng. Mater. **1**, 127–131 (1997)
33. Peng, Z., Liu, M.: Preparation of dense platinum-yttria stabilized zirconia and yttria stabilized zirconia films on porous $La_{0.9}Sr_{0.1}MnO_3$ (LSM) substrates. J. Am. Ceram. Soc. **84**(2), 283–288 (2001)
34. Ryan, W., Massoud, E.: Electrophoretic deposition could speed up ceramic casting. Interceram. **2**, 117–119 (1979)
35. Ryan, W., Massoud, E., Perera, C.T.S.B.: Fabrication by electrophoresis. Trans. Brit. Ceram. Soc. **80**, 46–47 (1981)
36. Tabellion, J., Clasen, R.: Electrophoretic deposition from aqueous suspension for near-shape manufacturing of advanced ceramics and glasses-applications. J. Mater. Sci. **39**, 803–811 (2004)
37. Tang, F.Q., Sakka, Y., Uchikoshi, T.: Electrophoretic deposition of aqueous nano-sized zinc oxide suspensions on a zinc electrode. Mater. Res. Bull. **38**(2), 207–212 (2003)
38. Ferrari, B., Farinas, J.C., Moreno, R.: Determination and control of metallic impurities in alumina deposits obtained by aqueous electrophoretic deposition. J. Am. Ceram. Soc. **84**(4), 733–739 (2001)
39. Winkle, M.R.: Elimination of film defects due to hydrogen evolution during cathodic electrodeposition. US Patent No. 5066, (1991)
40. Sakurada, D., Suzuki, K., Miura, T., Hashiba, M.: Bubble-free electrophoretic deposition of aqueous zirconia suspensions with hydroquinone. J. Mater. Sci. **39**, 1845–1847 (2004)
41. Tang, F.Q., Uchikoshi, T., Sakka, Y.: Electrophoretic deposition behavior of aqueous nanosized zinc oxide suspensions. J. Am. Ceram. Soc. **85**(9), 2161–2165 (2002)
42. Uchikoshi, T., Hisashige, T., Sakka, Y.: Stabilisation of Yttria aqueous suspension with polyethyleneimine and electrophoretic deposition. J. Ceram. Soc. Japan. **110**, 840–843 (2002)
43. Uchikoshi, T., Ozawa, K., Hatton, BD., Sakka, Y.: Dense, bubble-free, ceramic deposits from aqueous suspensions by electrophoretic deposition. J. Mater. Res. **16** (2), 321–324 (2001)
44. Wang, Y., Liu, FQ., Duboust, A., Neo, SS., Yuh Chen, L., Hu, Y.: Hydrogen bubble reduction on the cathode using double cell design. US Patent No. 7229535 B2 (2007)
45. Besra, L., Uchikoshi, T., Suzuki, TS., Sakka, Y.: Bubble-free aqueous electrophoretic deposition (EPD) by pulse potential application. J. Am. Ceram. Soc. **91**(10), 3154–3159 (2008)
46. Besra, L., Uchikoshi, T., Suzuki, T.S., Sakka, Y.: Application of constant current pulse to suppress bubble incorporation and control deposit morphology during aqueous electrophoretic deposition (EPD). J. Eur. Ceram. Soc. **29**, 1837–1845 (2009)
47. Sarkar, P., Mathur, S., Nicholson, P.S., Stager, C.V.: Fabrication of Texture Bi-Sr-Ca-Cu-O Thick Film by Electrophoretic Deposition. J Appl. Phys. **69**, 1775 (1991)
48. Sarkar, P., Nicholson, P.S.: Magnetically-enhanced reactive-sintering of textured YBa_2C-u_3O_X", P. Sarkar and P.S. Nicholson. Appl. Phy. Lett. **61**, 492 (1992)
49. Zhitomirsky, I., Petric. A.: Electrolytic deposition of zirconia and zirconia organoceramic composites. Mater. Lett. **46**, 1–6 (2000)
50. Zhitomirsky, I., Petric, A.: Fabrication of organoceramic films by electrodeposition. Bull. Am. Ceram. Soc. **80**, 41–46 (2001)
51. Sarkar, P., De, D.: Unpublished work
52. Sarkar, P., Huang, X., Nicholson, P.S.: Structural ceramic microlaminates by electrophoretic deposition. J. Am. Ceram. Soc. **75**, 2907 (1992)
53. Sarkar, P., Huang, X., Nicholson, P.S.: Zirconia-Alumina functionally-gradiented composites by electrophoretic deposition techniques. J. Am. Ceram. Soc. **76**, 1055 (1993)

54. Nicholson, P.S., Sarkar, P., Datta, S.: Producing ceramic laminate composites by EPD. Am. Ceram. Soc. Bull. **75**, 48 (1996)
55. Prakash, O., Sarkar, P., Nicholson, P.S.: Structure and fracture behaviour of t-ZrO_2/Al_2O_3 lamellar composites. Fatigue Fract. Eng. Mater. Struct. **18**, 897 (1995)
56. Nicholson, P.S., Sarkar, P., Huang, X.: Potentially Strong and Tough ZrO_2-Based Ceramic Composites at 1300°C by Electrophoretic Deposition. In: Badwal, S.P.S., Bannistar, M.J., Hannik, R.H.J. (eds.) Science and Technology of Zirconia V, Technomic Publishing Company, Inc., Lancaster, Pennsylvania, USA, p. 503 (1993)
57. Illston, T.J., Ponton, C.B., Marquis, P.M., Urler, G.: Electrophoretic deposition of Silica/ Alumina colloids for the manufacture of CMC's. Ceram. Eng. Sci. Proc. **15**, 1052 (1994)
58. Boccaccini, A.R., Ponton, C.B.: Processing ceramic-matrix composites using electrophoretic deposition. JOM Oct: 34 (1995)
59. Ishihara, T., Sato, K., Mizuhara, Y., Takita, Y.: Preparation of yittria-stabilized zirconia films for solid oxide fuel cells by electrophoretic deposition method. Chem. Lett. **992**, 943–946 (1992)
60. Krkljuš I, BrankovićZ, Katarina_uriš, Vukotić, V., Brankovic, G.: The electrophoretic deposition of lanthanum manganite powders for a cathode-supported solid oxide fuel cell in planar and tubular configurations. Int. J. Appl. Ceram. Technol. **5**(6), 548–556 (2008)
61. Matsuda, M., Hosomi, T., Murata, K., Fukui, T., Miyake, M.: Direct EPD of YSZ electrolyte film onto porous NiO-YSZ composite substrate for reduced-temperature operating anode-supported SOFC. Electrochem. Solid State Lett. **8**(1), A8–A11 (2005)
62. Hosomi, T., Matsuda, M., Miyake, M.: Electrophoretic deposition for fabrication of YSZ electrolyte film on non-conducting porous NiO-YSZ composite substrate for intermediate temperature SOFC. J. Eur. Ceram. Soc. **27**, 173–178 (2007)
63. Besra, L., Liu, M. Electrophoretic deposition on non conducting substrates: the case of YSZ film on NiO-YSZ composite substrates for Solid oxide fuel cell application. J. Power Sources **173**(1), 130–136 (2007)
64. Sarkar, P., Yamarte, L., Rho, H. Johanson, J.: Anode-Supported Micro-Solid Oxide Fuel Cell. Int. J. Appl. Ceram. Technol. **4**, 103 (2007)
65. Tian, Z.R., Voigt, J.A., Liu, J., McKenzie, B., Mcdermott, M. J., Rodriguez, M.A., Konishi, H., Xu, H.: Compex and oriented ZnO nanostructures. Nature **2**, 821–826 (2003)
66. Youngblood, G.E., Gordon, R.S.: Texture-conductivity relationships in polycrystalline lithia-stabilized β″-Alumina. Ceram. Int. **4**(3), 93–98 (1978)
67. Norton, D.P., Goyal, A., Budai, J.D., Christen, D.K., Kroeger, D.M., Specht, E.D., He, Q., Saffian, B., Paranthaman, M., Klabunde, C.E., Lee, D.F., Sales, B.C., List, F.A.: Epitaxial $YBa_2Cu_3O_7$ on Biaxially Textured Nickel (001): An Approach to Superconducting Tapes with High Critical Current Density. Science **274**, 755–757 (1996)
68. Uchikoshi, T., Suzuki, T.S., Okuyama, H., Sakka, Y., Nicholson, P.S.: Electrophoretic deposition of alumina suspension in a strong magnetic field. J. Eur. Ceram. Soc. **24**, 225–229 (2004)
69. Uchikoshi, T., Suzuki, T.S., Tang, F.Q., Okuyama, H., Sakka, Y.: Crystalline-oriented TiO_2 fabricated by electrophoretic deposition in a strong magnetic field. Ceram. Int. **30**, 1975–1978 (2004)
70. Uchikoshi, T., Suzuki, T.S., Iimura, S., Tang, F., Sakka, Y.: Control of crystalline texture in polycrystalline TiO_2 (anatase) by electrophoretic deposition in a strong magnetic field. J. Eur. Ceram. Soc. **26**, 559–563 (2006)
71. Sakka, Y., Suzuki, T.S., Uchikoshi, T.: Fabrication and some properties of textured alumina-related compounds by colloidal processing in high-magnetic field and sintering. J. Eur. Ceram. Soc. **28**(5), 935–942 (2008)
72. Habibovic, P., Barrere, F., van Blitterswijk, C.A., de Groot, K., Layrolle, P.: Biomimetic hydroxyapatite coating on metal implants. J. Am. Ceram. Soc. **85**(3), 517–522 (2002)
73. Ma, J., Liang, C.H., Kong, L.B., Wang, C.: Colloidal characterization and electrophoretic deposition of hydroxyapatite on titanium substrate. J. Mater. Sci.: Mater. Med. **14**, 797–801 (2003)

74. Wei, M., Ruys, A.J., Milthorpe, B.K., Sorrell, C.C., Evans, J.H.: Electrophoretic deposition of hydroxyapatite coatings on metal substrates: a nanoparticulate dual-coating approach. J. Sol-gel Sci. Technol. **21**, 39–48 (2001)
75. Mayr, H., Ordung, M., Ziegler, G. Development of thin electrophoretically deposited hydroxyapatite layers on TiAl6V4 hip prosthesis. J. Mater. Sci. **41**, 8138–8143 (2006)
76. Stoch, A., Brozek, A., Kmita, G., Stoch, J., Jastrzebski, W., Rakowska, A.: Electrophoretic coating of hydroxyapatite on titanium implants. J. Mol. Struct. **596**, 191–200 (2001)
77. Hamagami, J.I., Ato, Y., Kanamura, K.: Fabrication of highly ordered macroporous apatite coating onto titanium by electrophoretic deposition. Solid State Ionics **172**(1–4), 331–334 (2004)
78. Sridhar, T.M., Mudali, U.K., Subbaiyan, M.: Preparation and characterissation of electrophoretically deposited hydroxyapatite coatings on type 316L stainless steel. Corros. Sci. **45**, 237–252 (2003)
79. Kitabatake, T., Uchikoshi, T., Munakata, F., Sakka, Y., Hirosaki, N.: Electrophoretic deposition of Eu^{2+} doped Ca-a-SiAlON phosphor particles for packaging of flat pseudo-white light emitting devices. J. Ceram. Soc. Japan **116**(6), 740–743 (2008)
80. Bailey, R.C., Stevenson, K.J., Hupp, J.T.: Assembly of micropatterned colloidal gold thin film via microtransfer molding and electrophoretic deposition. Adv. Mater. **12**(24), 1930–1934 (2002)
81. Giersig, M., Mulvaney, P.: Formation of ordered two-dimensional gold colloid lattices by electrophoretic deposition. J. Phys. Chem. **97**, 6334–6336 (1993)
82. Giersig, M., Mulvaney, P.: Preparation of ordered colloid monolayers by electrophoretic deposition. Langmuir **9**, 3408–3413 (1993)
83. Yeh, S.R., Michael, S., Boris, I.S.: Assembly of ordered colloidal aggregates by electric-field-induced fluid flow. Nature **386**, 57–59 (1997)
84. Trau, M., Saville, D.A., Aksay, I.A.: Field induced layering of colloidal crystals. Science **272**, 706–709 (1996)
85. Tabellion, J., Clasen, R.: Electrophoretic deposition from aqueous suspension for near-shape manufacturing of advanced ceramics and glasses-applications. J. Mater. Sci. **39**, 803–811 (2004)
86. Boehmer, M.: In situ observation of 2-dimensional clustering during electrophoretic deposition. Langmuir **12**(4), 5747–5750 (1996)
87. Hayward, R.C., Saville, D.A., Aksay, I.A.: Electrophoretic assembly of colloidal crystals with optically tunable micropatterns. Nature **404**, 56–59 (2000)
88. Kamacheva, E., Golding, R.K., Allard, M., Sargent, E.H.: Colloid crystal growth on mesoscopically patterned surfaces: effect of confinement. Adv. Mater. **14**(3), 221–224 (2002)

Chapter 6
Preparation of High-Purity Glasses and Advanced Ceramics Via EPD of Nanopowders

Rolf Clasen

6.1 Introduction

For the preparation of advanced materials via a powder technological route the shaping process is most important after the synthesis of the powder. Therefore high production rates and the formation of a homogeneous compact are generally desirable. Furthermore, a high green density is necessary to reduce shrinkage during drying and sintering. This is of special importance for near net shaping or micro-structuring. For coatings an even thickness with a good coverage of the edges is an additional specification that has to be fulfilled. The typical shaping processes for fine or advanced ceramics with particle sizes in the µm or even sub-µm region are slip casting, forming of high-viscous masses, and dry pressing. The decreasing of particles size leads to the area of nanoparticles with sizes less than 100 nm. These nanopowders show higher sintering activity leading to a decrease of sintering temperature. Thus smaller final grain sizes can be achieved or even quantum size effects can be observed. For glasses new perspectives are opened with nanopowders. Due to the reduction of processing temperatures nanosized glass powders can be completely sintered to transparent glasses via viscous flow below the crystallization temperature. Thus no crucibles are needed any more and high-purity glasses can be prepared. For coatings on substrates with limited temperature stability glazes and enamels can be applied without additional low melting components. Theses fluxes generally reduce the chemical stability of the coating.

As these nanopowders tend to aggregate due to the increased influence of van der Waals attraction forces, these nanopowders have to be dispersed perfectly. For that suspensions are most suited and the formation of aerosols (dust of nanopowders) is prevented, which might be hazardous. In electrostatically stabilized suspensions the dispersed particles show a surface charge, which can be directly used for moving this particles in an applied external electric field leading to an electrophoretic deposition (EPD) [1–14]. But it should be kept in mind that the electrostatical as well as

R. Clasen (✉)
Saarland University, Saarbrücken, Germany
e-mail: r.clasen@nanotech.uni-saarland.de

J. H. Dickerson, A. R. Boccaccini (eds.), *Electrophoretic Deposition of Nanomaterials*, Nanostructure Science and Technology,
DOI 10.1007/978-1-4419-9730-2_6, © Springer Science+Business Media, LLC 2012

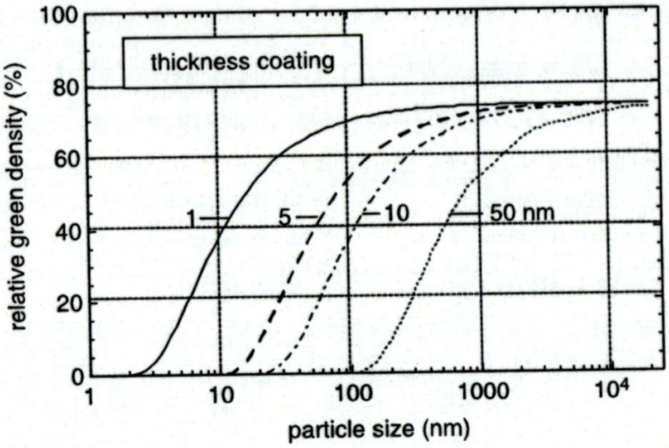

Fig. 6.1 Dependence of green density on the particle size for different coatings

the steric stabilization increases the effective particle diameter. The thickness of this coating of adsorbed ions or molecules leads to a decrease of the green density of the compact. This is schematically shown for a simple model of densely packed spheres in Fig. 6.1. Therefore there is a practical lower limit of 20–30 nm of the particle diameter for the preparation of mechanically stable compacts (green density≥40%). Furthermore it should be considered that the zetapotential decreases with particle size in electrostatic stabilization [15]. Very small nanopowders can only be sterically stabilized with applications in diluted suspensions like sun creams.

In contrast to slip casting or pressure slip casting EPD shows no particle size effect [16]. Therefore EPD seems to be a very promising shaping method especially for nanopowders or bimodal powder mixtures with nanopowders [10, 11]. Furthermore it could be proofed that the highest green densities could be obtained with EPD [17–19]. This is of high importance for the preparation of optoceramics.

For the EPD there seems to be a contradiction between the facts that on one side a stable suspension needs highly charged particles (high zeta potential) to prevent aggregation, on the other side a dense deposition can only be achieved from this stable suspension. For that, different theoretical models exist but all need further improvements. Thus the particles interactions during deposition were calculated [20, 21] and numerical simulations were performed [22, 23]. Experimentally the coagulation was observed and measured in-situ [22–25]. But a better understanding of this mechanism is necessary to understand the self-organization of monosized SiO_2 particle layers [26]. This would be of great importance for the self-healing of faults during the deposition of thin dense layers. Periodically arranged particles are of great interest for photonic materials. Thus highly-ordered monolayers of gold and latex particles could be achieved, but staple faults occurred in multilayer systems [27].

For the investigation of the electrokinetic effects it is also important to consider the electroosmotic effects besides the EPD. Electroosmosis leads to a movement of the liquid in the porous compact and influences he phase boundary liquid–solid [28]. Furthermore, the local pH value is changes through this transport reaction. This has to be considered for the investigation of the deposition mechanism.

One problem of the EPD is the bubble formation in aqueous suspensions at the electrodes. For non-aqueous suspensions these problems with the bubble formation are generally not observed. Therefore many working groups focused on these organic systems [29]. But he understanding of the mechanisms of surface charges are still a problem. For the future ionic liquids might also be of interest which gained much interest in the last years for chemical synthesis instead of their high price (typically 500 € for 100 mL). But there are a lot of reasons to focus on aqueous systems. Due to the high polarity of water high solid loadings can be achieved and it is generally much easier and cheaper to handle water instead of (flammable) organic suspensions. This is of importance especially for industrial applications. Therefore the focus of this paper is on aqueous systems and the problem with the decomposition of water will be discussed in detail. It is the objective to give an overview of the possibilities and advantages of shaping compacts and structured coating with the EPD of nanopowders. Both, glass and ceramics are considered. The processing of nanosized glass powders is state of the art for the preparation of optical fibers [30–34], but there are much more perspectives for sintered glasses. The term glass is used in its precise sense and should not be mixed up with organically modified glass-like materials. Sintered glasses behave like molten glasses, but the processing temperature can be significantly decreased when nanosized glass powders are used. Examples of advanced ceramics are alumina and zirconia. As nanopowders are more expensive than traditional fine powders, applications are expected in applications with high added value like optoceramics or dental ceramics. Furthermore, only the addition of small amounts of nanopowders as a sintering aid might be of interest. With these bimodal powder mixtures higher solid contents in the suspension and higher green densities of the compacts or coatings can be achieved. Finally, new ideas for micro-structuring with EPD are presented. This might be of interest for decoration of glass and ceramics as well as the layer-wise deposition of 3D structures for rapid prototyping.

6.2 Suspensions and Deposition Modes

A homogeneous compact with a small size distribution can only be obtained if the particles are perfectly dispersed [35]. This is usually done in diluted suspension, at least for the measurements of the particle sizes. But high deposition rates in EPD need high solids loadings of the suspensions. The decomposition of water during EPD can also be prevented with chemical additives, which might influence the colloidal system. Therefore the preparation of the suspension and some specific characterization methods for concentrated suspensions are briefly discussed. Evonik Degussa Aerosil® OX50 (fumed silica nanopowder, mean particle size 40 nm) was used as a model powder. After that a detailed presentation of all practicable methods for the prevention of disturbing bubbles is given.

6.2.1 Characterization of Suspensions

The properties of the suspensions and the behavior of the particles during deposition is a very fundamental problem of EPD, where all details are not yet completely understood. Depending on the number of aggregates inside the starting powder the dispersing apparatus and the dispersing time has to be chosen. In most cases high-speed stirrers, ball mills and ultrasonic dispersers are used and the final particle size has to be measured. Chemical additives are used for adjusting the pH value, as dispersing aids or binders. One example is shown in Fig. 6.2. It can be seen from the voltammetry measurements that complex electrochemical reaction occur which need further investigations. More experimental details are given in the cited publications in the examples in Chap. 3.

As the particle size measurements of nanopowders in real suspensions with typical solid contents of 50 wt.% and more are difficult, the cryo-SEM investigations gives at least a qualitative results of the state of aggregation. In this method a small amount of the suspension is shock-freezed and then observed in a SEM. In Fig. 6.3 Aerosil OX50 suspensions adjusted to different pH values are shown in 3 magnifications. It is obvious that the primary particle size of Aerosil OX50 (approx. 40 nm) was not reached [36]. Another powerful method for observing real suspensions is the ATR (attenuated total reflection) spectroscopy.

In Fig. 6.4 a schematic presentation of this method is given. The ATR-unit was placed in the sample compartment of a Bruker IFS66v FT-IR spectrometer. The incident beam enters the Ge-crystal on one side. Ge has only a limited spectral transmittance (500…5,000 cm^{-1}) compared to ZnSe (650…20,000 cm^{-1}), but offers a sufficient electrical conductivity to apply an electric field during the optical measurement. Due to the higher refractive index of the Ge-crystal compared to the suspension the IR beam is totally reflected inside the Ge-crystal, but an evanescent electric field enters the suspension. The penetration depth is only some μm. If there is an absorbing medium (e.g. the vibrational modes of the dispersed particles), the

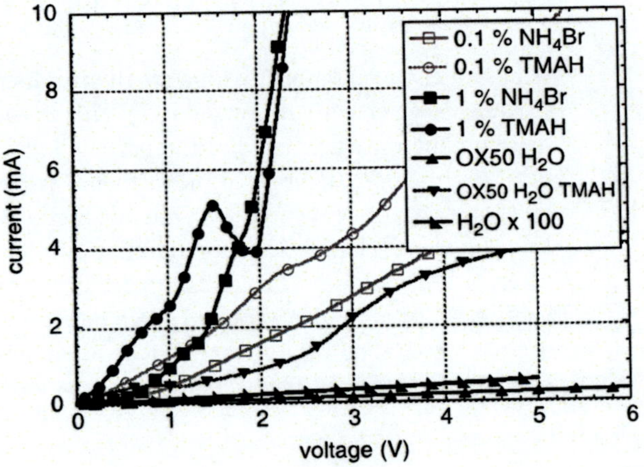

Fig. 6.2 Voltammetry of suspensions and solutions with different additions

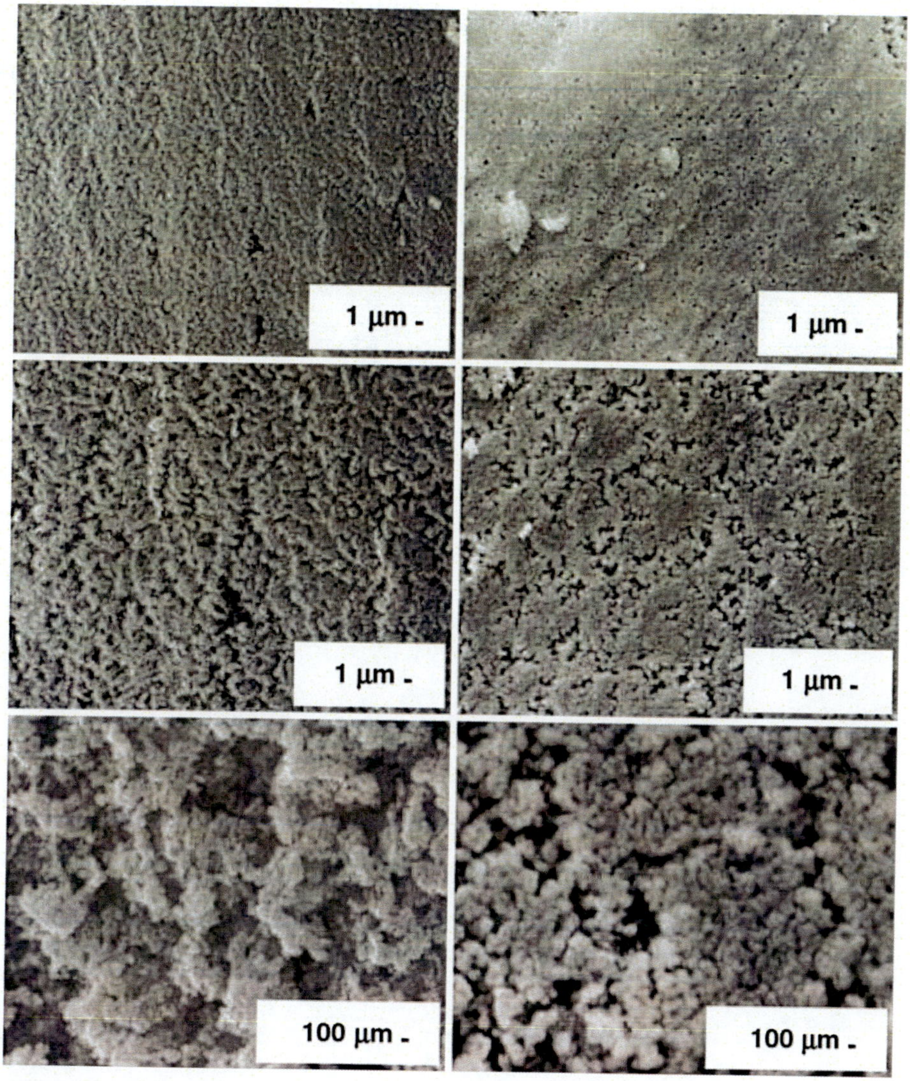

Fig. 6.3 Pictures of cryo-SEM of concentrated Aerosil OX50 suspensions, left row: pH=4, right row: pH=7.5. [36]

incident beam is not completely reflected and a transmission spectrum of the absorbing medium is obtained. An example is shown in Fig. 6.5. The vibrational mode of Si–O at 1120 cm^{-1} can be seen and peak height is depending on the concentration.

The advantage of the ATR method in comparison to the conventional transmission measurement is that a very thin layer can be measured without any preparation problems of a thin sample. It is known, that the shape of the absorption mode can also depend on the shape of the nanoparticles (topological effect) [38]. This explains the shift of 20 cm^{-1} of Aerosil A380 in the spectra shown in Fig. 6.6. Aerosil A380 has

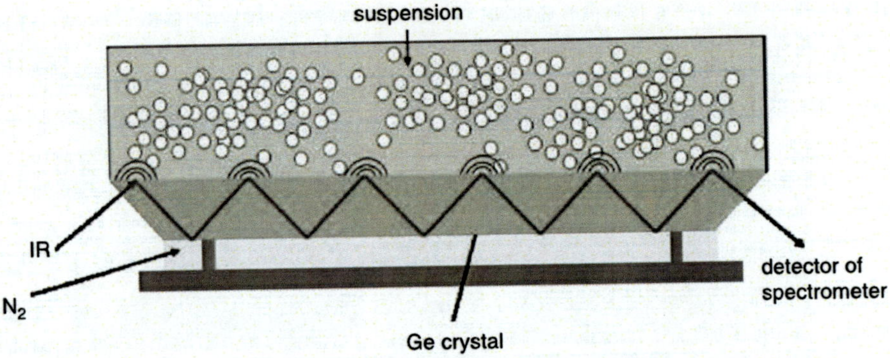

Fig. 6.4 ATR-spectrometry of a suspension on a Germanium crystal (schematic). [37]

Fig. 6.5 ATR spectra of aqueous Aerosil OX50 suspensions with different solid contents

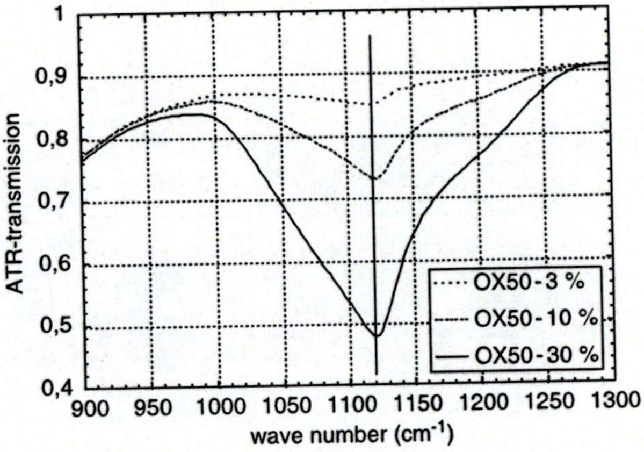

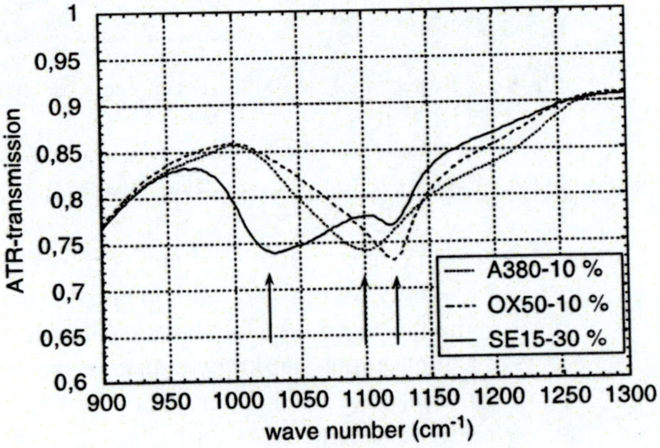

Fig. 6.6 ATR spectra of silica suspensions different particles sizes

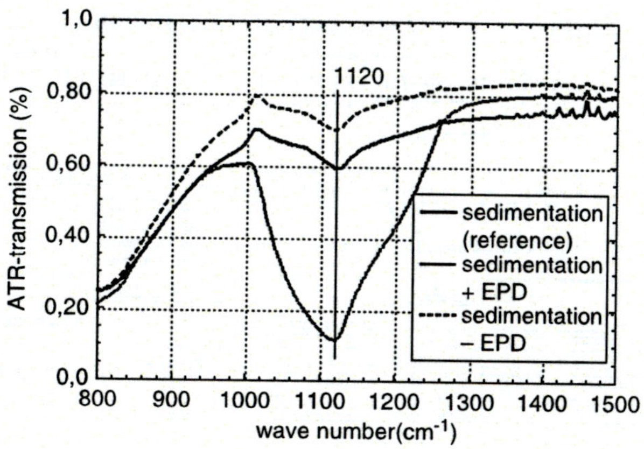

Fig. 6.7 ATR spectra of an EPD in-situ measurement of a diluted Aerosil OX50 suspension (1 wt.%). [37]

a much smaller particle size (12 nm) compared to Aerosil OX50 and Tokuyama Excelica SE15 (silica glass particle, Ø=15 μm). The additional absorption peak of SE15 at 1,025 cm^{-1} is not yet understood. With the electrical conducting Ge-crystal EPD effects close to the Ge-electrode can be measured in-situ. For that a Pt-electrode was placed as a counter electrode on top of the suspension. In Fig. 6.7 the results for moving the particles away from the Ge-electrode (sedimentation—EPD) and moving them towards the Ge-electrode (sedimentation + EPD) can clearly be seen. It would be very interesting if agglomerates of the dispersed powder could be detected in this way with improved sensitivity utilizing topological effects.

For the characterization of the real EPD system further in-situ measurements with high spatial (3D) resolution are necessary. It would be most important to measure the local electric field, the current, and the pH value. While it should be possible to measure the electric properties with fine, shielded electrodes, it is still a major problem to measure the local pH value with a spatial resolution of less than 1 mm. An attempt was made to use optical fibers and to detect the color change of a pH indicator. But this was not successful [37]. Integral measurements of the whole EPD cell are also possible with impedance spectroscopy or voltammetry.

6.2.2 *Solutions for the Gas Bubble Problem in Aqueous Suspensions*

For aqueous suspensions the main problem is to suppress the formation of gas bubble coinciding with the deposition of the ceramic particles. For practical applications four possibilities are presented to solve this problem for an anodic deposition of negatively charged particles. For a cathodic deposition of positively charged particles H_2-absorbing electrodes like Pd can be used. For EPD with microelectrodes close to the deposition area the formation of bubbles disturbs the electric field at the electrodes. This case is considered in Sect. 3.5.2.

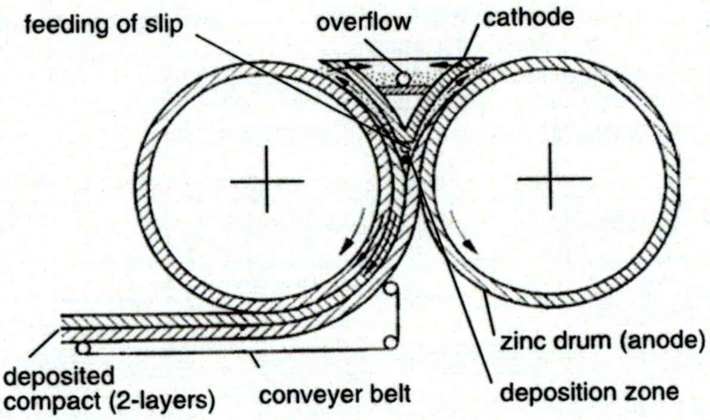

Fig. 6.8 Elephant process (schematic) for EPD of fine ceramic tiles

6.2.2.1 Reactive Electrode

In the Elephant-Process the electrode material (e.g. Zn) is oxidized instead of OH-groups [39, 40]. The deposition of clay particles takes place on two rotating zinc electrodes, see Fig. 6.8. Both deposited layers were fed together forming a ceramic tape from which tiles were cut. But this process was not competitive to dry pressing for he production of ceramic tiles. Another problem is the contamination of the slip with Zn-ions, which may influence the properties of the slip or at least the final product. Further approaches with EPD of clay slips were made for sanitary ware [41, 42] and high deposition rates up to 1 mm/s could be obtained. But in the field of fine ceramics dry (isostatic) pressing seems to be more promising to reduce production cost and time because the time consuming drying can be avoided.

6.2.2.2 Reactive Electrolyte

In this case the electrolyte is oxidized instead of OH-groups. Therefore a compound is needed which shown different states of valence. In the Miele enameling ETE-process (Elektro-Tauch-Emaillierung), which is discussed in more detail in Sect. 4.1, sodium bromide (NaBr) is used [43–45]. Then the following reaction takes place at the anode

$$NaBr + 4OH^- \rightarrow NaBrO_3 + 2H_2O + 4e^- \quad (1)$$

On the cathode the reverse reaction might occur so that NaBr is formed again. The oxidizing of the electrolyte was also successfully demonstrated for hydroquinone [46] and a 10% ethanol addition to water [47].

6.2.2.3 Membrane Process

This process is a technical development of the well-know diaphragms in physical chemistry. Due to the spatial separation of water decomposition and electrophoretic

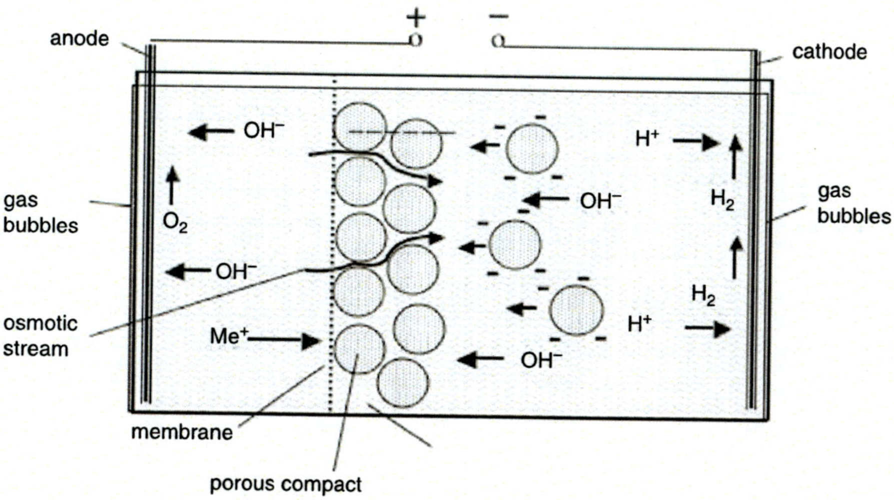

Fig. 6.9 Schematic presentation of the EPD membrane process

deposition dense compacts can be deposited on the ion-permeable membrane without included bubbles and is applied on a laboratory scale since a long time [48]. In Fig. 6.9 the EPD membrane process is schematically shown. On the left hand side of the membrane there is the compensation compartment, on the right hand side is the suspension compartment. The voltage is applied between the electrodes (Pt, stainless steel, carbon). Beside the movement of the dispersed particles an osmotic stream is observed, which leads to a change of the pH value in both compartments and generally to an elevation of the level of the suspension compartment. As the electroosmosis is observed in any EPD process, the membrane process offers additional free parameters like the variations of the concentration and the species of the electrolyte in the compensation compartment.

With the vertical membrane arrangement there might be some problems with a homogeneous deposition at longer deposition times and high depletion in the suspension compartment. Due to the difference in density in the enriched and depleted parts compensatory current can occur. This is shown in Fig. 6.10 experimentally (left hand side) and schematically (right hand side). This problem can be circumvented in a horizontal cell. In this case some precautions have to be made to prevent disturbances with the bubbles rising from the electrodes (e.g. wire-gauze top electrode, deflector bar in the lower compartment). This horizontal cell can be operated in different modes: electrically driven movement (1) in the direction or (2) counterflow of the sedimentation [49]. In the latter case bigger agglomerates settle to the ground and only well-dispersed particle are deposited on the membrane.

This process was applied for the deposition of a large variety of nanopowders to form silica glass plates and tubes, alumina and zirconia ceramics [10, 50]. Furthermore, this process was successfully applied for bimodal powders mixtures. With bi- or multimodal powder mixtures relative green densities higher than 74% (dense

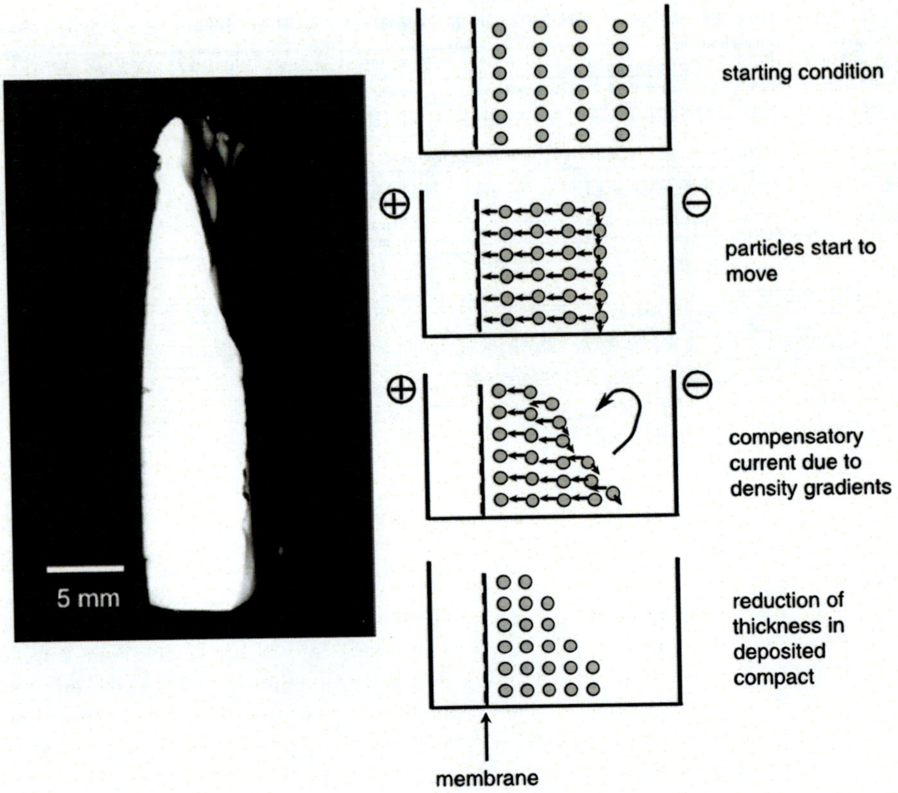

Fig. 6.10 *Left side:* cross section of an deposited Aerosil OX50 compact, *right side:* schematic presentation of the reduction of the thickness in the compacts deposited in a vertical cell

cubic packing) can be achieved. As the shrinkage during sintering is directly correlated to the green density this is of high importance for preparing near net-shaped compacts. Thus it was shown that with a bimodal mixture of coarse silica particles (Ø 15–30 μm) and fumed silica powders a relative density higher than 80% could be obtained leading to a linear shrinkage of less than 4% [51].

6.2.2.4 Pulsed or Alternating Currents

Most of the investigations on EPD were focused on the chemical reactions, especially for the reduction of the bubble formation due to the decomposition of water in aqueous systems. The new approach to suppress the decomposition of water in electrochemistry is the applications of AC fields. In symmetric AC fields a non-oscillating movement of a particle in one direction can only be observed as dielectrophoretic deposition (DEP) in an electric field gradient. This mechanism is the basis

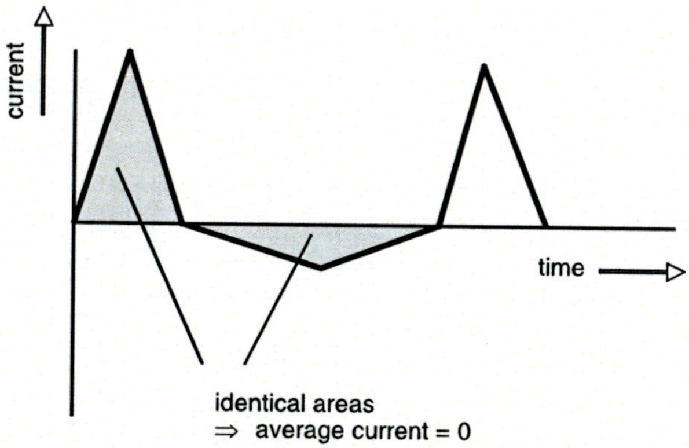

Fig. 6.11 Schematic presentation of the AC-EPD current

of electrorheological effects and can be used for a textured deposition [52–56]. As the dielectrophoretic deposition is dependent on particle size it will only be considered in this paper for coatings on sharp edges.

Very promising approaches are pulsed and non-symmetric electric fields. These offer the possibility to utilize dynamic processes to suppress the electrode reactions of the water. With square-wave pulses bubble-free alumina deposits could be achieved [57]. Another approach was the superposition of AC fields on the DC field [58] to improve the green density of the deposits. In this case the particles are oscillating in the AC field and friction effects are reduced. The alternating current electrophoretic deposition with asymmetric pulses (ACEPD) is a new approach to solve the problem with the evolution of gas bubbles in aqueous suspensions [9] and is schematically shown in Fig. 6.11. The positive and negative pulses differ in height in time, but the amount of the integral over the current (charge) is identical. Thus the average current is zero and no decomposition of water is observed. Due the non-linear response of the moving particles to the electric field, which is known for a long time [59] there is a preferential direction of movement. This leads to the deposition of a compact. This new method seems to be very attractive and much more work is needed in this field to recognize the full potential of this process.

As already mentioned in the introduction, there is still a big lack of knowledge of the microscopic mechanisms of EPD. Although the basic electrochemical reactions and influence of the height of the electric fields on the formation of the surface charges of the particles (e. g. Wien effect) are known for a long time [60], the real electrophoretic system is much more complex due to suspensions with high solid loadings and overlapping electrokinetic effects. Therefore much more effort is needed to understand all fundamental process and to get a solid basis for technical developments.

Finally, it should be pointed out that the bubble formation might also be used for generating defined pores. Thus it could be shown that a defined pore structures can be produced on a structured electrode [61].

6.3 Deposition of Compacts from Aqueous Suspensions

As the velocity of a single particle depends on the viscosity of the suspension, the viscosity should be low. On the other hand a high solid loading is necessary to improve the deposition rate. Both are opposed effects. Therefore an optimum has to be found which generally depends on the specific properties of each powder [62]. Consequently, a careful selection of the most appropriate dispersants and binders as well as a powerful dispersing apparatus is needed.

6.3.1 Homogeneous Compacts of Fumed Silica

Nanosized silica powders are produced since a long time on an industrial scale (fumed silica powders like Evonik Degussa Aerosil®, Wacker HDK®, Cabot Cabosil®). As these powders are synthesized in a hot flame, they are molten glass particles, which can be directly sintered to a transparent glass at temperatures significantly below the melting temperature. Furthermore, these powders are also ideal model materials because they are available at a moderate price (for a nanopowders), a high purity, and a wide range of particle sizes. In Fig. 6.12 two examples are shown. Due to the flame hydrolysis generation these powders show agglomerates. Therefore the effective particle diameter measured in a real suspension is generally higher than the diameter measured in a TEM (Aerosil A380: 12 nm, OX50: 40 nm). As already mentioned in the introduction, there are intrinsic problems to reach a high green density with very small nanopowders. This can also be seen in Fig. 6.13, where the deposition rate is plotted for different fumed silica powders. The solid content of silica had to be reduced for Aerosil A200 and A380 suspen-

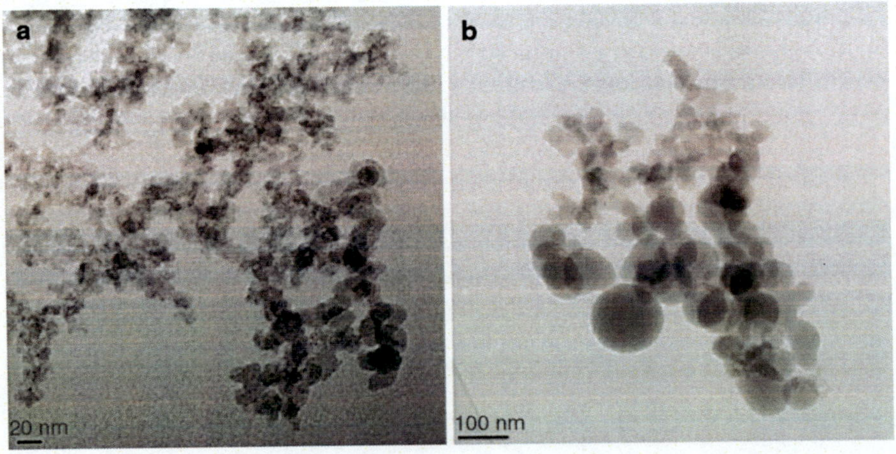

Fig. 6.12 TEM pictures of fumed silica glass powders Aerosil A380 (**a**) and OX50 (**b**)

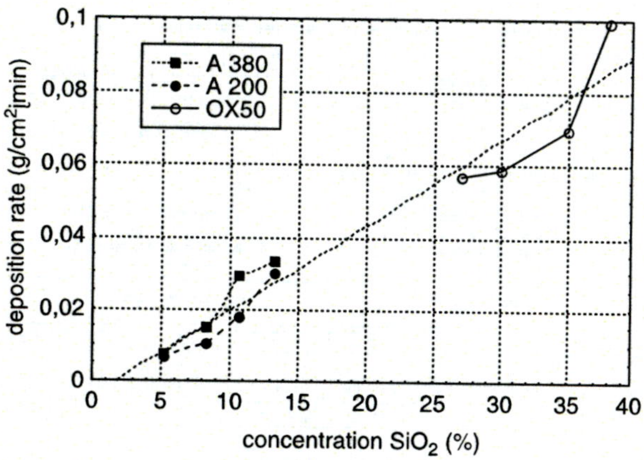

Fig. 6.13 Deposition rates of different Aerosil silica glass (3.75 V/cm, $2 \cdot 10^{-2}$% TMAH/m^2 Aerosil)

sions to reach the same viscosity. Furthermore, the linear shrinkage during drying of the compacts of these fine Aerosil powders exceeded 10% leading to problems with crack formation. Consequently, the most appropriate powder for preparing silica compacts with a sufficient green density is Aerosil OX50. As small amounts of silicic acid are soluble in water and the solubility depends on particle size, no additional binder is needed for silica compacts. During drying this solved silicic acid is precipitated at the necks between nanoparticles leading to compacts with good mechanical properties. Only the pH value was adjusted with tetra methyl ammonium hydroxide (TMAH).

In Fig. 6.14 a typical compact of Aerosil OX50 deposited via EPD is shown before and after sintering (zone furnace 1,500°C, 10 mm/min, He atmosphere). The linear shrinkage (27%) during sintering depends directly on the green density of the compact (40%, only one half of the compact was sintered).

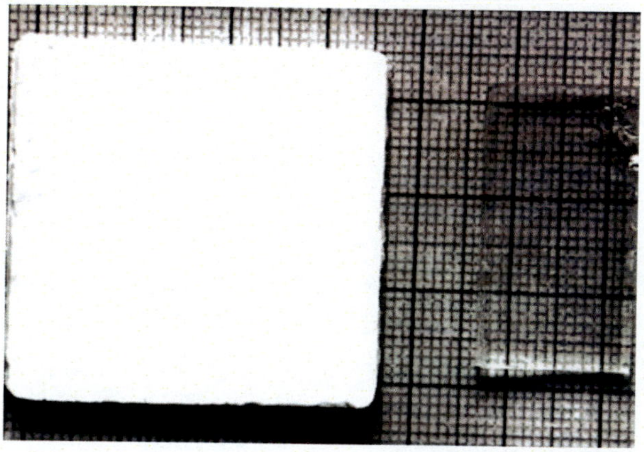

Fig. 6.14 Green compact of Aerosil OX before (*left*) and after sintering (*right*)

6.3.2 *Bimodal Mixtures of Silica Powders*

It is well known that the theoretical green density of densely packed monosized spheres of 74% can be surpassed with bimodal or multimodal powder mixtures. To decrease the shrinkage during sintering the green density has to be increased. This is of great importance for near net-shaping. The smaller amount of nanopowders should also reduce the costs of the final product. For sintering compacts of silica glass Excelica SE15 (Ø = 15 μm) and Aerosil OX50 was combined. If powder mixtures are used as starting materials it is essential that no separation takes place. As in EPD the deposition rate is independent of the particle size, EPD is an ideal process for preparing compact of powder mixtures. This can be seen in the comparison of compact made by slip casting and EPD in Fig. 6.15. In both cases the inner and outer surface of the compacts are shown. While the EPD compact is homogeneous and no differences can be seen, the slip casted compacts clearly shown separation of the coarser and finer powders.

The maximum of the green density of 82% was reached for 6 wt.% Aerosil OX50, which correspond with a linear shrinkage of 6%, see Fig. 6.16. The nanosized Aerosil OX50 powders also acts as an inorganic binder and increase the mechanical properties of the green body. In Fig. 6.17 the bending strength of a green body is plotted over the Aerosil OX50 content. The maximum of the bending strength coincides with the maximum of the green density.

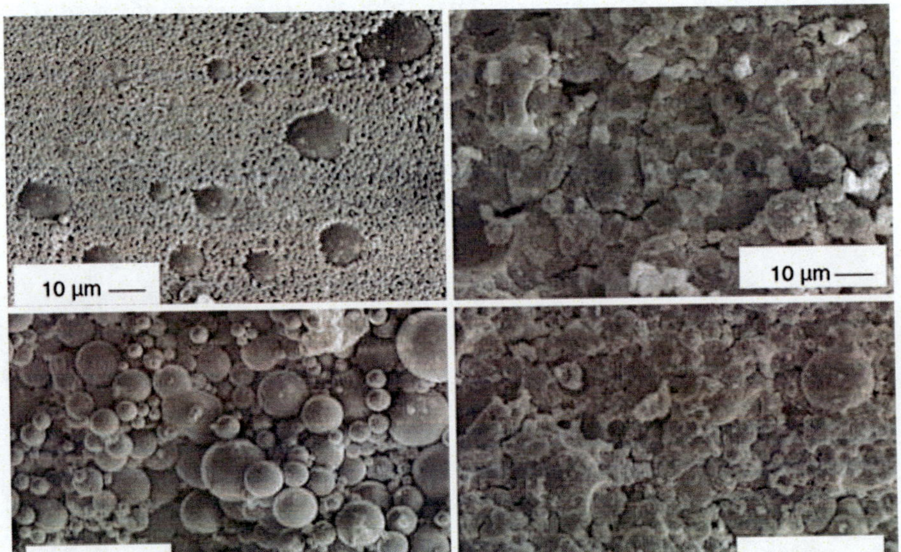

Fig. 6.15 SEM pictures of fracture planes of compacts made from Excelica SE15 and Aerosil OX50. *Left hand side:* slip casting, inner (*top*) and outer surface (*bottom*). *Right hand side:* EPD samples, inner (*top*) and outer surface (*bottom*). [36]

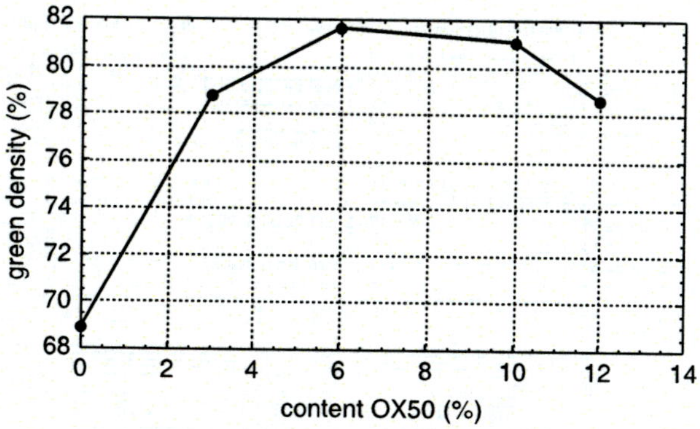

Fig. 6.16 Green density of a compact of bimodal powders (SE15 and OX50)

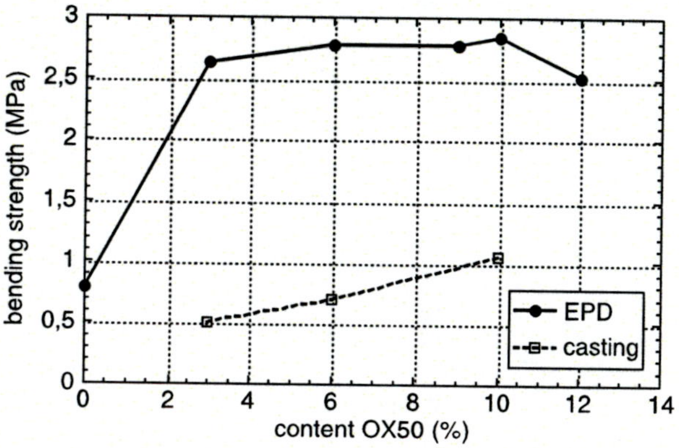

Fig. 6.17 Bending strength of compacts of mixtures of Excelica SE15 and Aerosil OX50

With these coarser particles the sintering temperature raises. But the increased green density facilitates sintering again. In total there was only a slight increase of sintering temperature about 50…70°C, because the nanopowder acts as an effective sintering aid. Thus these compacts made from bimodal powder mixtures could be sintered to transparent glass without crystallization. One example of a transparent silica glass crucible is shown in Fig. 6.18. The bigger crucible on the right hand side could not be sintered due to the lack of a sintering furnace of that size. This crucible should only demonstrate the possibilities of the EPD of bimodal powder mixtures on a laboratory scale.

6.3.3 Compacts Made by Electrophoretic Impregnation (EPI)

Up till now only single component systems were considered. The combination of different powders as well as the combination of EPD and electrochemical reactions offers nearly unlimited variations and possibilities. In Fig. 6.19 the modified process is shown, where a porous compact is filled up with much smaller particles

Fig. 6.18 Crucibles made from bimodal powder mixtures Excelica SE15 and Aerosil OX50. [36]

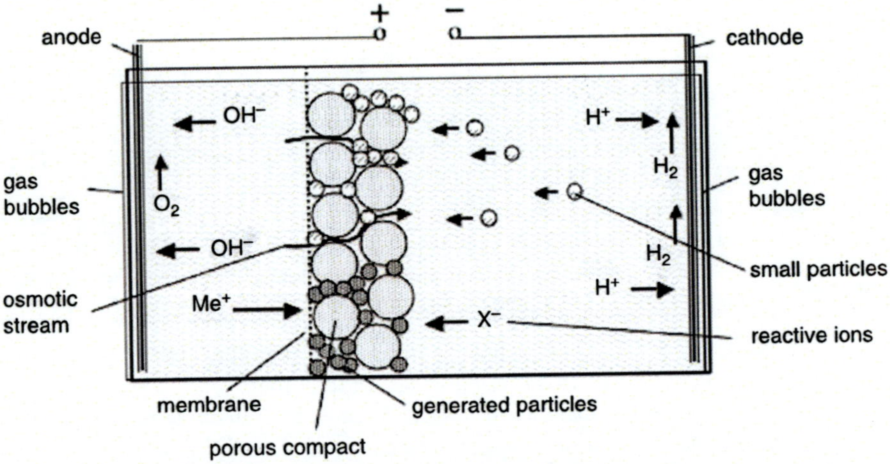

Fig. 6.19 Schematic presentation of the modified EPD processes inside a porous compact with an impregnation with smaller particles (*top part*) and a co-reaction of ions (*bottom part*)

(upper part) or ions coming from both sides react inside the compact forming a nanoparticle (lower part).

Typically, the solid loading of the suspension for the impregnation and the deposition rate has to be low to prevent particle sticking on the surface of the porous compact.

Furthermore, the particle size of the impregnating suspension should be much smaller than the pore size of the compact and the zeta-potentials of both (particles in the compact and in the EPI suspension) have to be considered. This electrophoretic impregnation (EPI) offers a lot of possibilities. Two examples of particle impregnation are shown in Fig. 6.20 and 6.21.

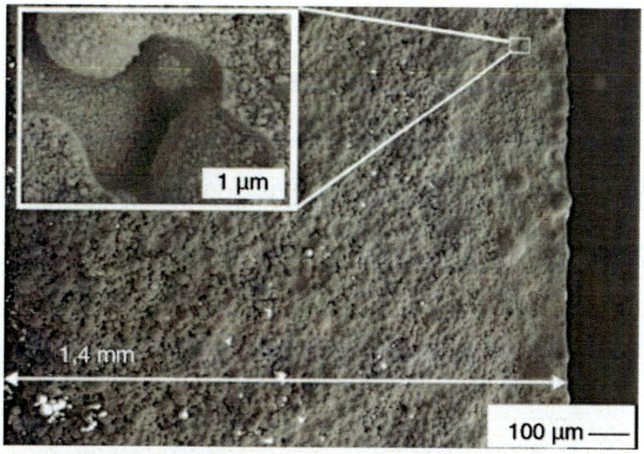

Fig. 6.20 EPI of Aerosil OX50 into a compact of Excelica SE15 from the *right* to the *left hand side*. [36]

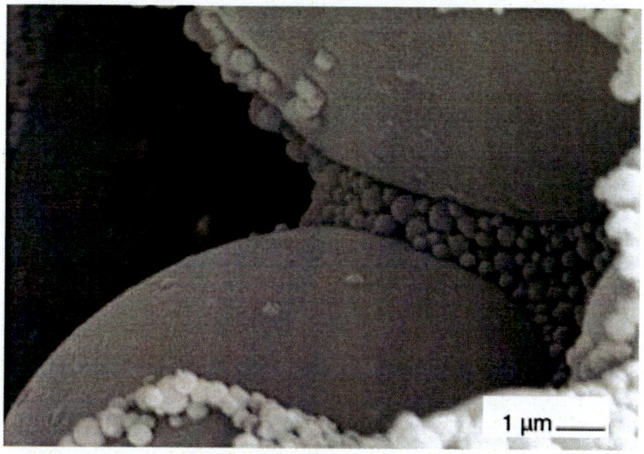

Fig. 6.21 EPI of a compact with coarse Excelica SE15 silica particles and submicron alumina particles

In Fig. 6.20 the porous compact of Excelica SE15 was impregnated with a diluted suspension of Aerosil OX50 from the right to the left side. The amount of the much smaller Aerosil OX50 particles decreases with increasing penetration depth leading to a density gradient. In this case the penetration depth was 1,4 mm and it should be possible to penetrated even bigger compacts by optimizing the EPI process [63]. In Fig. 6.21 a similar compact is impregnated with small alumina particles showing that the pores between the coarse particles can be completely filled up.

An example for the reaction of two components coming from both sides of the compact is shown in Fig. 6.22 [64, 65]. The Cd^{2+} ions and the S^{2-} ions from the other side react to CdS. The particles size of the CdS can be controlled by the pore size of the compact (template process). Unfortunately, the red orange-red color in the center can only be seen as a gray shadow in this picture. It would be most interesting to prepare red colored glasses in this way. But unfortunately the sintering temperature even of compact of nanosized silica particles is still too high for sintering such doped compacts. At 1300–1500°C most of the CdS would evaporate

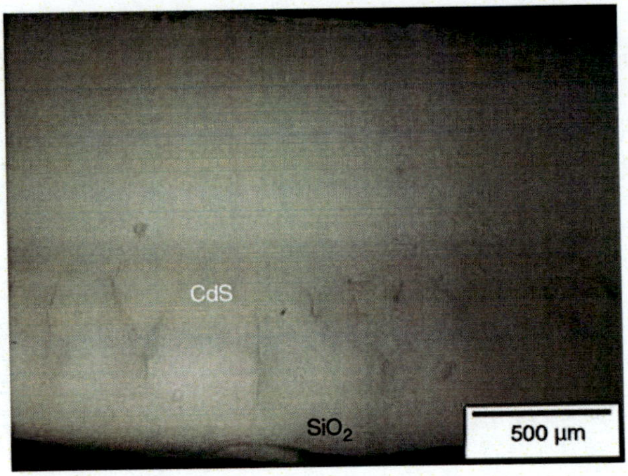

Fig. 6.22 Pictures of a green body of Aerosil OX50, where CdS was precipitated inside. [66]

and disappear though the open pore structure before the final transparent glass is achieved. The only chance to use this versatile template sintering process for the preparation of (annealed) red glasses is to reduce the sintering temperature and to prepare compacts of multicomponent nanosized powders. This will be discussed at the end of the next chapter. Nevertheless, it is possible to get nice gold ruby glasses using the template process [66–68] although the melting temperature of gold is exceeded during the sintering process of the silica glass matrix. With a sufficient low concentration of the salt solution (< 1% $AuCl_3$) the gold particles formed by reducing the metal salt cannot coagulate in the molten state and the size is preserved. As the size of the gold particles is controlled by the pore size of the compacts a mixture of Aerosil OX50 and A380 has to be taken to reduce the pore size. In a pure Aerosil OX50 compact it is not possible to get a bright red ruby color. In this case the gold particles are bigger than 50 nm and the red color is a little bit dull.

Another area with a high application potential of the EPI process are fiber composite materials [69, 70]. One example for a CFC material protected with a SiC coating is shown in Fig. 6.23. A 30 vol.% suspension of SiC powder (ESK SM15, d_{50}=0.77 µm), combined the sintering additives B_4C (ESK Tetrabor1500, d_{50}=1 µm) and carbon black (Evonik Degussa FW200, d_{50}=12 nm), Tween (Roth) as dispersing additive, with a pH adjusted to 7 was used. The negatively charged carbides and the positively charged carbon showed a heterocoagulation and a coating of thickness 50…100 µm was achieved after sintering in vacuum at 1750°C [14].

6.3.4 *Reactive Electrophoretic Deposition (REPD)*

The EPI process is a two-step process with a post-processing of an existing compact. Although the reactions are more complex it is also possible to add other components to the compact by an in-situ reaction in a single step process. It is possible

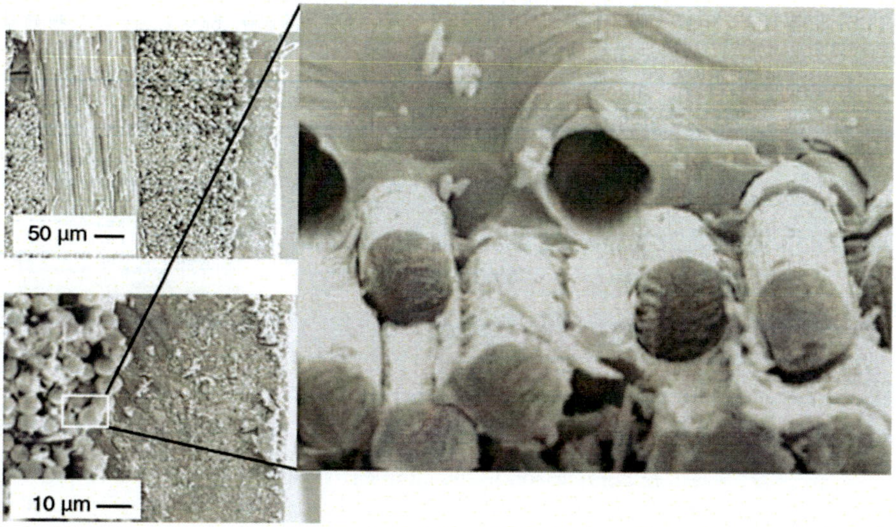

Fig. 6.23 EPI of fiber composite materials, *top left:* boundary layer 2D-CFC with EPD-SiC on the *right* surface, *bottom left and right hand side:* magnification of the embedded carbon fibers

to get a homogeneous as well as a heterogeneous doping of the compact. It is also possible to combine the EPD of ceramic (dielectric) particles and the electrodeposition (ED) of metals. As the reduction of metal salts takes place at the anode, the ceramic particles have to be positively charged for a codeposition. A lot of work was done in field of cermets with an ED of Ni and an EPD of nanosized alumina powder [71–76]. Thus the scratch resistance of corrosion protective Ni coatings is improved by the ceramic particles. Ni-alumina cermets are also attractive as selective solar absorbers. But in this case the final coating must be rather thin (typ. 200 nm thick). Another approach was the combination of silica and tungsten to get a gradient material with reduces residual stresses at the boundary layer [77]. This is important for a fed-through in high power lamps with a silica glass envelope. But much more work is needed in this field.

For compacts of nanopowders the small diffusion path in nanoparticles enables a homogeneous distribution of the incorporated materials, which can be obtained at least after annealing the compact at elevated temperatures. Thus it was possible to prepare doped silica and multicomponent glasses from SiO_2 nanoparticles by reactive electrophoretic deposition [65], but also gold ruby glasses with the heterogeneous formation of nano-sized colloids [78]. The incorporation of potassium and boron oxides in silica glass compacts is of very high technical importance to reduce the sintering temperature and to increase the coefficient of thermal expansion (CTE). Both are necessary for applying coatings of nanosized glass powders on substrates (see Chap. 4). In Fig. 6.24 the increase of the density of a pure (undoped) Aerosil OX50 compacts as a reference and compacts doped with boron and, additionally, potassium oxide is plotted versus temperature of the zone-sintering furnace. With these dopings a remarkable reduction of the sintering temperature

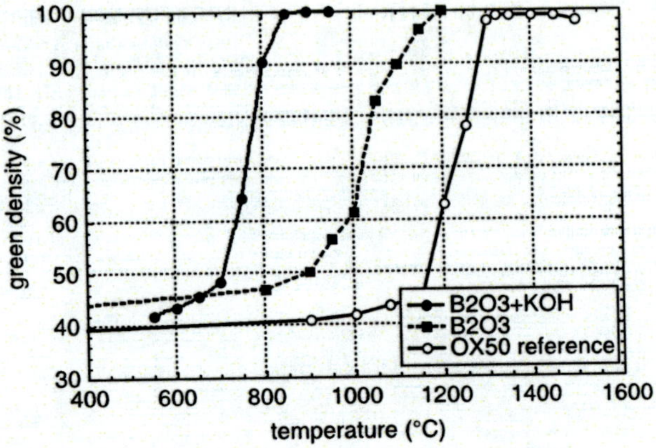

Fig. 6.24 Sintering shrinkage of pure and doped silica glass compacts. [36]

was achieved. In comparison to the simple doping of an Aerosil OX50 compact by soaking with a salt solution a higher amount of the dopant was incorporated into the silica glass network via REPD [36].

Due to the high importance of coatings the perspectives of nanosized multicomponent glass powders are discussed in more detail. Borosilicate glasses (BSG) are taken as an example because BSG shows a high chemical durability, a CTE close to many important materials like silicon, and a reduced tendency to crystallize in comparison to soda lime glass. Unfortunately, nanosized BSG powders are not commercially available and only very few attempts were made to produce borosilicate glasses via a sintering technique of nanopowders [79–86]. As no results are reported of the sintering kinetics of compacts of nano-BSG powders, the viscosities of bulk silica glass and BSG are plotted in Fig. 6.25. From the well-known sintering temperatures of Aerosil compacts it can be concluded that it should be possible to sinter compacts of nano-BSG powders between 600 and 700°C. This would be a dramatic reduction of processing temperature for a glass with high chemical durability without any "chemical compromise" like the addition of fluxes.

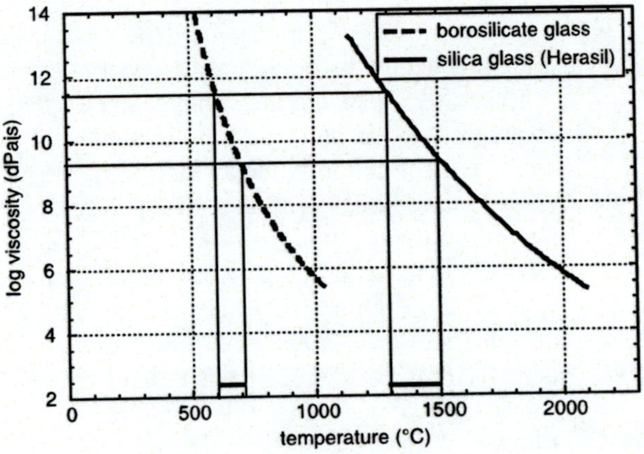

Fig. 6.25 Viscosity of borosilicate and silica glass with the sintering regime of silica (1,300…1,500°C) and borosilicate (600…700°C) nanopowders

Fig. 6.26 SEM pictures of borosilicate nanopowders synthesized via flame hydrolysis at the EMPA (*left*) and laser ablation (*right*)

The next problem that had to be solved was to synthesize at least a small quantity of nano-BSG powders for first experiments. Based on first results to the flame hydrolysis of nano-BSG powders [87] several grams of nano-BSG powders were synthesized by T. Graule and A. Heel from the EMPA (Dübendorf, Switzerland) via a flame hydrolysis in an ethine flame starting with alkoxide precursors of silica, boron and sodium (data not published yet). Due to the high flame temperature and a short duration of the molecules inside the flame a very high heating rate of 230,000 K/s was achieved. Thus it can be expected that no phase separation due to different reaction kinetics of the single components should occur and a homogeneous BSG powder should be obtained. In a second approach a BSG rod was ablated with a 100 W CO_2-laser. With both methods similar results of BSG nanopowders were obtained, see Fig. 6.26. The chemical analysis of silica and boron oxide shown that there were only minor changes in the chemical composition during ablation (composition before/after ablation): SiO_2 80/73.3%, B_2O_3: 15/13.5%.

From these powders small compacts were prepared and sintered in zone furnace (v = 8.3 cm/min) at different temperatures. The SEM pictures of fracture planes of the bulk materials are shown in Fig. 6.27. At 800°C most of the bulk material had already completely sintered to a transparent glass. Unfortunately, some bubbles were formed inside the glass, which still grow in size during further heating to higher temperatures (900°C). The mechanism for this formation of bubble is not yet understood and further investigations are necessary. One reason might be, that some free boron oxide is still present which evaporates at higher temperatures leading to the formation of these bubbles. If this problem can be solved a new and very interesting route for the preparation of borosilicate glasses is created. There would be a high potential for preparing Cd-free red glasses via the template process because a much wider variety of nanosized semiconductor materials can be tested and incorporated into the BSG matrix. For annealed red glasses CdS/Se has to be applied because this

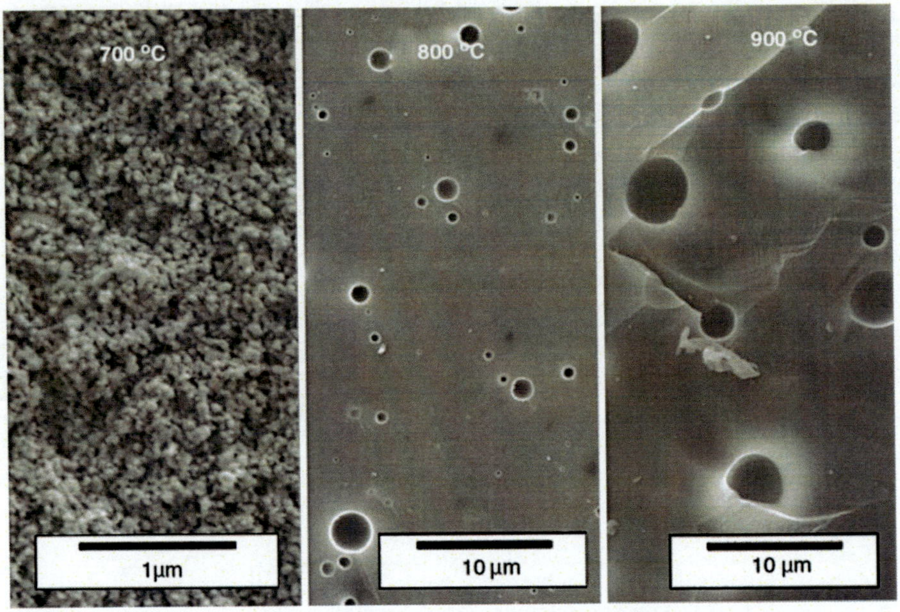

Fig. 6.27 SEM pictures of fractured planes of sintered nano-borosilicate compacts

is the only compound that fits with the desired band gap, is solvable in a glass melt and can be precipitated as nanoparticles without chemical alterations from the melt. This restriction is no longer valid for the template process, were all kinds of salts can be used for doping the porous glass compact.

6.3.5 Optoceramics

Optoceramics are polycrystalline materials, where the optical properties determine the application. In case of transmittance the optoceramic should be at least translucent, but for high-grade applications transparency is required. To reduce scattering on residual pores optoceramics have to be sintered to full (theoretical) density (typically > 99.99%), which is at the high-end of all advanced ceramics. Due to the polycrystalline structure scattering might also occur on grain boundaries. For non-isotropic (non-cubic) structures this influence can be very high. This is the reason why (non-absorbing) ceramic generally looks white. This scattering can only be reduced if the grain size is very small compared to the wavelength of the incident radiation. For visible light the grain size has to be smaller than 100 nm. Such small grain sizes can only be achieved with nanosized starting powders, which has to be sintered to full density without significant grain growth. Therefore it is very favorable to get a compact with a high green density.

The first interest in alumina optoceramics for lamps started almost 50 years ago [88–94] and it was sufficient to get translucent envelopes for high-pressure sodium lamps. In contrast to glasses alumina is stable against the chemical attack of hot so-

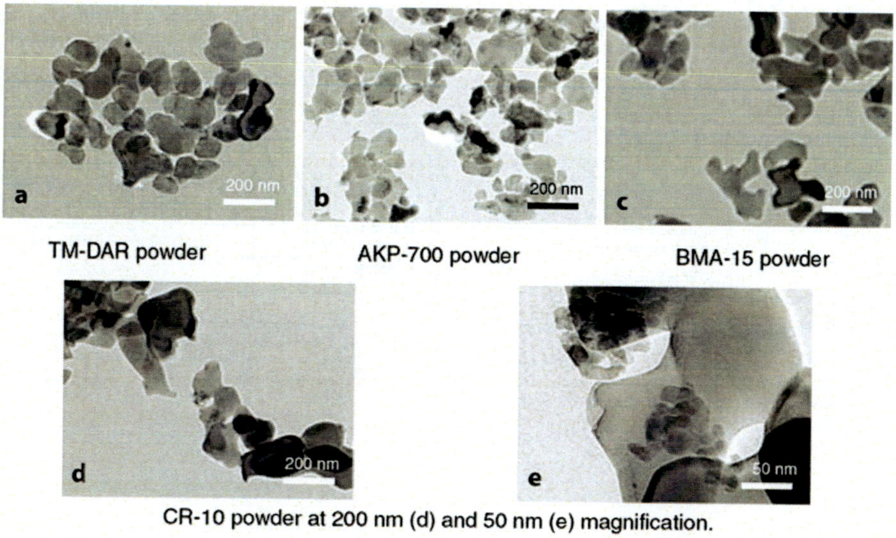

Fig. 6.28 TEM-images of submicron alumina powders used for the preparation of optoceramics. [18]

dium vapor. The interest for spot lamps for projection renewed the interest in better transparency and nearly transparent alumina ceramics were achieved [18, 95–101]. But at high operation temperatures and long operation times grain growth might occur again. Therefore other applications like large windows for ceramic amour (replacing expensive sapphire single crystals) and IR transparent windows of missiles might be more promising for application.

Like in all other cases the starting powder is the first important link in the long process chain and the alumina powder should have the α-phase. Although nano-sized α-Al_2O_3 was often announced most of the commercially available powders with reasonable prices are submicron powders like shown in Fig. 6.28. These powders have to be perfectly dispersed to get an optimum green density and additives like MgO or ZrO_2 have to be added to control the grain growth during sintering [102]. The complexity of this optimizing process can be seen in Fig. 6.29. Here

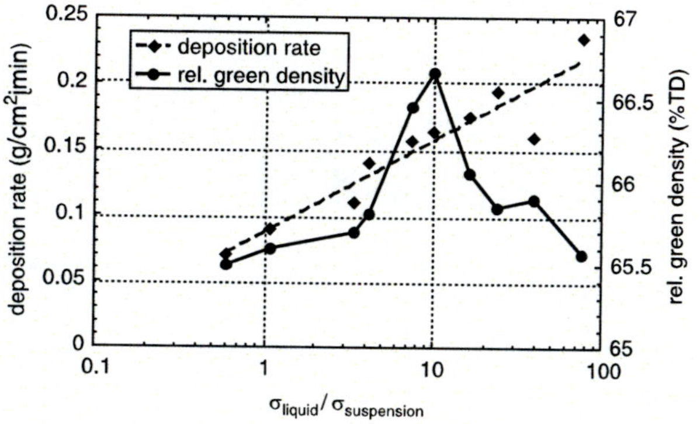

Fig. 6.29 Influence of the ratio of electrical conductivities of both compartments. [18]

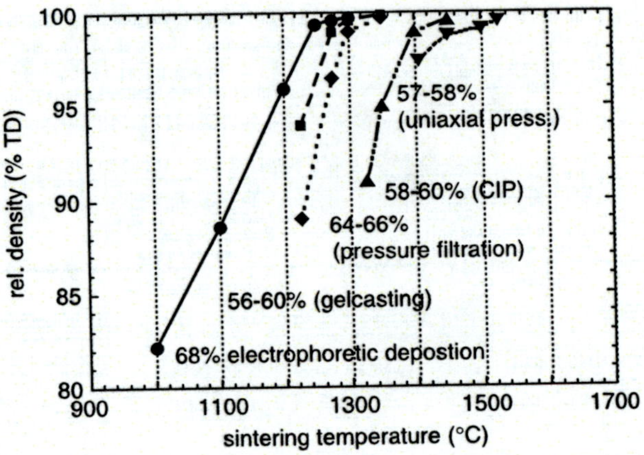

Fig. 6.30 Sintering density of compacts made via different shaping methods. [18]

it is demonstrated that the green density of an EPD compacts depends on the ration of the conductivities in the compensatory and suspension compartment of the membrane cell. The comparison of the density of compacts made via different shaping methods plotted in Fig. 6.30 reveals impressively the advantages of the EPD process.

In most cases the reduction of grain growth requires additional forces like hot isostatic pressing (HIP) or electric field-assistant sintering (FAST/SPS). Thus the grain growth could be limited to grain sizes to approximately 500 nm (see Fig. 6.31) and nearly transparent alumina ceramics could be obtained (Fig. 6.32).

For materials with a cubic structure things are a little bit easier because the grain size can be bigger and there is no necessity to use nanopowders. In spite of that, the EPD can be advantageous to get a homogeneous compact with high green density. Typical materials are Mg-Al-spinel [103, 104], Y_2O_3 [105–107], YAG [108], AlON [109] or PLZT [110], which might be applied for windows, scratch resistant coatings, laser host material and optical elements. The laser materials (e.g. Nd:YAG) are

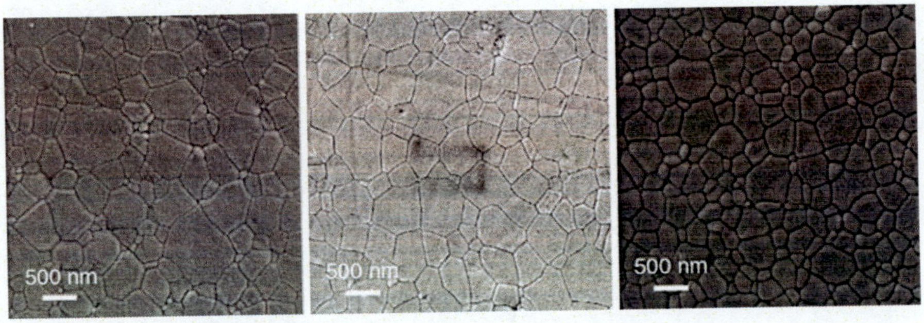

Fig. 6.31 HR-SEM images of non doped and doped samples after sintering (5 K/min) in air at 1,200°C with 48 h dwelling times. *Left:* undoped Al_2O_3 average grain size 890 nm, *middle:* Al_2O_3 with 250 ppm MgO average grain size 791 nm, *right:* Al_2O_3 with 250 ppm ZrO_2 average grain size 603 nm. [18]

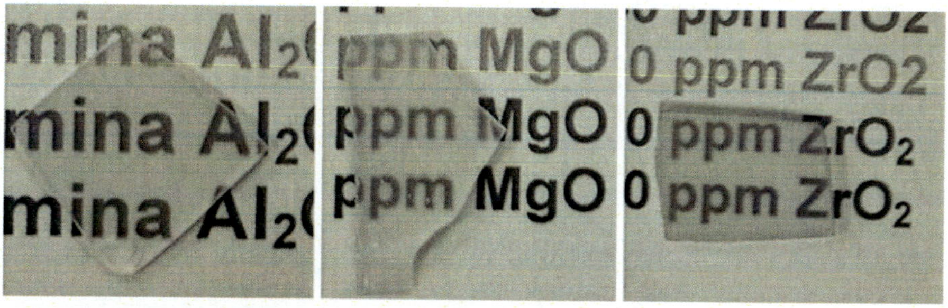

Fig. 6.32 Pictures of dense alumina optoceramics. *Left:* pure alumina, *middle:* alumina doped with magnesia, *right:* alumina doped with zirconia. [18]

of great interest, because higher Nd doping with better homogeneity and a lower price compared to YAG single crystals are possible. Optical element like lenses show—compared to optical glasses—a reduced dispersion combined with a high refractive index.

Zirconia ceramics is used in wide area of applications, mainly as a high-temperature ionic conductor and for devices for high mechanical loads. Cubic zirconia has similar properties like diamond. Therefore it is interesting to see if this material can also be prepared by sintering of nanosized or submicron powders [111–113]. In Fig. 6.33 some starting powders of stabilized zirconia are shown.

These powders were shaped via EPD and finally sintered. Due to the superior green density of these compact the activation energy for sintering calculated from experiments was smaller than for compacts made by dry pressing, see Fig. 6.34. To get a real transparent zirconia a lot of effort has to be spent for optimizing the sintering conditions. In this case a 2-step sintering process was used [111]. Like in most preparation routes for optoceramics a post treatment by hot isostatic pressing was necessary. Surprisingly, there was an optimum for the density of the presintering of the compact, see Fig. 6.35. After sintering the cubic zirconia ceramics was transparent and shown only a small scattering. This can be seen from the right hand side picture in Fig. 6.36. The center part is a view through the polished ceramic disc.

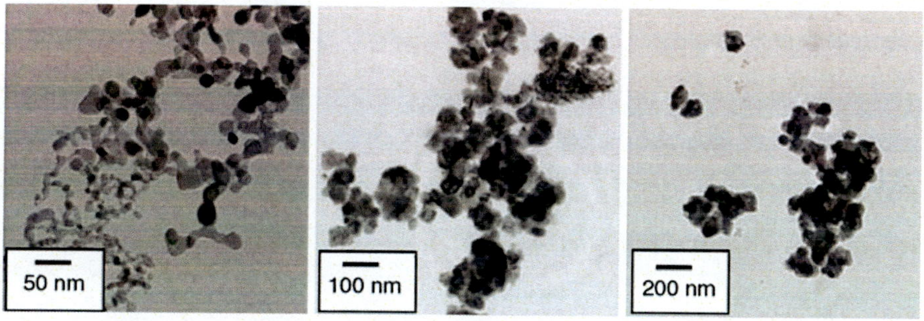

Fig. 6.33 TEM-images of the zirconia powders YSZ3, NA8Ym and TZ8Y. [111]

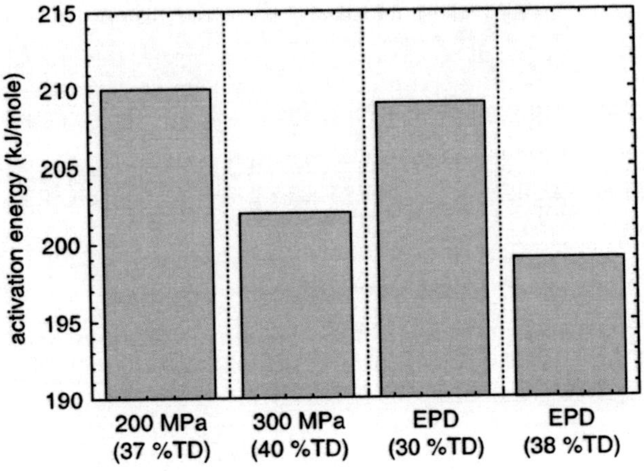

Fig. 6.34 Activation energy for sintering of compacts made via EPD and dry pressing (200 and 300 MPa). Surface diffusion was suggested as the sintering mechanism. [111]

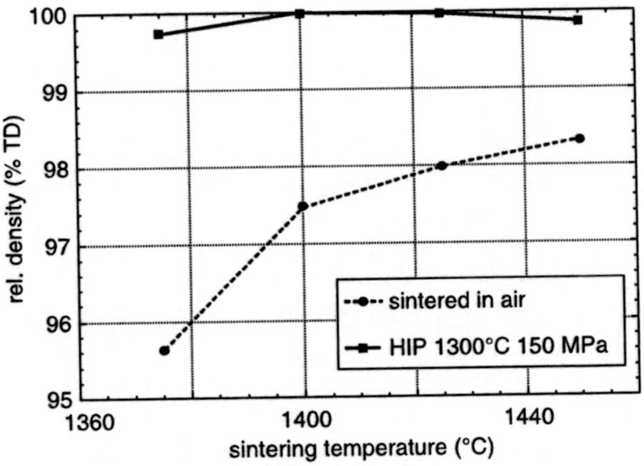

Fig. 6.35 Sintering density of a compact of TZ8Y after 2-step sintering and post-hipping. [111]

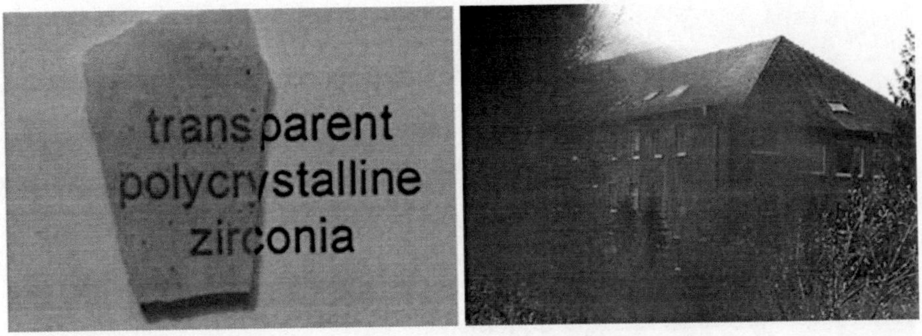

Fig. 6.36 Pictures of dense cubic zirconia (TZ8Y) optoceramics. *Left:* ceramic disc on substrate, *right:* view through ceramic disc. [111]

Alternatively to the hot isostatic pressing a microwave sintering [114] was applied and revealed non-thermal microwave effects with an increase of 2% of final sintering density combined with a significantly reduced grain growth.

6.3.6 *Deposition of Structured Compacts*

Up till now most of the sample shown were made on the laboratory scale with rather simple geometries. For all kinds of application a shaped body with more or less complex geometry is generally desired. The importance of a low shrinkage rate for near net-shaping was already mentioned, which can be facilitated with high green density obtainable with bimodal powder mixtures. With the EPD membrane process there are two possibilities of shaping:

1. Deposition on a shaped membrane
2. Local deposition with a single point electrode or an array of point electrodes

Both possibilities will be presented and discussed. The second choice is of special interest because alls kinds of structures could be generated by a computer controlled system for coatings (2D) as well for rapid prototyping (3D).

6.3.6.1 Shaped Membrane

Dental ceramics are one interesting field of application, where a high added value is still within easy reach. Here individual shaping of single items like crowns and bridges are necessary which require high skills and a lot of expensive manpower. Although the process development is moving in the direction of computer controlled rapid prototyping (CAD-CAM approach), a lot of work is still done with the traditional technique of making replicas. With this replica technique a solution has to be found for compensating the shrinkage during processing of ceramics. Fortunately, special kinds of gypsum exist which expand up to 10% during curing. The structure of these materials is shown in Fig. 6.37. Therefore a ceramic system had

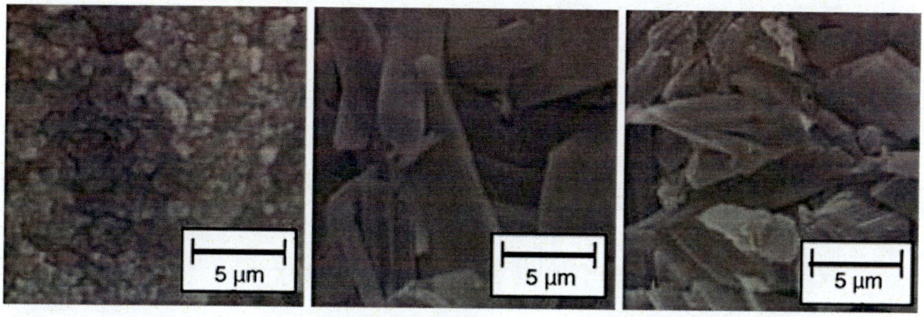

Fig. 6.37 SEM pictures of different plaster moulds. [116]

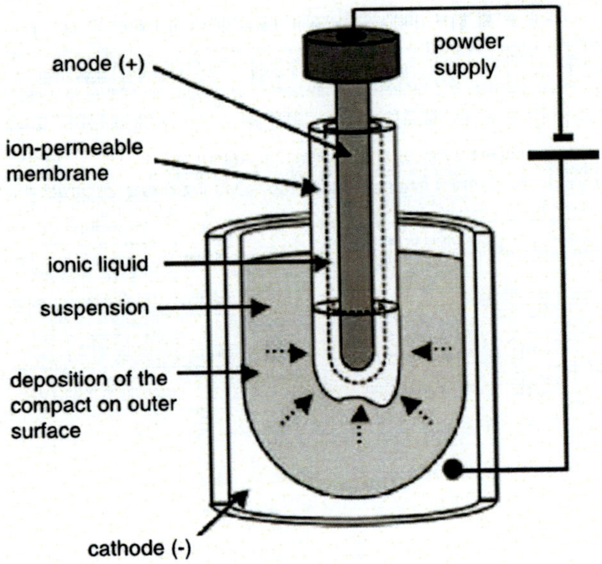

Fig. 6.38 Apparatus for deposition of dental model crowns. [116]

to be found where the total linear shrinkage does not exceed 10%. This means that a green density of 74% had to be reached, which is only achievable with a bimodal powder [115].

This gypsum moulds were used as a membrane for the EPD shaping. In Fig. 6.38 a schematic presentation of the deposition device is shown. The inner part corresponds to the compensation compartment, the outer part to the suspension compartment. The deposition took place on the outer side of the gypsum mould. After deposition the compact was carefully removed from the mould to prevent cracking during drying. This is a critical step, which needs some exercise but it was manageable.

In Fig. 6.39 the starting powders are shown. On the left hand side is a coarser Cer-stabilized zirconia powder and two much finer powders are in the middle and

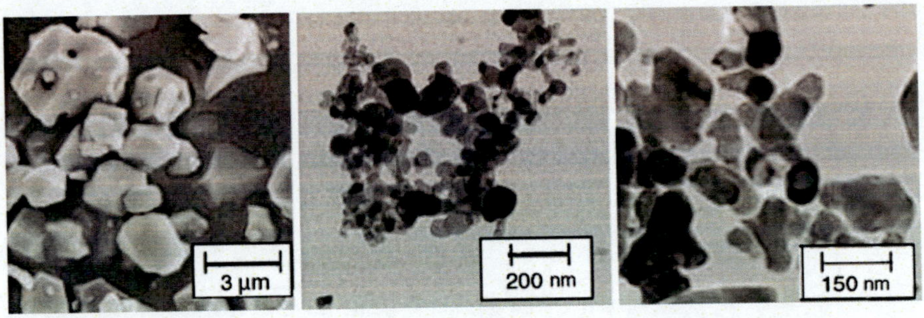

Fig. 6.39 SEM pictures and particle size distribution of starting materials Ce3,3, Y0,01 and TM-DAR. [116]

Fig. 6.40 H-REM-pictures of Ce3,3-TMDAR-compacts deposited with different amounts of the fine alumina powder (TMDAR). **a** 8 wt - % fine powder, **b** 8 wt - % fine powder, **c** 16 wt - % fine powder, **d** 16 wt - % fine powder. [116]

on the right hand side. One is Yttria-stabilized zirconia for a zirconia powder mixture and the second is an alumina powder for alumina-zirconia ceramic. Both powder bimodal mixtures were investigated [116, 117].

After drying a homogeneous compact was achieved, see Fig. 6.40. It was important to optimize the system in such a way that on one hand a high solid loading is achieved leading to a high green density and, on the other hand, the viscosity is still low enough to get a sufficient deposition rate. One successful example is shown in Fig. 6.41. In Fig. 6.42 some model caps of tetragonal zirconia are shown.

Similar to slip casting and pressure slip casting an improvement was made by replacing the gypsum with a polymer membrane. As all attempts to use the porous membranes developed for pressure slip casting failed. Polyether sulfon (PES) membrane were made by casting [68]. The SEM pictures in Fig. 6.43 show the pore structure in two magnifications. Furthermore it was the objective to deposit smaller structures with this membrane. One example is shown in Fig. 6.44.

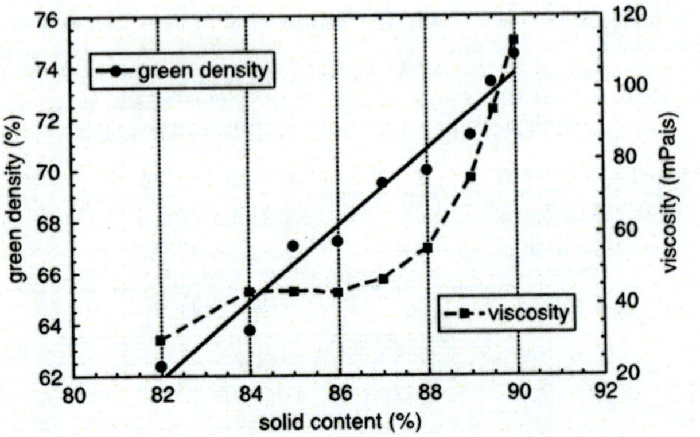

Fig. 6.41 Green density and viscosity of bimodal zirconia compacts. [116]

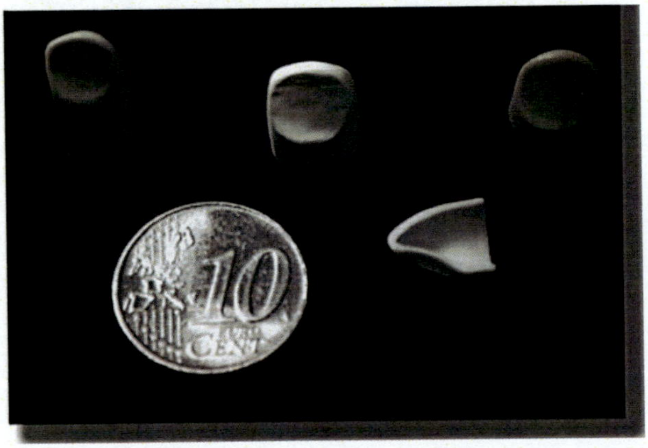

Fig. 6.42 Sintered model caps with different geometries, *top row*: complete caps, *right side*: cut cap. [116]

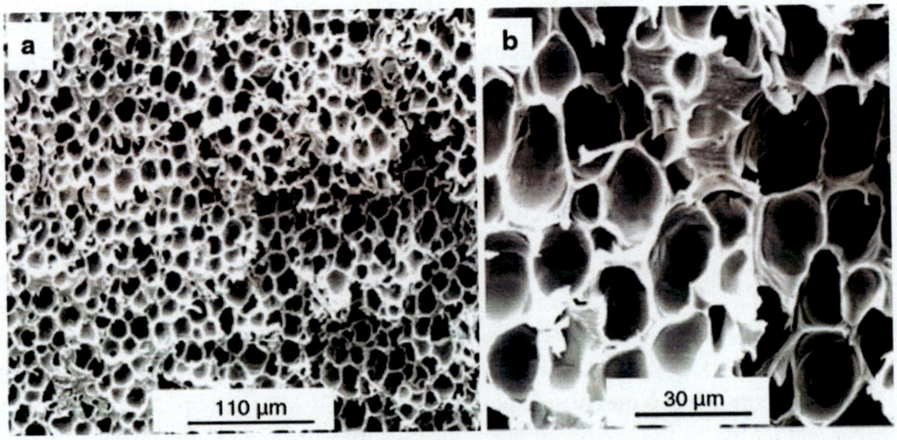

Fig. 6.43 SEM pictures of PES membrane (different magnifications)

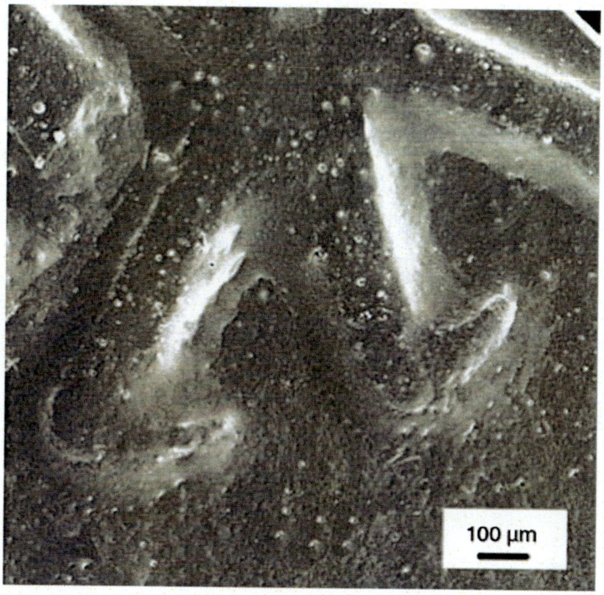

Fig. 6.44 Picture of compact deposited on a structures membrane. [68]

For porous compacts with very fine structure the mechanical properties—at least in the green state—might be a problem. Therefore some mechanical calculations were performed for the maximum aspect ratio, which limits the length of finely structured compact. For that the mechanical properties of the porous compacts had to be measured to get realistic parameters for the modeling. An example of the measured Young's modulus is shown in Fig. 6.45 demonstrating the superior properties of EPD compacts again.

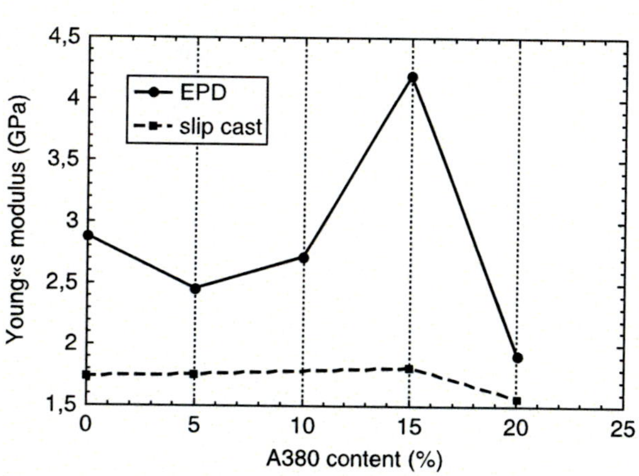

Fig. 6.45 Young-modulus for compacts deposited from slips with 50 wt.% via EPD and slip casting. [68]

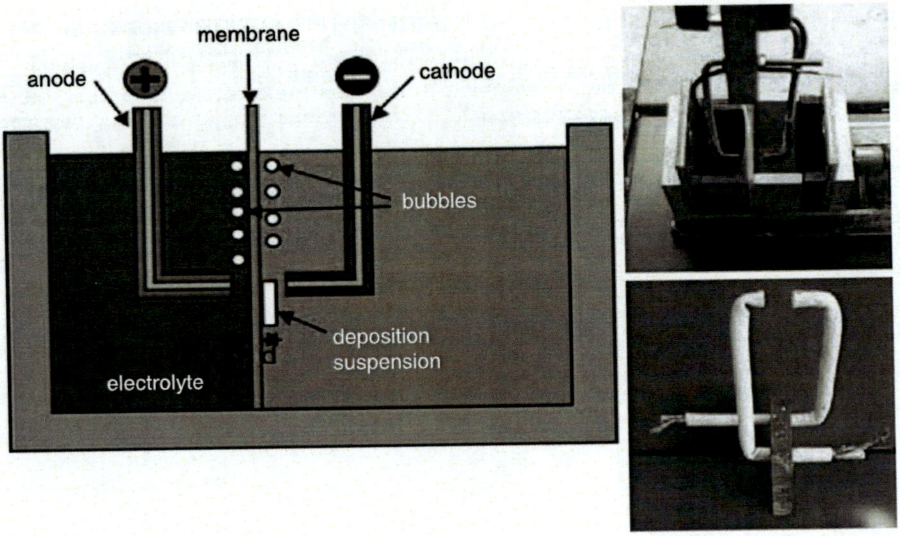

Fig 6.46 Schematic presentation of the apparatus for EPD wit moved point electrodes (*left*) and pictures of the laboratory device (*right side*). [68]

6.3.6.2 Point Electrodes

An alternative to the replica based structuring on shaped porous moulds is a controlled local deposition with one or two point electrodes. This can be a mechanically guided point single or double electrode shown in Fig. 6.46 or an electrically switched array of point electrodes shown in Fig. 6.47 [118–123].

In this case it is most important to focus the electric field to a small point, which is not trivial in an electrically conducting suspension. Therefore shielded coaxial electrodes were applied and model calculations with the program FlexPDE were made. One example is shown in Fig. 6.47. For the first experiments commercial coaxial cables were used, see the upper picture in Fig. 6.48. In the lower picture two examples for a deposition of Aerosil OX50 from an aqueous suspension with electrodes guided with a robot is shown. The corresponding results for an electrode array are shown in the lower part of Fig. 6.49. Further improvements have to be made in both cases to increase the aspect ratio of width and height. Therefore model calculations were performed to optimize the geometry of the coaxial electrodes. Furthermore, the point electrodes can be combined with the shaped membrane process opening a wide field of variations.

Another problem that arises with this configuration is the formation of gas bubbles at the electrodes. For focusing the electric field the diameter of the core wire and the distance to the deposition membrane has to be decreased. Therefore any gas bubble formed at the electrode disturbs the local electrical field significantly. A perspective might be the combination with the ACEPD process. This is presently under investigation. Finally, in Fig. 6.50 the measured aspect ratio is plotted.

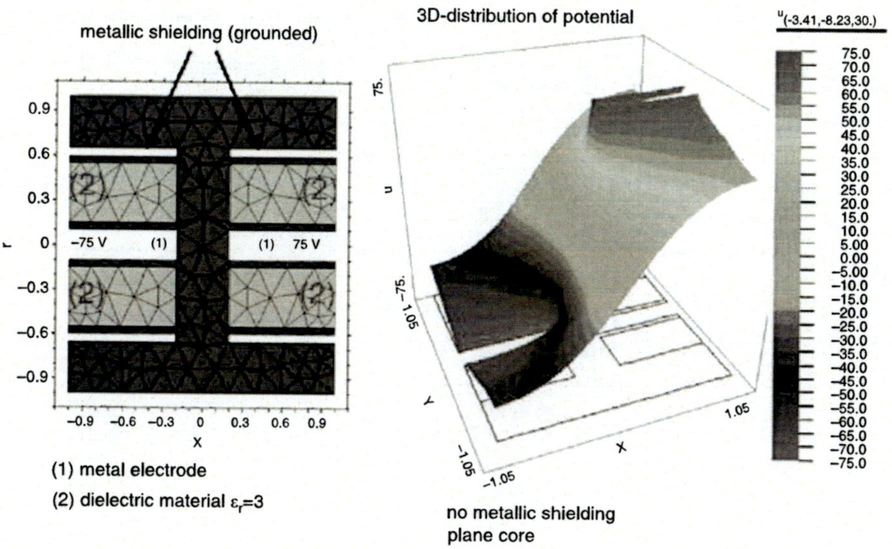

Fig 6.47 Calculated electric fields for a coaxial electrode: *Left:* configuration of the shielded electrode, *right:* calculated electric fields. [68]

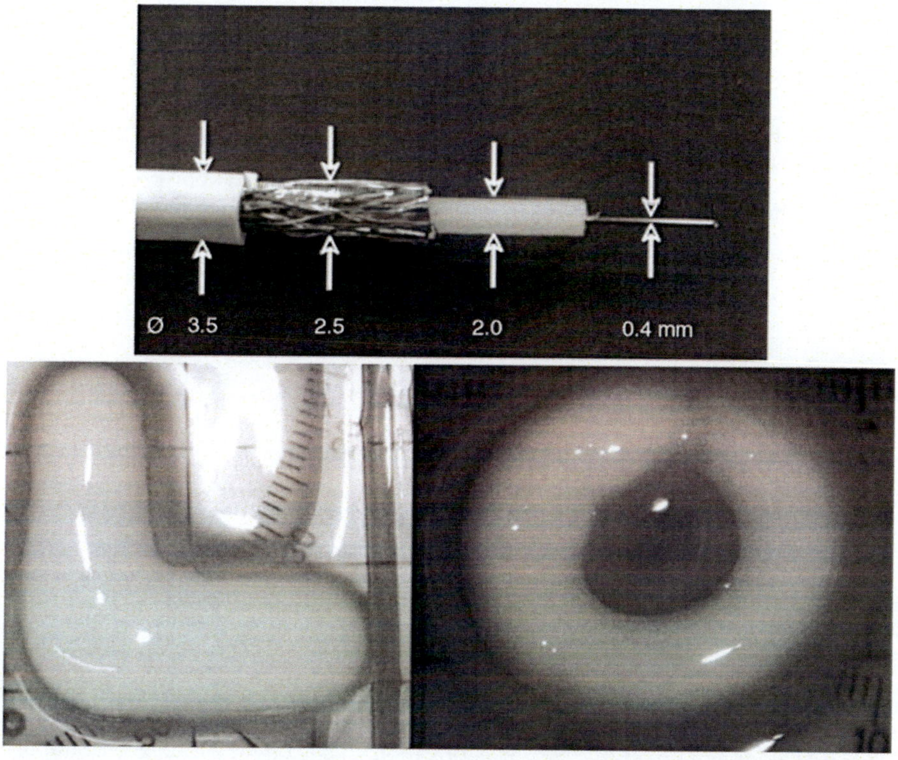

Fig. 6.48 Pictures of coaxial cable used for the electrode (*top*), and deposited Aerosil OX50 with a guided electrode (*bottom*). [68]

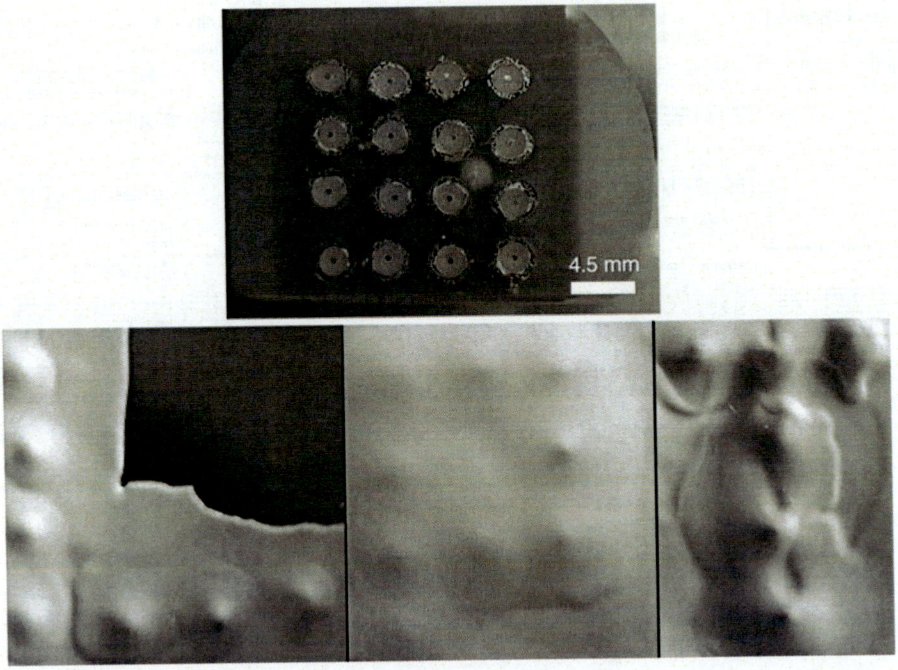

Fig. 6.49 Pictures of the array electrodes (top), and deposited Aerosil OX50 with 3 different combinations of switched-on electrodes of the electrode array

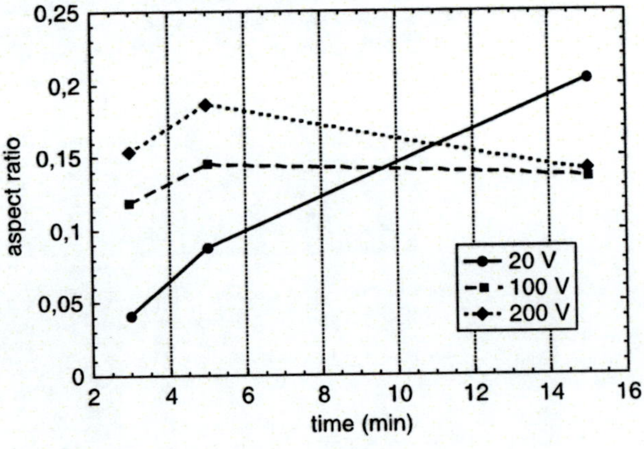

Fig. 6.50 Measured aspect ratio for the electrode array (electrode diameter 3 mm, electrode distance 2 mm) for different voltages. [68]

6.4 Deposition of Coatings

EPD is a process, which is applied for a long time in many fields. But most of these industrial applications are in the field of coatings. One of the oldest applications is enameling [124–128]. Newer publications focus on the efficiency of the deposition process [129] and glass-like coatings based on nanoparticles, which can be densified at 500°C [130]. But EPD has been proven to be a very versatile coating method for a large variety of materials [131, 132].

There a numerous publications on all kinds of EPD coatings (aqueous and non-aqueous systems) like hydroxylapatite layers [133–136], dense titania layers [47, 137], zirconia layers [138], silica sol-gel deposition [139], SiC oxidation protection layers [140], electrolyte layers for SOFC [141–146], MgO coatings [147, 148], 0.1–300 μm thick CeO_2 films [149], CdS/Cu_2S and TiO_2 for solar cells [150, 151], phosphors [152, 153], PZT layers [154], $BaTiO_3$ films 1–5 μm thick [155], quantum sized nano-ZnO films [156], up to 4 mm thick Ni-alumina cermet layers [157], abrasion resistant WC-Co layers [158], diamond films on silicon substrates [159–161], MoS_2 lubricant films [162], ordered gold colloid monolayers [27, 163], and alumina thermal insulations films [164].

Furthermore, it should be briefly noticed, that the cataphoretic deposition for corrosion protection lacquers is applied since a long time [165]. This technique utilizes one of the advantages of an E-field assisted processing, that sharp edges are efficiently coated. In other coating methods the surface tension of the liquid lacquer or suspension leads to a decrease of the coating thickness in this critical part. This is schematically shown in Fig. 6.51. The troublesome areas at the sharp edges of a conventionally coated substrate (upper part) are marked with arrows.

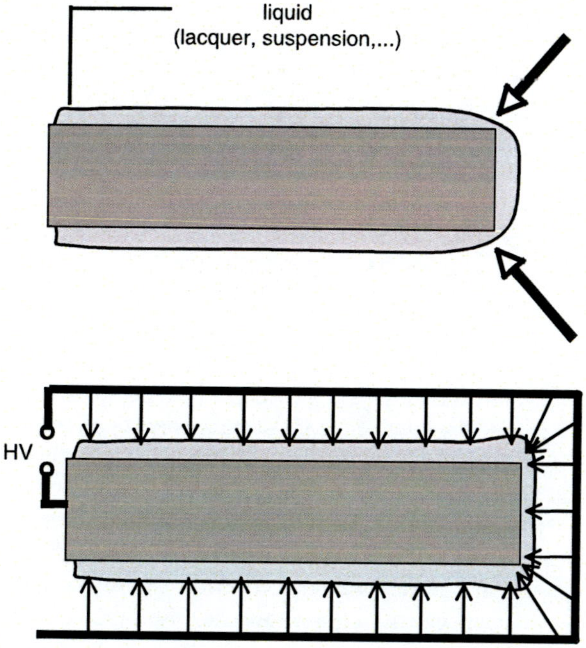

Fig. 6.51 Schematic presentation of the "sharp edge problem"

6.4.1 ETE Process

For inorganic coatings enamels are still one of the biggest applications. The ETE (Elektro-Tauch-Emaillierung) process was developed by Miele (Gütersloh, Germany) almost 30 years ago [43–45, 166] to surpass the problems with the bubble formation in aqueous systems. Apart of the edge enhancement of the EPD process another additive (sodium aluminate) is used for a self-adjustment of the thickness of the coating. This additive forms a gel-like structure in the pores of the deposited enamel particles, which increases the electrical resistance. At thinner parts of the coatings or wholes in the layer the local electric field is higher leading to the preferred deposition of particles. Thus a very smooth surface and a coating without faults can be obtained. The ETE process is applied on an industrial scale and probably one of the biggest industrial applications of EPD coatings.

6.4.2 Electrophotography

There are many perspectives for coatings. This is especially important for nanopowders, which are generally more expensive. For coatings only a small quantity is needed to improve or change the property of a component part. In many cases the coatings have to be structured. The most economic way is an additive (generic) process, because in subtractive processes a lot of material is lost. One of the most emerging markets of a structured EPD is electrophotography (EP). EP is the basis of all copiers and laser printers [167–170]. Most of the applications are in the office area, where black or colored toners are printed on paper. The EP process is schematically shown in Fig. 6.52. The photoconductor is charged in the dark. This is generally done with a high voltage corona. This might be a thin wire, which is moved over the photoconductor in a constant distance or a multi-wire plane corona. Alternatively, a volatile electrode can be used [171]. Thus it is possible to charge a large area in a short time. Most of the copiers and laser printers on the market use a rotating drum instead of the flat substrate plotted in Fig. 6.52 leading to a sequential processing. The pictorial illumination causes a corresponding charge pattern on the surface of the photoconductor.

This charge pattern is developed with charged toner particles. Most of the consumer laser printers now use dry toners, which consist of black or colored pigments and thermoplastic powders. The charged toner particles move in the electric field and are attached to the surface charges on the photoconductor. Afterwards these pigments are transferred on a paper and are glued on the surface via a thermal fixing. As the particle size is typically 10 μm, as can be seen in Fig. 6.53, only a limited resolution can be obtained with such dry toners.

Liquid toners (suspensions) are more troublesome to handle, but finer particles can be used and, consequently, the resolution can be improved [172]. A development electrode, as shown in Fig. 6.52, can improve the sensitivity. There was a

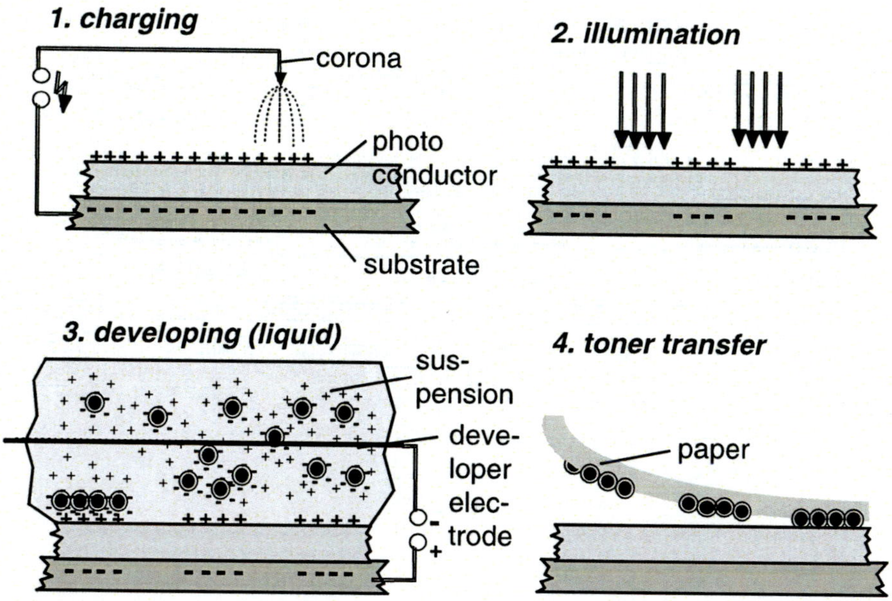

Fig. 6.52 Principle of electrophotography (EP) with a liquid developer (LEP)

Fig. 6.53 SEM picture of black toner for dry development of a laser printer

great interest in electroradiography to develop very sensitive developers. Therefore stable suspensions with toner particles with minimized surface charges were created to reduce the x-ray doses of the patient [59, 173]. First attempts to utilize EP for new applications were performed in the field of RGB phosphors for TV screens [174–177]. But these processes have never been applied in production.

6.4.2.1 Deposition of Glazes for Ceramic Decoration

The decoration of ceramics with glazes or ceramic powders is generally done with all kinds of printing techniques. This is an inexpensive process for large quantities. With the improvement of laser printers not only single prints or copies are interesting from the economic point of view, but also higher quantities are now made with laser printing. Thus laser printers replaced small printing machines. From this viewpoint it was not astonishing, that attempts were made to apply laser printing for ceramic decoration. Especially for small quantities or individual decors there is a big interest. As the ceramic surface is not electrically conducting, a transfer process seems to be most promising. Printed decals are state of the art in ceramic decoration. Therefore digital ceramic/glass decals made by laser printing provide an easy way to produce low-cost images with photo quality [178, 179]. This can be done on commercial laser printers where the toner is replaced by a ceramic powder.

Alternatively to the dry toners a suspension with charged pigments can be used. The effort for processing is higher, but the resolution is better due to the effect, that smaller particles can be used. Furthermore, the surface charge and, consequently, the electrophoretic motion can be better controlled. One disadvantage of these laser printers with a rotating photoconductor drum is the limited charge that can be stored on the photoconductor. This leads to a limitation of the amount of powders, which is deposited on the electric charge pattern. This can be seen in Fig. 6.54, where a cross section of a glaze on a ceramic substrate made with a commercial equipment is shown. Furthermore, these thin coating show only a limited scratch resistance. Another problem is a homogeneous coating of larger areas because more toner is deposited at the electric field gradient of the edges.

Therefore new concepts were developed for the deposition of structured coatings based on the EPD. For non-conductive ceramic or glass substrates a transfer process is needed. This means that the deposition has to be performed on a carrier close to the electrode and the bubble problem has to be solved for aqueous suspen-

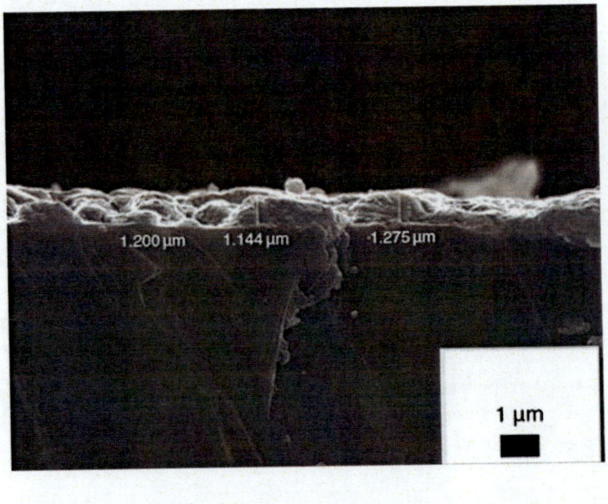

Fig. 6.54 Laser printed glaze powder transferred on a ceramic tile

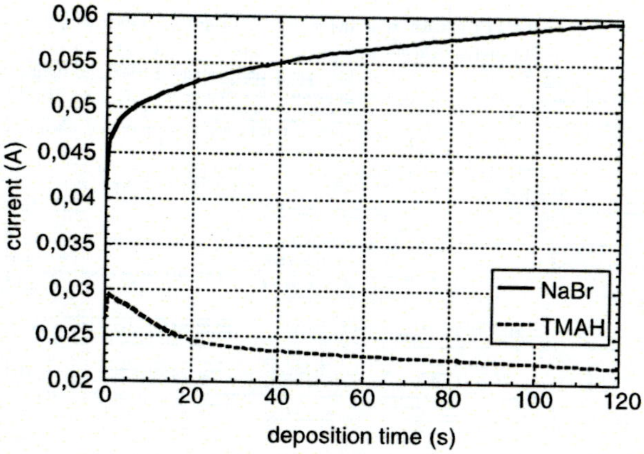

Fig. 6.55 Current with and without addition of NaBr

sions again. In this case the electrolyte oxidation was chosen according to the ETE process. A crude estimation of the charge needed to deposit particles for a coating of some 10 μm thickness shows that for a compensation of 7.2 As (60 mA, 2 min) 1,3 mg NaBr are needed. The influence on the electrical conductivity of this additional electrolyte in a model system consisting of water, Aerosil OX50 and TMAH for adjusting the pH value is plotted in Fig. 6.55. Although the electrical conductivity is increased by factor of 3 with the addition of NaBr, there was no significant influence on the deposition properties. In Fig. 6.56 pictures of deposited compacts are

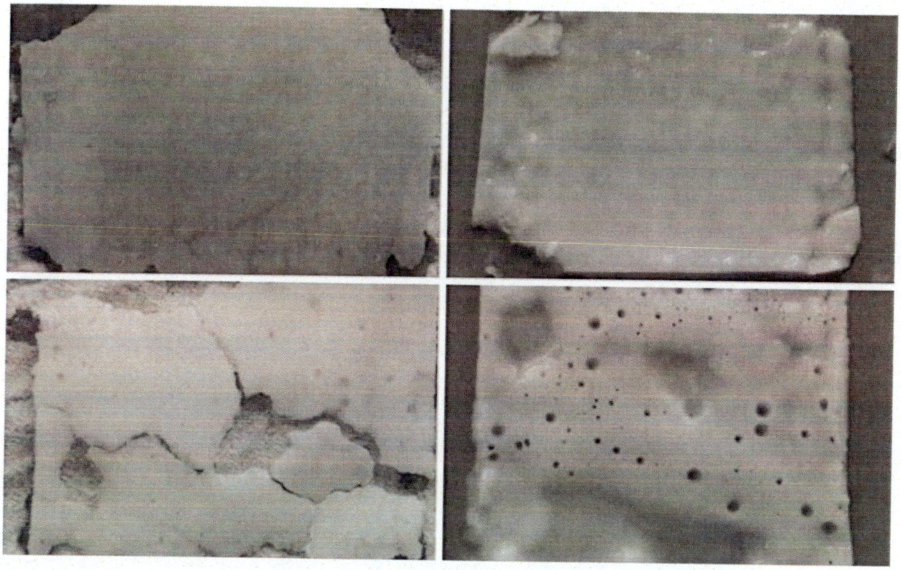

Fig. 6.56 Pictures of deposited compacts. *Left hand side*: electrode side, *right hand side*: suspension side. *Top row*: Deposition with addition of NaBr on a 5 μm sieve, which laid directly on the electrode. *Bottom row*: Deposition without the addition of NaBr

Fig. 6.57 Picture of a structured layer deposited with the addition of NaBr

shown. With the addition of NaBr no bubbles are formed inside the deposited layer. Another examples are shown in Fig. 6.57, where a deposition was performed on structured electrode. The thickness of the layer was approximately 500 µm showing, that there is no limitation in thickness. But is can be expected that the spatial resolution of the deposit and the maximum thick are correlated. Therefore more investigations are necessary to get a better understanding of the possible aspect ratio.

6.4.3 *Deposition of 3D Compacts Layer by Layer*

The forming process of ceramic powders should lead to the final shape as close as possible under the consideration of shrinkage. From the deposition of coatings to a 3D compact is a logic step forward in the field of rapid prototyping (RP), which is of great importance for individually shaped parts or small series. Most of the established rapid prototyping processes are based on a layer-wise processing [180–187]. The applications cover a broad field in rapidly growing markets. Examples are dental ceramics [188–191], functional gradient materials [192], and medical materials [193–195]. Also EPD generated layers are already applied for RP [196–200].

The disadvantage of the layer-wise structuring is the limited mechanical strength because delamination in the boundary region between the layers is a major problem. In EPD at least one side can be structured by depositing on a shaped electrode or membrane [68, 78, 118, 138, 201, 202]. Applications were demonstrated for ceramic microcomponents of PZT or alumina [201, 203], PZT microtubes [204], nanorods [205, 206], and artifical opal growth [207]. Structures can also be obtained by a codeposition of fibers [208] or deposition of carbon nanotubes for field emission displays [209, 210] or tissue engineering scaffolds [211].

6.5 Conclusions

The electrophoretic deposition is a versatile process for depositing structures coatings or compacts especially for nanopowders or mixtures with nanopowders. For aqueous suspensions, which are most favorable for industrial applications the formation of bubbles due to the decomposition of water is a problem. Apart of established solutions with chemical additives new approaches using non-linear effects with pulsed currents seem to be very promising for coatings. For non-conductive substrates the transfer technique known from electrophotography and laser printers offers new perspectives for the decoration of ceramics. More effort is needed to improve these processes. This should be done together with more fundamental research to understand the local mechanisms during deposition.

References

1. Gani, M.S.J.: Electrophoretic deposition—a review. Ind. Ceram. **14**(4), 163–174 (1994)
2. Nicholson, P.S., Sarkar, P.: The Electrophoretic Deposition of Ceramics, pp. 469–479. Am. Ceram. Soc., Westerville (1994)
3. Wittwer, H., Krüger, H.G.: Möglichkeiten und Grenzen der Elektrophorese, cfi/Ber. DKG **72**(9), 556–560 (1995)
4. Boccaccini, A.R., Zhitomirsky, I.: Application of electrophoretic and electrolytic deposition techniques in ceramic processing. Solid State Mater. Sci. **6**, 251–260 (2002)
5. Moritz, K., Tabellion, J., Clasen, R.: Elektrophoretische Abscheidung keramischer Feinst- und Nanopulver—Anwendungen als Formgebungsverfahren und Infiltrationsmethode, Fortschrittsber. Dtsch. Keram. Ges. **17**(1), 48–55 (2002)
6. Corni, I., Ryan, M.P., Boccaccini, A.R.: Electrophoretic deposition: from traditional ceramics to nanotechnology, J. Euro. Ceram. Soc. **28**, 1353–1367 (2008)
7. Moreno, R., Ferrari, B.: Advanced ceramics via EPD of Aqueous Slurries. Am. Ceram. Soc. Bull. **Jan 2000**, 44–48 (2000)
8. Boccaccini, A.R., Biest, O.v.d., Clasen, R.: Electrophoretic Deposition: Fundamentals and Applications II, p. 272. Trans Tech Publications Ltd., Ütikon (2006)
9. Neirinck, B., Fransaer, J., Biest, O.v.d., Vleugels, J.: Aqueous electrophoretic deposition in asymmetric AC electric fields (AC-EPD). Electrochem. Commun. **11**(1), 57–60 (2009)
10. Clasen, R., Tabellion, J.: Electric-field-assisted processing of ceramics. Part I: perspectives and applications, cfi/Ber. DKG **80**(10), E40–E45 (2003)
11. Oetzel, C., Clasen, R., Tabellion, J.: Electric-field assisted processing of ceramics. Part II: electrophoretic impregnation and use for manufacturing of glass and ceramic functionally graded materials, cfi/Ber. DKG **81**(4), E35–E41 (2004)
12. Boccaccini, A.R., Roether, J.A., Thomas, B.J.C., Shaffer, M.S., Chavez, E., Stoll, E., Minay, E.J.: The electrophoretic deposition of inorganic nanoscaled materials—a review. J. Ceram. Soc. Japan **114**(1), 1–14 (2006)
13. Conslik, A.T.: The Debye-Hückel approximation: its use in describing electroosmotic flow in micro- and nanochannels. Electrophoresis **26**, 1896–1912 (2005)
14. Tabellion, J.: Manufacturing of ceramics and glasses by aqueous electrophoretic deposition. Interceram **55**(5), 338–342 (2006)
15. Fendler, J.H.: Nanoparticles and nanostructured films, p. 468. Wiley, Weinheim (1998)
16. Clasen, R., Tabellion, J.: Electrophoretic deposition from aqueous suspensions—technical applications. In: Boccaccini, A.R., Biest, O.v.d., Talbot, J.B. (eds.) Electrophoretic Deposi-

tion: Fundamentals and Applications, pp. 138–145. The Electrochemical Society, Inc., Pennigton (2002)
17. Braun, A., Wolff, M., Oetzel, C., Tabellion, J., Falk, G., Clasen, R.: Alumina ceramics by means of electrophoretic deposition of submicron powders. Ceram. Eng. Sci. Proc. **25**(4), 591–596 (2004)
18. Braun, A.: Transparent Polycrystalline alumina ceramic by means of electrophoretic deposition for optical applications, Dissertation, p. 170. Universität des Saarlandes, Saarbrücken (2005)
19. Braun, A., Falk, G., Clasen, R.: Transparent polycrystalline alumina ceramic with sub-micrometre microstructure by means of electrophoretic deposition. Mat.-wiss. u. Werkstofftech. **37**(4), 293–297 (2006)
20. Nichols, S.C., Loewenberg, M., Davis, R.H.: Electrophoretic particle aggregation. J. Colloid Interface Sci. **176**, 342–351 (1995)
21. Solomentsev, Y., Guelcher, S.A., Bevan, M., Anderson, J.L.: Aggregation dynamics for two particles during electrophoretic deposition under steady fields. Langmuir **16**, 9208–9216 (2000)
22. Pérez, A.T., Saville, D., Soria, C.: Modeling the electrophoretic deposition of colloidal particles. Europhys. Lett. **55**(3), 425–431 (2001)
23. Greil, P., Cordelair, J., Bezold, A.: Discrete element simulation of ceramic powder processing. Z. Metallkd. **92**, 682–689 (2001)
24. Nadal, F., Argoul, F., Hanusse, P., Pouligny, P., Ajdari, A.: Electrically induced interactions between colloidal particels in the vicinity of a conducting plane. Phys. Rev. E **65**(067409), 1–7 (2002)
25. Solomentsev, Y., Böhmer, M., Anderson, J.L.: Particle clustering and pattern formation during electrophoretic deposition: a hydrodynamic model. Langmuir **13**, 6058–6068 (1997)
26. Sarkar, P., De, D., Yamashita, K., Nicholson, P.S., Umegaki, T.: Mimicking nanometer atomic processes on a micrometer scale via electrophoretic deposition. J. Am. Ceram. Soc. **83**(6), 1399–1401 (2000)
27. Giersig, M., Mulvaney, P.: Preparation of ordered colloid monolayers by electrophoretic deposition. Langmuir **9**, 3408–3413 (1993)
28. Roth, J.-E.: Grenzflächeneffekte bei der Fest/Flüssig-Trennung. Chem. -Ing. Tech. **63**(2), 104–115 (1991)
29. Boccaccini, A.R., Biest, O.v.d., Talbot, J.B.: Electrophoretic Deposition: Fundamentals and Applications, p. 307. The Electrochemical Society. Inc., Pennigton (2002)
30. Rabinovich, E.M.: Review: preparation of glass by sintering. J. Mater. Sci. **20**, 4259–4297 (1985)
31. MacChesney, J.B., DiGiovanni, D.: Materials development of optical fiber. J. Am. Ceram. Soc. **73**(12), 3537–3555 (1990)
32. Sakka, S.: Gel method for making glass. In: Tomozawa, M., Doremus, R.H. (eds.) Treatise on Materials Science and Technology, pp. 129–167. Academic Press, New York (1982)
33. Yan, M.F.: Optical fiber processing: science and technology. Am. Ceram. Soc. Bull. **72**(5), 107–119 (1993)
34. Clasen, R.: Preparation and sintering of high-density green bodies to high purity silica glasses. J. Non-Cryst. Solids **89**, 335–344 (1987)
35. Clasen, R., Zeiner, J., Oetzel, C.: Shaping of nanosized powder by electrophoretic deposition—optimisation of suspension and green body properties by dispersing methods, cfi/Ber. DKG **84**(13), 18–22 (2007)
36. Magnucki, K., Stasiewicz, P.: Elastic buckling of a porous beam. J. Theor. Appl. Mech **42**(4), 859–868 (2004)
37. Kastler, A.: Untersuchungsmethoden zur in-situ Charakterisierung der elektrophoretischen Abscheidung, Diplomarbeit, p. 73. Universität des Saarlandes, Saarbrücken (2004)
38. Bohren, C.F., Huffmann, D.R.: Absorption and Scattering of Light by Small Particles, p. 530. Wiley, New York (1983)
39. Schmidt, E.W.: Elephant-Experiment oder Zukunft. Ziegelindustrie Int. 217–220 (1978)

40. Chronberg, M.S., Händle, F.: Processes and equipment for the production of materials by electrophoresis ELEPHANT. Interceram **27**(1), 33–34 (1978)
41. Mihailescu, M., Emandi, M., Vancea, V., Marcu, M.: Electrophoretic behavior of ceramic sanitaryware slips. Interceram **40**(3), 165–170 (1991)
42. Hector, I., Clasen, R.: Electrophoretic deposition of compacts from clay suspensions. Ceram. Eng. Sci. Proc. **18**(2), 173–186 (1997)
43. Engelhardt, T.: Physikalische und chemische Eigenschaften wässriger Emailschlicker für die Elektrotauchemaillierung, PhD-Thesis. RWTH, Inst. f. Gesteinshüttenkunde, Aachen (1988)
44. Warnke, H., Kaup, F., Bersch, B., Mertinkat, W., Tenhaven, U.: Method for directly enameling steel parts using a single enamel coating, Patent, US 4707385
45. Warnke, H.: Die Elektrotauchemaillierung (ETE), p. 218. Eigenverlag Autor, Marienfelde (2001)
46. Sakurada, O., Suzuki, K., Miura, T., Hashiba, M.: Bubble-free electrophoretic deposition of aqueous zirconia suspensions with hydroquinone. J. Mater. Sci. **39**, 1845–1847 (2004)
47. Lebrette, S., Pagnoux, C., Abélard, P.: Fabrication of titania dense layers by electrophoretic deposition in aqueous media, J. Eur. Ceram. Soc. **26**, 2727–2734 (2006)
48. Clasen, R.: Forming of compacts of submicron silica particles by electrophoretic deposition. In: Hausner, H., Messing, G.L., Hirano, S. (eds.), 2nd Int. Conf. on Powder Processing Science Berchtesgaden, 633–640.12.–14. 10. Deutsche Keramische Gesellschaft, Köln (1988)
49. Bonnas, S., Tabellion, J., Ritzhaupt-Kleissl, H.-J., Haußelt, J.: Systematic Interactions of Sedimentation and Electrical Field in Electrophoretic Deposition. In: Menz, W., Dimov, S., Fillon, B., 4M 2006—Second International Conference on Multi-Material Micro Manufacture Grenoble, France, Elsevier
50. Tabellion, J., Clasen, R.: Advanced ceramic or glass components and composites by electrophoretic deposition/impregnation using nanosized particles. In: Lin, H.-T., Singh, M. (eds.) 26th Annual Conference on Composites, Advanced Ceramics, Materials and Structures Cocoa Beach, vol. 23, pp. 617–627. The American Ceramic Society, Florida (2002)
51. Tabellion, J., Clasen, R., Jungblut, E.: Near-shape manufacturing of ceramics and glasses by electrophoretic deposition using nanosized powders. Ceram. Eng. Sci. Proc. **24**(3), 46 (2003)
52. Pohl, H.A.: The motion and precipitation of suspensoids in divergent electric fields. J. Appl. Phys. **22**(7), 869–871 (1951)
53. Klass, D.L., Martinek, T.W.: Electroviscous fluids. I. Rheological properties. J. Appl. Phys. **38**(1), 67–74 (1967)
54. Klass, D.L., Martinek, T.W.: Electroviscous fluids. II. Electrical properties. J. Appl. Phys. **38**(1), 75–80 (1967)
55. Bowen, C.P., Shrout, T.R., Newnham, R.E., Randall, C.A.: Tunable electric field processing of composite materials. J. Intell. Mater. Syst. Struct. **6**, 159–168 (1995)
56. Bowen, C.P., Bhalla, A.S., Newnham, R.E., Randall, C.A.: An investigation of the assembly conditions of dielectric particles in uncured thermoset polymers. J. Mater. Res. **9**(3), 781–788 (1994)
57. Besra, L., Uchikoshi, T., Suzuki, T.S., Sakka, Y.: Bubble-free aqueous electrophoretic deposition (EPD) by pulse-potential application. J. Am. Ceram. Soc. **91**(10), 3154–3159 (2008)
58. Yue, C.-F.J., Kumar, D., Singh, R.K.: Fabrication of Ag-sheathed Bi-Sr-Ca-Cu-O thik films by a novel a.c.-electric field assisted electrophoretic deposition method. Physica C **314**, 291–298 (1999)
59. Stotz, S.: Field dependence of the electrophoretic mobility of particles suspended in low-conductivity liquids. J. Colloid Interface Sci. **65**(1), 118–131 (1978)
60. Kortüm, G.: Lehrbuch der Elektrochemie, p. 564. Verlag Chemie GmbH, Weinheim (1957)
61. Moritz, K., Moritz, T.: Electrophoretically deposited porous ceramics and their characterization by X-ray computer tomography. Key. Eng. Mater. **412**, 255–2670 (2009)
62. Santillán, M.J., Membrives, F., Quaranta, N., Boccaccini, A.R.: Characterization of TiO_2 nanoparticle suspensions for electrophoretic deposition, J. Nanopart. Res. **10**, 787–793 (2008)

63. Tabellion, J., Oetzel, C., Clasen, R.: Densification of porous ceramics and glasses by electrophoretic impregnation. Electrochem. Soc. Proc. **2002–21**, 31–38 (2002)
64. Clasen, R., Smeets, K., Tabellion, J.: Incorporation of functional secondarys phases by electrophoretic impregnation. Electrochem. Soc. Proc. **2002–21**, 279–286 (2002)
65. Tabellion, J., Zeiner, J., Clasen, R.: Manufacturing of pure and doped silica and multicomponent glasses from SiO_2 nanoparticles by reactive electrophoretic deposition, J. Mater. Sci. **41**, 8173–8180 (2006)
66. Smeets, K.: Pulvertechnologische Herstellung und Charakterisierung von Grünkörpern und Gläsern mit nanoskaligen Zweitphasen, Dissertation, p. 161. Universität des Saarlandes, Saarbrücken (2004)
67. Clasen, R.: Preparation of coloured silica glasses made by sintering of particulate gels. Glastech. Ber. **66**(11), 299–304 (1993)
68. Zeiner, J.: Untersuchungen zur Herstellung von Mikrostrukturen aus nanoskaligen Pulvern mittels elektrophoretischer Abscheidung, Dissertation, p. 156. Universität des Saarlandes, Saarbrücken (2007)
69. Moritz, K., Müller, E.: Electrophoretic infiltration of woven carbon fibre mats with SiC powder suspensions. Key Eng. Mater. **206–213**, 193–196 (2002)
70. Dittrich, R.: Neuartige Technologieansätze zur Hrestellung von C/SiS-Verbundwerkstoffen, Dissertation, p. 147. TU Bergakademie Freiberg, Freiberg (2007)
71. Müller, B., Ferkel, H.: Properties of nanocrysatlline Ni/Al_2O_3 composites. Z. Metallkd. **90**(11), 868–871 (1999)
72. Müller, B., Ferkel, H.: Al_2O_3-nanoparticle distribution in plated nickel composite films. Nanostruct. Mater. **10**(8), 1285–1288 (1998)
73. Vidrich, G., Moll, O., Ferkel, H.: Verhalten von SiO_2- und Al_2O_3-Nanopartikeln in wässrigen Elektrolyten bei der Koabscheidung. In: 9. Werkstofftechnisches Kolloquium Chemnitz (2006)
74. Müller, B.: Elektrolytische Koabscheidung partikelverstärkter Nanoverbundschichten und deren Eigenschaften, p. 108. VDI, Düsseldorf (1999)
75. Krüger, H.G., Knote, A., Schindler, U., Kern, H., Boccaccini, A.R.: Composite ceramic-metal coatings by means of combined electrophoretic deposition and galvanic methods. J. Mater. Sci. **39**, 839–844 (2004)
76. Castagnet, J.-F.: Herstellung und Charakterisierung von nanopartikelverstärkten elektrolytisch abgeschiedenen Nickelschichten, Dissertation, p. 136. Technische Universität Clausthal, Clausthal-Zellerfeld (2005)
77. Jung, D.: Herstellung gradierter Glas-Metallübergänge mittels elektrophoretischer Abscheidung zur Reduzierung thermsicher Spannnungen, Dissertation, p. 151. Universität des Saarlandes, Saarbrücken (2008)
78. Zeiner, J., Clasen, R.: Fabrication of colored silica glasses by incorporation of nano-sized colloids via reactive electrophoretic deposition (REPD). Key Eng. Mater. **314**, 115–120 (2006)
79. Dislich, H.: Darstellung von Mehrkomponenetengläsern ohne Durchlaufen der Schmelzphase. Glastechn. Ber. **44**(1), 1–8 (1971)
80. Dislich, H.: New routes to multicomponent oxide glasses. Angew. Chemie. **10**(6), 363–370 (1971)
81. Dislich, H.: Glassy and crystalline systems from gels: chemical basis and technical application. J. Non-Cryst. Solids **57**, 371–388 (1983)
82. Mukherjee, S.P.: Sol-gel process in glass science and technology. J. Non-Cryst. Solids **42**, 477–488 (1980)
83. Mukherjee, S.P.: Homogenity of gels and gel-derived glasses. J. Non-Cryst. Solids **63**, 35–43 (1984)
84. Villegas, M.A., Navarro, J.M.F.: Characterization and study of Na_2O-B_2O_3-SiO_2 glasses prepared by the sol-gel method. J. Mater. Sci. **23**, 2142–2152 (1988)
85. Villegas, M.A., Navarro, J.M.F.: Characterization of B_2O_3-SiO_2 glassses prepared via sol-gel. J. Mater. Sci. **23**, 2464–2478 (1988)

86. Jabra, R., Phalippou, J., Zarzycki, J.: Synthesis of binary glass-forming oxide glasses by hot-pressing of gels. J. Non-Cryst. Solids **42**, 489–498 (1980)
87. Blasig, R.: Herstellung von Borosilikatglasschichten auf Glasformkörpern über einen Sinterprozeß nanoskaliger Pulver, Dissertation. Universität des Saarlandes, Saarbrücken (1997)
88. Coble, R.L.: Transparent alumina and method of preparation, Patent, US 3 026 210
89. Coble, R.L., Song, H., Brook, R.J., Handwerker, C.A., Dynys, J.M.: Sintering and grain growth in alumina and magnesia. In: Kingerly, W.D. (ed.) Advances in Ceramics Structure and Properties of MgO and Al_2O_3, pp. 839–852. The American Ceramic Society, Inc., Columbus (1984)
90. Peelen, J.G.J.: Influence of MgO on the evolution of the microstructure of alumina. In: Kuczynski, G.C. (ed.) Sintering and Catalysis, pp. 443–453. Plenum Press, New York (1975)
91. Peelen, J.G.J.: Zusammenhang zwischen Mikrostruktur und optischen Eigenschaften von polykristallinem Aluminiumoxid. Sci. Ceram. **6**, 1–13 (1973)
92. Peelen, J.G.J.: Transparent hot-pressed alumina. I: Hot pressing of alumina. Ceram. Int. **5**(2), 70–75 (1979)
93. Peelen, J.G.J., Metselaar, R.: Light scatering by pores in polycrystalline materials: transmission properties of alumina. J. Appl. Phys. **45**(1), 216–220 (1974)
94. Peelen, J.P.J.: Die Lichtdurchlässigkeit von gesintertem Aluminiumoxid. Philips Techn. Rdsch. **36**(2), 53–58 (1976/1977)
95. Apetz, R., Bruggen, M.P.B.v.: Transparent alumina: a light scattering model. J. Am. Ceram. Soc. **86**(3), 480–486 (2003)
96. Bruggen, M.v., Apetz, R., Kop, T., Krell, A., Hutzler, T.: Transparent Polycristalline Aluminum Oxide, Patent, WO 2004/007398 A1
97. Krell, A., Blank, P., Ma, H.W., Hutzler, T., Bruggen, M.P.B.v., Apetz, R.: Transparent sintered corundum with high hardness and strength. J. Am. Ceram. Soc. **86**(1), 12–18 (2003)
98. Krell, A., Ma, H.W.: Sintering transparent and other sub-mm alumina: the right powder, cfi-Ceram. Forum. Int. **80**(4), E41–E45 (2003)
99. Krell, A., Hutzler, T., Klimke, J.: Transparent ceramics for tructural applications: part 1. Physics of light transmission and technological consequences, cfi/Ber. DKG **84**(4), E41–E50 (2007)
100. Krell, A., Hutzler, T., Klimke, J.: Transparent ceramics for structural applications: part 2: fields of applications, cfi/Ber. DKG **84**(6), E50–E56 (2007)
101. Krell, A., Hutzler, T., Klimke, J.: Transmission physics and consequences for materials selection, manufacturing, and applications, J. Eur. Ceram. Soc. **29** 207–221 (2009)
102. Braun, A., Falk, G., Clasen, R.: Transparent polycrystalline alumina ceramic with submicrometre microstructure by means of electrophoretic deposition, cfi/Ber. DKG **82**(13), 161–165 (2005)
103. Dericioglu, A.F., Kagawa, Y.: Effect of grain boundary microcracking on the light transmittance of sintered transparent $MgAl_2O_4$, J. Eur. Ceram. Soc. **23**(6), 951–959 (2003)
104. Patterson, M.C.L., Caiazza, J.E., Roy, D.W.: Transparent spinel development, In: Arthurs, E.G. (ed.) Inorganic Optical Materials II vol. 4102, pp. 59–68. (2000)
105. Skandan, G., Hahn, H., Parker, J.C.: Nanostructured Y_2O_3: Synthesis and relation to microstructure and properties. Scripta Metall. Mater. **25**, 2389–2393 (1991)
106. Saito, N., Matsuda, S., Ikegami, T.: Fabrication of transparent yttria ceramics at low temperature using carbonate-derived powder, J. Am. Ceram. Soc. **81**(8), 2023–2028 (1998)
107. Ikegami, T., Li, J.-G., Mori, T., Moriyoshi, Y.: Fabrication of transparent yttria ceramics by the low-temperature synthesis of yttrium hydroxide, J. Am. Ceram. Soc. **85**(7), 1725–1729 (2002)
108. Ikesue, A.: Polycrystalline Nd:YAG ceramics lasers. Opt. Mater. **19**, 183–187 (2002)
109. Cheng, J., Agrawal, D., Zhang, Y., Roy, R.: Microwave reactive sintering to fully transparent aluminium oxynidride (ALON) ceramics, J. Mater. Sci. Lett. **20**, 77–79 (2001)

110. Snow, G.S.: Fabrication of Transparent PLZT Ceramics by Atmosphere Sintering, J. Am. Ceram. Soc. **56**(2), 91–96 (1973)
111. Wolff, M.: Untersuchungen zur Herstellung von transparentem Zirkonoxid, Dissertation, p. 164. Universität des Saarlandes, Saarbrücken (2005)
112. Wolff, M., Clasen, R.: Fabrication of transparent polycrystalline zirconia ceramics, cfi/Ber. DKG **82**(9), E49 (2005)
113. Wolff, M., Falk, G., Clasen, R.: Sintering behavior of dense nanocrystalline zirconia ceramics: a comparative investigation, Ceram. Eng. Sci. Proc. **25**(4), 19–24 (2004)
114. Link, G., Wolff, M., Takayama, S., Thumm, M., Falk, G., Clasen, R.: The densification behavior of zirconia ceramics during millimeter-wave sintering, cfi/Ber. DKG **13**, 312–316 (2005)
115. Braun, A., Clasen, R., Oetzel, C., Wolff, M.: Hochleistungskeramik aus Nanopulvermischungen für optische und dentale Anwendungen, Fortschrittsberichte der DKG, Beiheft cfi/Ber. DKG **20**(1), 131–136 (2006)
116. Oetzel, C.: Herstellung vollkeramischer Zahnkronen aus Zirkonoxid und Zirkonoxid-Aluminiumoxid mit Hilfe der elektrophoretischen Abscheidung, Dissertation, p. 202. Universität des Saarlandes, Saarbrücken (2005)
117. Oetzel, C., Clasen, R.: Preparation of zirconia dental crowns via electrophoretic deposition, J. Mater. Sci. **41**(24), 8130–8137 (2006)
118. Zeiner, J., Clasen, R.: Fabrication of microstructures by means of electrophoretic deposition (EPD). Key Eng. Mater. **314**, 57–62 (2006)
119. Zeiner, J., Clasen, R.: Electrophoretic deposition as shaping technique—a CAM approach. Adv. Sci. Technol. **45**, 714–719 (2006)
120. Zeiner, J., Nold, A., Clasen, R.: Electrophoretic deposition as shaping technique—a CAM approach. Adv. Sci. Technol. **45**, 714–719 (2006)
121. Nold, A., Zeiner, J., Clasen, R.: Local electrophoretic deposition with coaxial electrodes. Key Eng. Mater. **412**, 307–312 (2009)
122. Nold, A., Zeiner, J., Assion, T., Clasen, R.: Electrophoretic deposition as a rapid prototyping method. J. Euro. Ceram. Soc. **30**(5), 1163–1170 (2010)
123. Nold, A., Zeiner, J., Clasen, R.: Fast deposition of surface structures by two-step EPD. Key Eng. Mater. **412**, 3–8 (2009)
124. Krannich, R.: Elektrophoretisches Emailauftragen. Silikattechnik **18**(7), 229–357 (1967)
125. Hoffmann, H.: Der elektrophoretische Emailauftrag. Sprechsaal **6**, 221–224 (1969)
126. Vliet, W.E.v.d.: Verhalten farbiger Emails bei elektrophoretischer Abscheidung. Mitteilungen VDEfa **25**(4), 49–52 (1977)
127. Salge, H.: Emailauftrag im elektrischen Feld. Mitteilungen VDEfa **25**(3), 29–40 (1977)
128. Hoffmann, H.: Theory and practice of electrocoating of porcelain enamel. Ceram. Bull. **57**(6), 605–608 (1978)
129. Yatsenko, E.A., Selivanov, V.N., Shchepeleeva, M.S.: Efficiency of depositing glass enamels by electrophoresis. Glass Ceram. **61**(9–10), 352–354 (2004)
130. Nguyen, H.Q., Fürbeth, W., Schütze, M.: Nano-enamel: a new way to produce glass-like protective coatings for metals. Mater. Corros. **53**, 772–782 (2002)
131. Gutierrez, C.P., Mosley, J.R., Wallace, T.C.: Electrophoretic deposition: a versatile coating method. J. Electrochem. Soc. **109**(10), 923–927 (1962)
132. Koura, N., Tsukamoto, T., Shoji, H., Hotta, T.: Preparation of various oxide films by an electrophoretic deposition method: a study of the mechanism. Jpn. J. Appl. Phys. **34**(3), 1643–1647 (1995)
133. Zhitomirsky, I., Gal-Or, l.: Electrophoretic deposition of hydroxyapatite. J. Mater. Sci: Mater. Med. **8**, 213–219 (1997)
134. May, H., Ordung, M., Ziegler, G.: Development of thin electrophoretically deposited hydroxyapatite layers on $TiAl_6V_4$ hip prosthesis. J. Mater. Sci. **41**, 8138–8143 (2006)
135. Wei, M., Ruys, A.J., Milthorpe, B.K., Sorrell, C.C., Evans, J.H.: Electrophoretic deposition of hydroxyapatite coatings on metal substrates: a nanoparticulate dual-coating approach. J. Sol-Gel Sci. Technol. **21**, 39–48 (2001)

136. Wang, R., Hu, Y.X.: Patterning hydroxyapatite biocoating by electrophoretic deposition. J. Biomed. Res. A **67A**(1), 270–275 (2003)
137. Yanagida, S., Nakajima, A., Kameshima, Y., Yoshida, N., Watanabe, T., Okada, K.: Preparation of crack-free rough titania coating on stainless steel mesh by electrophoretic deposition. Mater. Res. Bull. **40**, 1335–1344 (2005)
138. Pfrengle, A., Both, H.v., Knitter, R., Haußelt, J.: Electrophoretic deposition and sintering of zirconia layers on microstructured steel substrates. J. Eur. Ceram. Soc. **26**, 2633–2638 (2006)
139. Nishimori, H., Tatsumisago, M., Minami, T.: Preparation of thick silica films by the electrophoretic sol-gel deposition on a stainless steel sheet. J. Ceram. Soc. Jpn. **103**(1), 78–80 (1995)
140. Damjanovic, T., Argirusis, C., Borchardt, G., Leipner, H., Herbig, R., Tomandl, G., Weiss, R.: Oxidation protection of C/C–SiC composites by an electrophoretically deposited mullite precursor. J. Eur. Ceram. Soc. **25**, 577–587 (2005)
141. Zhitomirsky, I., Petric, A.: Electrophoretic deposition of electrolyte materials for solid oxide fuel cells. J. Mater. Sci. **39**, 825–831 (2004)
142. Will, J., Hruschka, M.M., Gubler, L., Gauckler, L.J.: Electrophoretic deposition of zirconia on porous anodic substrates. J. Am. Ceram. Soc. **84**(2), 328–332 (2001)
143. Basu, R.N., Randall, C.A., Mayo, M.J.: Fabrication of dense zirconia electrolyte films for tubular solid oxide fuel cells by electrophoretic deposition. J. Am. Ceram. Soc. **84**(1), 33–40 (2001)
144. Ishihara, T., Shimose, K., Kudo, T., Nishiguchi, H., Akbay, T., Takita, Y.: Preparation of yttria-stabilized zirconia thin films on strontium-doped $LaMnO_3$ cathode substrates via electrophoretic deposition for solid oxide fuel cells. J. Am. Ceram. Soc. **83**(8), 1921–1927 (2000)
145. Chen, F., Liu, M.: Preparation of yttria-stabilized zirconia (YSZ) films on $La_{0.85}Sr_{0.1}5MnO_3$ (LSM) and LSM-YSZ substrates using an electrophoretic deposition (EPD) process. J. Euro. Ceram. Soc. **1–2**, 127–134 (2001)
146. Zhitomirsky, I., Petric, A.: Electrophoretic deposition of ceramic materials for fuel cell applications. J. Eur. Ceram. Soc. **20**(12), 2055–2061 (2000)
147. Hosseinbabaei, F., Raissidehkordi, B.: Electrophoretic deposition of MgO thick films from an acetone suspension. J. Euro. Ceram. Soc. **20**, 2165–2168 (2000)
148. Ferrari, B., Moreno, R., Sakai, P., Nicholson, P.S.: Electrophoretic deposition of MgO from organic suspensions. J. Eur. Ceram. Soc. **20**, 99–106 (2000)
149. Zhitomirsky, I., Petric, A.: Electrolytic and electrophoretic deposition of CeO_2 films. Mater. Lett. **40**, 263–268 (1999)
150. Williams, E.W., Jones, K., Griffiths, A.J., Roughley, D.J., Bell, J.M., Steven, J.H., Huson, M.J., Rhodes, M., Costich, T.: The electrophoresis of thin Film CdS/Cu_2S solar cells. Solar Cells **1**, 357–366 (1979/1980)
151. Matthews, D., Kay, A., Grätzel, M.: Electrophoretically deposited titanium dioxide thin films for photovoltaic cells. Aust. J. Chem. **47**, 1869–1877 (1994)
152. Shane, M.J., Talbot, J.B., Schreiber, R.D., Ross, C.L., Sluzky, E., Hesse, K.R.: Electrophoretic deposition of phosphors. I. Conductivity and zeta potential measurements. J. Colloid Interface Sci. **165**, 325–333 (1994)
153. Shane, M.J., Talbot, J.B., Kinney, B.G., Sluzky, E., Hesse, K.R.: Electrophoretic deposition of phosphors. II. Deposition experiments and analysis. J. Colloid Interface Sci. **165**, 334–340 (1994)
154. Kaya, C., Kaya, F., Su, B., Thomas, B., Boccaccini, A.R.: Structural and functional thick ceramic coatings by electrophoretic deposition. Surf. Coat. Technol. **191**, 303–310 (2005)
155. Zhang, J., Lee, B.I.: Electrophoretic deposition and characterization of micrometer-scale $BaTiO_3$ based X7R dielectric thick films. J. Am. Ceram. Soc. **83**(10), 2417–2422 (2000)
156. Wong, E.M., Searson, P.C.: ZnO quantum particle thin films fabricated by electrophoretic deposition. Appl. Phys. Lett. **74**(20), 2939–2941 (1999)

157. Windes, W.E., Zimmerman, J., Reimanis, I.: Electrophoretic deposition applied to thick metal-ceramic coatings. Surf. Coat. Technol. **157**, 267–273 (2002)
158. Henker, A., Jahn, D., Pursche, G.: Zu Problemen des elektrophoretischen Abscheidensvon WC-Co Teilchen zum Zwecke der Herstellung verschleißmindernder Schichten, Wiss. Z. d. Techn. Hochsch. Karl-Marx-Stadt **XXI**(1), 21–32 (1979)
159. Lee, D.-G., Singh, R.K.: Synthesis of (111) oriented diamond thin films by electrophoretic deposition process. Appl. Phys. Lett. **70**(12), 1542–1544 (1997)
160. Valdes, J.L., Mitchel, J.W., Mucha, J.A., Seibles, L., Huggins, H.: Selected-area nucleation and patterning of diamond thin films by electrophoretic seeding. J. Electrochem. Soc. **138**(2), 635–636 (1991)
161. Affoune, A.M., Prasad, B.L.V., Sato, H., Enoki, T.: Electrophoretic deposition of nanosized diamond particles. Am. Chem. Soc. **17**(2), 547–551 (2000)
162. Panitz, J.K.G., Dugger, M.T., Peebles, D.E., Tallant, D.R., Hills, C.R.: Electrophoretic deposition of pure MoS_2 dry film lubricant coatings. J. Vac. Sci. Technol. A **11**(4), 1441–1446 (1993)
163. Zhao, S.-Y., Lei, S.-B., Chen, S.-H., Ma, H.-Y., Wang, S.-Y.: Assembly of two-dimensional ordered monolayers of nanoparticles by electrophoretic deposition. Colloid. Polym. Sci. **278**, 682–686 (2000)
164. Narisawa, T., Arato, T., Koganezawa, N., Shibata, M., Nonaka, Y.: Formation of alumina insulation films for a CRT heater by constant current electrophoretic deposition method. J. Ceram. Soc. Jpn. **103**(1), 53–57 (1995)
165. Benjamin, M., Osborn, A.B.: The deposition of oxide coatings by cataphoresis. Trans. Faraday. Soc. **36**, 287–295 (1940)
166. Thielmann, C.: Elektrotauch-Emaillierung—Vom Versuchsmodell zur Großserienanlage. JOT **1**, 20–24 (1998)
167. Schaffert, R.M.: Electrophotography, p. 989. Focal Press Ltd., London (1975)
168. Comizzoli, R.B., Lozier, G.S., Ross, D.A.: Electrophotography—a review. Proc. IEEE **60**(4), 348–369 (1972)
169. Hays, D.A., Ossman, K.R.: Electrophotographic copying and printing (xerography). In: The Optics Encyclopedia, pp. 587–607. Wiley, Weinheim (2004)
170. Pai, D.M., Springett, B.E.: Physics of electrophotography. Rev. Mod. Phys., **65**(1), 163–211 (1993)
171. Clasen, R.: Low-dose electroradiographic multilayer system with PbO binder layers. J. Photografic Sci. **28**(6), 226–230 (1980)
172. Diamond, A.S., Weiss, D.S.: Handbook of Imaging Materials. Marcel Dekker, Inc., New York (2002)
173. Busselt, W., Junginger, H.-G., Peschmann, K.R., Stotz, S.: Verfahren zur Herstellung von elektrophotografischen Suspensionsentwicklern, Offenlegungsschrift, DE 2826127
174. Kaplan, S.H., Libman, P.C., Wainscott, W.E.: Process for making color television screens by electrophoretic deposition, Patent, US 4130472
175. Junginger, H.-G., Littmann, W., Loosdrengt, P.C.v.: Verfahren zur Herstellung eines Bildschirmes für Farbbildröhren auf elektrophotografischem Wege, Offenlegungsschrift, DE 3008419 A1
176. Junginger, H.-G., Littmann, W., Loosdrengt, P.C.v.: Verfahren zur Herstellung eines Bildschirmes für eine Fernsehbildwiedergaberöhre, Offenlegungsschrift, DE
177. Talbot, J.B., Sluzky, E., Kurinec, S.K.: Electrophoretic deposition of monochrome and color screens for information displays. J. Mater. Sci. **39**, 771–778 (2004)
178. Zimmer, M.: Verfahren zur Herstellung dekorierter Keramik- und Glaserzeugnisse und keramische Farbzusammensetzungen zur Durchführung des Verfahrens, Patent, DE 44 13 168 C2
179. Weinberg, W., Auchter-Krummel, P., Zimmer, M.: Verfahren und Vorrichtung zum Aufbringen von Dekors und/oder Zeichen auf Glas-, Glaskeramik- und Keramikerzeugnisse, Patent, DE 19849500A1

180. Cawley, J.D., Heuer, A.H., Newman, W.S., Mathewson, B.B.: Computer-aided manufacturing of laminated engineering materials. Am. Ceram. Soc. Bull. **75**(5), 75–79 (1996)
181. Cima, M.J., Yoo, J., Rynerson, M., Nammour, D., Giritlioglu, B., Grau, J., Sachs, E.M.: Structural ceramic components by 3D printing. In: Proceedings of the Symposium on "Solid Freeform Fabrication", Austin, 479–488
182. Zhang, Y., He, X., Han, J., Du, S., Zhang, J.: Al_2O_3 ceramics preparation by LOM (laminated object manufacturing). Int. J. Adv. Manuf. Technol. **4**, (2001)
183. Heinrich, J.G.: New developments in the solid freeform fabrication of ceramic components, cfi/Ber. DKG **76**, 29–35 (1999)
184. Mott, M., Evans, J.R.G.: Zirconia/alumina functionally graded material made by ceramic ink jet printing. Mater. Sci. Eng. **A 271**, 344–352 (1999)
185. Slade, C.E., Evans, J.R.G.: Freeforming ceramics using a thermal jet printer. J. Mater. Sci. Lett. **17**, 1669–1671 (1998)
186. Deitz, D.: Stereolithography automates prototyping. Mech. Eng. **112**(2), 34–39 (1990)
187. Griffith, M.L., Halloran, J.W.: Freeform fabrication of ceramics via stereolithography. J. Am. Ceram. Soc. **79**(10), 2601–2608 (1996)
188. Rekow, E.D.: Dental CAD-CAM systems—what is the state of the art?. J. Am. Dent. Assoc. **122**, 43–48 (1991)
189. Rekow, D.: Computer-aided design and manufacturing in dentistry: A review of the state of the art. J. Prosth. Dent. **58**(4), 512–516 (1987)
190. Duret, F., Blouin, J.-L., Duret, V.: CAD-CAM in dentistry. J. Am. Dent. Assoc. **117**, 715–720 (1988)
191. Mörmann, W.H., Brandestini, M., Lutz, F., Barbakow, F.: Chairside computer-aided direct ceramic inlays. Quintessence Int. **20**(5), 239–339 (1989)
192. Ogawa, T., Ioroi, T., Uchimoto, Y., Ogumi, Z., Takehara, Z.: A novel method for preparation of Ni/YSZ cermet by vapor-phase processes. In: Third Inter. Symp. on Solid Oxid Fuel Cells, Honolulu, Hawaii, 479–483
193. Sanghera, B., Naique, S., Papaharilaou, Y., Amis, A.: Preliminary study of rapid prototype medical models. Rapid Prototyping J. **7**(5), 275–284 (2001)
194. Sumeel, J., Lewis, J., Doraiswamy, A., Deravi, L.F., Sewell, S.L., Gerdon, A.E., Wright, D.W., Narayan, R.J.: Piezoelectric ink jet processing of materials for medical and biological applications. Biotechnol. J. **1**(9), 976–987 (2006)
195. Seitz, H., Rieder, W., Irsen, S., Leukers, B., Tille, C.: Three-dimensional printing of porous ceramic scaffolds for bone tissue engineering. J. Biomed. Mater. Res. **74B**(2), 782–788 (2005)
196. Manjooran, N.J.: Development of an alpha silicon carbide based liquid toner for electrophotographic solid freeform fabrication, Master Thesis. Univeristy of Florida (2003)
197. Manjooran, N.J., Kumar, A., Sigmund, W. M.: Development of a liquid toner for electrophotographic solid freeform fabrication. J. Eur. Ceram. Soc. **26**, 2459–2465 (2006)
198. Mahale, T.R.: Three dimensional electrophotographic printing through layered manufacturing: an exploration into personal fabrication, Master Thesis. North Carolina State University, Raleigh (2003)
199. Mahadevan, P.: Analysis of layer development and fusing for 3D laser printing pHD thesis, p. 87. North Carolina State University, Raleigh (2003)
200. Kumar, A.V., Dutta, A., Fay, J.E.: Electrophotographic printing of part and binder powders. Rapid Prototyp. J. **10**(1), 7–13 (2004)
201. Both, H.v., Hausselt, J.: Ceramic microstructures by electrophoretic deposition of colloidal suspensions. In: Boccaccini, A.R., Biest, O.v.d., Talbot, J.B. (eds.) Electrophoretic Deposition: Fundamentals and Applications, pp. 78–85. The Electrochemical Society, Inc., Pennigton (2002)
202. Both, H.v.: Elektrophoretische Herstellung keramischer Mikrostrukturen. Keram. Z. **56**(5), 298–303 (2004)

203. Laubersheimer, J., Ritzhaupt-Kleissl, H.-J., Haußelt, J., Emig, G.: Electrophoretic deposition of sol-gel ceramic microcomponents using UV-curable alkoxide precursors. J. Eur. Ceram. Soc. **18**, 255–260 (1998)
204. Yoo, J.H., Gao, W.: Electrical properties of PZT micro tubes fabricated by electrophoretic deposition. In: Boccaccini, A.R., Biest, O.v.d., Talbot, J.B. (eds.) Electrophoretic Deposition: Fundamentals and Applications, pp. 146–150. The Electrochemical Society, Inc., Pennigton (2002)
205. Limmer, S.J., Seraji, S., Wu, Y., Chou, T.P., Nguyen, C., Cao, G.: Template-based growth of various oxide nanorods by sol-gel electrophoresis. Adv. Funct. Mater. **12**(1), 59–64 (2002)
206. Wang, Y.C., Leu, I.C., Hon, M.H.: Preparation and characterization of nanosized ZnO arrays by electrophoretic deposition. J. Cryst. Growth **237–239**, 564–568 (2002)
207. Holgado, M., García-Santamaría, F., Blanco, A., Ibisate, M., Cintas, A., Míguez, H., Serna, C.J., Molpeceres, C., Requena, J., Mifsud, A., Meseguer, F., López, C.: Electrophoretic deposition to control artificial opal growth. Langmuir **15**, 4701–4704 (1995)
208. Zhitomirsky, I.: Electrophoretic hydroxylapatite coatings and fibers. Mater. Lett. **42**, 262–271 (2000)
209. Gao, B., Yue, G.Z., Qiu, Q., Cheng, Y., Shimoda, H., Fleming, L., Zhou, O.: Fabrication and electron field emission properties of carbon nanotube films by electrophoretic deposition. Adv. Mater. **13**(23), 1770–1773 (2001)
210. Boccaccini, A.R., Cho, J., Roether, J.A., Thomas, B.J.C., Minary, E.J., Shaffer, M.S.P.: Electrophoretic deposition of carbon nanotubes. Carbon **44**, 3149–3160 (2006)
211. Thomas, B.J.C., Shaffer, M.S.P., Freeman, S., Koopman, M., Chawla, K.K., Boccaccini, A.R.: Electrophoretic deposition of carbon nanotubes on metallic surfaces. Key. Eng. Mater. **314**, 141–146 (2006)

Chapter 7
Electrophoretic Deposition of Phosphors for Information Displays and Solid State Lighting

Jan B. Talbot

7.1 Introduction

The trend for information displays has been towards larger and thinner devices with higher resolution. The performance of emissive display screens depends upon the uniformity, density, and adhesion of the phosphor layer. Phosphors are chemical compounds that emit visible light when excited either by a photon (photoluminescence), electron (cathodoluminescence), or electric field (electroluminescence). Emissive display technologies include cathode ray tubes (CRTs), field emission displays (FEDs) and plasma display panels (PDPs). On a CRT display, the phosphors are on the front glass and are excited by a beam of electrons from a thermionic cathode device (electron gun) using high accelerating voltages of about 20 kV. An FED uses an array of tiny electron emitters (cold cathodes) operating at lower voltages of 1–8 kV to provide electrons that strike the phosphor to produce an image. For PDPs the phosphors are excited by UV light produced by plasma of a mixture of argon, neon and xenon. For all of these display devices, a thin layer of phosphor (either as a thin film or powder) is laid down upon a glass substrate, making the screen.

All phosphor screens must meet a number of requirements for use in information displays [1]. The deposit thickness must be optimized to ensure a pin-hole free coverage, yet must not reduce light emission due to internal absorption. The packing density should be optimized for the best light output at specific excitation conditions. The screen must be uniform to ensure consistent optical performance. The amount of non-luminescent material needs to be minimized. Finally, the deposit must have sufficient adhesion strength to withstand handling during manufacturing, as well as during use. Typical phosphor powder coating methods include slurry coating, screen printing, and settling [2].

Electrophoretic deposition (EPD) of powder phosphors, typically 0.5–10 μm diameter, is an alternate method used in the manufacturing of displays, particularly

J. B. Talbot (✉)
Department of NanoEngineering, University of California, San Diego,
9500 Gilman Drive, La Jolla, CA 92093-0448, USA
e-mail: jtalbot@ucsd.edu

J. H. Dickerson, A. R. Boccaccini (eds.), *Electrophoretic Deposition of Nanomaterials*,
Nanostructure Science and Technology,
DOI 10.1007/978-1-4419-9730-2_7,

high-resolution screens, in which a thinner coating is desired [2]. EPD has been used for the preparation of specialized CRTs since the 1950s [3–5]. In 1970 Grosso et al. [6] reported on the deposition of luminescent materials from isopropyl alcohol (IPA) and the effects of water in the bath on the deposition. Sluzky and Hesse [7] in 1989 reported on that an EPD screen can demonstrate brightness equal to coatings made by standard settling methods and were capable of very high resolution. More recently, EPD has been used to prepare flat panel emissive displays, such as FEDs, PDPs, and vacuum fluorescent displays (VFDs).

To improve the brightness and resolution of information displays, phosphors with high quantum efficiency, controlled morphology, and fine particle sizes are needed. Multiple scattering of the emission from micrometer-sized phosphor powders leads to poor beam collimation, substantial backscattering of the emission, and absorption losses in the phosphor itself. As particle scattering scales as the square of the particle mass, reducing the particle size to the nanoscale would help to eliminate scattering. It is also advantageous that the phosphor particle size is as small as possible as this potentially leads to higher screen resolution, lower screen loading, and a higher screen density. With the ability to produce nanocrystalline and nanometer-sized phosphors [8], EPD is a useful coating method. For example, a film of the spherical Sr_2CeO_4 particles with an average diameter of 70–80 nm was formed by EPD [9]. Nanocrystalline Y_2O_3:Eu (70–100 nm) was synthesized and films were made by EPD with an optimum loading of about 1.8 mg/cm^2 [10]. Spherical, 300 nm Y_2O_3:Eu phosphor particles were deposited at 150 V for ~2 min, resulting in a coating thickness of 3 mg/cm^2 [11]. Thin films of nanometer size ZnO particles were fabricated by EPD which exhibited optical properties characteristic of the quantum size particles [12]. Both the band-to-band and visible photoluminescence (PL) were progressively blue-shifted with decreasing particle size in the film. Recently, EPD was used to deposit soft X-ray scintillator phosphor (Gd_2O_2S:Pr) crystals (120 × 150 nm average size) onto an indium-tin oxide (ITO)-coated polyethylene terephthalate (PET) transparent substrate so that it can be used to detect X-rays generated by a laser generated plasma [13].

Light emitting diodes (LEDs) offer promise of energy efficient, low cost white solid state lighting. One approach for generating white light combines UV or blue light emitting diodes with phosphors that down-convert the LED emission to longer wavelength light. In such devices, the UV or blue light emission from an active region of a LED excites a phosphor that receives the LED-emitted light. In turn, the excited phosphor emits light at a longer wavelength. The net result is a device that emits light having a plurality of wavelengths over the visible spectrum with which for an appropriate combination of different wavelengths can be perceived as white light by the human eye. For example, yttrium aluminum garnet ($Y_3Al_5O_{12}$ or YAG) cerium-doped (YAG:Ce) phosphor strongly absorbs in the blue region and converts this radiation into a broad yellowish spectrum, which combines with the blue light emission from an InGaN or GaN LED to produce white light. In some cases, more than one phosphor is used, wherein each phosphor emits at a different wavelength. The commercial technique typically employed in phosphor deposition on LEDs involves simply painting or dispersing phosphor powders blended in a liquid polymer onto the LED die, which is then dried or cured. However, a conformal coating improves the spatial color distribution of LEDs. More recently, EPD has been used to deposit phosphor onto the LED.

Our group has systematically investigated the fundamentals of the EPD process for phosphors, in particular, the dissociation of nitrate salts in IPA [14, 15], the zeta potential of charged phosphor particles [14, 16], as well as the formation of the adhesive agents [15, 17]. The deposition rates of the phosphor and a binder were modeled [18, 19]. The factors, which affect the adhesion strength of the deposited phosphors, were identified [15, 20]. The optical performance of phosphor screens was tested and was found not to be affected by the process itself nor by the conditions, which enhance phosphor adhesion under 1–4 kV excitation voltage applicable to FEDs [21]. New screening methods were developed for color displays by combining EPD and photolithography [22–24]. EPD in a thermoreversible gel was explored [25]. More recently, EPD has been used to deposit phosphors for LEDs. Our research results, as well as others, concerning the EPD of phosphors will be reviewed here. The descriptions of the experimental apparatus and procedures with details of the results can be found in the references.

7.2 Fundamentals of the Process

To understand the EPD of phosphors, the process was divided into the following fundamental steps: (a) the charging of the particles in suspension, (b) the transport of the particles under the influence of an electric field, and (c) the deposition and adherence of the particles onto a substrate. The standard EPD bath of interest in our and many other studies is a suspension of phosphor particles in isopropyl alcohol (IPA) (1–4 g/L), which contains dissolved nitrate salts ($\sim 10^{-3}$ M $Mg(NO_3)_2$) and small amounts of water (~1 vol.%). The nitrate salt dissociates slightly, providing ions to charge the particles positively. Therefore, the approach taken was to (a) investigate the dissociation behavior of nitrate salts in IPA, (b) study the effects of phosphor chemistry and suspension medium on the zeta potential of the particles, and (c) study the effects of the EPD process conditions and model the deposition rates.

The conductivity of various nitrate salts in IPA was measured and analyzed using the Ostwald dilution law to determine the conductivity at infinite dilution and dissociation constants as shown in Table 7.1 [14]. The mobility of the ions can be determined with the use of the limiting conductivity, and concentration of ions can

Table 7.1 Limiting conductivities (Λ_o) and dissociation constants (K_D) of nitrate salts in IPA. [14]

Reaction	Λ_o(cm^2/mol Ω)	K_D(M)
$NaNO_3 \rightarrow Na^+ + NO_3^-$	17	3×10^{-4}
$Mg(NO_3)_2 \rightarrow Mg(NO_3)^+ + NO_3^-$	18	6×10^{-5}
$Mg(NO_3)^+ \rightarrow Mg^{2+} + NO_3^-$	120	2×10^{-7}

be calculated from the dissociation constant [14, 18]. The dissociation constants are very low and in the concentration range of 10^{-4} to 10^{-3} M $Mg(NO_3)_2$ typically used in EPD baths, $Mg(NO_3)^+$ is the predominant cation available to charge the phosphor and to form the binder.

The zeta potentials of several oxide, sulfide, silicates, and oxysulfide phosphors were measured in IPA and in IPA containing nitrate salts and water [16]. The zeta potentials of phosphors in pure IPA were negative. With the addition of 5×10^{-4} M nitrate salt, the zeta potentials of nearly all the phosphors became positive. A more focused study measured the zeta potentials of Zn_2SiO_4:Mn (P-1) and ZnS:Ag (P-11) phosphor particles in IPA under a wide range of nitrate salt concentrations and pH values [14]. The zeta potential was negative (~−50 mV) at salt concentrations less than 10^{-6} M. As the salt concentration increased, the zeta potential increased and became positive, reaching a maximum (~50 mV) at 10^{-5} M. In IPA or IPA with 10^{-5} M magnesium nitrate, the zeta potential was positive at $pH < 6$, but became negative for $pH > 6$. At higher salt concentrations of 10^{-4} and 10^{-3} M, the zeta potential remained positive for all pH values. Therefore, the zeta potential is dependent upon the nitrate salt concentration and pH. Figure 7.1 shows the zeta potential as a function of magnesium concentration in IPA for ZnS:Ag, Zn_2SiO_4:Mn and YAG:Ce [26].

During EPD, electrolysis of the water present in the IPA creates a basic environment at the cathode. Thus, $Mg(NO_3)^+$ reacts with the hydroxide ions to produce magnesium hydroxide [14, 17], as follows:

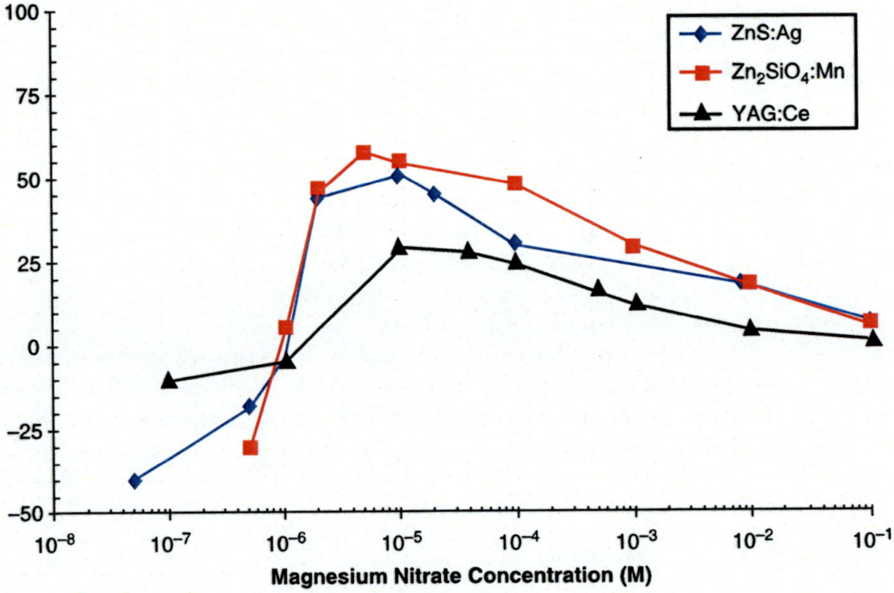

Fig. 7.1 Zeta potential of ZnS:Ag, Zn_2SiO_4:Mn, and YAG:Ce as a function of magnesium nitrate concentration in isopropanol

$$2H_2O + 2e^- = H_2(g) + 2OH^-$$

$$Mg(NO_3)^+ + 2OH^- = Mg(OH)_2 + NO_3{}^-$$

By carefully controlling the amount of water in the IPA bath, it was determined that alkoxide formation can also occur, which can contribute to the binder formation [15]:

$$Mg^{2+} + 2C_3H_7OH = Mg(C_3H_7O)_2 + 2H^+$$

$$Mg(C_3H_7O)_2 + 2H_2O = Mg(OH)_2 + 2C_3H_7OH$$

When the water concentration in the EPD bath is low (<1 vol.%), alkoxide predominately forms, whereas at high water content (>5 vol.%), all of the alkoxide is converted to the hydroxide. At intermediate water concentrations, the binder is a mixture of the two materials. Thus, the role of the magnesium nitrate in the EPD bath is to charge the phosphor particle positively, to maintain the positive zeta potential at high pH at the cathode, and to form the adhesive material.

The amounts of deposited phosphor and magnesium hydroxide binder were simply modeled by integrating the flux of material over time multiplied by the fraction of material adhering [18, 19]:

$$M = \int \alpha ACvdt$$

where M is the mass of material deposited (mg) in time t (s), α is the fraction of material reaching the cathode that adheres, C is the concentration of the material in solution (mg/cm^3), v is the velocity under the influence of the electric field (cm/s), and A is the area of the cathode (cm^2). The velocity is due only to electrical migration and is the product of the mobility (u) and the electric field strength (E), which can be calculated from the average current density (i) during deposition by $v=uE=u\ (i/kA)$, where k is the specific conductivity of the solution. For the phosphor particles, the mobility was determined from the measured zeta potential (ζ) using the Smoluchowski equation, $u = \xi\varepsilon/\mu$, where ε is the permittivity (18.3) and μ is the viscosity (2 cP) of IPA. The mobility of $Mg(NO_3)^+$ was calculated from $u=\Lambda_0 t_+/F$, where t_+ is the cation transference number (~ 0.4) and F is Faraday's constant. The concentration of $Mg(NO_3)^+$ was multiplied by the molecular weight of $Mg(OH)_2$, as this is the species that deposits. With conductivity measurements of $Mg(NO_3)_2$ in IPA, the specific conductivity, the limiting conductivity, and the concentration of $Mg(NO_3)^+$ were determined [14]. The deposition rates of both the phosphor and magnesium hydroxide predicted from this simple model agreed with experimental results provided that the $Mg(NO_3)_2$ concentration was greater than $\sim 10^{-4}$ M [19]. This minimum concentration maintains a positive zeta potential near the cathode and provides the necessary amount of binder to adhere the particles (α equal to 1).

7.3 Adhesion of Deposits

Adequate adhesion of the phosphor on the screen is important as the phosphor coating must endure several processing steps before final display assembly, and during use the display must tolerate shocks and vibrations. As discussed, hydroxides and alkoxides formed via cathodic precipitation are the binder materials for the phosphor particles in the deposited layer [15, 17] with a minimum amount of binder necessary for particle adhesion [19]. A limitation of phosphor screens made by EPD is the low adhesion strength of the deposit. Therefore, the effects of the inherent EPD processing conditions on the adhesion strength and ways to enhance adhesion of deposited phosphors were investigated [15].

The adhesion strength of a deposit was tested by removal of particles by applying a jet of nitrogen gas perpendicular to the coating [27]. The adhesion strength was determined by one of two methods. The first method is qualitative for the weak deposits, using the weight percent of particles remaining on the substrate after testing as the adhesion strength. The second method provides a quantitative measurement for the strong deposits by relating the width of the ring of material removed from the substrate to the shear stress on the substrate, which equals the adhesion strength (in the range of 0.08–0.4 N) [28].

The variables investigated were phosphor concentration, type of phosphor, particle size distribution, water content in the bath, salt concentration and the type of salt (e.g. $Mg(NO_3)_2$, $Y(NO_3)_3$ and $La(NO_3)_3$) in the bath, and chemical additives (e.g. cellulose and glycerin). Unless otherwise stated, all depositions in our studies were made from an IPA bath with 10^{-3} M $Mg(NO_3)_2$ with a ZnS:Ag phosphor particle loading of 4 g/L. Samples were baked at 425°C for 1 h after deposition; deposit densities were 2 mg/cm^2. The baking temperature was limited to the softening point of the glass substrate which was coated with a conductive layer. After baking the binder is completely converted to MgO [18]. Table 7.2 lists those parameters which had no effect on the adhesion strength of EPD phosphor deposits and these have been discussed elsewhere [20]. The location of the binder was determined to be the interstitial region between the particles. Thus, the adhesion strength of the EPD phosphor deposits is due solely to the contact points between the particles. Additional binder material, which may fill the interstitial region or cover the deposit, does not enhance the adhesion strength. Table 7.3 lists those variables, in order of decreasing importance, which did affect the adhesion strength of phosphor depos-

Table 7.2 Parameters which did not affect adhesion strength of phosphor deposits. [27]

Parameter	Range
Applied voltage	50–800 V
Substrate coating	Al, indium tin oxide
Deposit thickness	>1 mg/cm^2 <7.4 mg/cm^2
Water soaking	1, 2, 3 min
MgO overcoating	0.5–2 mg
Phosphor concentration	0.5–8 g/L

Table 7.3 Parameters affecting adhesion strength of phosphor deposits. [20]

Parameter	Range	Effect on adhesion
Glycerin	1–2 vol.%	Very large increase
Water	0–5 vol.%	Large increase
Baking	425°C, 1 h	Increase
Particle size distribution	0–100% 3 and 6 μm dia.	Increase
Salt identity	$Mg(NO_3)_2$, $La(NO_3)_3$, $Y(NO_3)_3$	Increase
Phosphor identity	P-11, P-1, P-43	Increase/decrease
Added cellulose	1–3.3 wt.%	Decrease

its. The results as listed in Table 7.3 will be summarized, but the adhesion data are presented elsewhere [20, 27].

Glycerin, which is commonly used industrially to disperse particles, had the largest effect on the adhesion strength. Virtually no particles were dislodged by the adhesion test on the deposits with 2 vol.% added glycerin to the bath. This increase in adhesion strength is likely due to dispersion of smaller particles and polymerization of the glycerin upon baking. Added water, up to 5 vol.%, to the deposition bath greatly enhances the adhesion strength of EPD phosphor deposits. This increase in adhesion is due to the preferential deposition of $Mg(OH)_2$ instead of $Mg(C_3H_7O)_2$ as the binder with increasing water concentration in the bath. With water concentrations higher than 5%, the deposits became irregular and of poor quality due to excessive H_2 gas evolution at the cathode, which disrupts particle deposition, yielding coarse layers with pin holes. Hydrogen also can cause transparent oxide electrodes, such as In_2O_3 and SnO_2:Sb, to brown on the glass substrate, due to reduction of the oxide [29]. Since industrial systems typically do not control absorption of atmospheric moisture into the deposition bath or onto the deposit, there may always be high concentrations of water. While water content should be monitored, since excessive water can lead to irregular deposits, adsorbed water onto the deposit will increase the adhesion strength during handling prior to mounting and evacuation in a display device.

Post-deposition baking at 425°C for 1 h converts the $Mg(OH)_2$ and alkoxide binder to MgO. The MgO has a higher bond strength than the $Mg(OH)_2$ by 60%. The amount of MgO increased from 0.39 to 1.34 wt.% in samples deposited from baths with 10^{-3} M and 5×10^{-3} M $Mg(NO_3)_2$, respectively. The particle size distribution also can affect the adhesion strength of phosphor deposits. It was found that a mixture of 20% 6 μm and 80% 3 μm mean diameter particles enhanced the adhesion strength, as well as increased the packing density. The change in porosity and packing of the deposit allows for more contact points for each particle. With more contact points, there are more adhesive points, leading to increased adhesion strength.

Three different salts, $Mg(NO_3)_2$, $La(NO_3)_3$ and $Y(NO_3)_3$, which are all commonly used in EPD processes [7], and combinations thereof, were studied. Using $Y(NO_3)_3$ in the deposition bath instead of $Mg(NO_3)_2$ was found to increase the adhesion strength. While the initial deposits from a $Y(NO_3)_3$ bath were weaker than those from a $Mg(NO_3)_2$ bath, the adhesion strength increased significantly after priming of the bath. The deposition bath is "primed" by depositing an initial sample, which usually has lower density than predicted by about 20% [27].

The type of phosphor used can increase or decrease adhesion. For ZnS:Ag, the salt concentration in the bath was optimized to control the zeta potential of the particle and the amount of binder. For the same deposition conditions, changing the phosphor material radically changed the adhesion strength of the deposit. Adhesion strengths for ZnS:Ag (P-11, 3 μm) and Zn_2SiO_4:Mn (P-1, 2.7 μm) deposits were approximately twice that of Gd_2O_2S:Tb (P-43, 2.3 μm). This difference in adhesion strength most likely can be attributed to the different particle size distributions of these materials. High adhesion strength most likely can be achieved for any particle material, provided that the deposition conditions are optimized.

While cellulose is a useful additive for anodic EPD [29], it was found unsuitable for cathodic EPD of phosphor as it decreased adhesion. Also, the morphology of the deposits was rougher. Other additives that have been used for enhancing adhesive strength are nitrocellulose, acryl resin or nitrate groups. Adhesion has been enhanced also by spraying an aqueous solution of polyvinyl alcohol (PVA) and ammonium dichromate (ADC) photoresist onto an EPD phosphor layer followed by curing under UV light [30]. This top coating may be advantageous if baking is not possible, such as with the EPD of phosphors onto plastic or flexible substrates.

7.4 Optical Performance of Phosphor Films

The optical performance of phosphor screens is defined in terms of resolution, brightness, spatial noise, and degradation. Apart from the intrinsic properties of the phosphor material, such as efficiency and saturation, the optical properties of the screen significantly influence its overall performance, especially resolution and brightness. Spatial noise arises from inhomogeneity in the packing density and from irregularities in the thickness of the screen. The optical properties include light absorption in the phosphor and at the aluminum film (if used), as well as scattering due to the microstructure of the screen, which is determined by the particle size distribution, screen density, and packing density. The resolution of screens increases with decreasing layer thickness and increasing packing density. However, the factors that enhance resolution often decrease brightness and vice versa. Generally, high resolution screens require a processing technique to produce a densely packed thin screen composed of small particles. It is also important that no contamination of the phosphor surface occurs during the screening process to maintain the efficiency. Particularly, luminescent intensities of phosphors are sensitive to the surface state of the phosphor for low voltage excitation. Therefore, the deposition of a screen must be optimized for the best optical performance for its particular application.

In 1966, McGee et al. [4] compared the performance of P-11 (ZnS:Ag, ~1 μm size) phosphor screens made by gravity settling and EPD. Optimum efficiency was found for screen densities of ~1 mg/cm^2 for settled screens and 0.5–0.7 mg/cm^2 for EPD. The light output was slightly higher for EPD screens than for those made by settling; however, the resolution was comparable. The optical performance of screens of P-20 ((Zn,Cd)S:Cu, 2.2 μm mean particle diameter) phosphor of various thicknesses prepared by centrifugal settling and by EDP were compared by Prener and Swank in 1978 [31]. The packing density, spatial resolution (described by the modulation transfer function, MTF) and the spatial noise were found to be the same for screens made by the two methods. Gibilini compared the performance of screens of 2 μm mean size particles made by gravity settling, gravity settling from an alcohol solution, centrifugal settling, and EDP in 1986 [32]. The resolutions (MTF) that increased with packing density were ~45, 50, 56, and 58% for standard settling, settling in alcohol, EDP, and centrifugal settling, respectively. In 1988, Sluzky and Hesse developed a technique by using a CRT with half of its faceplate made from a single-crystal and the other half coated with a powder phosphor for determining the contribution of the phosphor screen to the overall spot size of a CRT [33]. It was determined that, for a given electron beam size smaller than 50–60 μm, resolution is influenced by the screen quality. They compared the quality of an EPD powder phosphor screen using 0.9 and 2.5 μm mean particle size P-53 ($Y_3Al_5O_{12}$:Tb) with sputtered and single-crystal screens under 13 kV irradiation [34]. The single-crystal and sputtered screens were shown to have spot size capabilities better than 4 mm (full width at half maximum) compared to 6 mm for the 0.9 μm particle screens. However, brightness of the powder screens was significantly better (>40%) than for the sputtered screens, which in turn was better than for single-crystal screens.

The effects of screening on degradation were studied by Raue et al. [35]. Phosphor screens are subject to Coulombic aging, a decrease in light output from an accumulation of charge per area. The degradation of phosphors was found to depend on the screen density. At low screen densities (<3 mg/cm^2) the increased degradation was ascribed to an increase in glass browning. At high screen densities, the degradation increased due to the longer path of the photons leaving the screen. It was also shown that high packing densities can improve the lifetime of screens.

Busselt and Raue in 1988 [36] developed a theoretical model for the simulation of light propagation in cathodoluminescent (CL) phosphor screens that predicts light output as well as resolution. The model takes into account scattering and absorption of the phosphor, reflectance of the Al film, and the layer thickness of the phosphor powder. The model predicted an optimum layer thickness for screen efficiency of 2S, where S is the scattering length. The scattering length was determined from an empirical correlation $S = d_{50}/P^{1/3}$, where d_{50} is the mean particle size and P the packing density, calculated from $P = W/(\rho D)$, where W is the screen weight and ρ the phosphor material density. It was determined that the optimum screen weight with respect to brightness can be estimated from $W^* = 2d_{50}\rho P^{2/3}$. If the resolution is specified in terms of line width, an approximate value for the maximum layer thickness can be calculated from $D^* = L/4.5$, where L is full line width at 5% of maximum.

For the ultimate use in information displays, the optical performance of the phosphor screens produced by EPD must be tested [1, 2]. Therefore, the effects of the EPD processing conditions on the film thickness, packing density, and optical performance of phosphor deposits were investigated [21]. The optical performance was evaluated by measurement of chromaticity and CL efficiency at low voltages (1–4 kV) applicable for FEDs as well as high voltages (13–20 kV) in CRTs. Of particular interest were the EPD processing conditions that were found to enhance adhesion in Table 7.3.

The packing density of deposits made at different applied voltages and with different deposit densities from 1 to 9 mg/cm^2 (5–40 μm thick) was constant at very high value of ~55%. Other studies have reported larger packing densities over 70% [37]. This suggests that the deposits with different thicknesses are homogeneous with a similar packing structure. Due to their similar structure, all of the deposits with different thicknesses made at 200 V demonstrated a similar CL reflection efficiency for deposit densities greater than 1 mg/cm^2. However, from transmission efficiency, an optimal thickness was found to be about 7 μm; this thickness corresponds to a deposit of about two layers of phosphor particles.

The EPD process does not affect the intrinsic efficiency or chromaticity of the phosphor [21]. The introduction of glycerin or water into EPD baths, which dramatically increases the deposit adhesion strength, does not show any deleterious effect on the optical performance. Also, other conditions shown to enhance phosphor adhesion, such as the use of lanthanum nitrate in the bath or the particle size distribution, do not affect the optical performance of the phosphor deposit.

The influence of the EPD deposition parameters on luminescence of low voltage $ZnGa_2O_4$ phosphor (~2–3 μm size) films for FEDs [38] was investigated. The optimal deposition parameters for electrode distance, stirring speed, electrode bias, and deposition time were found to be 20 mm, 450 rpm, 50 V, and 40 s, respectively. The optimum phosphor film thickness was 8–10 μm. As the film thickness increased, the luminance intensity improved and a maximum intensity was obtained for an 8.6 μm thick film (about 3–4 particle layers). The luminescence of the as-deposited phosphor film was improved by annealing in a 25%O_2/Ar atmosphere due to the improved crystallinity. The luminance efficiency of phosphor films prepared by EPD, which were smoother and denser, was reported to be ~15 times higher than that prepared by the sedimentation method.

In another study the optical characteristics of the phosphor screen made by EPD in FED environments were measured [39]. One of the issues for a phosphor screen which is excited at high current density and low voltage in an FED is the Coulombic degradation, due to the accumulated charge dosage. It depends not only on the material properties of phosphor, but also on the morphology of the phosphor screen, that is influenced by the screening process. The maximum CL occurs with a very thin phosphor layer approximately two particle layers. The deposition rate was found to be the important to obtaining the highest packing density for maximum luminance. The process variable examined were electric field strength, deposition time, and $Mg(NO_3)_2$ concentration for $ZnGa_2O_4$:Mn phosphor deposition. Also, a conductive powder, such as WO_3 and V_2O_3, was deposited with Y_2O_3:Eu phosphor to increase the screen conductivity. Alternatively, the phosphor screen was treated in a solvent with metallic salts such as $SnCl_4$ to improve conductivity. The CL intensity was increased and degradation reduced when the conductive materials were incorporated into a Y_2O_3:Eu phosphor screen.

It has been noted that the choice of anode is important in the EPD of phosphors as corrosion products can become impurities in the phosphor that can degrade the optical performance [40]. $YNbO_4$:Bi phosphor screens were examined using laser Raman spectroscopy and it was found that the screens were contaminated with Fe^{3+} from the stainless steel anode used in EPD. Also, poisoning of ZnS:(Cu, Al) due to corrosion of the anode during EPD altered the PL emission spectrum and thus, the chromaticity of the phosphor [41].

EPD with the standard IPA bath is routinely used to deposit phosphor coatings for study of their luminescent behavior [42–44]. For example, spectral properties of films of red Y_2O_2S:Eu [42, 43, 45], Y_2SiO_5:Ce [46], and ZnO:Zn [47] prepared by EPD have been measured. EPD offers easy control of the film thickness and a high packing density.

7.5 Full Color Displays

Techniques for fabricating high-resolution, full-color screens are desirable for information displays. Several techniques to pattern phosphor screens using EPD into triads of stripes were developed in the 1970s [48–50]. However, inherent problems, such as large resistance drops and poor phosphor adhesion, associated with each of these processes have made them unacceptable as a viable alternative to conventional phosphor screening techniques [2]. Color VFDs have been produced using EPD by selecting a conductive line for each color phosphor. Ultra-high resolution color VFD screens for microdisplays on silicon chips were deposited with tricolor pixels as small as 18 μm × 25 μm [51].

By combining a photolithographic technique similar to that developed by Mooney [51] and EPD, we developed a process shown in Fig. 7.2 to deposit triads

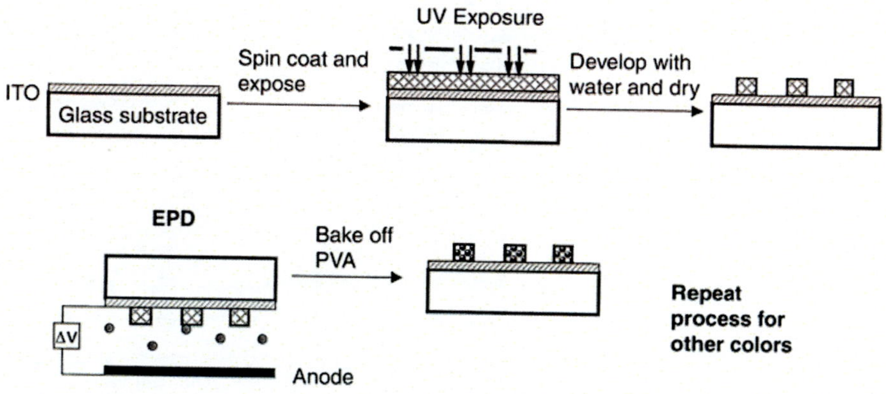

Fig. 7.2 Schematic of the process steps for full color screens of Chang et al. [22]

of phosphor stripes with a line resolution of 100 triads of 75 μm stripes per inch [22]. The photoresist used was a mixture of 4.5 wt.% PVA and 0.45 vol.% ADC in distilled water. The PVA solution was spin-coated upon an ITO-covered glass substrate to a film thickness of ~6 μm. Then the PVA was patterned using a mask aligner and a shadow mask to yield alternating 175 μm stripes of cross-linked PVA and 75 μm uncross-linked PVA stripes. The coating was then sprayed with warm water to dissolve and remove any uncross-linked PVA from the substrate. Next, the PVA-covered substrate was heated for several minutes to dry excess water and to complete the cross-linking reaction.

Small particle size (1–3 μm) green $ZnSiO_4$:Mn (P-1), red Y_2O_3:Eu (P-56), and blue ZnS:Ag,Cl (P-11) phosphors were deposited onto the 75 μm conductive striped regions of the substrate using EPD [22]. Successive depositions of the phosphors were performed by repeating the aforementioned steps. The shadow mask had to be aligned in precise registry with the edge of the previously coated stripes during each of the successive lithography steps. Also, the concentration of the ingredients within the solution bath was optimized according to the specific phosphor that was deposited [22]. The duration of exposure of the PVA-ADC film to UV light was optimized in order to fabricate PVA stripes with a high degree of uniformity, edge straightness, water insolubility, and adhesion. The EPD conditions were optimized in order to comply with several requirements. A screen thickness of 4–6 μm was desired in order to maximize phosphor efficiency and resolution, while minimizing phosphor cross-contamination. Also, ample phosphor particle adhesion was required. The optimum deposition time and applied voltage for each phosphor were determined based upon the stripe thickness, continuity, edge resolution, and particle packing density. The phosphors were deposited in sequential order of P-1, P-56, and P-11 with deposition times of 20, 35, and 33 s, respectively. Figure 7.3 shows a SEM micrograph of a deposited triad, each stripe approximately 75 μm wide with very distinct, straight edges. Also, little cross-contamination of phosphors was observed. It must be noted that higher line resolutions may be attainable with this

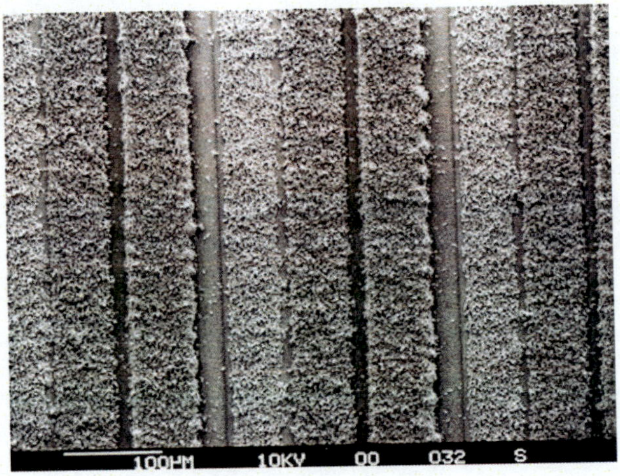

Fig. 7.3 SEM image of P-1, P-56, and P-11 stripes. [22]

method. The line resolution was constrained only by the shadow mask which had a resolution of 100 triads of lines per inch.

A similar method of patterning phosphor screens was developed by Kang et al. [52] using a PVA-ADC photoresist, that is insoluble in IPA necessary for the subsequent EPD process. A three-color phosphor screen was made with a triad pitch of 210 μm with three 50 μm phosphor lines and 20 μm spaces between each line. It was noted that small amounts of phosphor were deposited on the PVA, especially when the deposition time was long. The phosphor luminance decreased with repeated patterning due to the carbon residue from photoresist ashing. The first phosphor deposited undergoes three spin coating and ashing processes. The phosphor efficiency was examined for each phosphor:Y_2O_3:Eu for red, ZnO:Zn for green, and Y_2SiO_4:Ce for blue. The CL intensity of phosphor screens decreased by 25% during the first coating and ashing process and 18% during the second process. Optimization of the screen intensity and color gamut requires consideration of each phosphor luminance and the effects of one or more photoresist ashing cycle.

A similar, but somewhat more complicated, method for fabricating tri-color phosphor screens by combining photolithography with EPD was developed by Kurinec and Sluzky by the trilayer process [23]. A cross-linked thick photoresist (AZ P4620, baked at 250°C) was employed to act as a mask. It was spin-coated over the conductive seed layer to a thickness of ~10 μm. The second layer of the trilayer scheme was an inorganic spin-on glass (SOG) (Accuspin 311) that acted as an etch stop in subsequent processing. For patterning, a thin (~1 μm) imaging positive resist was applied and exposed with the desired pattern. Following the imaging resist development, SOG was etched in regions where phosphor deposition was required. To make the process efficient and manufacturable, dry etching was chosen. Reactive ion etching of SOG was carried out using a CF_4/H_2 plasma chemistry such that a reasonable selectivity was achieved between the resist and SOG. After etching SOG, the plasma chemistry was switched to oxygen to etch the hard-baked thick resist. The etching stopped on the conductive aluminum layer. The plasma residue was cleaned using a buffered HF-glycerin solution before carrying out the EPD of phosphor. The process was further improved by the development of a single chamber plasma process [53]. The commercial phosphors ($ZnSiO_4$:Mn, ZnS:Ag, and Y_2O_3:Eu) used had particle sizes of 2–6 μm. A timed deposition was performed to fill the trench. The process was repeated for the other two color phosphors as depicted in the process flow diagram given in Fig. 7.4. The red, green and blue (RGB) lines were separated by a black material ($MnCO_3$) also deposited by EPD. After EPD of the three phosphor lines, the process for black lines in between becomes self-aligned. Trial pitches of 15-μm color-line width and 5-μm spacing between colors were achieved. The exposure tool utilized for this work was a 5× reduction standard wafer stepper commonly employed in semiconductor technology. The advantage is that the mask is not in contact with the screen and higher lithographic resolution is easily attained. However, the resolution is only limited by the phosphor particle size.

Kwon et al. [54] developed a full-color screen for an FED application with a pixel size of 400 μm. For the fabrication of the patterned anode plate, first a mixture of commercial positive photoresist and hexamethyldisilane was spin-coated on the

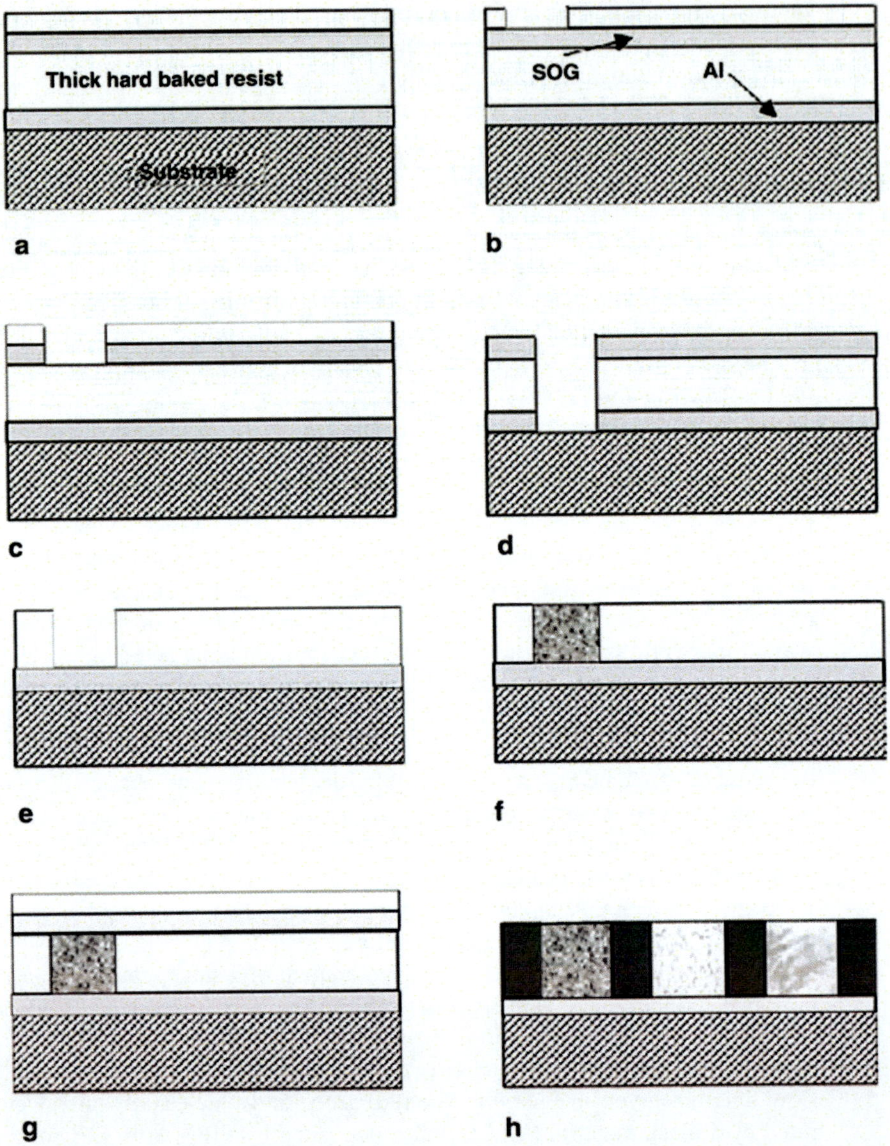

Fig. 7.4 Schematic of process steps of the trilayer process for full color screens of Kurinec and Slusky. **a** Trilayer deposition, **b** Imaging resist exposure, **c** RIE of SOG in CF_4/H_2, **d** RIE of resist in O_2, **e** RIE of SOG in CF_4/H_2, **f** EPD of first phosphor, **g** Re-application of SOG and IR, **h** Repeat for other two colors. [23]

substrate to about 1 μm thick. After baking for 25 min at 958°C, the photoresist was placed on a mask aligner and exposed to UV light for 30 s. It was developed, baked and then etched in a solution of $HCl:H_2O:FeCl_3$ of 10:10:1 for about 2.5 min. The EPD process sequentially deposited Y_2O_3:Eu, $ZnGa_2O_4$:Mn, and Y_2SiO_5:Ce as red, green, and blue phosphor, respectively, onto the patterned ITO-glass at 400 V for 10 s, while a reverse bias voltage in the range of 150–100 V was applied to the area where the phosphors were not to be deposited. Each 400 μm pixel was composed of 100, 100, and 2 × 50 μm for the red, green, and blue lines, respectively. The etched line was 33 μm. Applying a reverse bias on the dormant electrodes while depositing a phosphor onto the stripe pattern was critical in preventing cross-contamination of each phosphor in a pixel. Also, the resistivity due to the long electrodes used in the pattern could be effectively reduced by dividing it into segments.

For the process developed by Yum and Sung [37, 55], a commercial positive photoresist based on novolac resin and naphthoquinone diazide was spin coated onto an substrate, which became insoluble in alcohol by optimizing the baking temperature (190°C for 20 min) as outlined in Fig. 7.5. This photoresist could then be directly used as a mask for deposition with EPD in IPA without a post-plasma etching process. Then the photoresist film was simply removed by annealing at 400°C for 1 h. The adhesion strength of the patterned color screens was improved by coating with a PVA-ADC photoresist cured under UV light. The red, green and blue phosphors used were Y_2O_3:Eu (2.9 μm), $LaPO_4$:Ce,Tb (2.8 μm), and $(SrCaBa)_5(PO_4)_3$ Cl:Eu (4.0 μm), respectively. For the deposition of the RGB phosphor layers in which each thickness was ~6 μm, the process conditions for each phosphor were 200 V for 60 s for red, 200 V for 70 s for green, and 250 V for 60 s for blue from a standard bath. The phosphor layers had a high packing density above 70% and the thickness or uniformity of the layers could be easily controlled. A full color screen with 65 μm wide lines was fabricated by this method.

7.6 Flat Panel Displays

As discussed, it has been demonstrated that phosphor screens made by EDP can have optical performance equal to or even better than other coating methods. Hence, EPD has been used to prepare flat panel emissive displays, such as FEDs, PDPs, and VFDs. EPD of carbon nanotubes as the field emitters for FED applications also has been explored (e.g., [56]), but will not be discussed here.

7.6.1 FEDS

Contrary to CRT phosphors with particle sizes ranging from 5 to 20 μm, FED operating at low voltage require small phosphor particles due to the high resolution of the screen and efficiency requirements. The display resolution is not only limited by the beam size, but also by phosphor particle size. Thin phosphor screens, which allow the maximum transmittance, are required to offset the low luminance of low voltage FEDs. Other issues of concern in FEDs are phosphor adhesion in the presence of high electric fields and out-gassing from the phosphor particles that

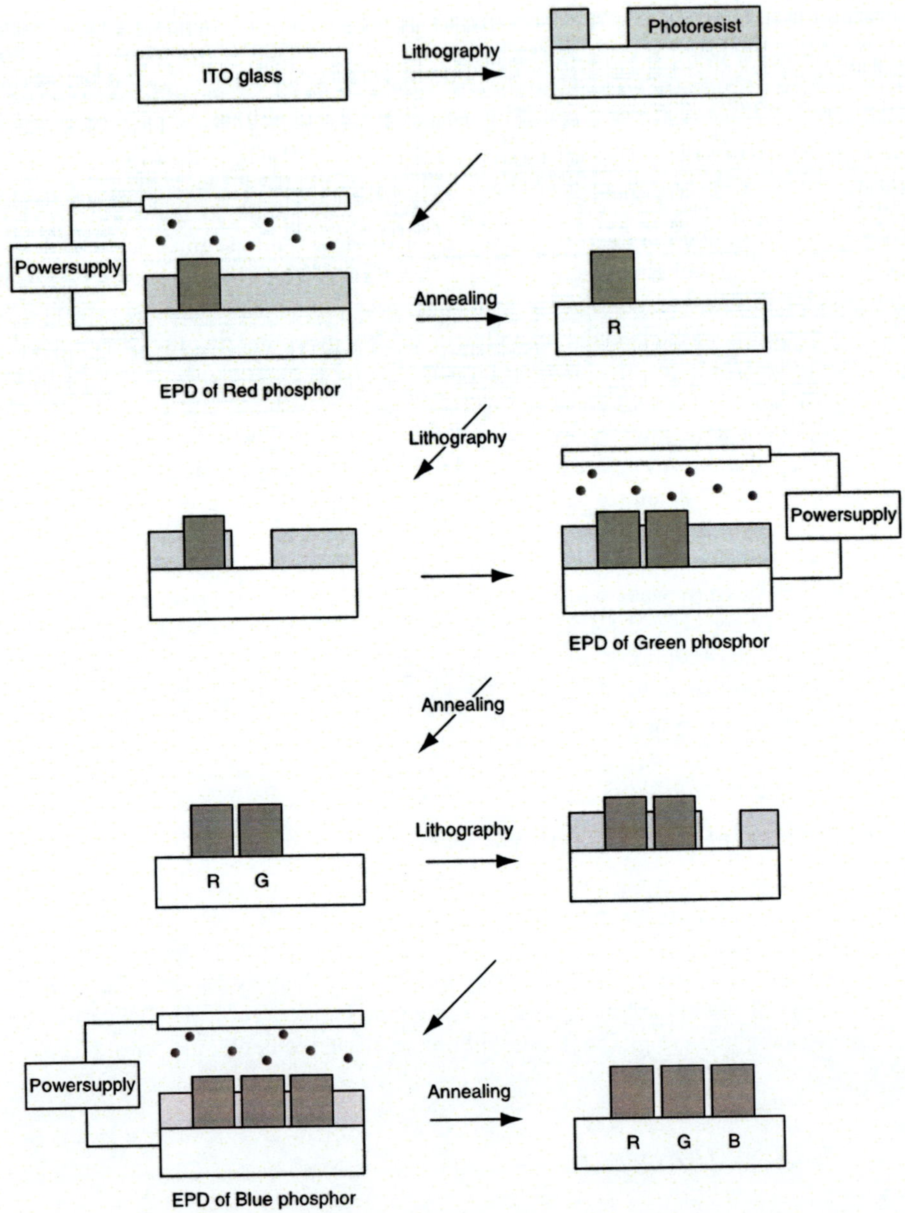

Fig. 7.5 Schematic diagram of EPD and photolithography method of Yum and Sung. [37]

may result in contamination of field emission tips. Therefore, typical factors to be considered when FED screens are prepared include the transmittance of a phosphor screen, charge accumulation on the screen which causes thermal degradation and flicker, and adhesion of phosphors onto the glass. Since phosphors are dielectric materials, electron bombardment in a FED results in a surface charge that may deflect the electron beam. In high voltage CRT displays, an aluminum layer is coated on the phosphor layer to prevent charge accumulation on the screen and also reinforce the adhesion of phosphors to the screen. However, in FED applications due to the low voltages employed an aluminum layer cannot be used, so strong adhesion and reduction of surface reactivity of the phosphors become important. Also, conductive materials, such as indium oxide or tin oxide, often are used to discharge the accumulated charge on the phosphor surface.

A patent issued to Kiyomiya et al. of Sony in 1995 [57] for screens for an FED application used EPD (although the process is ascribed as electrodeposition). A suspension of green phosphor particles with aluminum nitrate, lanthanum nitrate and glycerol in IPA was used. The method describes using a transparent electrode with a guard electrode to which reverse bias voltage was applied to prevent deposition in undesired areas.

One method for improving the luminescent characteristics of low voltage phosphors is to increase the conductivity in order to remove the surface charge of the insulating phosphor. The EPD of green ZnS:Cu,Al phosphor (1–10 μm) mixed with a conductive In_2O_3 powder onto ITO-coated glass for FEDs was studied [58]. Then, the luminance properties of the phosphor mixed with and without conductive In_2O_3 powder (less than 1 μm size) from 3 to 15 wt.% were compared. The CL intensity increased linearly with increasing indium oxide content and a maximum was reached at about 15 wt.%. The improvement in the brightness was due to pronounced reduction of the charging effect, but above 15% the amount of non-emitting material had a detrimental effect. The aging performance was clearly enhanced with the addition of In_2O_3 (optimum concentration of 10 wt.%) at 1,000 V excitation voltages with a current density of 1 mA/cm^2. These researchers [59] also used a laser treatment on EPD ZnS:Cu,Al films as a way to remove contaminants, such as oxides or carbides, from the phosphor surface. A Nd:YAG laser was used as an irradiation source and by adjusting laser power and treatment time, brightness was enhanced by about 10%.

In another study, Y_2O_3:Eu phosphor screen deposited by a standard EPD process was subsequently electroplated with tin from a solution of $SnCl_4 \cdot 5H_2O$ dissolved in IPA [60]. Then the screen was heat-treated to attain a SnO coating on the phosphor screen. Maximum brightness was found when the deposited phosphor weight was 2.0 mg/cm^2, corresponding to 8.0 μm thickness. Considering that the particle size was 3.5 μm, the optimized phosphor thickness was about two layers of phosphors particles. It is found that the tin oxide coating slightly enhanced luminescence output. These effects were ascribed to an increase in the screen conductivity by SnO incorporation that prevents the coulombic charging of a phosphor screen, which could lead to a longer lifetime. Also, it was observed that the adhesion increased twofold with SnO without any loss in brightness.

Phosphors have also been coated with an In_2O_3 conductive layer by the hydrolysis of indium chloride [61] which at an optimum thickness improves the luminescent properties. As the conductive coating affected the surface chemistry of the phosphor particles, the zeta potential of the coated ZnS:Cu,Al in IPA solution increased to 32 mV as compared to 1.2 mV of that of the uncoated phosphor. Consequently, the EPD rate of the phosphors onto the substrate increased.

The influence of the panel sealing process environment on the CL degradation of ZnS:Ag,C1 phosphor screens made by the standard EPD method was investigated [62]. The surface reactions caused by the electron bombardment and the bonding states of adsorbed components on the surface of the phosphor screen before and after heat treatment at the panel sealing temperature were discussed.

An FED prototype was made using diamond-like-carbon thin film with millions of rectangular microholes as the emitter (cathode) and EPD ZnO:Zn phosphor (PI5) patterned into 50 anode electrodes [63]. The ZnO:Zn phosphor (1.5 g/L) was mixed with 0.15 g/L indium oxide powder in a standard bath. During EPD the voltage was 200 V and the deposition time was 10 min. After deposition the samples were dried at 80°C and then annealed at 450°C for 30 min. To improve the adherence of phosphor layers, the screens were treated in 0.2 M Na_2SiO_3 solution for 24 h to add additional binder. This matrix-addressed diode FED prototype was packaged and tested up to 3 kV.

In another design, a matrix-addressable FED was fabricated using a carbon nanotube-epoxy composite as the electron emission source and EPD phosphor stripes on the anode plate [64]. For anode processing, red Al_2O3:Eu was deposited by EPD at 400 V in a standard IPA bath onto ITO-patterned stripes 200 μm wide with a pitch size of 300 μm. The 32 × 32 matrix-addressable FED prototype was tested and showed steady emission in vacuum with well-defined and switchable pixels.

Full color screens have been made by EPD as discussed previously, some with specific application to FEDs [60] A 4-in. diagonal full color prototype FED was successfully made using EPD [65, 66]. The suspension consisted of Y_2O_2S:Eu, ZnS:Cu,Al, ZnS:Ag,Cl phosphors as red, green, and blue colors, respectively, with a bath of $La(NO_3)_2$ and $Al(NO_3)_2$ salts in IPA. Since these commercially-available phosphors had particle sizes of up to 7–9 μm, a sieving procedure was necessary to attain a smaller size of 2–3 μm. Conductive In_2O_3 powder at 10 wt.% was mixed with each color phosphor in order to provide high conductivity of the phosphor surface [66]. The deposition rate of each color phosphor was systematically investigated by changing deposition time and applied voltage, which varied from 50 to 140 V. Deposits had a uniform thickness of ~4–9 μm over the entire ITO-coated glass substrate. The brightness of each deposited phosphor was analyzed in a fully-sealed vacuum chamber utilizing Spindt-type field emission arrays.

In another study, process variables to improve the reliability of full color FEDs were investigated [67]. A full color FED panel consisted of pixels of 500 μm×120 μm, 100 μm, and 120 μm for red (Y_2O_2S:Eu), green (ZnS:Cu,Al), and blue (ZnS:Ag,Cl) phosphor, respectively, each electrophoretically deposited. The phosphors also were coated with In_2O_3. The width of the green color pixel was designed to be narrower than those of red and blue colors due to its higher efficiency. However, in order to avoid cross-talk, cleaning and drying of the phosphor lines after EPD were optimized. Unwanted deposition of phosphor particles on the unbiased line (cross-contamination) was observed with a cleaning step time of more than 15 s in acetone. But by reducing the cleaning and drying step to less than 15 s immediately following EPD significantly improved the cross contamination. Emission properties of the panel were analyzed in a fully-sealed environment that indicated that phosphor aging after sealing was required. Full color images were demonstrated with a 4 in. FED device.

Upon the deposition of phosphors by EPD on patterned substrates, a weak, non-uniform deposit of phosphor on the adjacent unbiased ITO patterns has been observed in addition to an expected uniform phosphor film on the biased patterns [68]. In practice, for EPD of phosphors the bias is only applied between a single group of biased ITO patterns (for example red lines) and the anode at a time. During the deposition, small amounts of phosphor are deposited on the unbiased neighboring ITO patterns, which affect the color purity of the screen. This occurs regardless of the pattern size, from micrometers to millimeters, and the non-uniform deposition appears to develop near the outer side of neighboring unbiased ITO patterns during EPD. The origin of the nonuniform deposition was explained by solving Laplace's equation with the boundary conditions simulating EPD, which showed an electric field with a parallel component from the unbiased ITO to the biased ones, in addition to the major vertical components of the electrical field, that increase as the resistance of the biased pattern increases as it was covered with the phosphor during EPD. It was suggested that the nonuniform deposition on the unbiased ITO electrodes may be avoided by connecting the unbiased ITO to the counter electrode through an appropriate resistor.

7.6.2 PDPS

In 1997, Schermerhorn et al. [69] proposed a grooved structure for large high-resolution color ac PDP using EPD of phosphor powders. The back plate of discharge cell was fabricated by selective deposition of red, green, and blue phosphors on all recessed surfaces of the groove pattern, which were prepared by isotropically etching and successively coating with a conductive material.

EPD process variables have been systematically studied and the optical characteristics of the screen investigated under vacuum ultraviolet (VUV) excitation for high-resolution PDP screens [70]. A suspension of green Zn_2SiO_4:Mn powder (~ 1–2 μm diameter) in a standard bath was used. The PL intensities of the phosphor screens were dependent upon the applied voltage, the screen density, and the binder material. The PL intensity of the screen increased linearly with screen weight, reaching a maximum at 4 mg/cm^2 (20 μm thickness) under VUV excitation. The PL intensity changed as a function of magnesium nitrate concentration in the bath for the same screen weight, exhibiting a maximum at $\sim 5 \times 10^{-4}$ M, indicating the influence on the binder content as it is a nonluminescent material. Also, an analytical expression for the optical performance of the phosphor screen was derived as a function of UV intensity, the reflectivity of back plate, and other material properties, which compared well with the experimental data.

A transparent ac-PDP test panel was prepared by EPD of nanosized phosphor powders onto on a metal mesh [71]. The optical transmittance and luminance of the panel indicated that a full color transparent ac-PDP is feasible with this approach.

7.7 EPD in a Thermoversible Gel

Another novel technique for depositing and patterning phosphors developed is EPD in a thermoreversible gel at specific points on the substrate [25]. A thermoreversible gel will melt upon heating and will resolidify when cooled. Thus, phosphors suspended in the gel matrix migrate in an applied electric field only when spot melting of the gel allows the phosphors to deposit on the substrate at that point. The melting and gelling may be repeated numerous times due to the physical bonds, unlike gels that are formed by chemical bonds that do not gel again once melted. The process as shown in Fig. 7.6 is described as follows.

Initially, a heated (liquid) solution of the polymer-solvent-phosphor system is coated on top of the substrate (e.g., ITO-coated glass). After the system is solidified (cooled), voltage is applied, creating an electric field. Spot-melting of the gel (e.g. by laser radiation) allows selective EPD of the phosphor to occur only at the desired spot or pattern. The electric field is removed and the gel is removed by melting, leaving only the deposited phosphor. This process may be repeated to deposit patterned multicolor screens. A possible advantage of utilizing EPD in a thermoreversible gel over the current methods which use organic photosensitive resins as lithographic structures [2] is its relative simplicity and avoiding the introduction of chromium (which decreases the phosphor brightness) and other chemicals (which may outgas and decrease the cathode performance) into the system.

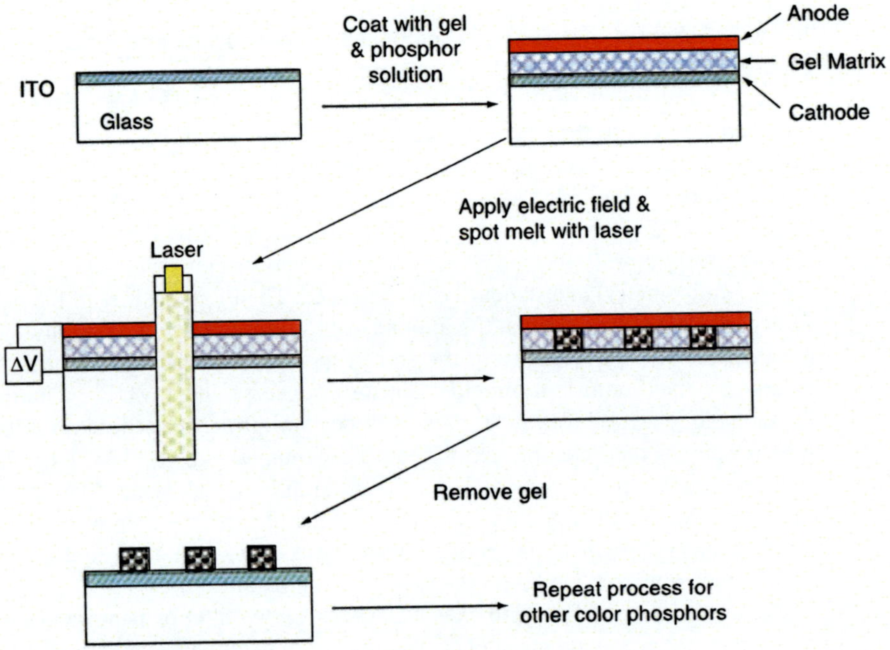

Fig. 7.6 Schematic of the process for EPD in a thermoreversible gel. [25]

First, an appropriate polymer-solvent system was required. Of the many polymer-solvent systems available [72], only a few solvents are suitable for EPD [73]. The system investigated was polybutyl methacrylate (PBMA), which forms a thermoreversible gel with isopropanol as the solvent [74]. PBMA does not form a homogeneous gel at concentrations less than 30 wt.% [74]. At lower concentrations and temperatures below 15°C the mixture separates into two phases. However, at the higher concentrations of polymer, the viscosity is very high, which inhibits the movement of phosphors in the electric field. The viscosity for various mixtures of PBMA in isopropanol and 4 g/L phosphors decreased significantly at higher temperatures. Therefore, a gel system must be optimized to gel at an appropriate temperature, yet not be too viscous as to impede EPD of phosphors.

A deposit density of ~1–2 mg/cm^2 of phosphor is required to ensure adequate optical performance [1, 2]. EPD in homogenous solutions of various concentrations of PBMA (10, 20, 30, and 45 wt.%) with and without added water (3 vol.%) and at various temperatures from 22–60°C was performed. The addition of water did not have a large effect on the viscosity. The addition of Zn_2SiO_4:Mn (P-1) phosphor did not significantly alter the gelation or the viscosity. For the 20 wt.% PBMA mixture, water appeared to be a necessary ingredient to have an acceptable phosphor loading. Without the addition of water, the loading was less than 0.50 mg/cm^2. This increased to > 1.0 mg/cm^2 for the first deposit with the addition of 3 vol.% water. However, the addition of water caused the polymer to precipitate into a sheet-like solid. For the higher concentrations of PBMA the phosphor loading was low. However, successive depositions (using a new cathode) from the same bath resulted in a reduction of phosphor loading after the first deposition. The decrease in deposit weight was determined to be caused by an increase in the local pH at the cathode.

Although EPD of phosphor in a thermoreversible gel has been demonstrated, it does not appear likely that this PBMA system will allow for adequate phosphor deposition due to the high viscosity at the required concentration of PBMA for gelation. Therefore, either a lower molecular weight PBMA [75] or a new polymer-solvent system with better gelation characteristics needs to be found for this proposed deposition method of phosphor.

7.8 Solid State Lighting

White light has been obtained by mixing blue light from the emission of GaN and yellow light by the fluorescence of a $Y_3Al_5O_{12}$:Ce (YAG:Ce) yellow phosphor. A uniform coating and an optimized thickness of phosphor on a GaN chip are necessary for achieving an efficient white LED. A patent from Lumileds in 2003 describes the use of EPD to produce a conformal phosphor coating on an LED for a uniform white light without colored rings [76]. A voltage was applied to a submount on which an LED chip was attached and placed into a suspension of phosphor particles for EPD. A successful deposit that we have made using a standard bath is shown

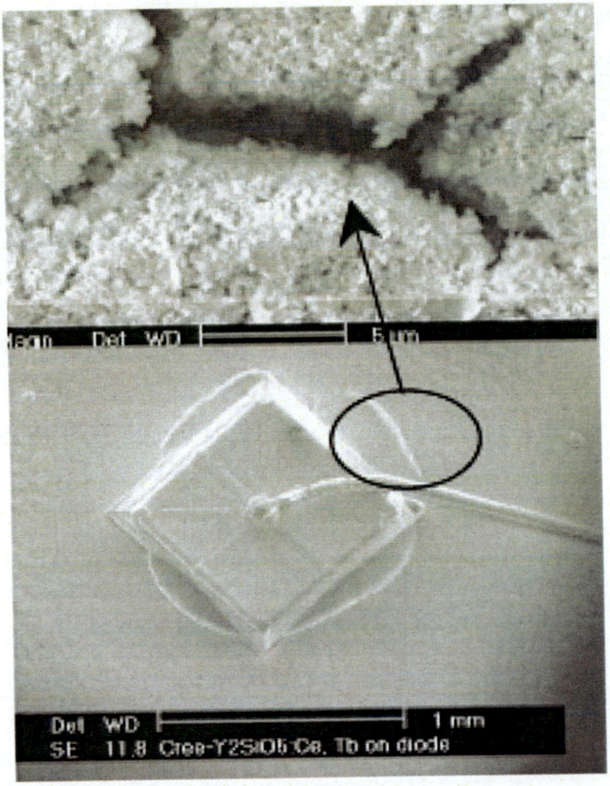

Fig. 7.7 EPD phosphor coating on an LED with a crack shown off the chip

in Fig. 7.7, which follows the contour of the LED chip. The arrow points out some cracking around the diode, but not on the chip, in an otherwise uniform coating.

Several methods for coating YAG:Ce (2.7 μm) particles, such as the slurry method, the settling method, the standard EPD, and a modified EPD were compared for preparing the phosphor layer on GaN [77]. For the modified EPD method, post-annealing was not done, but the deposit was sprayed with a PVA photoresist and UV cured to enhance adhesion [30]. For a homogeneously uniform phosphor layer, the EPD conditions were a bath concentration of 3 g/L of phosphor, deposited at 700 V for 240 s, yielding 9.88 mg/cm^2 with a thickness about 30 μm, which was almost one-third of the thickness compared with those coated by the slurry or settling methods. The packing density of the EPD layer was 69%, whereas those fabricated by slurry or settling methods were below 50%. The morphologies of the phosphor layer fabricated by the slurry and settling methods showed a lower uniformity than that for EPD. The adhesion strength of the EPD deposit was lower than for the phosphor layers formed with the other coating methods. However, the modified EPD method improved adhesion to be as strong as achieved with the slurry method. The intensity and color of white light was dependent on the thickness of the phosphor layer. It was found that the properties of the phosphor layer, such as packing density, thickness and uniformity, could be more easily controlled by EPD than by the slurry and settling methods. Furthermore, the high packing density from EPD could allow a thinner layer to be fabricated.

The YAG:Ce phosphor particles were coated onto a flexible ITO-coated PET polymer by EPD for use in fabrication of a white LED without directly coating onto a GaN chip [26]. The YAG:Ce phosphor particles (2.7 μm) were deposited from a standard bath at 400 V for 3 min. The packing density of the phosphor film on ITO-PET was lower than that of the film on rigid ITO because of the low conductivity of the ITO-PET. This was overcome by applying compression to the phosphor film at 200 bar pressure for 3 min. The packing density of the phosphor layer (26.6 μm thick) without compression was 53% and after compression to a thickness of 22.9 μm was 62%. To enhance adhesion, the deposits were also spray-coated with a PVA photoresist and UV cured [30].

Yellow YAG:Ce powder (~5–10 μm) was coated by EPD onto ITO-coated glass and a surface mounted device (SMD) LED [78]. A standard bath with 0.75 g/L phosphor was used with a 200 V/cm applied field for 20 min. After EPD, the deposit was rinsed with IPA and baked at 150°C for ~30 min. It was found that the average deposition rate increased with time with saturation at about 20 min yielding 3.0 mg/cm^2. The deposition process was more stable with stirring of the suspension at a rate of 60 rpm. EPD of phosphor was done on two die-attached SMD LEDs, one directly and one filled with a thin epoxy layer around the LED chip in which the thickess was less than that of LED die. These LEDs were compared to the conventional method of dispersing the phosphor in the epoxy. With the thin epoxy layer EPD LED, the light color was more uniform, and the white color value was well controlled; however, the luminous efficiency was lower than that of without the thin epoxy layer. It was found the both EPD LEDs had a lower efficiency that the dispersed phosphor LED. It was suggested that this was due to larger difference in refractive index between the phosphor and gallium nitride compared to the difference between epoxy and gallium nitride, such that blue light extraction efficiency was decreased.

A patent [79] describes an EPD method for coating on an LED having a p-side and an n-side. An anode is place in the bath with the LED, with a voltage applied between the anode and the p-side to be positive with respect to the p-side and a second voltage applied between the p-side and the n-side to cause the suspended particles to deposit onto the LED.

The EPD of yellow Eu^{2+}-doped Ca-α-SiAlON phosphor onto ITO-coated glass was investigated for its use in white solid state lighting [80]. The powder (0.2 g/L) was dispersed in ethanol with small amounts of phosphate ester and polyethylene imine as dispersants and polyvinyl butyral as a binder. The particle size was about 1 μm, but the powder also contained relatively large aggregates of ~20 μm. The thickness of the phosphor films was controlled by altering the deposition time up to 5 min at 30 V. The light emission from the deposit surfaces was characterized by comparing the intensities of the excited yellow light and the transmitted blue light, both which decreased with an increase of the film thickness from 2 to 6 μm. Therefore, deposit density greatly affected the color tone that was controllable from blue-white to yellow-white.

7.9 Summary

The EPD process is a general phosphor screening procedure applicable to a host of information display applications. Newer display technologies, such as FEDs and PDPs, have placing more stringent requirements on the phosphor screen. Practices of lacquering and then aluminizing the phosphor deposit are often inappropriate for the newer, advanced displays. EPD of phosphor particles is well suited to deposit the fine (nanometer to micrometer diameter) particles needed for high resolution displays. Additionally, EPD can be used as a method to make samples to test new phosphors, as the process does not alter the inherent optimal performance of the phosphor [81].

To electrophoretically deposit a wide variety of phosphors (and other powders), the fundamentals of the process in a standard IPA bath were studied. By investigation of the dissociation behavior of nitrate salts in IPA, measurement of the effects of pH and nitrate salt concentration on the zeta potential of the particles, and by studying the EPD process conditions and modeling the deposition rates, the fundamentals of this EPD have been well-characterized. The electrochemical precipitation reactions that form the adhesive agents were identified and even utilized (without phosphor) to form hydroxide precursors for superconducting films [82]. These fundamental studies not only allow better design of the process for the EPD of phosphors, but for other powdered materials (e.g. zeolites [83] and CNTs [56]).

As discussed, a major limitation of EPD phosphor deposits has been the low adhesion strength of the deposit. Our research has provided a better understanding of the adhesion of EPD phosphor particles and techniques to enhance the adhesion strength to meet the requirements of these new display technologies. The single greatest effect on the adhesion strength was the added 2 vol.% glycerin to the deposition bath. Adhesion has been also significantly enhanced by spray coating a photoresist layer on top of the EPD phosphor film [30].

The ability to improve the adhesion strength of an EPD phosphor deposit is useless if the optical performance is degraded. The color and brightness of screens made with these deposition conditions were unchanged by the processing conditions. EPD phosphor screens also only outgas H_2, CO, CO_2 and low level hydrocarbons under electron bombardment in FED tests [84], which make them attractive for use.

New methods to fabricate full-color have been proposed. Several methods that combine EPD and photolithography were developed to produce screens with a high resolution of triads of red-green-blue phosphor stripes per inch. Even higher resolution could be achieved with a higher resolution mask and smaller sized phosphor particles. The EPD in a thermoreversible gel was explored but needs further work to find a polymer-solvent system with better gelation characteristics than the PBMA polymer used.

A new application of EDP of phosphor is white solid state lighting which has used similar processing conditions as for information displays. The ability to coat a uniformly thin, high packed, conformal phosphor layer on the LED has been the advantage of EPD to control color and efficiency.

Acknowledgements The author acknowledges all of the students, post-doctoral researchers and industrial colleagues, in particular Esther Sluzky, who contributed to the many years of research on the EPD of phosphors. The research was supported by Hughes Aircraft Co., Electroplasma Inc., Candescent Technology Corp., the UC MICRO Program, the American Display Consortium, the DARPA Phosphor Technology Center of Excellence, Sony Electronics, Inc., and the UC SMART Program. The facilities of SRI International, Menlo Park, and Coloray, Fremont, CA were used for our photolithography work.

References

1. Ozawa, L.: Cathodoluminescence. Kodansha, Tokyo (1990)
2. Sasaki, K.Y., Talbot, J.B.: Deposition of powder phosphors for information displays. Adv. Mat. **11**(2), 91 (1999)
3. Cerulli, N.F. (1958) US Patent 2851408
4. McGee, J.D., Airey, R.W., Aslam, M.: The evaluation of cascade phosphor-photocathode screens. Adv. Electr. Elect. Phys. **22A**:57 (1966)
5. Libman, P.C. (1990) US Patent 4891110
6. Grosso, P.F., Rutherford, R.E. Jr., Sargent, D.E.: Electrophoretic deposition of luminescent materials. J. Electrochem. Soc. **117**, 1456 (1970)
7. Sluzky, E., Hesse, K.: Electrophoretic preparation of phosphor screens. J. Electrochem. Soc. 136, 2724 (1989)
8. Yen, W.M., Shionoya, S., Yamamoto, H.: Practical Applications of Phosphors. CRC Press, Boca Rotan (2007)
9. Xing, D., Gong, M., Qiu, X., Yang, D., Cheah, K.: Characterization of superfine Sr_2CeO_4 powder prepared by microemulsion-heating method. J. Rare Earth **24**(3), 289 (2006)
10. Wakefield, G., Holl, E., Dobson, P.J., Hutchison, J.L.: Luminescence properties of nanocrystalline Y_2O_3:Eu. Adv. Mater. **13**(20): 1557 (2001)
11. Silver, J., Withnall, R., Lipman, A., Irel, T.G., Fern, G.R.: Low-voltage cathodoluminescent red emitting phosphors for field emission displays. J. Lumin. **562**, 122–123 (2007)
12. Wong, E.M., Searson, P.C.: ZnO quantum particle thin films fabricated by electrophoretic deposition. Appl. Phys. Lett. **74**, 2939 (1999)
13. Fern, G., Irel, T., Silver, J., Withnall, R., Michette, A, McFaul, C., Pfauntsch, S.: Characterisation of Gd_2O_2S:Pr phosphor screens for water window X-ray detection. Nucl. Instrum. Methods Phys. Res. Sect. A **600**(2), 434 (2009)
14. Shane, M.J., Talbot, J.B., Schreiber, R.D., Ross, C.L., Sluzky, E., Hesse, K.R.: Electrophoretic deposition of phosphors I. Conductivity zeta potential measurements. J. Colloids. Interf. Sci. **165**, 325 (1994)
15. Russ, B.E., Talbot, J.B.: Analysis of the binder formation in electrophoretic deposition. J. Electrochem. Soc. **145**(4), 1253 (1998)
16. Shane, M.J., Talbot, J.B., Sluzky, E., Hesse, K.R.: Zeta potential of phosphors. Colloids Surf. **96**, 301 (1994)
17. Siracuse, J.A., Talbot, J.B., Sluzky, E., Hesse, K.R.: Adhesive agent in cataphoretically coated phosphor screens. J. Electrochem. Soc. **137**, 346 (1990)
18. Siracuse, J.A., Talbot, J.B., Sluzky, E., Avalos, T., Hesse, K.R.: Cataphoretic deposition of phosphor. J. Electrochem. Soc. **137**, 2336 (1990)
19. Shane, M.J., Talbot, J.B., Kinney, B.G., Sluzky, E., Hesse, K.R.: Electrophoretic deposition of phosphors II. Deposition experiments and analysis. J. Colloids. Interf. Sci. **165**, 334 (1994)
20. Russ, B.E., Talbot, J.B., Sluzky, E.: Study of the variables affecting adhesion strength of electrophoretically deposited phosphors. J. Soc. Inform. Displays. **4**(3), 207 (1996)
21. Luo, S., Talbot, J.B.: Optical characterization of electrophoretically deposited phosphors. J. Electrochem. Soc. **148**(7), H73 (2001)

22. Chang, D.C., Talbot, J.B., Rao, R.P., Holland, C.: Electrophoretic deposition process for screening phosphor powders for color displays, J. Soc. Inform. Displays **8**(1), 51 (2000)
23. Kurinec, S.K., Sluzky, E.: Fabrication of ultra-high-resolution three-color phosphor screens. J. Soc. Inform. Displays **4**(4), 371 (1996)
24. Talbot, J.B., Sluzky, E., Kurinec, S.K.: Electrophoretic deposition of monochrome color phosphor screens for information displays. J. Mater. Sci. **39**(3), 771 (2004)
25. Choi, Y., Talbot, J.B.: Electrophoretic deposition of phosphor in a thermoreversible gel. J. Soc. Inform. Displays **10**(3), 256 (2002)
26. Yum, J.H., Kim, S.S., Sung, Y.E.: $Y_3Al_5O_{12}:Ce_{0.05}$ phosphor coatings on a flexible substrate for use in white light-emitting diodes. Colloids Surf. A **251**, 203 (2004)
27. Russ, B.E., Talbot, J.B.: Study of the adhesion of electrophoretically deposited phosphors. J. Electrochem. Soc. **145**(4), 1245 (1998)
28. Russ, B.E., Talbot, J.B.: Method for measuring the adhesion strength of powder coatings. J. Adhesion. **68**, 257 (1998)
29. Mizuguchi, J., Sumi, K., Muchi, T.: Highly stable nonaqueous suspension for the electrophoretic deposition of powdered substances. J. Electrochem. Soc. **130**, 1819 (1983)
30. Yum, J.H., Choi, K.H., Sung, Y.E.: Adhesion improvement of phosphor layer by combining electrophoretic deposition and UV curing. J. Electrochem. Soc. **150**(2), H43 (2003)
31. Prener, J.S., Swank, R.K.: Studies of X-ray image intensifier output. J. Electrochem. Soc. **125**, 583 (1978)
32. Gibilini, D.: Projection cathode-ray tubes for HDTV. SID 86 Digest 236 (1986)
33. Sluzky, E., Hesse, K.: High-resolution phosphor screens. J. Electrochem. Soc. **135**, 2893 (1988)
34. Sluzky, E., Hesse, K.: Characteristics of ultrafine grain P53 screens. J. Electrochem. Soc. **138**, 2418 (1991)
35. Raue, R., Vink, A.T., Welker, T.: Phosphor screens in cathode-ray tubes for projection television. Philips. Tech. Rev. **44**(11/12):335 (1989)
36. Busselt, W., Raue, R.: Optimizing the optical properties of TV phosphor screens. J. Electrochem. Soc. **135**, 764 (1988)
37. Yum, J.H., Sung,Y.E.: Full color screen by EPD combined with photolithography for flat panel displays. J. Electrochem. Soc. **151**(2), H27 (2004)
38. Yang, S.H.: Electrophoretic prepared $ZnGa_2O_4$ phosphor film for FED. J. Electrochem. Soc. **150**(10), H250 (2003)
39. Kang, S.W., Jeon, B.S., Yoo, J.S., Lee, J.D.: Optical characteristics of the phosphor screen in field-emission environments. J. Vac. Sci. Technol. B. **15**(2), 520 (1997)
40. Gibbons, C., Jing, X.P., Silver, J., Vecht, A., Withnall, R.: Danger of poisoning by stainless steel anodes used for the electrophoretic deposition of phosphors. Electrochem. Solid. State Lett. **2**(7), 357 (1999)
41. Davies, D.A., Lipman, A.L., Silver, J., Tseung, A.C.C.: On the effect of anode material in electrophoresis on the emission color of a zinc sulfide phosphor. Electrochem. Solid-State Lett. **4**(5), H12 (2001)
42. Cho, S.H., Yoo, J.S., Lee, J.D.: Synthesis and low-voltage characteristics of $CaTiO_3$:Pr luminescent powders J. Electrochem. Soc. **143**(10), L231 (1996)
43. Cho, S.H., Kwon, S.H., Yoo, J.S., Oh, C.W., Lee, J.D., Hong, K.J., Kwon, S.J.: Cathodoluminescent characteristics of a spherical Y2O3:Eu phosphor screen for field emission display application. J. Electrochem. Soc. **147**(8), 3143 (2000)
44. Jeon, B.S., Kang, S.W., Yoo, J.S., Lee, J.D.: Characterization of low-voltage phosphor screens for FED applications. Materials Research Society Symposium—Proceedings. **424**, 421 (1996)
45. Tseng, Y.H., Chiou, B.S., Peng, C.C, Ozawa, L.: Spectral properties of Eu^{3+}-activated yttrium oxysulfide red phosphor. Thin Solid Films **330**(2), 173 (1998)
46. Marsh, P.J., Silver, J., Vecht, A., Newport, A.: Cathodoluminescence studies of yttrium silicate: Cerium phosphors synthesised by a sol-gel process. J. Lumin. **97**(3/4), 229 (2002)

47. Jeon, B.S., Yoo, J.S., Lee, J.D.: Electrophoretic deposition of ZnO:Zn phosphor for field emission display applications. J. Electrochem. Soc. **143**(12), 3923 (1996)
48. Gupton, J.A.: US Patent 3681223 (1970)
49. Philips, D.M.: US Patent 3904502 (1973)
50. Kaplan, S., Libman, C., Wainscott, D.: US Patent 4130472 (1978)
51. Mooney, J.B.: US Patent 4990416 (1991)
52. Kang, S.W., Yoo, J.S., Lee, J.D.: Effects of SnO_2 on the phosphor screen behaviors under low energy electron excitation. J. Electrochem. Soc. **145**(2), 648 (1998)
53. Sluzky, E., Kurinec, S., Hesse, K., Ternullo, L.: US Patent 5582703 (1996)
54. Kwon, S.H., Cho, S.H., Yoo, J.S., Lee, J.D.: Fabrication of full-color phosphor screen by electrophoretic deposition for field emission display application. J. Electrochem. Soc. **147**(8), 3120 (2000)
55. Yum, J.H., Sung, Y.E.: US Patent 6627060 (2003)
56. Quale, S., Talbot, J.B.: Electrophoretic deposition of substrate-normal-oriented single-walled carbon nanotube structures. J. Electrochem. Soc. **154**, K25, (2007)
57. Kiyomiya, T., Ohoshi, T., Okita, M.: US Patent 5466358 (1995)
58. Kim, C.O., Seo, D.S., Hong, J.P.: Analysis of electrophoretically deposited low-voltage phosphors mixed with In_2O_3 conducting powders for field emission displays. J. Electrochem. Soc. **147**(6), 2394 (2000)
59. Hong, J.P., Seo, D.S., Kim, C.O., Song, B.K., Char, S.N., Lee, N.S., Kim, J.M., Chee, J.K.: Effect of laser irradiation on electrophoretically coated phosphors for field emission displays. J. Electrochem. Soc. **148**(4), H45 (2001)
60. Kang, S.W., Yoo, J.S., Lee, J.D.: Photolithographic patterning of phosphor screens by electrophoretic deposition for field emission display application. J. Vac. Sci. Technol. B. **16**(5), 2891 (1998)
61. Chang, C.H., Chiou, B.S., Chen, K.S., Ho, C.C., Ho, J.C.: The effect of In_2O_3 conductive coating on the luminescence and zeta potential of ZnS:Cu, Al phosphors. Ceramics Intern. **31**, 635 (2005)
62. Park, Z.M., Jeon, D.Y., Jin, Y.W., Cha, S.N.: Degradation behavior of low-voltage cathodoluminescence of a ZnS:Ag,Cl phosphor screen under a panel sealing environment. J. Vac. Sci. Technol. B. **21**(1), 527 (2003)
63. Li, W., Feng, T., Mao, D.S., Wang, X., Liu, X.H., Zou, S.C., Zhu, Y.K., Li, Q., Xu, J.F., Jin, S., Zheng, J.S.: A FED prototype using patterned DLC thin films as a cathode. Int. J. Mod. Phys. B5 16(6/7), 993–997 (2002)
64. Wang, Q.H., Setlur, A.A., Lauerhaas, J.M., Dai, J.Y., Seelig, E.W., Chang, R.P.H.: A nanotube-based field-emission flat panel display. Appl. Phys. Lett. **72**, 2912 (1998)
65. Hong, J.P., Jang, J.E., Jin, Y.W., Jung, J.E., Ryu, Y.S., Lee, H.W., Kim, J.M., Milhalova, V.: Optimization and analysis of low voltage phosphors deposited electrophoretically for the FED applications. Proceedings of IEEE International Vacuum Microelectronics Conference, IVMC, 692 (1997)
66. Jin, Y.W., Jang, J.E., Yi, W.K., Jung, J.E., Lee, N.S., Kim, J.M., Jeon, D.Y., Hong, J.P.: Performance of electrophoretic deposited low voltage phosphors for full color field emission display devices. J. Vac. Sci. Technol. B. **17**(2), 489 (1999)
67. Kim, J.M., Hong, J.P., Choi, J.H., Ryu, Y.S., Hong, S.S.: Parameters for improving reliability of full color field emission display devices. J. Vac. Sci. Technol. B. **16**(2), 736 (1998)
68. Shin, S.H., Jeon, D.Y., Jin, Y.W., Kim, J.M. Investigation of nonuniform deposition of field emission displays phosphors on unbiased indium-tin-oxide coated glass during electrophoresis. Japn. J. Appl. Phys. A **39**(12), 6743 (2000)
69. Schermerhorn, J.B., Sweeney, R.B., Wang, W., Park, B.Y., Park, M.H., Kim, J.S.: A grooved structure for a large high-resolution color ACPDP. SID Digest. **28**, 229 (1997)
70. Jeon, B.S., Hong, K.Y., Yoo, J.S., Whang, K.: Studies on the phosphor screen prepared by electrophoretic deposition for plasma display panel applications. J. Electrochem. Soc. **147**(11), 4356 (2000)

71. Choi, H.N., Lee, S.Y., Kim, Y.S.: Developments of Transparent ac-PDPs. Proceedings of the International Meeting on Information Display, iMiD/IDMC/Asia Display **8**, 1621 (2008)
72. Guenet, J.: Thermoreversible gelation of polymers and biopolymers. Academic Press, San Diego (1992)
73. Powers, R.W.: The electrophoretic forming of beta-alumina ceramic. J. Electrochem. Soc. **122**, 490 (1975)
74. Schneider, T., Wolf, B.A., Kasten, H., Kremer, F.: Thermoreversible gelation vitrification of highly concentrated polymer solutions under poor thermodynamic conditions. Macromolecules **24**, 5387 (1991)
75. Jelich, L.M., Nunes, S.P., Paul, E., Wolf, B.A.: On the cooccurrence of demixing thermoreversible gelation of polymer solutions 1. Experimental observations. Macromolecules **20**, 1943 (1987)
76. Collins, W.D., III, Krames, M.R., Verhoeckx, G.J., van Leth, N.J.M.: US Patent 6576488 (2003)
77. Yum, J.H., Seo, S.Y., Lee, S., Sung, Y.E.: $Y_3Al_5O_{12}:Ce_{0.05}$ phosphor coatings on gallium nitride for white light emitting diodes. J. Electrochem. Soc. **150**(2), H47 (2003)
78. Chen, W.R., Bai, J.A., Meen, T.H., Huang, C.J.: Electrophoretic deposition of YAG phosphor on SMD LED. ECS Trans. 19(12), 27 (2009)
79. Summers, C.J., Menkara, H., Chua, J.B.Y. US Patent 6864110 (2005)
80. Kitabatake, T., Uchikoshi, T., Munakata, F., Sakka, Y., Hirosaki, N.: Electrophoretic deposition of Eu^{2+} doped Ca-α-SiAlON phosphor particles for packaging of flat pseudo-white light emitting devices. J. Ceramic. Soc. Jpn. **116**(6), 740 (2008)
81. Shea, L.E.: Low-voltage cathodoluminescent phosphors. Interface **7**(2), 24 (1998)
82. Abolmaali, S.B., Talbot, J.B.: Synthesis of superconductive thin films of $YBa_2Cu_3O_{7-x}$ by a nonaqueous electrodeposition process. J. Electrochem. Soc. **140**(2), 450 (1993)
83. Ahlers, C.B., Talbot, J.B.: Fabrication of zeolite-modified electrodes via electrophoretic deposition. J. Electrochem. Soc. **146**(9), 3259 (1999)
84. Malinowski, M.E., Stewart, K.D., Ohlberg, D.A.A., Felter, T.E., Chakhovskoi, A.G., Hunt, C., Shea, L., Russ, B.E., Talbot, J.B., McKittrick, J.: Gas desorption from FEA—phosphor screen pairs. Proceedings of 8th Intern Conf on Vacuum Microelectronics, IVMC-95. Electr. Device Soc. IEEE 202 (1995)

Chapter 8
Electrophoretic Deposition in Production of Ceramic Matrix Composites

Saša Novak, Katja König and Aljaž Iveković

Acronyms

ATZ	Alumina Toughened Zirconia
CFCMC	Continuous Fibre-reinforced Ceramic Matrix Composite
CMC	Ceramic Matrix Composite
CNT	Carbon Nanotube
CTAB	Cetyl-trimethylammonium bromide
CVD	Chemical Vapour Deposition
CVI	Chemical Vapour Infiltration
EPD	Electrophoretic Deposition
EPI	Electrophoretic Infiltration
FCI	Flow Channel Insert
IEP	Isoelectric Point
NITE	Nano-powder Infiltration and Transient Eutectoid
PEI	Polyethylene-imine
PIP	Polymer Infiltration and Pyrolysis
PZC	Point of Zero Charge
SDOSS	Sodium-dioctyl-sulfosuccinate
SiCf/SiC	Silicon Carbide Fibre-reinforced Silicon Carbide Composite
SITE	Slip Infiltration and Transient Eutectoid
TMAH	Tetramethylammonium hydroxide
ZTA	Zirconia Toughened Alumina

S. Novak (✉)
Department for Nanostructured Materials, Jožef Stefan Institute,
Jamova c. 39, SI-1000, Ljubljana, Slovenia
e-mail: sasa.novak@ijs.si

J. H. Dickerson, A. R. Boccaccini (eds.), *Electrophoretic Deposition of Nanomaterials*, Nanostructure Science and Technology,
DOI 10.1007/978-1-4419-9730-2_8,

8.1 Introduction to Ceramic Matrix Composites [1, 2]

Ceramic matrix composites (CMCs) are ceramic-based materials that by a combination of ceramic matrix and a reinforcing phase form a new material with superior properties. The main goal of introduction of the reinforcing phase is to provide toughness to an otherwise brittle ceramic matrix. The ceramic matrix composites posses properties that allows unique engineering solutions and are therefore of high technological interest. Continuous development of new, improved and less expensive processing routes is permanently leading to increasingly varied applications.

The ceramic matrix composites are available in a variety of forms where the reinforcement phase is either discontinuous or continuous. The reinforcements in the so-called discontinuously reinforced CMCs are particulates such as particles, whiskers, platelets or short fibres (Fig. 8.1). The continuous fibre-reinforced ceramic matrix composites (CFCMCs) represent a special group of CMCs, where the reinforcing phase-continuous fibres, are typically woven into ceramic fabric, while the matrix may be a monolithic ceramics or a composite with a discontinuous phase.

The continuous fibre-reinforced ceramic matrix composites represent the most advanced group of ceramic materials in which the continuous fibres provide unique performance and ability to meet requirements for highly demanding applications.

Fig. 8.1 Schematic of ceramic matrix composites: **a**, **b** and **c** discontinuous phase reinforced composites with particles, short fibres, whiskers or nanotubes and with platelets, respectively; **d** continuous fibre-reinforced composite

Due to their specially tailored structure they can significantly reduce susceptibility to catastrophic failure that is in particular important in large structures. On the other hand, their production is usually expensive that, to a great extent, limits their application. In contrast to the composites with discontinuous reinforcements that can be produced by the same shaping techniques as monolithic ceramics (slip casting, pressing, etc), the CFCMCs require an infiltration, which represent major bottleneck in production and a great technological challenge.

In a brief introduction, the main characteristics of both major types of the ceramic matrix composites will be given, while in continuation basic principles for production by means of EPD and some case studies will be presented. The main stress in this chapter will be given to fibre-reinforced CMCs and among them to the SiC_f/SiC composites that are among the most challenging products.

8.1.1 Discontinuous Phase (particulate) Reinforced CMCs

As already stated above, the reinforcement in this group of CMCs is achieved by incorporation of either particles, whiskers, platelets or short fibres. To achieve sufficient green density of the composite, several traditional shape forming techniques can be employed from dry-pressing to forming from colloidal suspensions. The main advantages of the EPD in comparison with other shaping techniques is that it enables not only the production of complex shaped composite parts or coatings with homogeneously distributed reinforcement, but also the distribution of the materials in a desired manner, as for example into layered structures (laminates, functionally gradient composites etc.). Not the least, when the reinforcements are anisotropic particles, such as needles, platelets or nanotubes, EPD has the potential to arrange those species into textured structures. This, however, remains an open field for further research.

The most important requirement to be followed during processing of the composite material is good control of microstructure and homogeneity. The suspensions used in EPD contain two (or more) sorts of particles, typically with different chemical composition that may also significantly differ in surface charge. Therefore, a great care must be taken to prepare and to retain the suspension with homogeneously distributed all components throughout the process.

Electrophoretic deposition of a complex suspension containing two- or more particulate phases requires a control of zeta potential (ζ-potential) of all particulates in order for all of them to migrate to the same electrode and consequently to co-deposit homogeneously or to form a controlled microstructure. In the case of mixed suspensions of two oppositely charged species, as for example alumina and silica in alkaline region, heterocoagulation may appear. This results in the formation of particles clusters, which migrate in a form of composite particles and not as single particles. Therefore, the co-deposition is preferably performed in such medium that both types of particulates are similarly charged.

Although the EPD offers a simple and low-cost fabrication technique, not many bulk composites with discontinuous reinforcements have been prepared by means of electrophoresis. Besides a few successful trials to prepare particulate composites by EPD, there have been some sporadic reports on EPD of whiskers- and nanotubes- reinforced composites. It is interesting that most of the papers in this specific field present the final results of their experiments, while they rarely present the electrokinetic properties of the used suspensions, though it is well known that the surface charge plays a major role. Many of the reports are based upon trial-and-error approach, which are often effective by chance but do not lead to optimal results and in particular do not serve as a good scientific basis for development of the field.

One of the most well known examples of the two-component systems is a composite of Al_2O_3 and ZrO_2 that exhibit opposing surface charge in a certain range of pH values in aqueous suspensions. By control of the ζ-potential, Hadraba et al. successfully deposited particulate composites from a mixture of Al_2O_3 and ZrO_2 in isopropanol suspensions stabilized with monochloroacetic acid [3]. They presented that by selecting the conditions where electrophoretic mobility of Al_2O_3 and ZrO_2 particles are similar, it is possible to prepare ceramic composites with controlled microstructure, such as particulate, layered or functionally gradient composite with smooth concentration transition (Fig. 8.2a, b, c). Pioneer work toward industrial production of alumina-zirconia composite with a functionally graded structure has been done by the group of Van der Biest [4] that presented the technique for shape-forming of composite hip-joint with significantly improved performance in comparison with the particulate zirconia toughened alumina (ZTA) material. In this case suspensions were based on methyl-ethyl-ketone and n-butylamine was used to disperse the suspension.

EPD was also confirmed to be a versatile technique for forming metal-free ceramic dental crowns. In deposition of alumina toughened zirconia (AZT) composites Moritz et al. [6] used a commercial composite powder dispersed in water-free ethanol and stabilised by 4-hydroxybenzoic acid. This work well illustrates the applicability of EPD technique in production of complex-shaped parts with dense and defect-free microstructures.

In 1994 Zhang et al. [7] presented a kinetic EPD model for the fabrication of SiC-whisker reinforced TZP composite with a uniform structure. The model predicts the decrease of the forming rate as an exponential function of time. Another early work on EPD of Al_2O_3 particles and SiC whisker in water-free isopropanol suspension was presented by Jean and co-authors [8]. They confirmed that during deposition Al_2O_3 and SiC whiskers migrate and deposit at approximately the same rate resulting in a nearly unchanged composition of the composite green parts.

The CMC composites reinforced with carbon nanotubes (CNTs) are of increasing technological interest due to the extraordinary behaviour of the CNTs. In the last decade an increasing number of publications report on significant improvement of the materials properties when the CNTs were incorporated. The main challenge is based upon the enormous elastic modulus and toughness of the CNT as well as upon its high thermal conductivity. The wide range of successful application of electrophoretic deposition in preparation of CNT reinforced SiO_2, TiO_2 or MnO_2 ceramic matrix nanocomposites was recently reviewed by Boccaccini et al. [10]

CNTs have also opened new opportunities to significantly improve the performance of the SiC-based composites for structural applications in future fusion

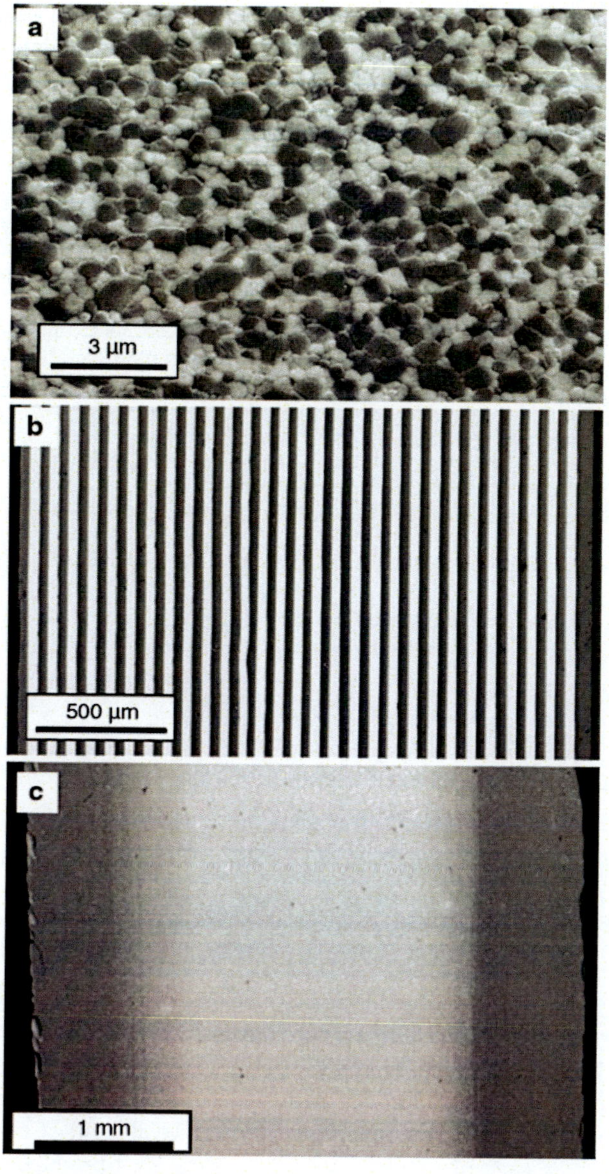

Fig. 8.2 Microstructures of alumina/zirconia composites prepared by EPD: **a** particulate, **b** layered [3] (Reprinted in part with permission from Hadraba et al. © 2004 Elsevier) and **c** gradient composites (Reprinted in part with permission from Hvizdos et al. © 2007 Elsevier). [5]

reactor, where high thermal conductivity in combination with high toughness is of outmost importance. Since both, SiC powder and CNTs have highly negative ζ-potential in alkaline region, it is possible to prepare a relatively stable mixed SiC-CNT suspension either in water or in ethanol [9, 11]. In the case of long "spaghetti-like" CNTs, however, the entanglement of the relatively rigid nanotubes in SiC particles limits the packing density in the composite. Figure 8.3a and b show microstructure of the green and sintered SiC-CNT deposit with 1 wt.% CNTs where the CNTs are homogeneously distributed within the SiC matrix; however, the density

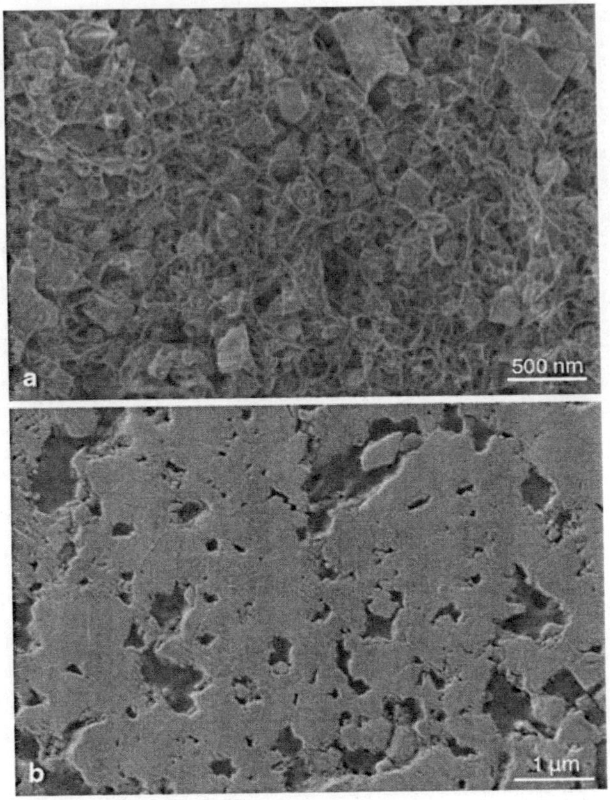

Fig. 8.3 SEM micrographs of the **a** fracture surface of green SiC-CNT deposit; **b** sintered and polished SiC-CNT deposit

of the deposit is apparently lower than the SiC deposit prepared under the same conditions.

8.1.2 Continuous Fibre-reinforced CMCs (CFCMCs)

Continuous fibre-reinforced CMCs represent a great challenge for technologists in the production of the advanced composite materials for application under extreme conditions. Although these materials with a highly complex structure are known and have been used for decades, their production typically involves very long and costly procedures. Therefore, their use has never been extended from the leading edge applications to a wider range of applications where they could contribute to a solution of many industrial issues. However, the principles of colloidal powder processing and the implementation of EPD process has recently opened new possibilities for production of CMCs with tailored properties at moderate price.

The continuous fibre-reinforced CMCs have been designed for production of structural parts to be used at high temperatures under high loads and in corrosive atmospheres. In contrast to their "forerunners", "fibreglass", the ceramic-matrix composites may have high-temperature strength, relatively high fracture tough-

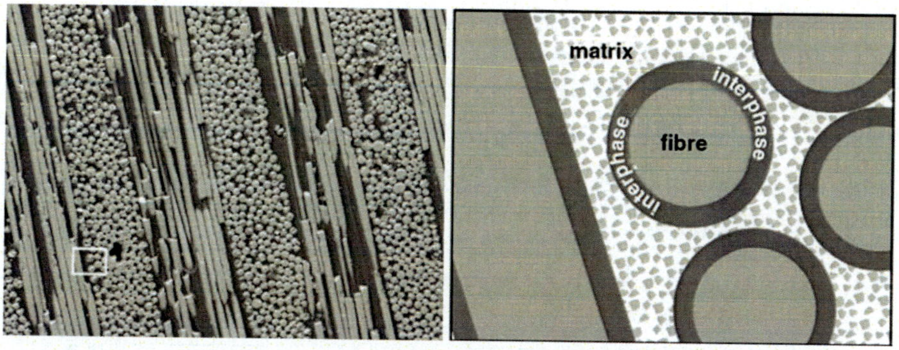

Fig. 8.4 Microstructure of continuous ceramic matrix composite (CFCMC)

ness, damage tolerance, and thermal shock resistance. The main challenge in this field is the durability of the materials when exposed to severe environments. These rather unique properties proceed from their specific macrostructure; they comprise a ceramic matrix in which long, typically ceramic fibres coated with an interphase layer are embedded (Fig. 8.4). The role of the fibres is to provide the material with toughness not being given by the matrix, while the interphase is an essential component that leads a propagating crack to lose a part of its energy and hence results in increased toughness and reliability. By tailoring the fibres properties, the fibre/matrix bonding and matrix microstructure the designed materials can thus achieve a non-brittle mechanical behaviour through different damaging phenomena occurring at the micrometer scale under loading. The stress-displacement diagram in Fig. 8.5

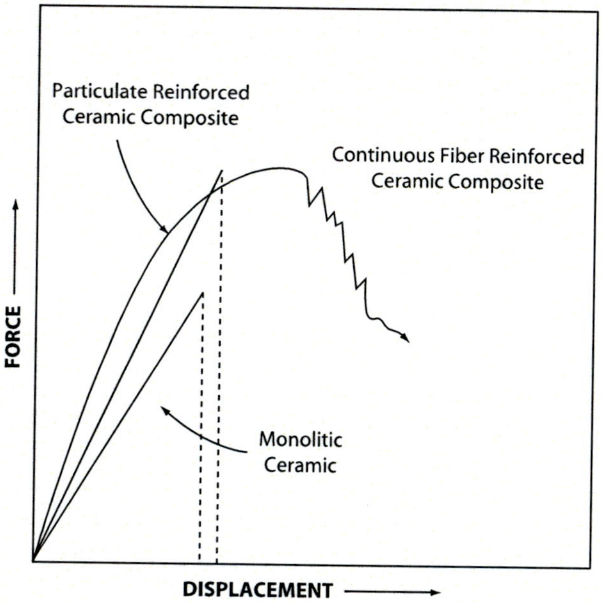

Fig. 8.5 Stress-displacement curves for fibre-reinforced CMC in comparison with particulate-reinforced composites and with monolithic ceramic

schematically illustrates the comparison in toughness for CFCMC with particulate CMC and monolithic material.

The matrices in CFCMCs are made of monolithic ceramics (oxide or nonoxide), glass-ceramics or particular composite, while fibres can be ceramic (oxide, e.g. alumina, mullite or non-oxide, e.g. carbon and SiC) or metallic (stainless steel, special alloys, titanium, etc.).

Fibres have the largest effect on composite properties. They are almost never used as single filaments in composites; since they are usually fragile, they are coupled in tows that are then weaved into a fabric. The fabric properties depend upon several parameters such as type of weaves, openness, thickness, etc. To improve the performance of a composite for some specialty applications, the fabrics are woven in three-dimensional fabrics. Thus, the complex architecture thus represents the main issue in infiltration with a ceramic matrix material.

In most applications of the CFCMCs, interfacial properties are critical to the performance. If there is a reaction between the matrix and the fibre, the two materials are generally strongly bonded to each other. This produces a ceramic composite with a low toughness because, when the fibres remain firmly bonded to the matrix, no energy is consumed in pulling the fibres out of the matrix as a crack propagates. Under these conditions, the mechanical behaviour of the composite is similar to that of the reinforced matrix. On the other hand, if the fibres have a non-reactive coating, there is little or no bonding at the interface. As a crack propagates through the stressed material, energy is consumed as fibres pull or slide out of the matrix.

Thus, an interphase layer between the fibre and the matrix plays a key role in the CMCs. Although it constitutes a small fraction of the total volume of a composite, the effects of its properties on the bulk properties are large because of the large surface area of the interfaces (Fig. 8.4b). It is typically thinner than 1 μm and has three main functions: load transfer, crack deflection, and diffusion barrier. The properties of the interphase layer depend on whether or not there has been a reaction at the interface, the type of bonding and on the nature (rate of crystallinity) and morphology of the interphase material. According to literature data, the most efficient interphase is anisotropic pyrolytic carbon deposited by chemical vapour deposition (CVD) from a gaseous hydrocarbon. It will be shown later that EPD represents a versatile alternative to this usually time-consuming and costly process.

According to the type of the matrix, interphase and fibres, most common CMC are non-oxide composites based on SiC or C fibres and oxide composites based on Al_2O_3 or mullite fibres. For high performance applications carbon and SiC fibres are the most prevalent. Since most applications take place in oxidizing atmospheres, preferred reinforcements are SiC-based fibres. Oxide-based CMCs with alumina and mullite matrices have also been successfully demonstrated in applications where a high tolerance to oxidation or salt corrosion is required in combination with the high toughness, light weight, and high-thermal-shock resistance.

8.1.3 Main Application Areas of CMCs

The great commercial importance of composites proceeds from complementary nature of matrix and reinforcements. In addition to the characteristics of constituent materials, the microstructure of the composite, e.g. orientation and amount of the reinforcement, as well as processing conditions, result in variety of composite's properties that enable various potential applications (Fig. 8.6). For example, with their good thermal and wear properties ceramic matrix composites are an excellent candidate for high temperature applications and applications where high stiffness is required. CMCs with dispersed particles, whiskers or other particulate reinforcement have been widely used in a variety of structural parts. In contrast, mainly due to the higher complexity usually associated with higher production costs, the use of continuous fibre-reinforced CMCs is limited to a few specific fields.

8.1.3.1 Armours

CFCMCs are being used in armours mainly due to their performance against high impact projectiles and light weight. They usually consist of alternating layers of extremely hard layers, which are designed to shatter an impinging projectile and ductile layers designed to yield under the force of the projectile. Hard layers consist of ceramic matrix produced form pre-ceramic polymer and hardness-producing filler materials (alumina, silicon carbide, silicon nitride, tungsten carbide, chrome carbide, chrome oxide, mullite, silica or boron carbide), reinforced with ceramic fibres (alumina, silicon nitride, silicon carbide, graphite or carbon). Intermediate ductile layers consist of at least one sheet of woven fibres.

When penetrated by the projectile CFCMCs expand in volume, producing a counterforce, which protects the second layer. Such armour has been shown to offer unique protection against high-explosive, anti-tank (HEAT) rounds. Light weight of

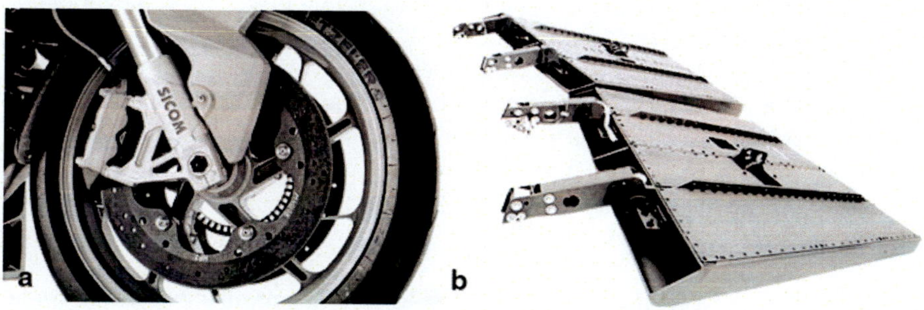

Fig. 8.6 Examples of fibre-reinforced CMC applications: **a** C/SiC (SICOM brake systems -MS Production), **b** CVI-C/SiC body flaps for the X-38, joined with C/SiC screws (MAN-T). [1]

makes it an ideal material for airplane and helicopter protection. Similar composites are also used for body armour as inserts in bullet-proof vests [2].

8.1.3.2 Advanced Friction Systems

The potential for using carbon fibre/carbon (C_f/C) materials in brake discs and pads has been known for a relatively long time. To overcome some of the materials disadvantages, such as a low coefficient of friction at temperatures below 450°C and a high rate of wear for the pads, a short-fibre-reinforced SiC material (C_f/C-SiC) was developed for lightweight brake discs, which were able to withstand such demanding conditions. The C_f/C-SiC composites are usually manufactured by an infiltration process. The most common are Liquid Silicon Infiltration and Liquid Polymer Infiltration. Brake systems using these materials were first introduced in aircraft, and then later in racing cars and motorcycles, high speed trains and emergency brakes for elevators and cranes [12, 13].

8.1.3.3 Automotive Applications

Advanced automotive applications are increasing toward higher temperature requirements. Automotive engines operate more efficiently at higher temperatures and therefore metal parts are being replaced by CMCs based on C_f/SiC, SiC_f/SiC. Initially, a few parts were made from CMC, such as piston heads and exhaust valves, however, as the trend to higher temperature engines continues, more and more metal components will be converted to CMCs [2].

8.1.3.4 Aerospace

The aerospace industry requires structural materials for high temperatures and aggressive environments, where silicon carbide offers great potential as a structural material. For example, C_f/C-SiC composites are proposed to be used as thermal protection systems of spacecrafts, e.g. the nose cap of X-38 crew return vehicle. Composites with much shorter operating times (only few seconds) and considerably higher thermo-mechanical stresses are needed for jet vanes which are used to divert the direction of thrust in solid fuel rockets of missiles. Advanced jet vanes are made of C_f/C-SiC composites and coated with ceramic surface protection, such as CVD-SiC, in order to withstand the immense blast of solid particles [14].

SiC-based ceramic matrix composites are also promising materials for advanced jet engines. They are being used for the fabrication of non-rotating parts in military jet engines, including combustion chambers and afterburner parts (exhaust cones, flame holders, or exhaust nozzle flaps [12].

8.1.3.5 Advanced Nuclear Applications for Energy Production

Most of the advanced energy production systems are concerned by the use of materials under extreme environment. In the advanced nuclear systems such as Generation IV and fusion reactors, the most important criterion for structural material is combination of high thermal stability and sufficient fracture toughness. Following neutron irradiation, the fracture toughness of most engineering alloys significantly drops. This is not the case for engineering ceramics, which, however, have critically low fracture toughness in monolithic form. Therefore, continuous ceramic fibres reinforced ceramic matrix composites seem to be the most promising option and has therefore attracted significant attention of researchers.

Fission Generation IV: Gas Fast Reactor

Fast reactors have superior ability to burn recycled nuclear fuel and hence by closing the fuel cycle by recycling they will reduce quantity and radiotoxicity of nuclear waste and increase uranium fuel utilization. They will be exposed to high temperature, neutrons and corrosive environments and therefore a development of materials with improved properties is needed. Currently, SiC_f/SiC and C_f/C seem to be the only composites of sufficient maturity for short term application; however, composites with a TiC or ZrC in the nanostructured matrix are expected to present better thermal properties.

Fusion

Next to fission reactors, SiC-based composites (SiC_f/SiC) are proposed to be used as structural material in the most advanced concept of the future fusion reactor—for high temperature tritium-breeding blanket (Model D). This is mainly due to the low neutron activation, radiation resistant strength and high heat resistance of the silicon carbide. Hence, the use of SiC_f/SiC composites for the blanket structural component is expected to enable a significant increase in the maximum operating temperature and to consequently contribute to a higher efficiency. Moreover, the material would not decay under neutron irradiation to produce long-lived radioactive wastes.

Continuous SiC-fibre-reinforced SiC-matrix composites are also proposed to be used in the Dual Coolant design (Model C) of fusion reactors as a functional material for the Flow Channel Inserts (FCI). A SiC_f/SiC-FCI would act as an insulating layer to separate electrically and thermally molten Pb-Li alloy from the structural, load-bearing ferritic steel channel walls of the blanket and allow significantly higher coolant outlet temperature. The required properties for SiC_f/SiC-FCI materials are low transverse thermal and electrical conductivity, as well as Pb-Li compatibility and radiation stability [15].

The development of most of the above mentioned structural parts is forced to take into account an economic criterion that favors the use of simple and cheaper techniques. The EPD offers an alternative option that may replace many costly and time-consuming processes in their production.

8.2 Processing Routes for Production of CMCs

The particulate reinforced CMCs can be manufactured by simple blending the powders and/or whiskers into the matrix powder or, using more advanced colloidal processing, the reinforcement is admixed into slurry of ceramic powder in a water or organic liquid carrier.

Conversely, the processing of continuous fibre-reinforced ceramic matrix composites (CFCMCs) is a complex multi-stage process. The fibres (filament tows) are woven in a number of ways into a preform, usually 2-dimensionsinal but ideally 3-dimensional fabric (Fig. 8.7), which is then infiltrated with a ceramic matrix, leading to a certain degree of densification. The composites incorporating two- or three-dimensional fibre reinforcements are particularly prone to exhibiting uncontrolled microstructures and residual porosity. This is because it is extremely difficult to achieve complete infiltration of the matrix material into the fibre tows, where the intra-tow openings may be down to the order of magnitude below 100 nm. To design properties via controlled microstructures, a reliable, simple, and cost-effective processing method is needed to completely infiltrate the fabric with matrix precursors.

For example, SiC-based composites reinforced with continuous carbon fibres (C_f/SiC) are currently prepared by various techniques: chemical vapour infiltration (CVI), liquid or vapour silicon infiltration (LSI or VSI), in-situ reaction sintering, or hot pressing (HP). Liquid silicon infiltration is a fast (short manufacturing cycles), low-cost manufacturing process that involves firstly the preparation of a C/C composite core, normally produced by pyrolysis of a carbon fibre-reinforced preform of the desired geometry, which is then subjected to a liquid-silicon-infiltration (LSI)

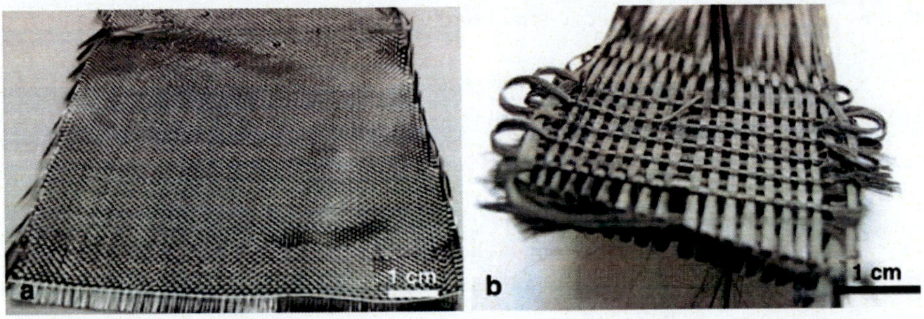

Fig. 8.7 2-D and 3-D SiC-fibre fabric

process based on the impregnation of porous carbon/carbon composite by molten silicon and its reaction to silicon carbide. The operation temperatures are normally in the range of 1600°C (at least beyond the melting point of silicon – 1415°C).

Selection of a suitable technique for production of SiC_f/SiC composites is based upon the target application of the material that determines the structural and compositional requirements for the material. In particular for the most demanding grades, such as that for fusion-applications, special attention is required. Although many methods have been attempted in the past, no adequate solution has yet been demonstrated that produces a low-activation composite with adequate mechanical and thermal properties and low (closed) porosity. The current method of choice for production of SiC_f/SiC composites is chemical vapour infiltration (CVI). This method enables the production of pure SiC with very low neutron activation, but it suffers from some processing drawbacks, notably the very long time required to build up sufficient material for the matrix, resulting in very high costs, and an inability to eliminate the porosity at the micro as well as the macro level (Fig. 8.8). Although for most applications of SiC_f/SiC made with CVI such drawbacks are not crucial, they are critical in the case of the proposed fusion applications, where the porosity must be minimised in order to prevent gas-permeability and to maximise thermal conductivity [16].

Similarly, polymer infiltration and pyrolysis (PIP) that is based on the infiltration of a SiC fibre-preform with a liquid preceramic polymer precursor, also fails to completely fill the gaps between the fibres and hence results in a material with an unacceptable residual open porosity. Attempts have been made to combine the PIP technique with pressure-assisted slurry infiltration prior to the polymer infiltration [17]; yet suitable material properties could not be achieved.

The third main group of processing techniques follows a ceramic route. Due to covalent bonding being responsible for the extraordinary properties of the SiC, the densification represents a demanding task. It is achieved either by solid-state sintering using boron as an additive, which is undesirable in fusion-relevant materials, or liquid-phase sintering using metal oxides as the sintering additives, typically Al_2O_3-

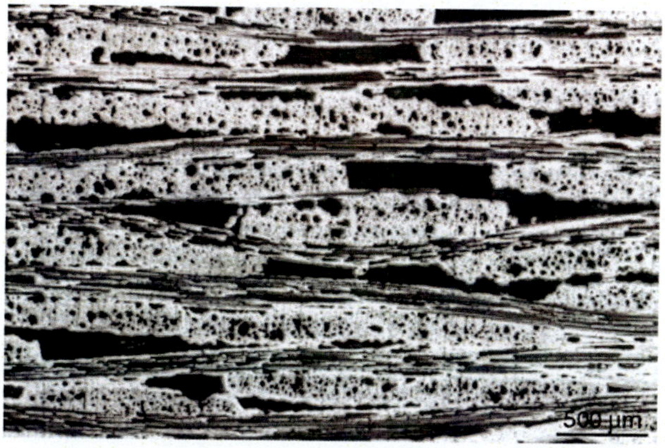

Fig. 8.8 Optical micrograph of polished cross-section of the SiC_v/SiC (CVI) composite

Y_2O_3. The temperatures needed for both the solid-state and liquid-phase sintering, even if combined with high pressures, exceed the temperature range that the present grades of SiC-fibres can withstand without serious damage. The pressure-less high-temperature densification typically performed above 1900°C, is also associated with an undesired β- to α-SiC transformation and with certain shrinkage that limits the possibility to prepare a gas-impermeable composite material.

Apparently, one of the most developed techniques for production of SiC_f/SiC composites is the so-called "NITE" process (nano-powder infiltration and transient eutectoid) [18], which may, according to the literature data, produce a material with a low porosity and a relatively high mechanical strength. In the NITE process, alumina and yttria are used as the sintering aids. In this process, the carbon-coated SiC fibre-fabric is first impregnated by polymer precursor containing nanoparticulate SiC, alumina and yttria powders. In this stage the SiC and oxide particles are introduced into narrow gaps between the fibres in the tows. After polymer pyrolysis, the preform is infiltrated by dipping into a slurry of SiC powder with sintering aids and final densification is achieved by hot pressing.

In the recently introduced "SITE" process (slip-infiltration and transient eutectoid) [19] the fibre preform is infiltrated by aid of electric field. In the next step, the green part is infiltrated with a liquid precursor for the subsequent thermal densification. The electrophoretic infiltration used in this process will be described in more details later in this chapter.

In addition, a few hybrid techniques have also been reported. In numerous trials to fully infiltrate the large voids between the fibre tows as well as the narrow gaps between fibres within the tows, a variety of approaches have been used. In order to avoid the time-consuming and expensive CVI process, they usually involve infiltration of the fibre fabric with ceramic slurry. However, neither the overpressure nor the vacuum infiltration led to the filling of the voids throughout the fabric thickness. Thus, the electric-field assisted infiltration has opened a new opportunity for improved and inexpensive production of CFCMCs.

8.3 EPD in Production of CMCs

Electrophoretic deposition of monolithic and particulate composite materials has been intensively explored for decades. Lately, it has been also recognised as a novel, simple and inexpensive method for fabricating particulate as well as fibre-reinforced ceramic composites with a great potential of achieving complete infiltration of tightly woven fibres performs. The main advantages of EPD over conventional processing routes are mainly the reduced processing times and improved control over green body microstructure. Aqueous or non-aqueous suspensions of ceramic submicron- or nano-sized particles are considered for forming the matrix or particulate reinforcement in composites; both conductive (e.g. SiC, carbon and stainless steel) and non-conductive (mullite) continuous fibres have been used to produce ceramic matrix composites.

In shaping of ceramic matrix composites by means of EPD more parameters have to be taken into account than in the case of monolithic materials. For example, the particles in suspensions for particulate CMCs may differ in shape, size and composition, may have different surface net charge and therefore they may move with different rates resulting in inhomogeneous composition of deposit. Further, in production of continuous fibre-reinforced CMCs, an electrically conductive fibre fabric may play a role of an electrode that is not always desired as it will be presented later. In the following section, the major parameters affecting the EPD in production of various CMCs will be presented. As the CFCMCs represent a greater challenge, a large part will be devoted to these composites.

8.3.1 Suspensions for EPD

With regard to the properties of suspensions for production of CMCs, a variety of parameters must be considered, such as the physicochemical nature of both suspended particles and the liquid medium, surface properties of the powders, and the influence of the type and concentration of the additives, mainly dispersants. In order to achieve the best possible result, these parameters should first be verified in EPD of bulk deposit as was in the case of electrophoretic infiltration.

8.3.1.1 Liquid Media

The selection of the liquid media, sometimes also called a "solvent", is usually the basic decision that determines most of the further parameters. Namely, a solvent must dissolve the surfactant, but not the powders, must provide sufficient conductivity to suspension (high dielectric constant is preferred), and must wet the particles in suspension well.

A variety of solvents have been successfully used in EPD. The effect of a solvent's polarity on adsorption of an anionic surfactant and consequently on ζ-potential of titania was studied by Kosmulski et al. [20] who presented that depending on the nature of the solvent, the same surfactant may adhere on the particles in two different ways resulting in positive or negative surface charge. Ethanol appears in literature to be a quite frequently used liquid in EPD. Its main drawback is the hygroscopicity that makes it difficult to control water content. Notably, a small amount of water may significantly alter the overall properties of ceramic suspension that hinders control over the EPD process [21].

Due to a high dielectric constant, water may result in relatively high ζ-potential values (usually higher than that in ethanol, Fig. 8.9) and, in addition to its more environment friendly nature is the preferred medium in EPD. In water, however, gas formation due to electrolysis must be considered. There are at least two ways of avoiding the problem of bubbles formation: using a membrane placed between the electrodes or using an electrode that will absorb the gas formed during electrolysis.

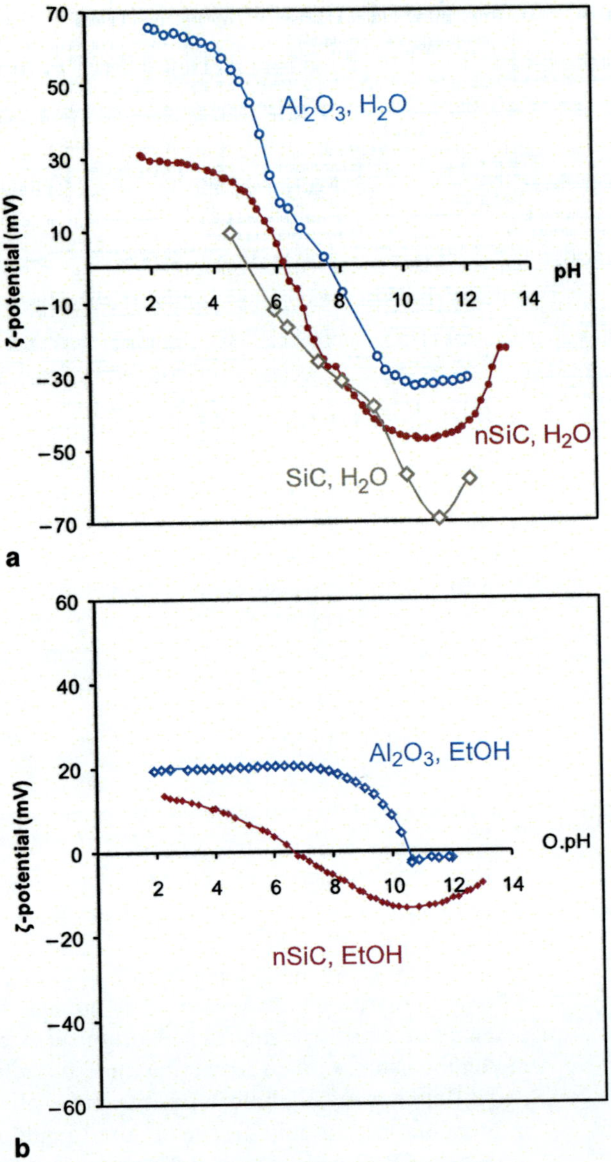

Fig. 8.9 ζ-potential of nano-sized Al_2O_3 and SiC in **a** aqueous and **b** in ethanol suspensions as a function of pH or operational pH (O.pH), respectively (pH is adjusted by HCl or NaOH)

8.3.1.2 Powder

One of the main goals of EPD is usually to achieve high particle packing density which is prerequisite to obtain a dense sintered material. For monodispersed spheres the maximum possible density is 74 vol. %; this may be further increased by the addition of smaller particles that fill the remaining voids. In a real world, i.e. in ceramic processing, particles typically do not have a regular shape and size distribution. Therefore, the packing density may significantly vary due to several parameters.

Conventional processing (dry pressing, slip-casting, injection moulding) usually result in up to 60% total density (TD). In EPD, favourable particle shape and size distribution in combination with high ζ-potential can result in relatively high densities of deposits. Literature review indicates that green densities well above 60 vol.% can be achieved, with the highest reported value obtained by EPD of bimodal particles size distribution as high as 84%TD [22]. In EPD of composites, beside the particles size the nature of the involved particulates is of decisive importance. The particulates may significantly differ for example in surface charge that may result in heterocoagulation.

8.3.1.3 Dispersants

To attain high stability of the colloidal suspension for EPD and to increase the particles mobility, high net-surface charge is needed. In many cases, highly positive or highly negative ζ-potential can be achieved simply by pH adjustment to more acidic or alkaline region, respectively. In some cases, the reactivity or solubility of solid particles at low and high pH values must be avoided. Therefore, the addition of an appropriate amount of charge-controlling agent—ionic surfactant (dispersant, deflocculant) is usually a better option.

Numerous commercial dispersants are available at the market among which the number of anionic dispersants is much higher than that for the cationic. The selection of a suitable dispersant depends on the liquid media used and the desired pH range of the suspension for the EPD. In general, a good dispersant is characterised with high solubility and high dissociation rate in liquid media. Water soluble surfactants show limited solubility in hydrocarbons and vice versa. For example, Na-dioctyl-sulfosuccinate (SDOSS) is only weakly soluble in water but is soluble in a wide range of organic liquids from methanol to hexane. As presented by Kosmulski et al. [20], SDOSS in polar or non-polar liquids results in opposite surface charges and, therefore, can be used either as cationic or as anionic surfactant.

In a composite suspension, a surfactant capable to effectively disperse all the present solids phases must be used. As a result, precisely determined optimal amount of a suitable surfactant yield both in the highest possible absolute value of ζ-potential and, accordingly, stable colloidal suspension. This assures that individual particles rather than agglomerates travel toward the oppositely charged electrode in the EPD cell. Figure 8.10 presents an example of SiC aqueous suspensions with addition of cationic or anionic deflocculant. It is evident that both result in high absolute values of ζ-potential, while the selection of the most suitable one depends on the desired particles polarity that will be discussed in continuation.

8.3.1.4 ζ-potential and Conductivity

Although many reports and reviews on the desired properties of suspensions for electrophoretic deposition and electrophoretic infiltration (EPI) have been published, no general rule has yet been agreed. It has been generally accepted, however,

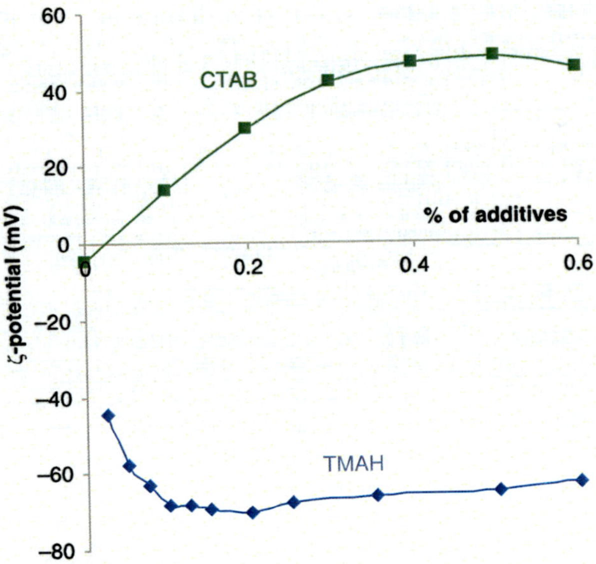

Fig. 8.10 ζ-potential of aqueous SiC suspensions as a function of the addition of cationic (CTAB—Cetyl-tri-methyl-ammonium bromide) or anionic (TMAH—Tetra-methyl-ammonium hydroxide) surfactant

that the most important characteristics of a suspension for EPD and EPI are high net-charge on the particles surface, suitable conductivity of the suspension, its high stability and low viscosity. To attain suitable combination of these properties, each particular system requires individual study.

Typical point-of-zero-charge (PZC) values for numerous oxide and non-oxide materials can be found in literature [23] that in principle enables at least a rough estimation of the particle's polarity in acidic and alkaline pH regions. As the ζ-potential determines the stability and viscosity of colloidal suspensions, in some cases it can be roughly estimated also by observing time-dependent sedimentation or by measuring the viscosity (the relative viscosity exhibits the minimum value at the highest absolute value of ζ-potential and vice versa). However, none of these is sufficient in preparation of suspension for EPD, in particular not in the case of heteroparticulate suspensions in EPD of composites. Only a direct and reliable measurement of the particles electrophoretic mobility or ζ-potential of the suspension to be used in EPD adequately and quantitatively describes the suspension.

Since EPD is an electrically driven process, the electrophoresis seems to be the most relevant component of the technique. This is true when very low-concentrated suspension is used. Such suspensions are, however, rarely used in production of CMCs, except for the coatings. The analysis of suspensions with relatively high powder loading are the best described by electroacoustic measurements.

Figure 8.11 shows comparison of the ζ-potential vs. pH curve for a SiC-powder suspension in water using two different principles: the electrophoresis and electroacoustic techniques. As is evident, a relatively wide range of ζ-potential values are obtained by measuring the same powder in aqueous suspensions. In the case of measuring electrophoretic mobility, the suspension must contain a low particle concentration to allow the laser beam to detect their movement properly. The dilution of the suspension re-

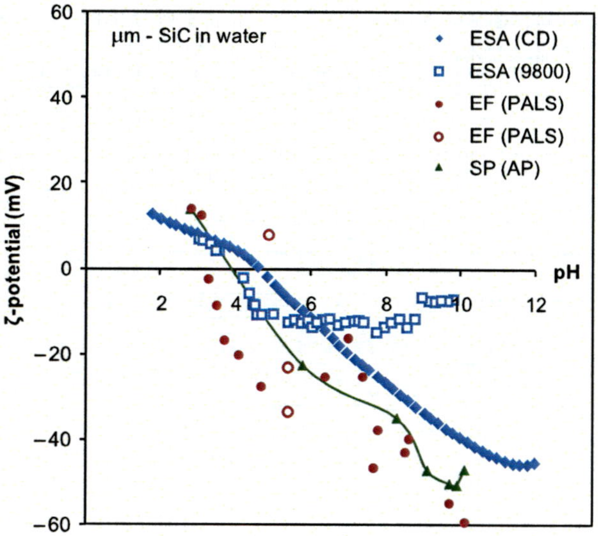

Fig. 8.11 pH related ζ-potential of a SiC-powder suspension determined with different techniques (*EF:* electrophoresis, *ESA:* electro-kinetic sonic amplitude, *SP:* streaming potential)

sults in significant pH change; therefore, to determine the ζ-potential of the suspension at its natural pH, an acid or base must be added. Hence, the ionic strength is increased. On the contrary, the electroacoustic technique enables direct characterisation of actual suspension for EPD without dilution, pH adjustment, or addition of an electrolyte.

Conductivity of the suspension is another important characteristic of suspension, which affects not only the deposition rate but also the quality of the deposit, i.e. particle packing density in deposit and its visco-elastic properties (firmness). Conductivity is significantly influenced by the intrinsic properties of the liquid media (dielectric constant), on the amount and properties of the powder and in particular by the amount and properties of (poly)electrolytes added. Conductivity is usually significantly higher in aqueous than in non-aqueous suspensions. Additionally, the results of the analysis of submicron and nano-sized SiC powders show that the ζ-potential is not significantly affected by the particles size (Fig. 8.12a), while the difference in conductivities appears larger. As presented in Fig. 8.12b, the conductivity of nanoparticulate suspension is significantly higher than for the submicron or mixed suspension. Excess acid or base also causes an abrupt increase in conductivity.

The addition of surfactants, i.e. (poly)electrolytes, can cause significant effects in the conductivity. Even with small addition, the suspension conductivity usually remains constant or slightly decreases until the complete adsorption of the electrolytes onto the particles' surfaces has been reached. Afterwards, the conductivity may increase with increasing amount of free molecules in suspension. This was observed for alumina and silicon carbide suspensions with addition of polyethylene-imine (PEI) or citric acid as surfactants, respectively (Fig. 8.13a, b). In both cases the onset of the conductivity increase nearly coincided with the optimum amount of the surfactant.

While the ζ-potential does not practically depend on solid loading in suspension (blue curve in Fig. 8.14a), the conductivity typically increases with solids loading

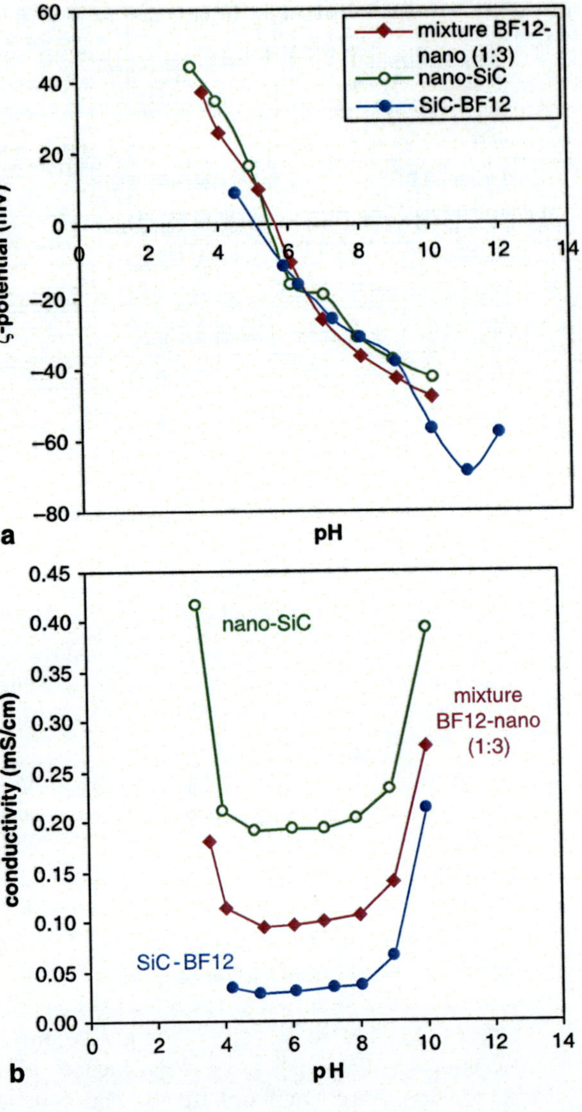

Fig. 8.12 The comparison of ζ-potential **a** and conductivity **b** vs. pH for 5 wt.% aqueous suspensions of two SiC powders with average particle size of 50 nm (nano-SiC) or 500 nm (SiC-BF12), respectively

in suspensions of conductive (e.g. SiC) as well as of non-conductive particles (e.g. alumina) (Fig. 8.14a and 8.14b). These characteristics were observed for suspensions at the optimum surfactant addition.

Hence, the conductivity of the suspensions is determined by several parameters; therefore, its values may vary in a very broad range. In contrast to ζ-potential (absolute) values that are considered as "high" above 60 mV and "low" below 20 mV, there is no recommended "appropriate" conductivity level that characterise a suspension suitable for the EPD. Hence, the conductivity must be determined for a particular system.

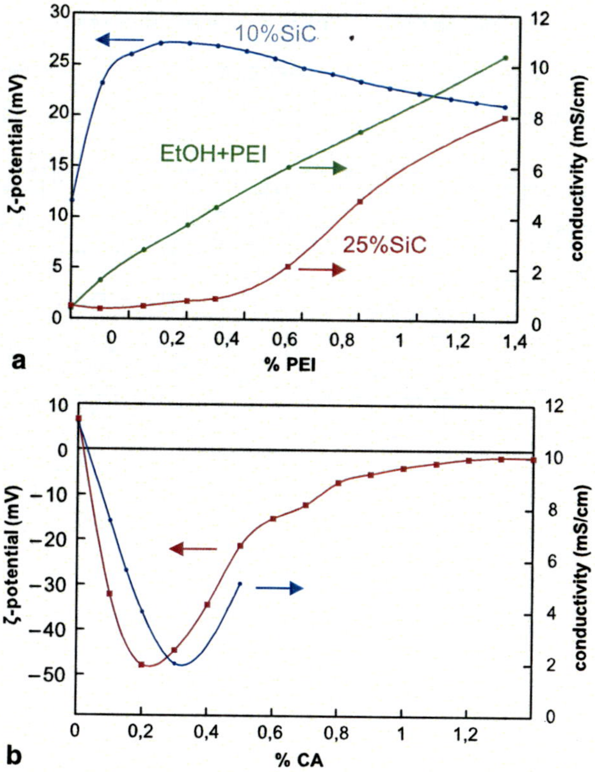

Fig. 8.13 Conductivity and ζ-potential for SiC and Al_2O_3 suspensions as a function of surfactant addition

The effect of conductivity of alumina-ethanol suspensions on EPD was studied by Stappers et al. [24], who claimed that high conductivity suspensions result in more uniform deposits than the low conductivity suspensions. This was attributed to the decrease in electric field over suspension as a result of the alumina deposit resistance. The increasing density of deposits with increasing conductivity of suspensions was observed also in the case of SiC, but only when the conductivity increased due to the increased solids content. In contrast, for the suspension with a certain solids loading the density of deposits decreased when the higher conductivity was caused by surplus of electrolyte or by lowering the particles size.

Numerous studies showed that too high a suspension conductivity does not lead to the formation of a firm deposit; naturally, an optimum value exists. Ferrari and Moreno [25] suggested that the conductivity was a key parameter for EPD. This was in part confirmed by Novak et al. [26] who analysed the effect of dispersant addition into alumina suspension in water and in ethanol. As expected, due to lower dielectric constant of ethanol, the values of the ζ-potential/conductivity ratio were two or three orders of magnitude larger than those for the aqueous suspensions (Fig. 8.15). Although the maxima appeared at different levels, they indeed corresponded to the highest densities of deposits.

Fig. 8.14 ζ-potential and conductivity of SiC and Al_2O_3 suspensions as a function of volume fraction of the powders in suspensions

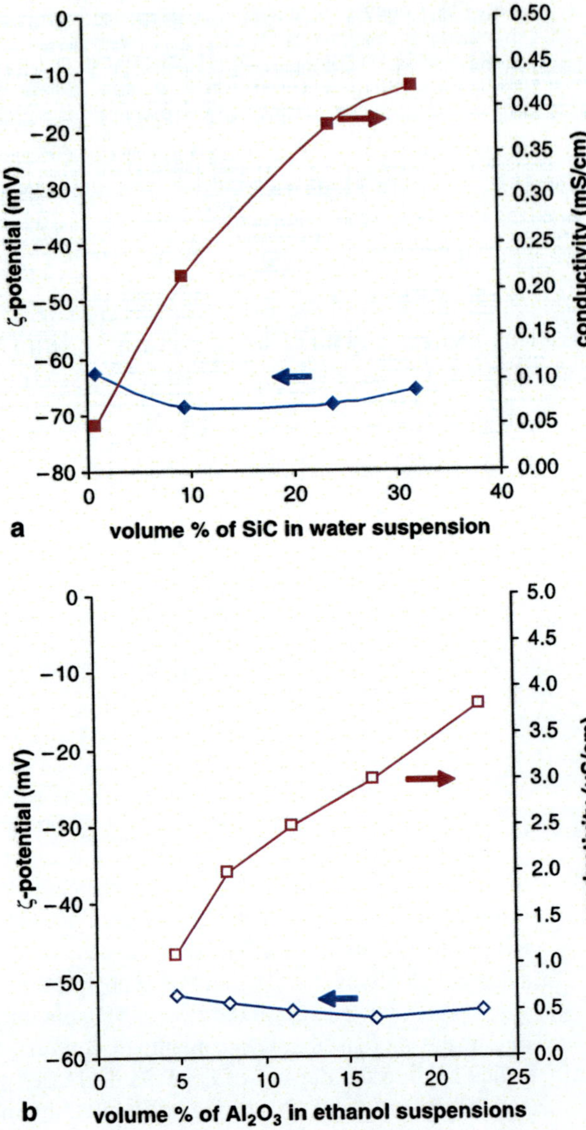

8.3.2 EPD Parameters

8.3.2.1 Electric Field Strength

The range of applied voltages during EPD, reported in literature, is rather wide, from millivolts to hundreds of volts. No general rule has been agreed on their affect, except on the well known linear increase in deposition yield in accordance to the

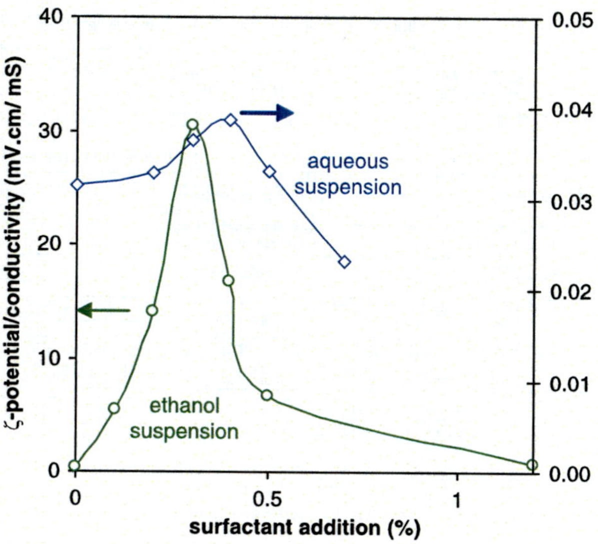

Fig. 8.15 ζ-potential vs. conductivity ratio for aqueous and ethanol alumina suspensions as a function of a dispersant (citric acid) addition

Hamaker equation. High voltages are usually needed for deposition from suspensions in organic liquids with low dielectric constant, while in water relatively low electric field strengths are also highly efficient.

An illustration of particles assembly during deposition at different voltages is presented in Fig. 8.16 showing monodispersed silica nanospheres with average diameter of 400 nm deposited on steel electrode from ethanol suspension in the range of field-strength 2.5–32 V/cm. The observation of the spheres array suggests that lower voltages enable better assembly and hence higher particle packing densities [27].

On the other hand, deposition of irregularly shaped polydispersed SiC- or Al_2O_3-particles of approximately the same average size reveals that the effect of applied voltage on particle packing density in the deposits is minor. As presented in Fig. 8.17, in the range of 5–35 V/cm, the densities of the submicron SiC and Al_2O_3 deposits are only approximately 60% TD. It is also evident from the graph that the EPD of well dispersed nanosized alumina and SiC powders resulted in even much lower green densities reaching only up to 30% TD.

8.3.2.2 Electrodes

For successful deposition, the properties of the electrode material have to be considered as well. Namely, the electrochemical reactions at the electrodes decrease the efficiency of the process. Polarisable electrodes result in significant potential drop near the electrode; therefore, non-polarisable materials are preferred. Graphite seems to be a good option in particular in the case of aqueous systems as it prevents formation of bubbles in the deposits. Evident from Fig. 8.18, the use of graphite

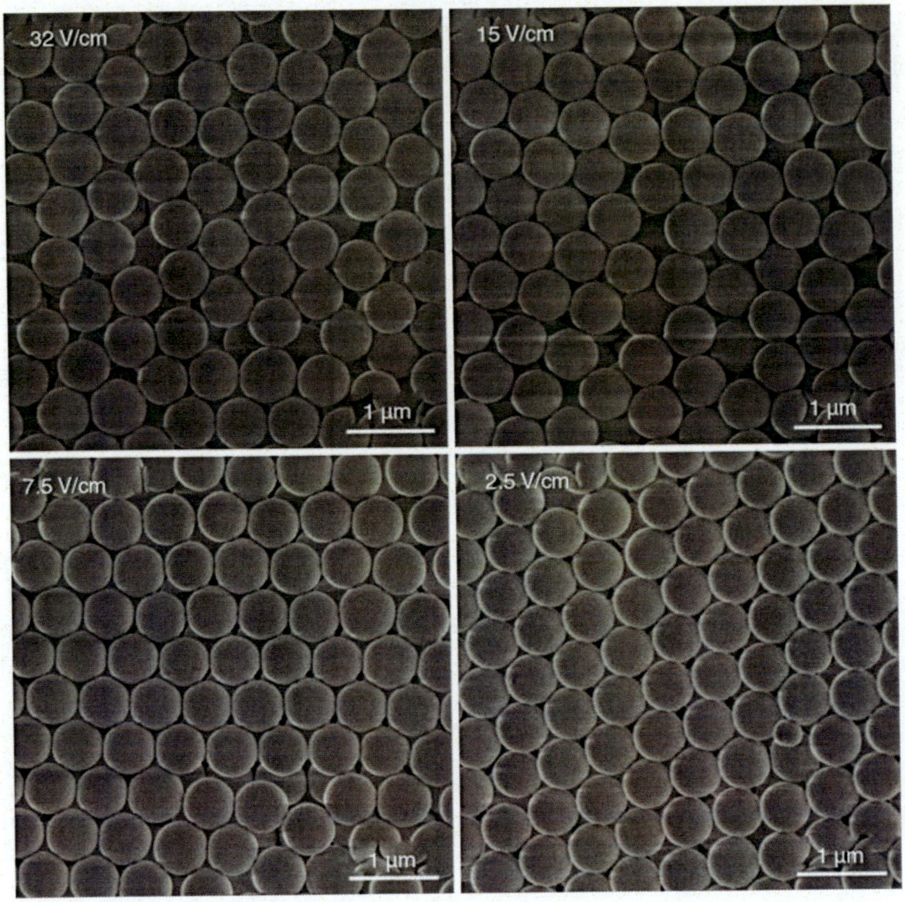

Fig. 8.16 The assembly of monodispersed silica spheres (d=400 nm) obtained by EPD of ethanol suspension at different voltages. [27]

electrode instead of stainless steel resulted in a complete elimination of pores in the thick SiC deposit. Some metals, e.g. Cu, Zn and Pd can also play a role of gas "absorbers" to prevent porosity of deposit due to water electrolysis; however, contamination of the deposit must be considered. This is, for example, in particular important in preparation of composite proposed for fusion application, where the contamination with any metal that could contribute to increase of neutron activation of the material during operation must be excluded.

8.3.2.3 Solids Content

A wide range of solids content is reported in literature to be used in EPD: from as low as 0.01 up to 40 vol.%. According to Hamaker's equation, deposition yield

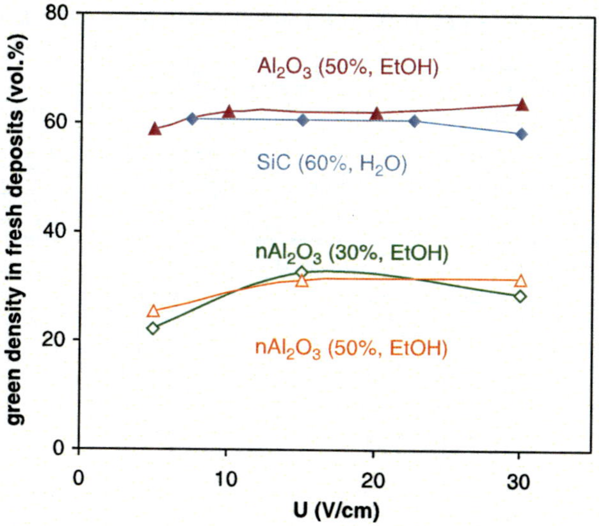

Fig. 8.17 Green density of nano-sized (nSiC, nAl_2O_3) and submicron Al_2O_3 and SiC deposits as a function of applied voltage

increases linearly with solids loading in suspension, which has been generally confirmed in practice. However, many researchers have noticed that in some systems no deposit is formed below a definite powder concentration independently on the applied voltage or deposition time. This phenomenon has been described by Radice et al. [28] who observed the threshold concentration for titania and polystyrene spheres below which the deposition yield does not follow the linear trend dictated by Hamaker's equation. Similar effect but at larger powder concentrations was observed also by the authors of this work: In EPD of aqueous SiC-suspension, no deposit was obtained from suspensions with less than 5 wt.%, while weak non-stable deposits were obtained from suspension containing below 10 wt.% powder. Above this concentration, green density of SiC deposits increases by increasing the solids content in suspension containing up to approx. 50 wt.%, where the highest green density was obtained (Fig. 8.19). The suspensions with such relatively high solids

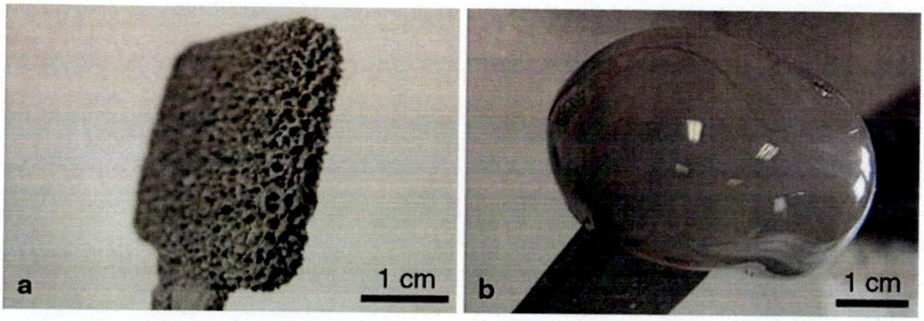

Fig. 8.18 Images of the SiC deposits prepared from aqueous suspension at steel **a** and on graphite **b** electrodes under the same EPD conditions

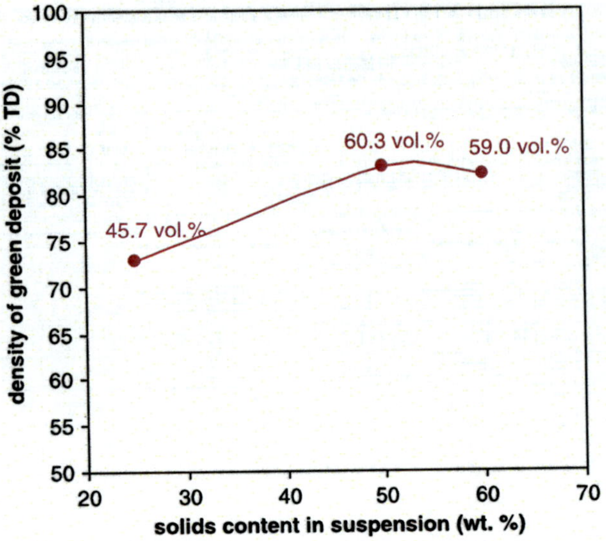

Fig. 8.19 Green density of SiC deposits as a function of solids loading in starting suspensions (aqueous suspension with TMAH addition)

loading were also used for electrophoretic infiltration of SiC-fibre perform and, as it will be presented later in this chapter, the result was also favourable.

8.3.3 Properties of Deposit

The apparent firmness of a deposit is the first observation after pulling the deposit from the EPD cell. Although this qualitatively describes a particle packing density does not sufficiently characterise the deposit. While for the monodispersed spheres assembly, the observation of the microstructure reveals the quality of the deposit, in ceramic systems, a more quantitative analytical tool is needed. Many reports refer to the density of green deposit measured by Archimedes method. Due to the open porosity, the sample must be either dipped into non-wetting liquid, or must be coated before measurement with a layer that prevents the liquid from penetrating the pores. While this is a relatively useful technique for analysis density of monolithic deposits, it is less accurate in the case of composite deposits, since due to the possible difference in deposition rate of chemically or structurally different particles the theoretical density is not exactly known. Hence, in this case, the percent of theoretical density preferably determined on the basis of the examination of the wet and dry deposit.

The evaluation of the deposit quality by measuring the fraction of powder in as-prepared, i.e. wet deposit can be done by weighing the deposit just after deposition and after drying. The weight loss equals the mass difference of the liquid suspension media after accounting for the solid content in the as-prepared deposit. This is a relatively good figure of merit for its quality and has been, after confirming good agreement with results of other techniques, accepted as the preferred technique used

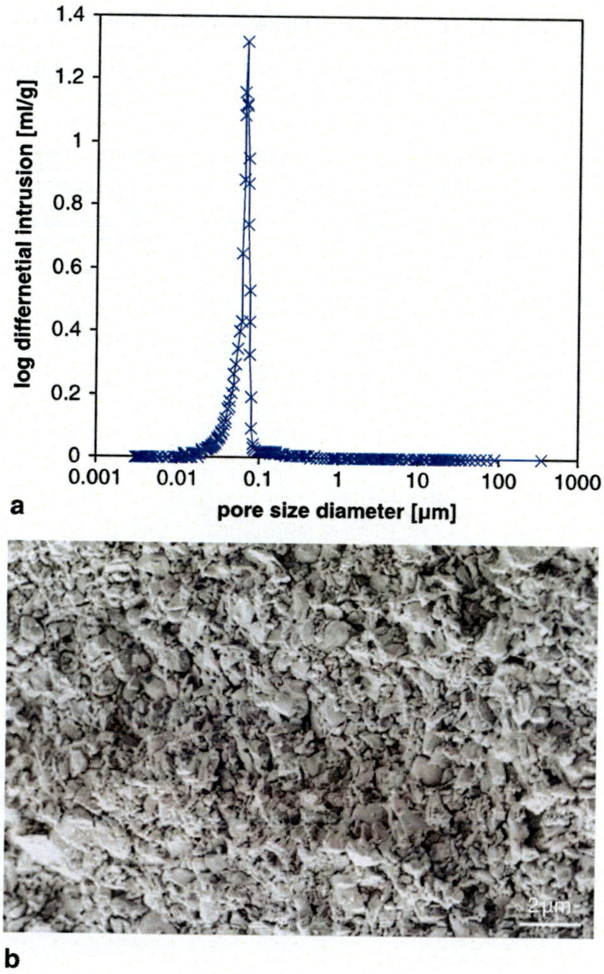

Fig. 8.20 **a** Pore size distribution of SiC deposit determined by high-pressure Hg-intrusion porosimetry; **b** SEM micrograph of fracture surface of the deposit

by the authors of this particular chapter. Hence, the "density of deposit" presented throughout this chapter is determined by measuring the solids content in wet deposit. The most reliable description of the quality of deposit is given by its porosity. As shown in Fig. 8.20, the deposition of aqueous SiC suspension containing 50 wt.% solids resulted in deposits with very small pores with average size of 70 nm and narrow size distribution.

8.3.4 Homogeneity of Composite Deposits

Figure 8.21 shows a fracture surface of a composite Al_2O_3-SiC deposit prepared by co-deposition of alumina and silicon carbide powders with approximately same

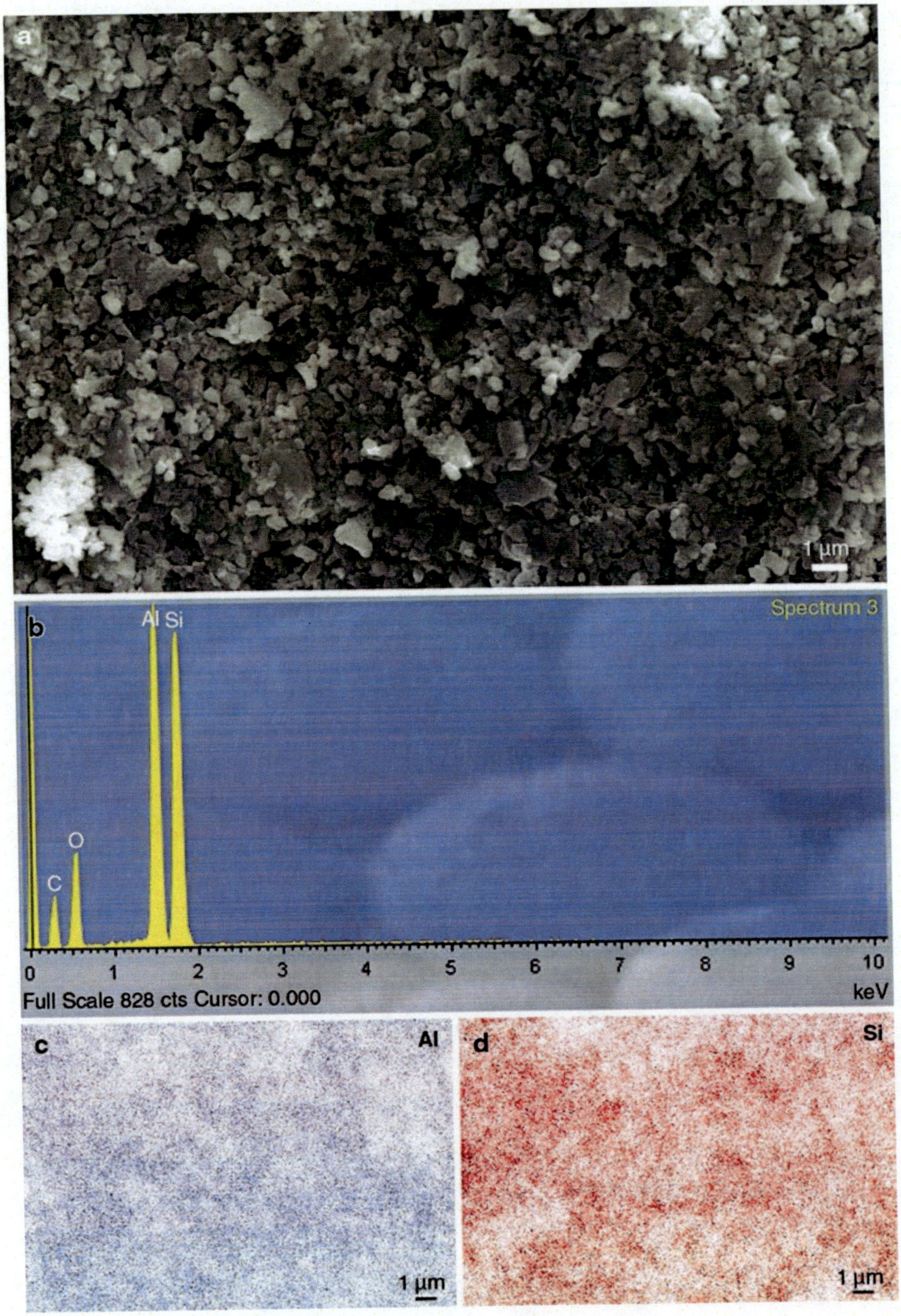

Fig. 8.21 FESEM micrograph **a** and EDXS **b** of a fracture surface of a composite deposit prepared by co-deposition of Al_2O_3 and SiC powder in aqueous suspension at pH 10 (EPD: 30 V/cm) elemental mapping of **c** Al and **d** Si

particles size in aqueous suspension. The suspension for the EPD experiment contained equal volume percents of the powders. Based on the results of the ζ-potential vs. pH behaviour shown in Fig. 8.9, the suspension with pH ~ 10 is presumably the most suitable for co-deposition since at this pH both powders posses highly negative surface charge. From quantitative standard-less EDXS analysis of the sample presented, we estimated that the concentrations of Al_2O_3 and SiC were similar, indicating that both types of particles travelled with the same rate and, hence, they formed a deposit with equal volume ratio of the powders as in the starting suspension (Fig. 8.21b, c, d). The analysis of aluminium and silicon distribution also confirmed the homogeneous distribution of the powders in green deposit.

8.3.5 Efficiency of Infiltration

In contrast to electrophoretic deposition, the efficiency of infiltration is difficult to quantify and is even tricky to illustrate. Since the reinforcing fibres resis the breaking of the infiltrated green parts to observe the fracture surface, the infiltrated part is frequently presented in SEM or optical microscopy images of polished epoxy resin-infiltrated cross-sections of green parts. However, this is sometimes misleading, since the resin perfectly wet the green part and fills the gaps and voids remaining after infiltration. Consequently, the apparent density estimated by observation of such a microstructure can be overestimated. This is illustrated in Fig. 8.22a–d, where the optical and SEM micrographs of the same region of infiltrated SiC_f/SiC are presented. The comparison of Fig. 8.22a and 8.22b indicates that polarised light microscopy reveal some resin-reach parts of microstructure where the amount of ceramic particles is quite low. In polarised light the SiC particles appears white, while the resin is almost black that helps to distinguish between ceramic from resin-infiltrated areas. As illustrated in Fig. 8.22c and 8.22d, back-scatter SEM results in even more pronounced effect. In general, a good option for realistic and correct presentation of the efficiency of fibre fabric infiltration is by observation of fracture surfaces of sintered composites. However, fracture surfaces are relatively difficult to prepare due to the intensive crack deflection at the fibres and consequently extremely rough surfaces.

8.4 EPD in Production of CFCMCs

As discovered at the beginning of this chapter, among the major requirements in the production of continuous fibre-reinforced CMCs are complete infiltration of the fibre preform and coating of the fibres with an interphase layer. EPD can be employed both to form protective coatings on the fibres and to infiltrate the preform with a matrix material. Several techniques have been used to introduce ceramic matrices into fibre-fabric performs. As mentioned before, the main drawbacks of

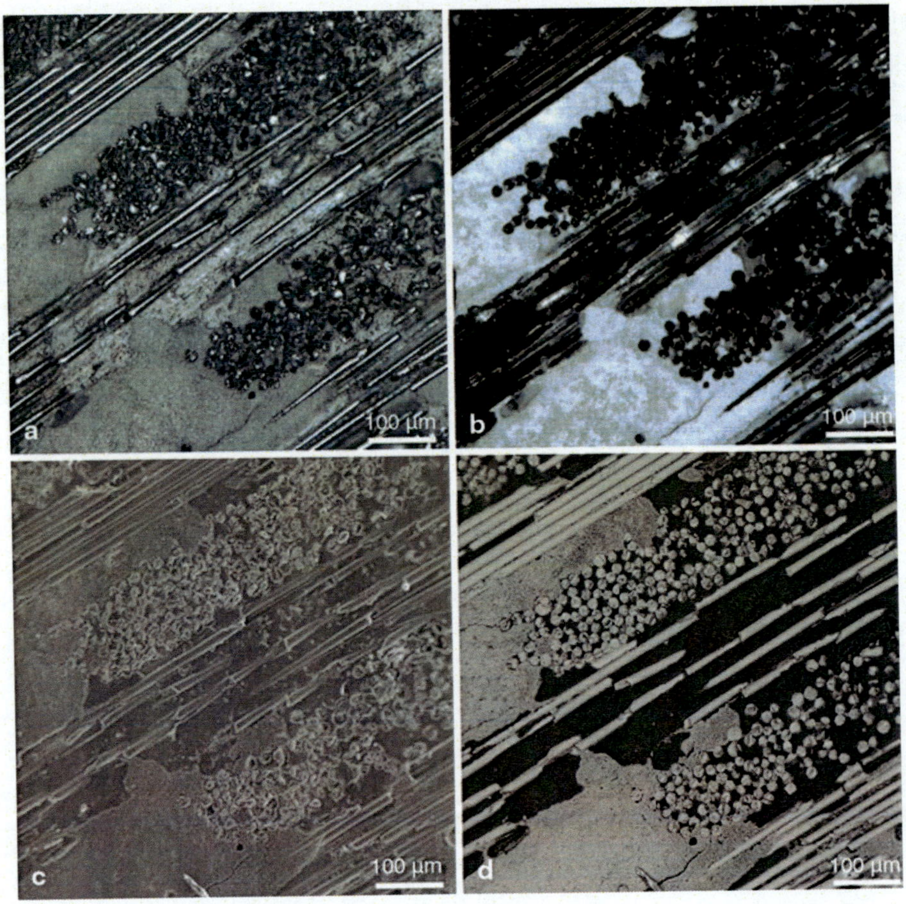

Fig. 8.22 Optical and scanning electron micrographs of epoxy resin-infiltrated and polished SiC_f/SiC composites taken by: **a** optical light microscopy; **b** polarised light microscopy, crossed prisms; **c** secondary electrons SEM and **d** back-scattered SEM

chemical vapour infiltration (CVI) and polymer impregnation technique (PIP) are their inability to completely fill the large voids between fibre bundles and to produce sufficiently dense composites. In the case of CVI, the required price and time needed are great drawbacks of an otherwise useful technique. Further, hybrid techniques that involve high- or low- pressure-assisted powder infiltration have relatively low penetration efficiency; hence, it is difficult if not impossible to obtain a dense green body.

Generally, it can be stated that it is extremely difficult to achieve complete infiltration of the matrix material into the fibre tows where the gaps between the tows may be in the range of submicron size. In spite of that, feasibility of fibre fabric infiltration by aid of electric field has been demonstrated for a variety of fibres, such as carbon, silicon carbide, alumina, mullite, etc. A literature survey suggests that the use of EPD in production of continuous fibre-reinforced ceramic-matrix composites (CFCMCs) has been rather limited. Most of the research experimented with two-dimensional (2-D) fabrics, and only a few reports were published on infiltration of 3-D fabric or coating of fibres by means of EPD. Therefore, this technique remains an attractive, but not well resolved, research field.

The key parameters affecting the efficiency of the infiltration can be classified into three main groups: (1) Parameters related to the properties of the suspensions (solvent properties, ζ-potential and conductivity, solids content in the suspension); (2) Parameters related to the inherent properties of the fibres (electrical conductivity, perform architecture, surface charge of the fibres at working pH); and (3) Parameters related to the process (applied voltage, time, cell configuration, vacuum-degassing).

Several requirements for successful infiltration of the fibre fabric exist, for example the selection of a suitable liquid and particle surfactants to achieve sufficient electrical net-charge at the particle surface to facilitate the particle's penetration and efficient void filling within the fabric. The resulting deposit (matrix) within the fibre mats must exhibit high particle packing density, which is connected to the absence of bubbles and highest possible homogeneity.

8.5 Electrophoretic Infiltration (EPI)

In addition to the general requirements for the production of high quality monolithic or particulate CMC deposits, other conditions must be met for fibre-reinforced CMCs: (a) the conductivity of the fibres must be considered and the cell configuration adjusted to the desired process (coating or infiltration); (b) the surface charge of the particles must be adapted accordingly; and (c) the ceramic particles size must be low enough to penetrate the gaps between the fibres within tows. Theoretical analysis of the influence of liquid media on penetration of particles into porous substrate (valid also for the voids between the fibres) was performed by Haber and Gal-Or [29]. They found that the penetration is enhanced at large values of Peclet number, Pe, which is a dimensionless parameter that in fluid dynamics determines the relative significance of the convective process vis-à-vis the diffusive process:

$$P_e = U \cdot b / D \tag{8.1}$$

where U is the particle velocity (in this case electrophoretic velocity [cm/s]), b is the mean pore radius [cm], and D the diffusion coefficient of the particle [cm^2/s]. The higher is electrophoretic velocity U of the particle and the larger the pore size, more efficient penetration of the particles is expected. The electrophoretic velocity U is proportional to electric field strength E [V/cm].

$$U = \mu \cdot E \tag{8.2}$$

The electrophoretic velocity per unit applied electric field is electrophoretic mobility μ [cm^2/sV], which is proportional to the ratio of the dielectric constant to the vis-

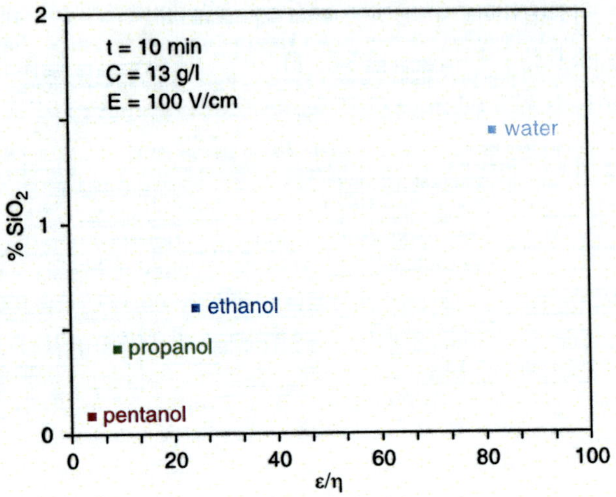

Fig. 8.23 Influence of ratio dielectric constant to viscosity (ε/η) of the liquid media on amount of penetrated SiO_2 particles into porous graphite substrate (Reprinted in part with permission from J. Electrochem. Soc., 139, 1078(1992). Copyright 1992, The Electrochemical Society). [30]

cosity ratio (ε/η) of the liquid. Figure 8.23 shows that the amount of SiO_2 particles penetrated into porous conductive (graphite) substrate increases with this ratio. This suggests water is the most suitable liquid media for fabrication of CFCMCs.

Beside the ratio of dielectric constant to viscosity of the solvent, Haber and Gal-Or suggested particle concentration and field intensity were very important parameters. They also stated that prevention of surface deposition on conductive fibres would favour an increased extent of the penetration.

8.5.1 Fibres Properties

Fibres interact with primary particle suspension in way that is similar to that of the secondary phase's (particles, whiskers, nanotubes) interaction within a composite suspension. Due to the continuity of the fibre-fabric in the composite, its conductivity also plays an important role. Additional complication in CFCMCs is that better mechanical properties are achieved by using three-dimensionally woven fabrics with highly dense texture (>30 vol.%), which are the most difficult to infiltrate with solid particles. For example, the current grades of SiC-fibre preforms used in production of SiC_f/SiC composites for fusion application contain up to 40 vol.% of fibres with diameter around 10 μm and with gaps between the fibres usually smaller than 1 μm.

8.5.1.1 Architecture of Fibre Mats

The architecture of the fibre mats plays an important role in the infiltration of conductive fabrics, where the probability that the infiltration process stops due to the deposition of the particles onto the front layer is high. Such findings were reported for example by Bao and Nicholson [31] who observed that no particles deposited in the central region of the conductive fibre bundle. They suggested that only single 2-D conductive fibre mats could be efficiently infiltrated, while 3-D mats or multiple 2-D mats resulted in high remnant porosity due to clogging effect. It has to be mentioned that the experiments were performed by using fibre mat as deposition electrode. The significant influence that the architecture and properties of conductive fibre mat on success of infiltration also observed Boccaccini et al. [31], who concluded that stainless steel filters were difficult to infiltrate due to its rigid structure.

Due to the structural complexity of 3-D woven fibre fabrics and the necessity for the particles to penetrate deeper into the spaces between fibres, many researchers used an alternative ways to fabricate 3-D CFCMCs; they infiltrated single 2-D fabrics and, thereafter, compacted them. However, the properties of composites prepared by stacking of 2-D fabrics can not attain the properties of the 3-D woven fabric that are conclusively preferred in most applications.

8.5.1.2 Electrical Conductivity of Fibres

Table 8.1 lists the electrical conductivity of the fibres most frequently used as reinforcement in CFCMCs.

Based on literature review, uniform composites were mostly achieved with non-conductive fibres, e.g. alumina or mullite. For example, as reported by Bao et al. [31], green composite was fabricated by EPI using 1 vol.% submicron alumina suspension in ethanol dispersed with 0.5 wt.% PEI. Alumina particles uniformly infiltrated through the inter-bundle voids to form a well-infiltrated matrix with only a few large pores. The authors also modelled electrophoretic infiltration of non-conductive fibre mats via a porous polyethylene (PE) board with alumina. Their result, illustrated in Fig. 8.24, shows that the interface of the PE board/deposited alumina is clear (the right side of the PE board is attached to the electrode). Alumina particles first infiltrated the inter-connected pores to deposit on the front of the electrode. Then they build backwards through the pores toward the board/suspension interface, i.e. the front pores did not clog as in the case of conductive fibres.

From these reports, one can conclude that the conductive fibre mats, e.g. carbon or silicon carbide, were usually used as a deposition electrode, which is possibly the main reason for the unsatisfactory results. Bao and Nicholson claimed that conductive fibres can not be efficiently infiltrated, especially not 3-D and several stacked

Table 8.1 Electrical conductivity of various fibres

	Electrical conductivity
graphite	3.10^4–2.10^5 Ωm^{-1}
Al_2O_3	$<10^{-13}$ Ωm^{-1}
SiC	80–120 Ωm^{-1}
mullite	$>10^{13}$ Ωm^{-1}
SiO_2	$>10^{10}$ Ωm^{-11}

Fig. 8.24 Porous polyethylene (PE) board infiltrated with alumina particles (Reprinted in part with permission from Bao et al. © 2007 J. American Ceramics Society). [31]

2-D fibre mats, due to the surface deposition that disables further penetration of the particles through the fabric (so called clogging effect) [31]. This was supported by the fact that conductive fibres suffer electric field shielding, which severely interferes with the deposition process. According to their explanation, the particles initially deposit on the surface of the outer fibres. If the deposit resistance is much higher than the suspension resistance, the particles deposit on the outer fibres and their surface becomes insulating. The electric field can now penetrate further and particles deposit on the fibres inside the preform. In the case of large fibres, the outer layer becomes clogged before the internal voids are filled that results in a central cavity. Bao et al also compared the electrophoretic deposition through conductive fibres with electrophoretic infiltration deposition through non-conductive fibres under constant current conditions. Conductive fibre mat acted as a deposition electrode, while non-conductive fibre mat was placed in front of the electrode. In their work the infiltration of non-conductive fabrics was called modified "electrophoretic infiltration deposition".

Lately, carbon and SiC fibre mats, both electrically conductive, became the most desired continuous reinforcement in composites that are proposed to be used under extreme conditions. Therefore, intensive investigations have been performed to enable the usually tightly woven carbon or SiC fabric to be infiltrated with SiC-based matrix. Novak et al have studied the infiltration of the conductive fibre-fabric perform using different experimental conditions [33]. They found that in the case when the SiC-fabric was fixed at the electrode, the fibres acted as an electrode and the dispersed SiC particles attached on the front side of the fabric that stopped the infiltration. The observation of the fibres after the experiment reveals firmly attached SiC particles. On the contrary, if the fabric was separated from the electrode either by a non-conductive spacer (e.g. Teflon) or with a cellulose membrane, the particles penetrated the fabric to reach the electrode (Fig. 8.25b). Also in this case, the central part of thicker fabric remained non-impregnated by particles. This happened because the particles found a way easier than that through the tiny gaps in the fabric, i.e. from the front and back side of the fabric. This was prevented by sealing the fabric, forcing the particles to penetrate through the fabric and to gradually form a deposit within the fabric.

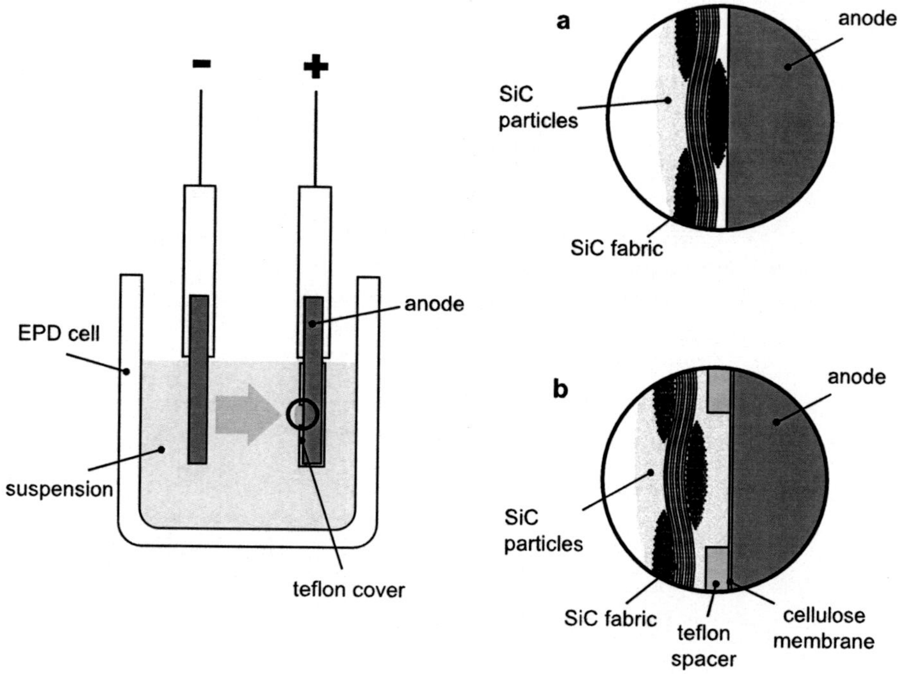

Fig. 8.25 Schematic of the electrophoretic infiltration cell configuration for fabrication of SiC fibre-reinforced ceramic matrix composites

8.5.1.3 ζ-potential of the Fibres

Surface charge of the fibres plays a role in particular during infiltration of non-conductive fibre-mats and in the case of conductive mats when they are not in contact with the electrode. If the fibres and the particles have the surface charges of opposite polarity, the particles are not attracted only by the electrode behind the fibre-mat but also by the fibres themselves. Consequently, the effect may be similar as in the case of the fibres connected to the electrode: the particles deposit onto the fibres on the front side of the fabric causing the infiltration to stop.

Conversely, the particles and the fibres repel each other when they have surface charges of like polarity. As a consequence on their way to the electrode, the particles avoid fibres, but forcefully pass through the fabric to reach the electrode surface and to deposit there. They gradually fill the spaces between the fibre filaments until the matrix extends to the outer fibre preform surface.

In first approximation, we can assume that the fibres posses the surface charge similar to the powder with nominally the same chemical composition. Alumina fibres, for example, exhibit isoelectric point around pH 8–9, the SiC fibres are negatively charged above pH 3, etc. Due to the great importance of the ζ-potential in EPI, the fibres' properties are evaluated instead of using this simplification. The simplest

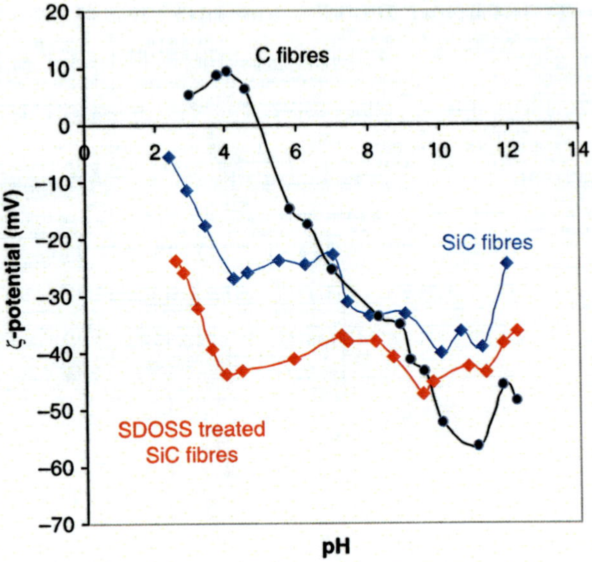

Fig. 8.26 ζ-potential of crashed carbon and SiC fibres in water

way is to analyse the electrokinetic properties of the crumbled fibres. Figure 8.26 illustrates ζ-potential of crumbled carbon and SiC fibres in water as a function of pH. The behaviour of SiC-fibres is quite similar to that for the aqueous SiC-powder suspensions (see Fig. 8.9a). Due to the relatively high repulsion between the SiC particles in alkaline suspension and SiC or C fibres, the particles do not form a deposit on the fibres (if separated from the electrode as described above). The particles do try to find a way between the fibres to reach the oppositely charged electrode.

The importance of mutual repulsion among the charged fibres and powder particles has been observed also by other authors [34, 38]. Kaya explored the applicability of EPD when the conductive fibres and the particles have the same surface charge polarity. Under these conditions, solely dipping of the fibres into suspension would not result in good adhesion between the particles and the fibres [35]. Specifically, in their work on silica particles in an aqueous solution and crushed stainless steel fibres were both negatively charged at the working pH value of 9.5, strongly repelling each other in absence of an external field. By employing electric field, thus forcing the negatively charged silica particles to move and to deposit onto the positive electrode, relatively effective infiltration was achieved.

In electrophoretic infiltration of non-conductive fibre mats, mullite fibres (Nextel) were most frequently used. Often single mats were infiltrated and subsequently pressed together. However, by considering the surface charge of the fibres and the particles, Stoll et al. [36] succeeded in the preparation of the high quality, simultaneous infiltration of multiple mullite fibre mats with alumina particles. Figure 8.27 shows that full intra-tow and inter-tow infiltration was achieved, and a very compact structure with high fibre volume fraction was obtained after simultaneous infiltration.

The authors also presented a phenomenological description of the mechanism of electrophoretic infiltration of oxide fibre mats by ceramic particles exhibiting the

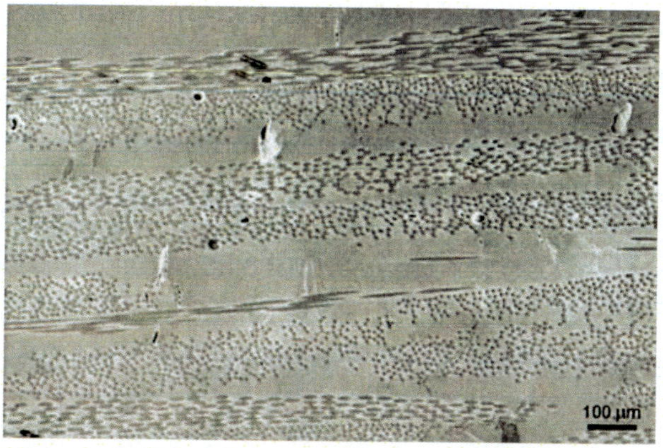

Fig. 8.27 SEM micrograph of a NextelTM 720 fibre-reinforced alumina matrix composite containing four fibre layers (fibre mats) fabricated by simultaneous electrophoretic deposition. The sample was sintered at 1300°C for 1 h (Reprinted in part with permission from Stoll et al. © 2006 J. American Ceramics Society). [36]

same surface charge polarity by using mullite fibre mats to reinforce alumina matrix. In their opinion it is important to achieve high electrostatic repulsion between the fibres and particles in order to induce that the charged particles infiltrate the non-conductive mat. The deposit grows in the opposite direction of particle motion until the non-conductive fibre mat is fully infiltrated. As presented in Fig. 8.28, they proposed that the total trajectory of the particles before reaching the electrode can be divided in two regions: the so-called "approaching trajectory" occurs in the suspension at a given distance away from the electrodes, where the particles move towards the deposition electrode only under influence of the externally applied electric field. The second region, the "infiltration trajectory", occurs close to the deposition electrode where the charge of the fibres influences the motion of the charged particles.

As discussed by Stoll, in the ideal case under external voltage particles in a stable non-agglomerated suspension move simultaneously in random manner, which can be understood as the "approaching trajectory". In the "infiltration trajectory", however, additional interaction forces appear between particles and the charged fibres. When fibres and particles possess the same charge, there are repulsion forces (F1) acting between particles and fibres as shown in Fig. 8.28. For each particle, these forces depend on the relative distance between particles and fibres. Due to the applied external electrical field, each positively charged particle is attracted to the fibre mat attached onto the cathode. However, the particles are repelled before they can reach the fibre surfaces (coagulation point) due to the positive charges on the fibres. The authors hypothesised that under the effect of the repulsive forces due to the surrounding fibres, the particles will follow the path with the fewest possible obstacles until reaching the next interstice between adjacent fibres. Thus, when the particles reach the electrode or the surface of previously deposited particles, they cannot move. So, the electrophoretic ceramic deposit grows with a high particle packing density. Conversely, when fibres and particles have opposite surface charge, coagulation should occur on the first layer of fibres encountered by the travelling particle. Consequently, the formation of a deposit on the outer fibre layer will block or, at least, will hinder the movement of the particles towards the interior of the fibre mat, which could result in poor infiltration and low quality microstructure of the green body [36].

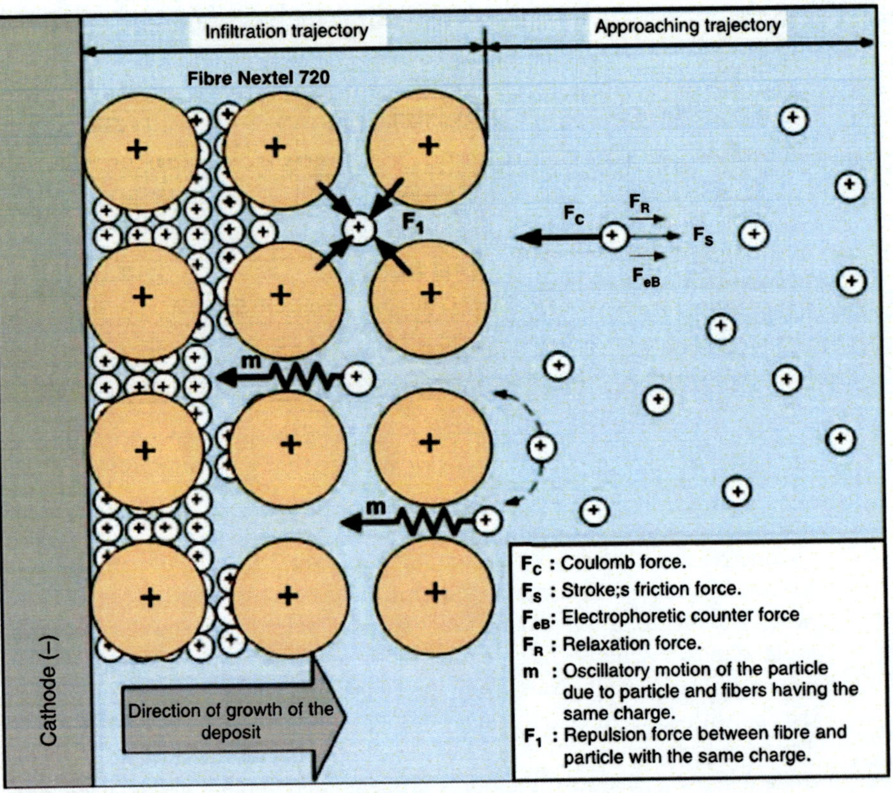

Fig. 8.28 Schematic diagram describing the motion of positively charged Al_2O_3 particles as they infiltrate a positively charged Nextel TM 720 fibre mat during EPD (Reprinted in part with permission from Stoll et al. © 2006 J. American Ceramics Society). [36]

8.5.1.4 Wetting of Fibres

Few authors have considered in their work the wetting of fibres with suspensions, although it plays a certain role in infiltration, in particular is aqueous-based systems. By examining the contact angles of the SiC-powder suspensions at the single SiC fibres, it was discovered that pre-treatment of the fibres with anionic surfactant sodium dioctylsulfosuccinate (SDOSS), can significantly improve the wetting with the suspension and hence the efficiency of infiltration (Fig. 8.29) [37].

8.5.2 Suspensions Properties

Müller et al. [38] reported that in their experiments the efficiency of the electrophoretic infiltration process was strongly influenced by the solids content of the suspension.

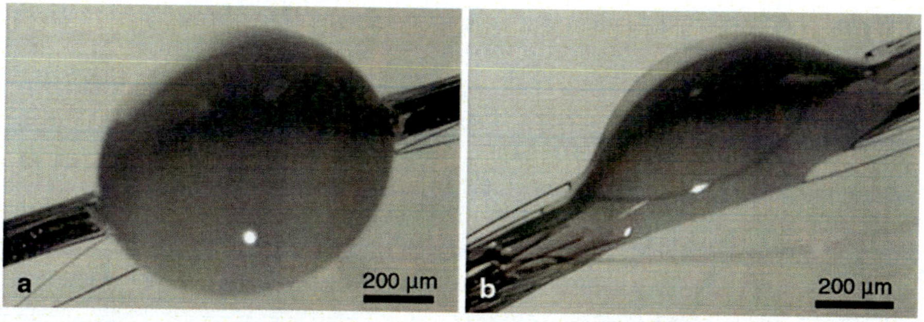

Fig. 8.29 Wetting of SiC-fibres with the SiC-suspension: **a** as-received fibres and **b** SDOSS pre-treated fibres

They performed infiltration of C-fibre mats by ethanol suspension of submicron SiC powder. By applying a slurry of a low solid content, the moving particles independently followed individual paths that led to the fibre surface where they deposited. Müller and colleagues suggested that narrow spaces between two fibres are gradually closed, restricting the path of the particles toward the interspace voids among the fibres. This left remnant pores between the C-fibres. However, in case of high powder content a collective movement of the particles was proposed. This meant that, although single particles could be caught by the charged fibres, additional particles could simultaneously pass through the space between the fibres. Gal-Or et al. [30] similarly reported that penetration increases with particle concentration up to a certain level. The amount of penetrated SiO_2 particles in the pores of graphite substrates increased with increasing particle concentration when using water and propanol as solvents.

8.5.2.1 Anionic or Cationic Deposition?

In general, both cationic and anionic depositions are feasible in infiltration, depending on the desired interaction between the fibers and powder particles. As presented above, equal surface charge polarities of particles and fibres are preferred. SiC particles can be positively or negatively charged in water through the addition of appropriate surface active agents. From either negatively or positively charged particles suspension bulk deposits with high particle packing density can be prepared. However, in accordance to the above mentioned findings, the use of positively charged SiC particles results in their deposition on the SiC fibres thus hindering the infiltration. Conversely, negatively charged particles successfully penetrate through a 3-D SiC fabric, which is detached from the electrode. The fibres may be also pre-treated with cationic surfactant, with positively charge particles presumably yielding the desired result. Another criterion that may be employed in infiltration with water-containing suspensions is gas formation due to electrolysis. In the case of anionic deposition the formed oxygen can be consumed for example by a graphite electrode, while in the case of cationic deposition Pd electrode may help to lower the porosity.

8.5.2.2 Particle Size

The small gaps in fibre tows are expected to be more easily infiltrated with nanosized particles than with larger ones. The nanosized particles are also highly attractive for electrophoretic infiltration not only because of their potential ability to form ceramic materials with improved properties but also since they have higher electrophoretic mobility. Nanosized powders usually exhibit higher ζ-potential than the submicron powders on the other side, their high tendency to agglomerate frequently represents a great problem in preparation of suspension for EPD. In addition, their high specific surface area reflecting in high viscosity of suspension does not allow for high solids loadings as for the submicron powders. For example, similar viscosities are observed for a suspension containing 60 wt.% of SiC powder with average particle size of 500 nm and for a suspension containing only 15 wt.% with average particles size of 50 nm. The comparison of electrokinetic properties of this two SiC powders reveals similar ζ-potential versus pH relationship within the pH range 4–9, but significantly higher conductivity for the nanosized powder than for the submicron-sized (see Fig. 8.12). This result in a significant difference in deposits prepared under the same conditions: while a thick dense deposit can be prepared from the submicron powder, the deposition of nano-SiC powder under the same conditions is not successful. This remains a topic of further research.

To allow the particles to penetrate though the narrow gaps in fibre tows, special attention must be paid to deagglomeration. It is well known that the effective size of particles in suspension is not given by the specified crystallite or "particle" sizes declared by the producer but by its state of dispersion, i.e. agglomeration. As illustrated in Fig. 8.30, the effective particle size is also reflected in ζ-potential. Its absolute values increase with addition of surfactant (till the optimum addition) but

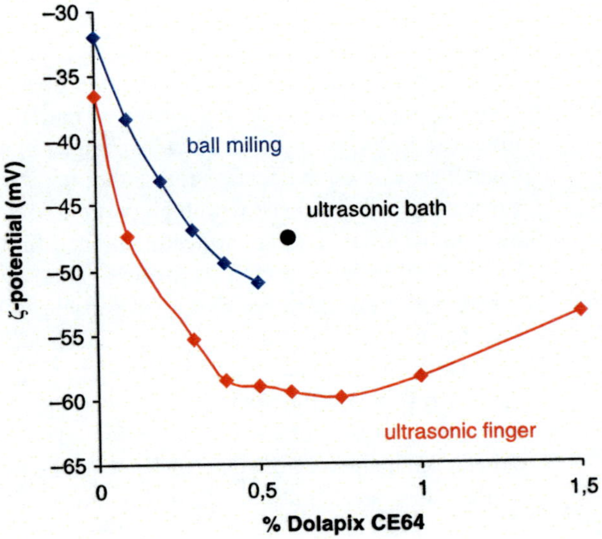

Fig. 8.30 Effect of homogenisation on ζ-potential of aqueous alumina suspension with solids loading 50 wt.%

the measured values increase with efficiency of homogenization. The ζ-potential of alumina in water with addition of 0.5% of Dolapix CE64 was −47 mV after 15 min of homogenisation in ultrasound bath. The value slightly increased after wet ball milling (−53 mV), while it reached the highest absolute value (−60 mV) only after vigorous homogenisation by aid of ultrasonic finger. This illustrates the great influence that a properly prepared suspension has on the formation of a dense homogeneous deposit and on infiltration efficiency.

8.5.3 EPD Parameters

According to Haber and Gal-Or [29], electric field has a dual effect on the extent of particle penetration into fibre-preform. Due to increase of the particle velocity and Peclet number with field strength, penetration is enhanced. However, it is proposed that in the case of conductive substrate the field may also enhance the buildup of an external deposit on the substrate that blocks penetration. An optimal field strength may exist, which decreases with increased particle concentration because of faster build-up of the external coating at higher concentrations. This effect was not observed in experiments performed by the authors of this work: the most effective infiltration was obtained with high solid loaded suspensions and relatively high field strength, i.e. up to 30 V/cm. This can be ascribed to the fact that in their experiments the conductive fibre mat was separated from the electrode by a Teflon spacer that prevented the above mentioned build-up effect.

8.5.4 The Electrodes

As previously mentioned, conductive fibre-fabric may serve as an electrode in EPD process. Though in this case, that the particles will infiltrate the fabric is difficult to expect. Hence, this configuration is more applicable for coatings than for infiltration. An electrically conductive material, such as different metals, alloys or graphite, is not necessarily passive in EPD. The electrode could consume oxygen or hydrogen that evolved due to water electrolysis, but could also emit metallic ions into the suspension. For example, when steel is used as an electrode material, it emits iron and alloying metal ions into the suspension, which travel in directions that oppose the motion of the particles and are ions are consequently incorporated in the deposit. This phenomenon is illustrated in the cross-section of alumina deposits prepared on zinc and copper electrodes, Fig. 8.31. The coloured core of the right hand side deposit reveals that the copper ions are incorporated in the deposit. Hence, when the absence of impurities in the final material is required, such as in the case of SiC_f/SiC composites for fusion application (the impurities are associated with too high neutron-induced radioactivity), proper selection of electrodes material is essential. In this case, graphite is proposed as a better option.

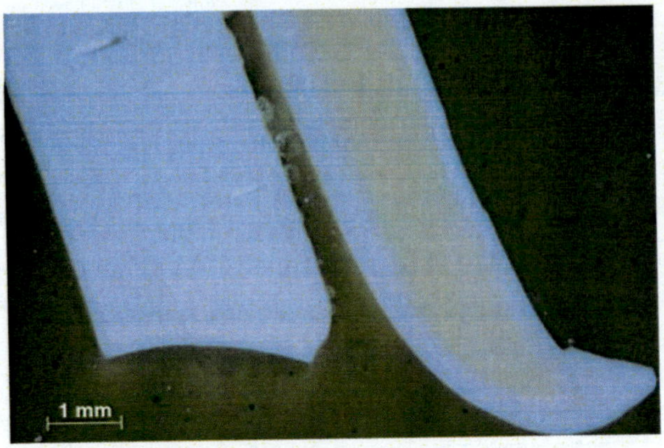

Fig. 8.31 Alumina deposit formed from alkaline aqueous suspension on zinc (*left*) and copper (*right*) electrode

8.5.5 Infiltration Rate and Penetration Depth

In EPD, the deposition rate usually decreases with deposition time, which is either due to the increased resistance of the deposit or to the altered properties of the remaining suspension. A similar rationale is valid in electrophoretic infiltration, in which deposit is formed in the spaces between fibres and fibre-tows. The penetration rate as a function of time has been studied by Gal-Or et al. [30], who observed a plateau in the amount of penetrated SiO_2 and explained it by the effect of external coating. They proposed that by prevention of surface deposition the extent of penetration should be increased.

8.6 State-of-the-art CFCMCs Prepared by EPI

In the last part of this chapter a few examples of electrophoretic infiltration employment to prepare continuous fibre-reinforced ceramic matrix composites (CFCMCs) will be reviewed. Most of the works concerned with oxide-based composites, in particular with 2-D, while less attention has been paid to non-oxide composites.

8.6.1 Oxide Fibres-reinforced CMCs

The main characteristic of the oxide fibres is their electrical resistance; thus, they are never used as an electrode. Since most of the oxide fibres are positively charged in acidic and negatively in alkaline region, both cationic and anionic depositions were used for infiltration. Stoll et al. [36] prepared alumina matrices with mullite fibres. They used 25 wt.% of α-alumina powder (d = 100–300 nm) dispersed in ethanol and 4-hydroxybenzoic acid as a dispersant agent that rendered alumina particles positively charged. Alumina-mullite composites were also prepared by Kaya et al. [39], who used mullite fibres coated with a crack-deflecting layer of $NdPO_4$. Nano-sized alumina powder (d_{50} = 160 nm) was dispersed in distilled water with addition of

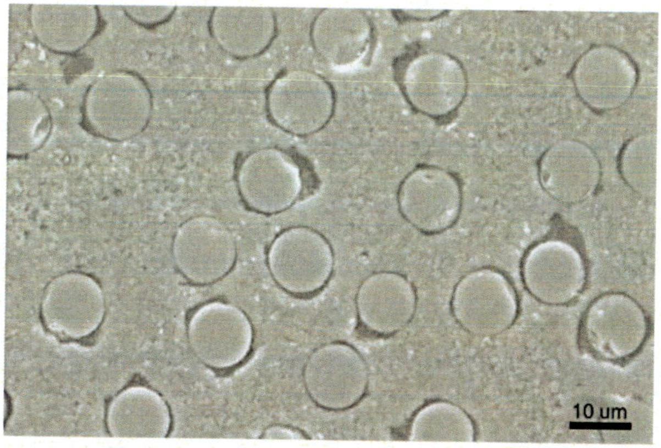

Fig. 8.32 Morphology of mullite fibre-reinforced reaction bonded mullite composite prepared by electrophoretic infiltration, pressure-less sintered at 1300°C for 2 h (Reprinted in part with permission from Bao et al. © 2008 Elsevier). [40]

water-based acrylic polymer (Duramax B1014), that acts as a dispersant and a binder. Furthermore, a mixture of boehmite and colloidal Y_2O_3 as a sintering aid was added.

Mullite-alumina fibre mats, previously coated with a layer of $AlPO_4$ (weak interphase layer), were used also to reinforce reaction bonded mullite matrix [40]. The alumina and silicon powders were dispersed in ethanol with addition of PEI that was protonated by glacial acetic acid. As the added PEI was also adsorbed onto the fibre surface, the fibres repelled the particles. This caused the particles to penetrate unimpeded into the fibre bundles. The resulting microstructure of sintered mullite fibre-reinforced reaction bonded mullite composite, illustrated in Fig. 8.32, reveals that the ceramic matrix filled the gaps between adjacent fibres.

Mullite matrices reinforced with mullite fibres (Nextel 720) were also fabricated by Kaya's group [41]. They used aqueous suspensions of 30 wt.% of nanosized alumina (72%) and silica (28%). At the working pH of 4.5, heterocoagulation of the positively charged alumina particles and negatively charged silica particles appear, hence resulting in negatively charged clusters. The authors report that under constant voltage conditions (12 V DC) woven mullite fibres were successfully infiltrated with the clusters although the fibres were separated only by 200–300 nm at some places. Kooner and Westby [42] used mullite fibres to reinforce alumina, silica and alumina/silica matrices. Alumina and silica powders were dispersed in ethanol. Tartaric acid and dibutylamine were added to alumina suspensions, while 2 wt.% of water was added to ethanol-based silica suspensions to obtain a negative charge on the silica. The fibre weave was placed in front of the deposition electrode, and the deposit formed both on the electrode and around the fibres.

8.6.2 *Metal Fibre-reinforced CMCs*

Besides a few reports on continuous ductile reinforcement of glass matrices, only a few research articles have been devoted to EPI of continuous metal-fibre reinforcement of ceramics, despite the advantages they may have over their ceramic-ceramic counterparts. These include an increased resistance to damage during composite processing due to the intrinsic ductility of metallic fibres and the possibility of exploiting their plastic deformation for composite toughness enhancement.

Since metallic fibres are mostly highly conductive, they can be used as a deposition electrode. As described above, although these are not the optimal conditions for the successful infiltration of 2-dimensional mats, examples of the successful implementation of metallic fibre mats exist in the literature. Stainless steel fibre mats were infiltrated with alumina, silica or cordierite particles [35, 43]. In most cases, the steel mats were used as deposition electrode. Alumina matrix for infiltration was prepared by using an acidic aqueous suspension of positively charged boehmite particles (d_{50}=20 nm) with 20 wt.% solids loading. Silica matrix was prepared by infiltration of commercial aqueous silica sol (d_{50}=20 nm) with addition of B_2O_3 and boehmite to enable sintering at 900°C. Silica suspension with 30 wt.% solids content was stable in the pH range 9-10, where the particles were negatively charged. Cordierite matrix was prepared from aqueous suspension of positively charged cordierite particles (d_{50}=120 nm) at pH 3 and solids content 20 wt.%.

8.6.3 *Carbon Fibre-reinforced CMCs*

Carbon fibres are also conductive and consequently, several experiments have been performed with the fibre-mats attached as electrode. This type of experiment was performed for example by the group of Müller [38], who used 2-D carbon fibre fabrics to reinforce submicron SiC matrix. SiC powder was dispersed in ethanol, with addition of PEI, addition of B_4C and carbon black (solids content from 1 to 10 vol.%). They observed that solids content of the suspension strongly influence on the success of infiltration. Using a suspension with 10 vol.% of powder yielded a much better result than by using a 1 vol.% suspension. Figure 8.33 shows the microstructure of the sintered composite, prepared by using a 25 vol.% suspension.

Kaya et al. [44] used nickel-coated 2-D carbon fibre mats to reinforce alumina and borosilicate matrices. The nickel coating was used to provide excellent conductivity as well as ease of fibre handling and adequate wettability. Alumina matrices

Fig. 8.33 SEM micrograph of a polished section of a carbon fibre mat, electrophoretically infiltrated with a 25 vol.% suspension of SiC powder

were prepared from aqueous suspension of positively charged boehmite particles, seeded with γ- and α-alumina, while borosilicate matrix was prepared by dispersion of negatively charged borosilicate powders in water, at pH 8.5. In both cases the fibre mat was used as an electrode and the infiltration was performed under vacuum.

8.6.4 SiC Fibre-reinforced CMCs

In comparison to other technical fibres the production of SiC fibres is small, but their excellent properties in particular at high temperatures, make them very attractive for some special applications where high elastic modulus and high corrosion resistance at high temperatures are required. SiC fibres are mostly used as reinforcement of SiC-matrix to form SiC_f/SiC composites. Silicon carbide fibres were introduced in 1960s. Since then, they have become a permanent subject of development, most notably in Japan. The currently produced SiC fibres exhibit significantly improved toughness and retained the properties at higher temperatures in comparison to the early grades. They differ in chemical composition, in diameter as well as in surface roughness, which all contribute significantly to the overall performance of the composite.

Silicon carbide is a semiconductor and therefore the fibres can be used in EPD as electrode. In addition, SiC-fibres in CMCs are usually coated with a thin carbon layer that, as described in the first part of this chapter, acts as a crack-deflecting barrier between the fibre and the matrix. Hence, in EPD the carbon-coated SiC fibres can be considered in the same way as carbon fibres. Since 3-D architecture enables much better performance than stacking 2-D fabric, it is preferred in particular for the extreme applications where transversal thermal conductivity and high shear strength are essential. This, however, requires rather demanding weaving procedure that in total results in overall very high-cost production of the SiC performs.

In 1993 Illston et al. [45] patented a method of reinforcing silica matrix with SiC-fibre mats coated with carbon. In their work several mats were stacked after EPD in a suitably shaped graphite dye. Later a similar procedure to produce a 3-D SiC/SiC composite was used by Kaya et al. [46]: In their work 20 wt.% of SiC powder (d_{50}=200 nm) was dispersed in distilled water at pH 9 with addition of boehmite and $Y_2Si_2O_7$ as sintering aids. The conductive carbon coated SiC fibres were used directly as an electrode. A 3-D composite was prepared by compacting 6 or 8 EPD-infiltrated layers together using pressure filtration.

2-D carbon-coated SiC-fibre mats (Nicalon, 5×5 cm^2) were also used to reinforce mullite matrix, prepared by aqueous suspension of fumed silica (28 wt.%) and alumina (72 wt.%) particles at pH 3.1. The fibre mat was employed as the deposition electrode (cathode), while the counter electrode was stainless steel electrode [34].

Based on systematic study of the electrokinetic properties of SiC-fibres and powders as well as the deposition and infiltration processes, Novak et al presented a method for control of the infiltration of 2-D and 3-D SiC-fibre mats [33]. They used Tyrano SA and Nicalon fibre-mats and for infiltration a submicron β-SiC powder. Due to the strict requirements for the target material proposed to be used in the first wall blanket of the future fusion reactor, graphite electrodes were used as the only acceptable to avoid any contamination with metals. In addition, due to the tough restrictions sintering additives were carefully selected and sintering temperature were kept low in order to prevent β- to α-SiC transformation and fibres properties deterioration.

In this phenomenological study, before infiltration the properties of the suspension for infiltration were optimised on the basis of a thorough analysis of the effects of various parameters on bulk deposit. Aqueous suspensions of submicron SiC particles were prepared with addition of tetramethylammonium hydroxide (TMAH) and dispersed by aid of vigorous ultrasonic agitation. The suspensions containing 20–60 wt.% solids expressed ζ-potential −60 to −70 mV and conductivity below 0.5 mS/cm. The deposit formed on graphite electrode contained above 83 wt.% solids (i.e. >60 vol.%) with average pore size as small as 70 nm. This implied that the successful infiltration of the fibre-preform with the SiC suspension will results in porosity of approximately the same range.

Before infiltration, wetting behaviour and ζ-potential of the crushed SiC fibres were analysed. The results showed that the SDOSS used as a surfactant beneficially affects the ζ-potential, resulting in highly negatively surface charged fibres (Fig. 8.26) that were well wetted by aqueous suspension of negatively charged SiC powder (see Fig. 8.29b). The 2-D SiC-fabric was then placed in the EPD cell in front of the graphite anode and separated from it by an insulating spacer. The infiltration was performed at field strength 30 V/cm for 2 min. The resulting green parts were dried, infiltrated with $Al(H_2PO_4)_3$ solution and sintered. The resulting microstructure illustrated in Fig. 8.34 reveals that the SiC-particles have filled the narrow gaps between the individual fibres.

The infiltration of 3-D SiC-fabric was performed in a similar way as for the 2-D fabric. The only difference was a longer time of deposition to allow particles to cross the 4 mm thick fabric. The SiC aqueous suspension for infiltration as well as the fabric pre-treated with anionic surfactant were characterised by highly negative ζ-potential (<−60 mV) at pH 10. The well dispersed suspension was poured into the cell where the pre-treated SiC fabric was fixed with Teflon sealing frame in front of a graphite anode that was separated from the fabric with a cellulose membrane. The configuration of the EPD cell, used in infiltration of the SiC fibre perform with SiC powder, is illustrated in Fig. 8.35.

The infiltration was performed at constant voltage of 30 V (the distance between electrodes was 2 cm). In 5 min, the preform was infiltrated and on both sides thin deposit (<1 mm) was also formed. After drying, subsequent infiltration with $Al(H_2PO_4)_3$ solution and sintering (SITE process [19]), thorough examinations of cross-sections of the composite samples at different depths was performed. Macroscopic image of the cross-section of the composite is presented in Fig. 8.36.

The examination of the polished cross-sections of the sintered composite revealed that the SiC particles penetrated throughout the 4 mm thick fabric, but a few relatively large non-infiltrated regions still remained, most probably as the consequence of presence of air pockets in the preform. This is supported by the following observations: 1) the particles appeared randomly throughout the thickness and not preferentially in the central region or near the electrode-side; 2) good penetration of the particles into the fibre tows could be observed at all parts of the composite (Fig. 8.37a). It is also obvious from Fig. 8.37b that the SiC particles filled the narrow gaps between fibres in all three directions.

Fig. 8.34 SEM micrographs of a low-angle cross-sections of the 2-D SiC-fabric infiltrated with an alkaline SiC suspension: **a** green composite and **b** impregnated with $Al(H_2PO_4)_3$ and sintered at 1300°C in Ar (Reprinted in part with permission from Novak et al. © 2008 Elsevier). [50]

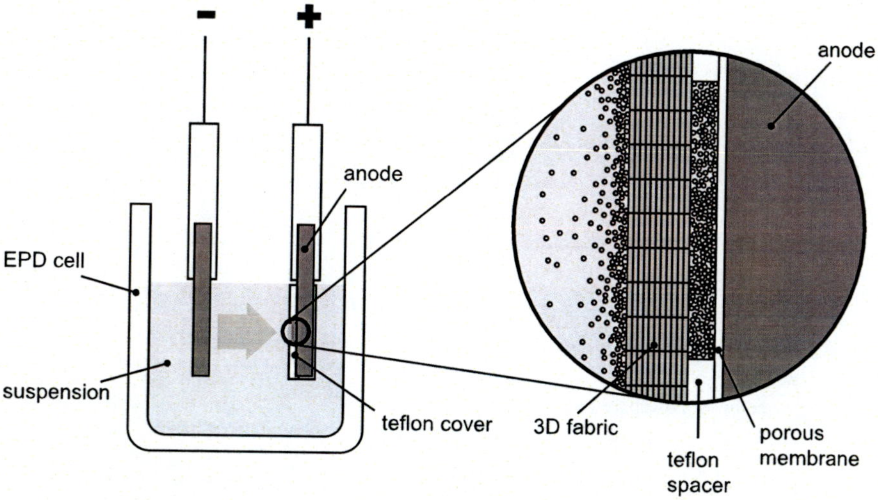

Fig. 8.35 Schematic presentation of cell configuration in electrophoretic infiltration of 4 mm thick SiC fibre-preform with SiC powder

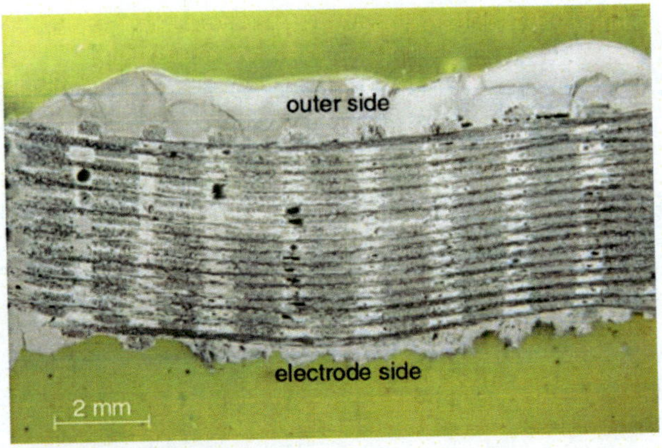

Fig. 8.36 Cross-section of the 3-D SiC-fibre fabric infiltrated in accordance to schematic presented in Fig. 8.35

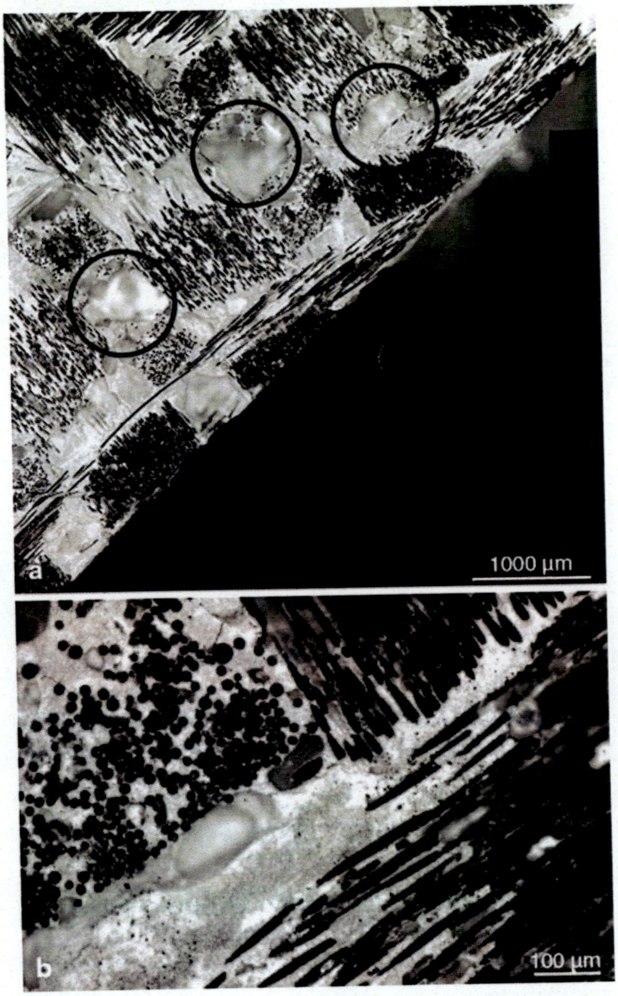

Fig. 8.37 Optical micrograph (polarised light) of polished cross-setion of the 3-D SiC_f/SiC composite prepared by EPI **a** composite picture; The circles show non-infiltrated regions and **b** at higher magnification showing infiltrated gaps between fibres in all directions

8.7 EPD of Interphase Layer on Fibres in CFCMCs

The electrical conductivity of some sorts of fibres can be also put into a good use to produce coatings (interphase layers) by aid of electric field. The coating thickness can be easily controlled by variation of deposition time, voltage or current density. For example, Boccaccini et al. [47] reported on fabrication of alumina and titania coatings on carbon and metallic fibres where the fibres were used as electrodes. Kaya et al. [48] recently reported when the use of EPD to fabricate an $NdPO_4$ or ZrO_2 interphase layer on alumina (Nextel) fibre mats that helped to produce an oxide-based composite with improved mechanical properties. Silicon carbide fibres were electrophoretically coated with carbon nanotubes (CNT) [11], which can be used as a more thermally conductive alternative to interphase layer than the pyrolytic carbon.

As stated at the beginning, the bonding between fibres and matrix is critical to the performance of the continuous fibre-reinforced ceramic matrix composites. Often a strong bond is desired so that the load is transferred efficiently from the matrix to the fibres. The ability to absorb energy may be increased by a weak bond that enables slipping between the fibres and the matrix and hence the energy absorption. Pyrolytic carbon and boron carbide are the most frequently used "weak" interphase layers. The deposition is usually done by chemical vapour deposition that enables a good control of the composition and thickness, but the process is time consuming and costly. As an alternative, it has been shown recently that coatings on the fibres can be also prepared by aid of electroporetic deposition by attaching the fibres to the deposition electrode.

In 2009 Yoshida and co-authors reported on deposition of interphase layer of carbon flakes on SiC fibres [49].The authors employed the negative charge of graphite particles at pH 9 to form a homogeneous coating on SiC-fibres connected as anode. In contrast to dip-coating and vacuum infiltration, the electrophoretic deposition on SiC fibres resulted in the surface of fibres wholly coated with flaky graphite particles including the fibres at the centre of the bundles. In a slightly different way König et al. [11] prepared carbon nanotubes (CNTs) coating on SiC fibres. To optimise the process, ζ-potentials of CNTs were first analysed. As they appeared to possess a highly negative surface charge, they were deposited on the SiC-fibres at the metal anode. In contrast to the Yoshida's approach, König and collaborators attached the SiC-fibre mats to the front side of the anode, so that due its conductivity it acted itself as the electrode. A similar result could be obtained by connecting the fibre fabric directly as an electrode, as shown elsewhere on the SiC/mullite system [49]. However, due to the lower conductivity of the SiC compared with a metal electrode, the applied voltage needed to be higher to achieve the same effect. Another reason for using the solid rectangular electrode behind the fabric in this study was that it results in a more homogenous electric field. The CNT-coating with a thickness approximately up to 400 nm was produced at an applied voltage of 2.8 V in 10 min. SEM micrographs of the SiC fibres before and after deposition with the CNTs are presented in Fig. 8.38a and 8.38b, respectively.

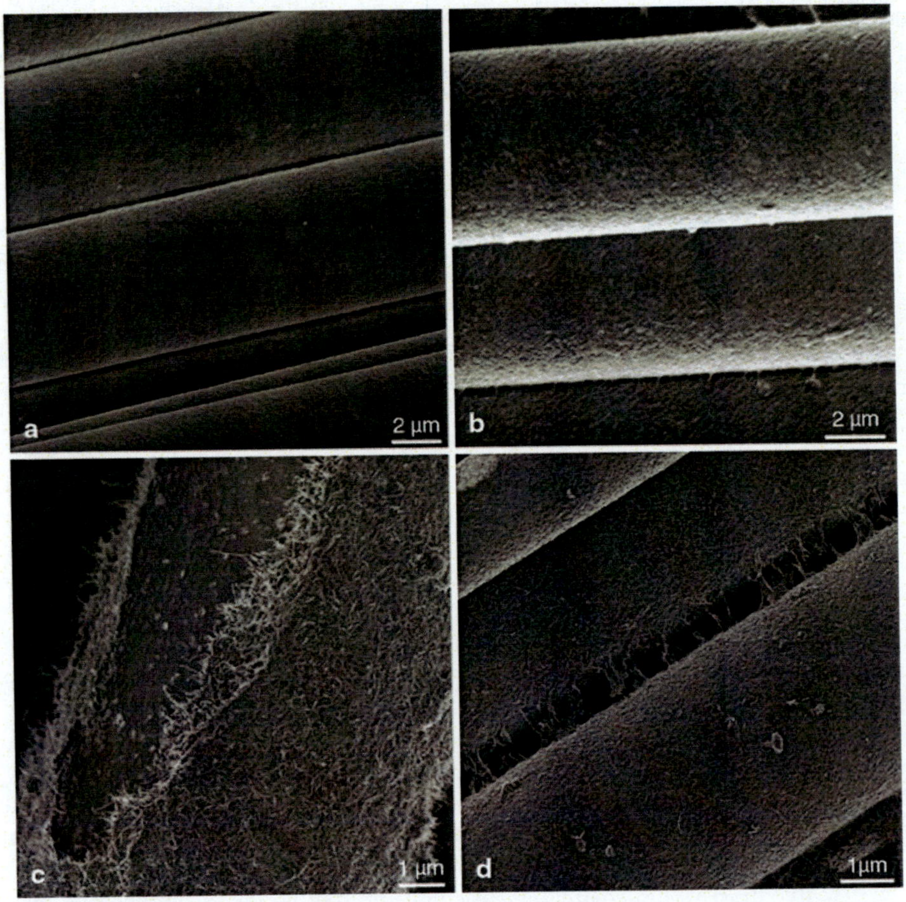

Fig. 8.38 **a** As-received SiC-fibres [11]; **b**, **c** and **d** the fibres coated with CNTs by aid of EPD (Reprinted in part with permission from König et al. © 2010 Elsevier). [11]

It is evident that individual fibres are completely covered with a fairly uniform layer of CNTs. Due to their very small size and well-dispersed state, the carbon nanotubes were able to penetrate into the spaces between the fibres in the 2-D fabric. By observing the cross-section of the sample it was confirmed that the SiC-fibres in the central region of the fibre mat were also efficiently coated by the CNTs. In some parts within the fabric, in particular where the fibres were very close to each other, the long spaghetti-like carbon nanotubes, however, could not penetrate and hence covered adjacent fibres. (Fig. 8.38c). This bridging of the adjacent fibres is undesired since in the further step, in infiltration, the particles can not penetrate deeper into the fabric. Hence, it could be assumed that in highly dense mats short nanofibres instead the long ones would presumably give a more favourable result.

Figure 8.38d presents the fibres with partly removed coating; the thickness of the coating can be estimated to less than 100 nm. It is obvious, however, that the thickness can be adapted in accordance to the need just by variation of the deposition

Fig. 8.39 CNT-coated fibres inside the central region of the sintered CNT-SiC_f/SiC composite (Reprinted in part with permission from König et al. © 2010 Elsevier). [11]

time. In the same study, König et al. prepared CNT-SiC_f/SiC composite, where the SiC-fibres mats where first electrophoretically coated by CNTs and then the fabric was infiltrated with SiC-powder aqueous suspension. In both cases well dispersed alkaline suspensions with ζ-potential approximately −60 mV were used. To assure the effectiveness, both stages have to be performed at different settings. While in the coating step the SiC-fibre mat was attached as anode, in infiltration it was separated from the electrode. This was necessary in order to prevent deposition of the SiC particles onto the front side of the fabric hence hindering further infiltration. The observation of sintered CNT-coated SiC-fabric infiltrated with SiC revealed that both steps, the coating and infiltration were relatively effective. Only some gaps between the adjacent fibres remained inaccessible for particles due to the bridging effect of long CNT (see Fig. 8.39). Consequently, the overall density of the CNT-SiC_f/SiC composite was lower in comparison with the SiC_f/SiC prepared at the same EPI conditions. This implies a need for further investigations in this field.

8.8 Conclusions

The electrophoretic deposition process known as an extremely versatile technique for shape forming of materials, has been compared against many other applications, as a potential tool in fabrication of ceramic matrix composites. EPD can be employed to form discontinuous- as well as for continuous phase-reinforced CMCs. In the first group of materials reinforced with particles of various size and shapes, the particulates can be controllably dispersed in the matrix either to form homogeneous, layered, or functionally graded composites. Here, the versatility of electrophoretic deposition and co-deposition of different species was demonstrated to be much larger than any other shaping technique. In addition, using optimal suspensions and EPD conditions, higher green densities can be achieved. In spite all of this, there is no indication yet of extensive industrial production of the CMCs. This implies a need for further optimisation of the process and technology to make EPD more attractive for industry. On the other hand, continuous fibre-reinforced composites still represent a great challenge for scientists, and the industrial production of CFCMCs is even much more distant. While numerous reports demonstrated applicability of the electrophoretic deposition principle for infiltration of various fibres woven into 2-dimensional mats, the electric-field assisted infiltration of 3-dimensional mats appears to be much less developed. Hence, in contrast to the widely investigated general field of EPD, currently there is still room for new findings that may result in new products and lower production costs.

Acknowledgements The work was financially supported by the Slovenian Research Agency within the National Research Programme P2-0084, project No. J2-7506-01016 and within the PhD studies of Mrs. Katja König and Mr. Aljaž Iveković (prj. No 1000-05-310034 and 1000-08-310084). A part of the work was also supported by the European Communities under the Contract of Association between EURATOM and the Ministry of Higher Education, Science and Technology of the Republic of Slovenia within the framework of the European Fusion Development agreement.

References

1. Bansal, N.P.: Handbook of composites. Kluwer, London (2005)
2. Strong, A. B.: Fundamentals of Composites Manufacturing, 2nd edn. Society of Manufacturing Engineers, Dearborn, Michigan (2007)
3. Hadraba, H., Maca, K., Cihlar, J.: Electrophoretic deposition of alumina and zirconia: II. Two-component sys-tems. Ceram. Int., **30**, 853–863 (2004)
4. Anne, G., Vanmeensel, K., Vleugels, J., Van Der Biest, O.: Electrophoretic deposition as a novel near net shaping technique for functionally graded biomaterials in: Electrophoretic deposition: Fundamentals and applications II, Book series: Key Engineering Materials, **314**, 213–218 (2006)
5. Hvizdos, P., Jonsson, D., Anglada, M., Anne, G., Van Der Biest, O.: Mechanical properties and thermal shock behaviour of an alumina/zirconia functionally graded material prepared by electrophoretic deposition, J. Eur. Cer. Soc. **27**, 1365–1371 (2007)
6. Moritz, T., Eiselt, W., Moritz, K.: Electrophoretic deposition applied to ceramic dental crowns and bridges Electrophoretic deposition applied to ceramic dental crowns and bridges, J. M. Sci. **41**, 8123–8129 (2006)
7. Zhang, Z., Huang, Y., Jiang, Z.: Electrophoretic Deposition Forming of SIC-TZP Composites in a Non-aqueous Sol Media. J. Am. Ceram. Soc. **77**(7), 1946–1949 (1994)
8. Jean, J.H.: Electrophoretic deposition of Al_2O_3-SiC composite, Mat. Chem. and Phys. **40**, 285–290 (1995)
9. König, K.: Electrophoretic deposition of advanced ceramic materials, PhD thesis, University of Ljubljana, 2011
10. Boccaccini, A.R., Cho, J., Kaya, C., Kaya, F.: Electrophoretic deposition of carbon nanotube-ceramic nanocomposites. J. Eur. Cer. Soc. **30**, 1115–1129 (2010)
11. König, K., Novak, S., Ivekovič, A., Rade, K., Meng, D., Boccaccini, A.R., Kobe, S.: Fabrication of CNT-SiC/SiC com-posites by electrophoretic deposition, J. Eur. Cer. Soc. **30**, 1131–1137 (2010)
12. Krenkel, W., Berndt, F.: C/C-SiC compo-sites for space applications and advanced friction systems, Mat. Sci. Eng. **412**, 177–181 (2005)
13. Stadler, Z., Krnel, Kristoffer, Kosmac, Tomaz: Friction behaviour of sintered metallic brake pads on a C-C-SiC composite brake disc. J. Eur. Cer. Soc. **27**, 1411–1417 (2007)

14. Naslain, R., Cristin, F.: SiC-Matrix Composite Materials for Advanced Jet Engines. MRS Bull. **9**, 654–658 (2003)
15. Youngblood, G.E., Kurtz, R.J., Jones, R.H.: SiC/SiC design for a Flow Channel Insert. http://www.ms.ornl.gov/programs/fusionmatls/pdf/dec2004/2_CERAMIC/YOUNGB1.PDF (2004). Accesed 29 March 2009
16. Hasegawa, A., Kohyama, A., Jones, R.H., Snead, L.L., Riccardi, B., Fenici, P.: Critical issues and current status of SiC/SiC composites for fusion, J. Nucl. Mater. **128**, 283–287 (2000)
17. Nannetti, C.A., Ortona, A., de Pinto, D. A., Riccardi, B.: Manufacturing SiC-fibre-reinforced SiC matrix composites vy improved CVI/slurry infiltration/polymer impregnation and pyrolysis, J. Amer. Ceram. Soc. **87**, 1205 (2004)
18. Katoh, Y., Kohyama, A., Nozawa, T., Sato, M.: SiC/SiC composites through transient eutectoid-phase route for fusion applications, J. Nucl. Mat. **587**, 329–333 (2004)
19. Novak, S., Dražić, G., König, K., Iveković, A.: Preparation of SiCf/SiC composites by the slip infiltration and transient eutectoid (SITE) process, J. Nucl. Mater. (2010), doi:10.1016/j.jnucmat.2010.01.014
20. Kosmulski, M., Próchniak, P., Rosenholm, J.B.: Solvents, in which ionic surfactants do not affect the ζ-potential, J. Coll. Int. Sci. **342**, 110–113 (2010)
21. Novak, S., König, K.: Fabrication of alumina parts by electrophoretic deposition from ethanol and aqueous suspensions, Ceram. Int. **35**, 2823–2829 (2009)
22. Tabellion, J., Clasen, R.: Electrophoretic deposition from aqueous suspensions for near-shape manufacturing of advanced ceramics and glasses—applications, J. Mat. Sci. **39**, 803–811 (2004)
23. Kosmulski, M.: pH-dependent surface charging and points of zero charge. IV. Update and new approach. J. Coll. Int. Sci. **337**, 439–448 (2009)
24. Stappers, L., Zhang, L., Van Der Biest, O., Fransaer, J.: The effect of electrolyte conductivity on electrophoretic deposi-tion, J. Coll. Int. Sci. **328**, 436–446 (2008)
25. Ferrari, B., Moreno, R.: Electrophoretic deposition of aqueous alumina slips. J. Eur. Cer. Soc. **17**, 549–556 (1997)
26. Novak, S., König, K.: Fabrication of alumina parts by electrophoretic deposition from ethanol and aqueous suspensions, Ceram. Int. **35**, 2823–2829 (2009)
27. König, K., Novak, Saša, Bele, M., Zorko, M., Kobe, S.: Influence of the suspension stability on the electrophoretic deposition of nanosized alumina and silica. V: D. Mihailovoć, S. Kobe et al. Hot nano topics 2008, 23 30 May, Slovenia: abstract book. Ljubljana, 2008, str. 249
28. Radice, S., Bradbury, C.R., Michler, J., Mischler, S.: Critical particle concentration in electrophoretic deposition. J. Eur. Cer. Soc. **30**, 1079–1088 (2010)
29. Haber, S., Gal-Or, L.: Deep Electrophoretic Penetration and Deposition of Ceramic Particles inside Porous Substrates: I. Analytical Model. J. Electrochem. Soc. **139**(4): 1071–1078 (1992)
30. Gal-Or, L.: Deep Electrophoretic Pene-tration and Deposition of Ceramic Particles inside Porous Substrates, II. Ex-perimental Model. J. Electrochem. Soc. **139**, 1078–1081 (1992)
31. Bao, Y., Nicholson, P.S.: Constant Current Electrophoretic Infiltration Deposition of Fiber-Reinforced Ceramic Composites. J. Am. Ceram. Soc. **90**(4): 1063–1070 (2007)
32. Boccaccini, A.R., Trusty, P.A.: Electrophoretic deposition infiltration of metallic fabrics with a boehmite sol for the preparation of ductile-phase-toughened ceramic composites. J. Mater. Sci. **33**, 933–938 (1998)
33. Novak, S., König, K., Iveković, A., Boccaccini, A.R.: Infiltration of 3-D fabric for the production of SiC/SiC composites by means of EPD. Key. Eng. Mater. **412**, 237–242 (2009)
34. Boccaccini, A.R., MacLaren, I., Lewis, M.H., Ponton, C.B.: Electrophoretic Deposition Infiltration of 2-D Woven Sic Fibre Mats with Mixed Sols of Mullite Composition. J. Eur. Ceram. Soc. **17**, 1545–1550 (1997)
35. Kaya, C., Boccaccini, A.R., Trusty, P.A.: Processing and Characterisation of 2-D Woven Metal Fibre-reinforced Multilayer Silica Matrix Composites Using Electrophoretic Deposition and Pressure Filtration. J. Eur. Ceram. Soc. **19**, 2859–2866 (1999)

36. Stoll, E., Mahr, P., Krüger, H.-G., Kern, H., Thomas, B.J.C., Boccaccini, A.R.: Fabrication technologies for oxide-oxide CMCs based on EPD. J. Eur. Ceram. Soc. **26**, 1567–1576 (2006)
37. Toplišek, T., Novak, S., Dražić, G., Kobe, S.: The preparation and characterization of continuous SiC fibres for infiltration with nanoparticle suspensions, 13th Conference on Materials and Technology, 10–12 October, 2005 Portorož, Slovenia.
38. Müller, E., Moritz, K., Dittrich, R.: Application of electrophoresis for fabricating SiC-matrix composites rein-forced by carbon fibres, Proceedings of 1st International Conference on Electrophoretic Deposition, 246–254 (2002)
39. Kaya, C., Butler, E.G.: Unidirectional all-oxide mini-composites with crack-deflecting $NdPO_4$ interface. J. Eur. Ceram. Soc. **29**, 363–367 (2009)
40. Bao, Y., Nicholson, P.S.: $AlPO_4$-coated mullite/alumina fiber reinforced reaction-bonded mullite composites. J. Eur. Ceram. Soc. **28**, 3041–3048 (2008)
41. Kaya, C., Gu, X., Al-Dawery, I., Butler, E.G.: Microstructural development of woven mullite fibre-reinforced mullite ceramic matrix composites by infiltration processing, Science and Technology of Advanced Materials **3**, 35–44 (2002)
42. Kooner, S., Westby, W.S., Watson, C.M.A., Farries, P.M.: Processing of NextelTM 720/mullite composition composite using electrophoretic deposition. J. Eur. Ceram. Soc. **20**, 631–638 (2000)
43. Kaya, C., Kaya, F., Boccaccini, A.R.: Electrophoretic deposition infiltration of 2-D metal fibre-reinforced cordierite, matrix composites of tubular shape. J. Mater. Sci. **37**, 4145–4153 (2002)
44. Kaya, C., Kaya, F., Boccaccini, A.R., Chawla, K.K.: Fabrication and characterisation of Ni-coated carbon fibre-reinforced alumina ceramic matrix composites using electrophoretic deposi-tion. Acta. Mater. **49**, 1189–1197 (2001)
45. Illston, T. J., Doleman, P.A., Butler, E., Ponton, C.B., Gilbert, M.J., Piramoon, R.: Method of manufacturing a com-posite material, patent WO 93/10056, 1993
46. Kaya, C., Chawla, K.K., Boccaccini, A.R.: Processing of oxide and non-oxide lightweight ceramic composites suitable for high temperature applications using EPD, Proceedings of the 1st Interna-tional Conference on Electrophoretic Deposition, 263–270 (2002)
47. Boccaccini, A.R., Schindler, U., Krüger, H.-G.: Ceramic coatings on carbon and metallic fibres by electrophoretic deposition. Mater. Letters **51**, 225–230 (2001)
48. Kaya, C., Kaya, F., Butler, E.G., Boccaccini, A.R., Chawla, K.K.: Development and characterisation of high-density oxide fibre-reinforced oxide ceramic matrix composites with improved mechanical properties. J. Eur. Ceram. Soc. **29**, 1631–1639 (2009)
49. Yoshida, K., Matsukawa, K., Imai, M., Yano, T.: Formation of carbon coating on SiC fibre for two-dimensional SiCf/SiC composites by electrophoretic deposition. Mat. Sci. Eng. **161**, 188–192 (2009)
50. Novak, S., Rade, K., König, K., Boccaccini, A.R.: Electrophoretic deposition in the production of SiC/SiC composites for fusion reactor applications. J. Eur. Ceram. Soc. **28**(14), 2801–2807 (2008)

Chapter 9
Electrophoretic Deposition of Nanostructured Electroactive Materials

Tao Li, Chen Yanhong and Jan Ma

9.1 Electrophoretic Deposition of Piezoelectric Materials

9.1.1 Introduction

Piezoelectric ceramics actuators have been widely applied in industries and attracted huge attention in academic research during the last decades due to their many advantages such as good displacement resolution, good mechanical durability, high speed, large output force, low power consumption and wide bandwidth of frequencies [1–5]. However, Most of the piezoelectric devices can only be of a very simple shape or structure, such as a plate or a disk. This is largely due to the various difficulties in the fabrication technique. Due to the brittle nature of the piezo ceramics, machining them into complicated shapes remains as a challenge.

However, the development of electrophoretic deposition (EPD) technique provides a choice for the fabrication of piezo devices of relatively complicated shapes and smaller dimensions [1, 5, 6]. EPD of piezo components is noted to be simple, cost effective, versatile and fast. EPD also allows for mass-production of devices, ranging in size from μm to cm in dimension. Technically, EPD is a colloidal process whereby ceramic bodies are shaped directly from a stable colloid powder suspension under a DC electric field. Generally, the colloid fabrication process consists of three stages: the formation of a charged suspension, the deposition of charged particles onto an electrode under a DC voltage and the final sintering of the deposited structure. Figure 9.1 shows the electrophoresis process schematically. A DC field causes the charged particles to move towards, and deposit on, the electrode of opposite charge. The depositing electrode reflects the shape of the ware required, and it is designed such that deposit release is facilitated.

The investigation of electrophoretic deposition on piezoelectric or ferroelectric materials has existed for more than a decade [6]; however, the shape of the deposit

J. Ma (✉)
School of Materials Science and Engineering, Nanyang Technological University, 639798 Singapore, Singapore
e-mail: asjma@ntu.edu.sg

J. H. Dickerson, A. R. Boccaccini (eds.), *Electrophoretic Deposition of Nanomaterials*, Nanostructure Science and Technology,
DOI 10.1007/978-1-4419-9730-2_9,

Fig. 9.1 Schematic drawing of electrophoretic deposition cell

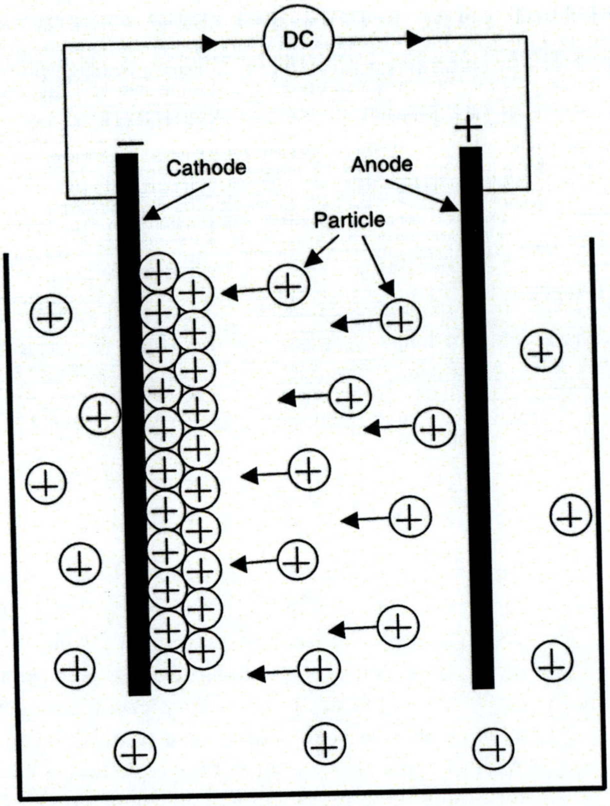

continues to be comprised of simple configurations, such as monolithic plates or thick layers. This paper introduces the application of EPD in the fabrication of several configurations, including tubular, helical and FGM monomorphs. Figure 9.2 shows the configuration of the piezo tube. It is normally coated with electrodes on its curved surfaces and poled along the thickness direction of the wall. The electrode configuration is flexible depending on the practical applications. The main advantage of a tubular configuration is its good structural rigidity. Its applications include scanning probe microscopy (SPM), ultrasonic motors and some medical diagnostics and therapeutics devices [7–11]. It is also noted that processes facilitating miniaturization are essential for the development of future generations of electronic devices. The EPD process meets this requirement very well and components of size in the mm order can be easily realized at a short time and a low cost.

Helical piezoelectric configurations are a good illustration of the capability of EPD in the fabrication of complicated shapes [12]. Figure 9.3 shows a piezo helix, which is a spring-shaped bimorph. The helix can generate both linear and rotational displacements, and has great potential for large displacement applications. The complicated configuration has posed difficulties in its manufacturing; however, the realization of such a complicated structure can be achieved easily using EPD.

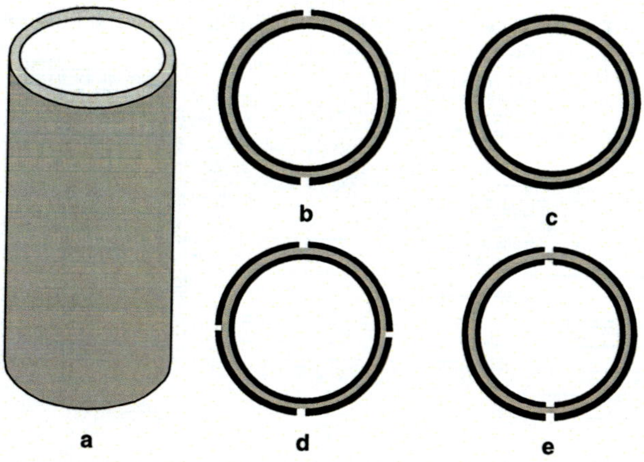

Fig. 9. 2 Configurations of the piezoelectric tubular transducers (**a**, piezoelectric tube; **b**–**e**, configuration of the electrodes)

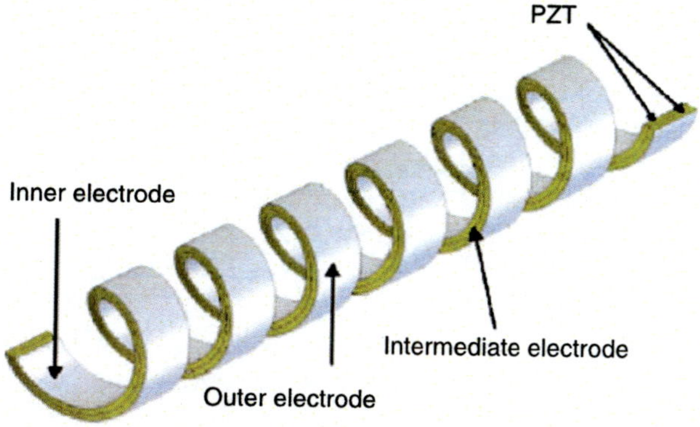

Fig. 9.3 Structure and working principle of the helical actuator

The FGM monomorph is another device configuration that can be fabricated using EPD [13–20]. Figure 9.4 shows the structure of such a configuration. It consists of a number of layers with different compositions of materials. From the bottom layer to the top layer, the composition of the materials changes gradiently, and all the materials layers are co-sintered to be an integrated monolithic piezo plate. The monomorph also is a bending device, and due to its integrated monolithic nature, the configuration is found to be more reliable than"plate-electrode-plate" sandwich-bonded configuration, such as the bimorph. Multiple electrophoretic deposition technique is found to be an efficient way to form such FGM monomorph structures.

9.1.2 Suspension Preparation and Characterization

In the EPD process of the various piezo configurations, the suspension was first prepared and characterized, then followed by deposition into respective piezoelectric actuators. Pure $Pb(Zr,Ti)O_3$ (PZT) and doped PZT material $0.95Pb\ (Zr_{0.52}Ti_{0.48})$

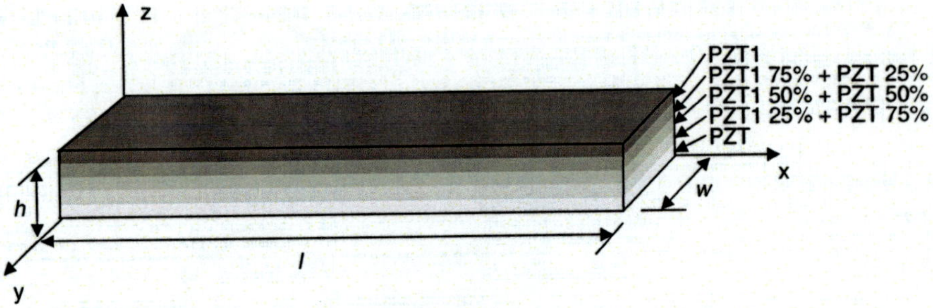

Fig. 9.4 Schematic structure of the FGM monomorph actuator

O_3·0.03$BiFeO_3$·0.02Ba($Cu_{0.5}W_{0.5}$)O_3+0.5wt%MnO_2 (PZT1) were used as the base materials for fabrication. Both base materials were synthesized by mixed-oxide method. The raw oxide powders of PbO(>99.9%), ZrO_2(>99.9%), TiO_2(>99.9%), BiO_3(>99.99%), Fe_2O_3(>99%), BaO (>99%), CuO(>99.99%), WO_2(>99%) and MnO_2(>99.99%) were mixed with the desired stoichiometrical composition and ball-milled for 24 h. To compensate the loss of PbO during sintering, an additional 5% of PbO was added into the raw mixtures. The mixed powders were calcined at 750°C for 2 h and then ground by a planetary ball milling machine at a speed of 150 rpm in ethanol for 8 h.

Suspensions were prepared by adding ball-milled powders in ethanol and then subjected to ultrasonic agitation for 6 min. The suspension was subsequently stirred for 3 to 6 h to ensure complete dissolution and uniform dispersion of the powders in the medium. The powder concentration used in the suspension was 50 g/l. The particle size distribution of the powders was confirmed using an acoustic particle sizer (DT-1200, Dispersion Technology, USA) to be ranging from 0.7 to 1.5 μm with a mean particle size was 1.0 μm (Fig. 9.5).

The zeta potential of the suspension as a function of pH was also measured as shown in Fig. 9.6. The pH was controlled by adding 10% HNO_3 into the suspension.

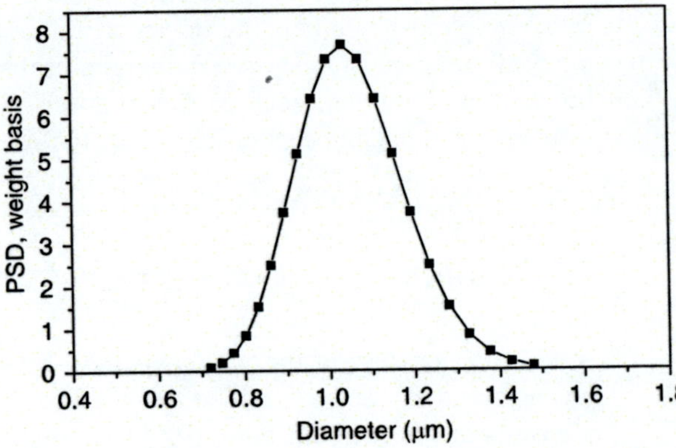

Fig. 9.5 Particle size distribution of PZT powders

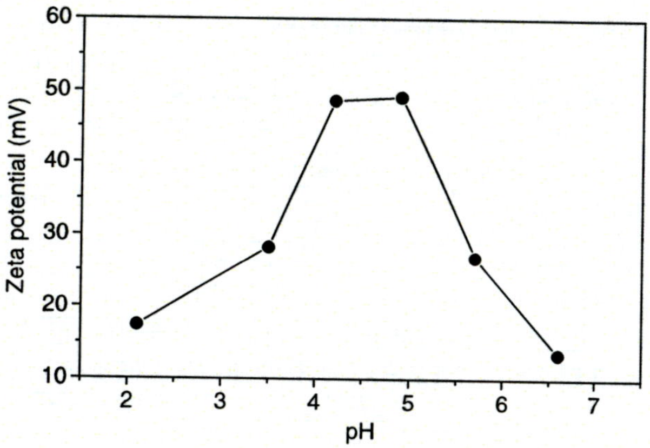

Fig. 9.6 Zeta potential vs. pH value of PZT suspension

A high zeta potential of around 50 mV was obtained when the pH is in the range of 4 to 5. Therefore, in the following deposition, pH of the suspension was controlled to be 4.5.

9.1.3 Fabrication of Various Configurations

9.1.3.1 Piezo Tube

Figure 9.7 depicts the electrophoretic deposition system used to fabricate the piezoelectric tube. The electrophoretic cell includes a cathodic graphite rod substrate positioned in the center of a stainless-steel counter-electrode. The deposition was performed at a constant voltage of 100 V for 3 to 8 min. The deposits obtained were dried for 12 h and then sintered at 1100°C for 1 h. Finally, the sintered product was cut into the designed dimension and painted with silver electrode. Poling was carried out in silicone oil at 100°C by applying a DC field of 2 kV/mm for half an hour.

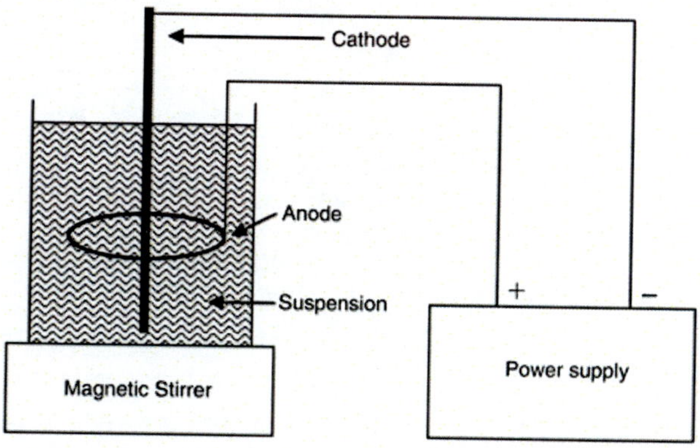

Fig. 9.7 Schematic drawing of electrophoretic apparatus

9.1.3.2 Helical Actuator

To fabricate the helical actuator, the same suspension and deposition procedure for EPD described for tube configuration was applied. The substrate electrode was machined into a screw shape as shown in Fig. 9.8a, and the deposition was performed in two stages. After the first stage deposition, a layer of platinum conductive paste was applied on the surface of the deposited structure to form the intermediate electrode. After the paste dried, the second stage deposition was carried out at a relatively higher voltage, and for a longer time, due to the decrease in conductivity from the platinum paste. The helical structure, after the removal of the graphite substrate, is shown in Fig. 9.8b. Figure 9.8c shows the products after sintering. The platinum electrode is about 7 μm thick and can be clearly seen in the cross section of the helix, as shown in the insert of Fig. 9.8c. After applying the electrodes, the helix is poled to complete the process (Fig. 9.8d).

9.1.3.3 FGM monomorph

Multilayer deposition is most suited for EPD technique. For the fabrication of FGM multilayered monomorph, five suspensions were prepared according to the ratio

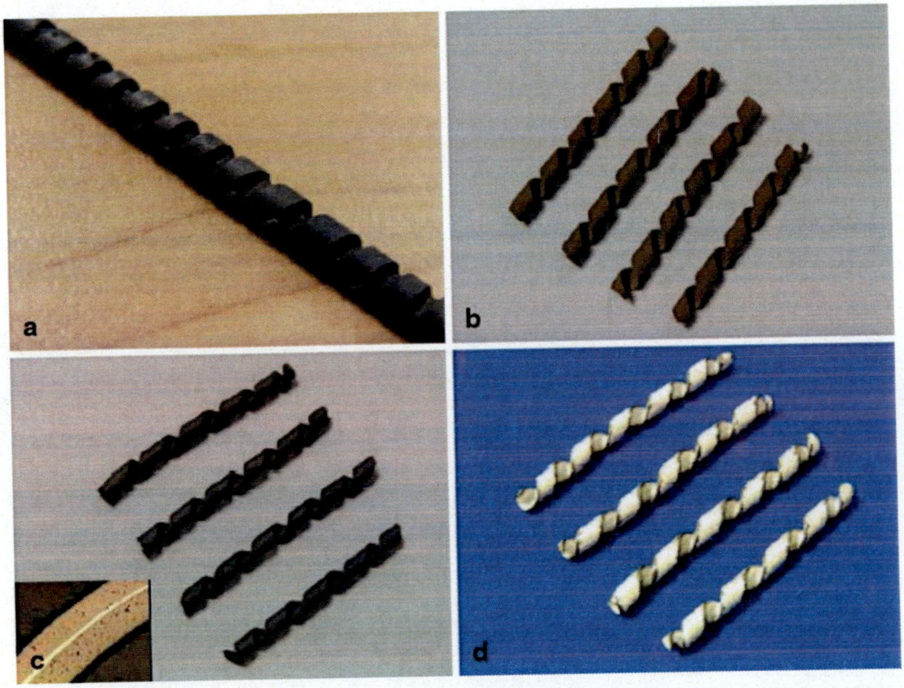

Fig. 9.8 Photographs of the samples, **a** graphite rod; **b** the deposits after burning out the graphite rod; **c** the deposits after sintering; **d** the deposits after poling

shown in Fig. 9.4. Each suspension was used to deposit 5 times to form one layer, with each deposition time of 1 min. The interval between each deposition was controlled to be 10 min. The time interval is important to allow the settling of the deposit to prevent crack or peel off. The initial applied voltage was 25 V from the first layer. Then the voltage was increased by 25 V while changing the suspension for subsequent layer since the resistance will increase with the thickness of the deposits. After deposition, the deposits were dried for 12 h.

The obtained green FGM plates were finally sintered at 1100°C for 1 h, and cut into rectangular-shaped piezoelectric plates, followed by coating with silver electrode and poling in silicone oil at 100°C for half of an hour under 2 kV/mm.

9.1.4 Characterization of Physical Properties

9.1.4.1 Piezo Tube

The electrophoretically deposited piezo tubes were characterized in various properties. Figure 9.9 shows the XRD (Shimadzu XRD 6000) pattern of the tube. It is confirmed that desired PZT perovskite structure has been obtained. Figure 9.10 is the SEM micrograph of the cross section of the sintered PZT tube. It shows that a dense component has been achieved. The density is measured to be about 7.54 g/cm^3, which is approximately 95% of the theoretical density. Figure 9.11 shows the X-ray radiography of the PZT tubes with different sizes. The structure is noted to possess reasonably uniform diameter and wall thickness. These results show that EPD is a good choice to fabricate the piezoelectric tubes with desirable physical properties.

9.1.4.2 FGM monomorph

Figure 9.12 shows the cross section of a FGM plate after sintering at 1100°C. The cross section shows a thickness of approximately 270 μm measured from Fig. 9.12a.

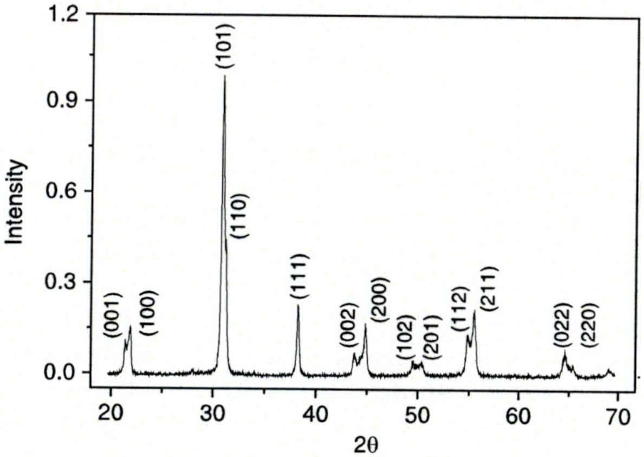

Fig. 9.9 XRD pattern of a tube sample

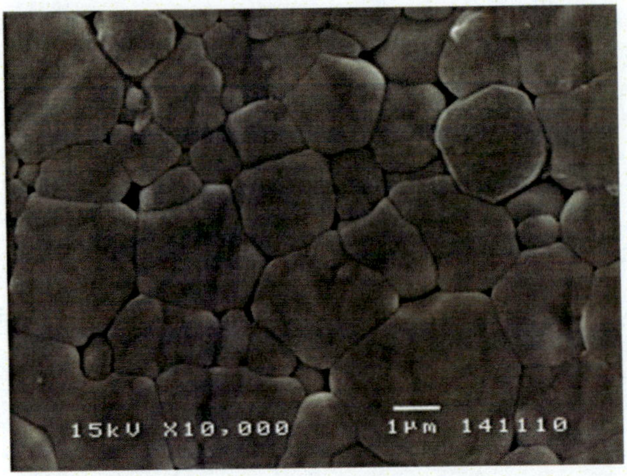

Fig. 9.10 Cross section of the sintered piezoelectric tube

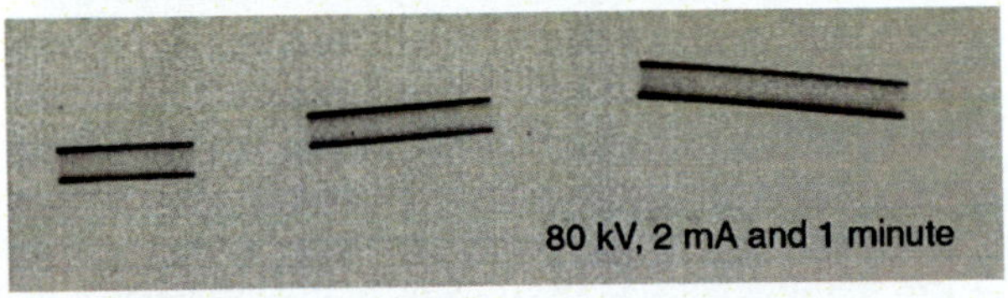

Fig. 9.11 X-ray radiography photo for different piezo tubes

Over this cross section, the gradient change in microstructure can be observed. From the pure PZT side (left) to the pure PZT1 side (right), it can be seen that the grain size gradually increases. This is due to the compositional gradient over the cross section (material PZT and PZT1 have different grain sizes). Figures 9.12b, c and d show the left, middle and right side of the cross section with higher magnifications, respectively. The variation of piezoelectric properties over the cross section induced by compositional and microstructural gradient is the driving source to actuate the monomorph. The observed microstructure gradient also confirms the effectiveness of fabricating an integrated FGM monomorph actuator.

9.1.5 *Characterization of Electromechanical Properties*

9.1.5.1 Piezo Tube

The electromechanical bending displacements of the piezo tube under both clamped-free and free-free boundary conditions were measured using MTI-2000 fotonic sensor. The displacement hysteresis loop (Fig. 9.13) shows a narrow shape, which indicates the low electromechanical losses of the piezoelectric tube.

The measured displacement data have also been compared with their theoretical values. The theoretical displacements under clamped-free condition and free-free

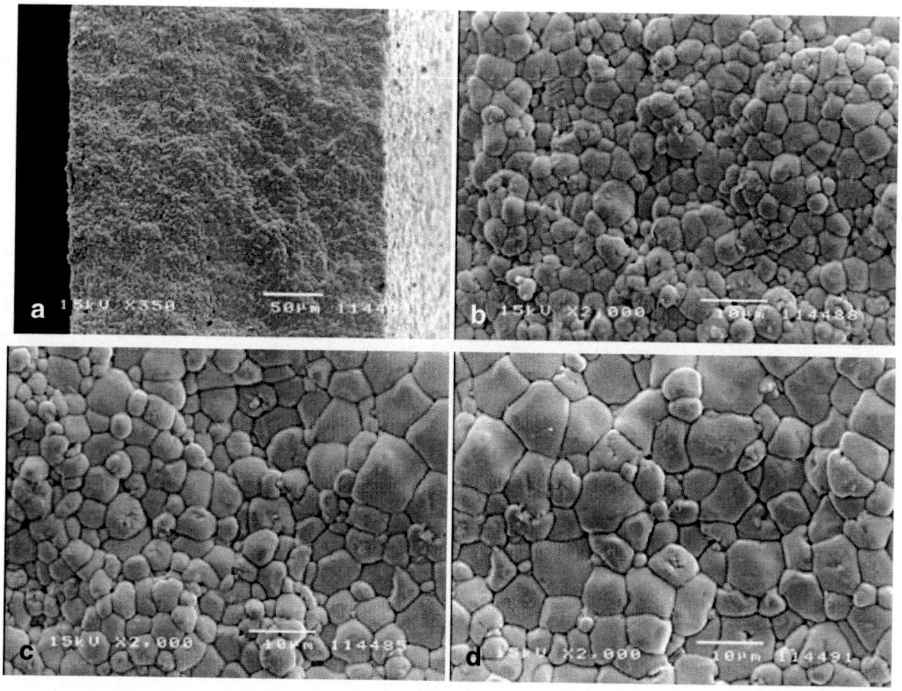

Fig. 9.12 Cross section of the actuator showing the grain structure from the left side to the right of the actuator. **a** Overview, **b** Left, **c** Middle, **d** Right

Fig. 9.13 Bending displacement in response to step voltage

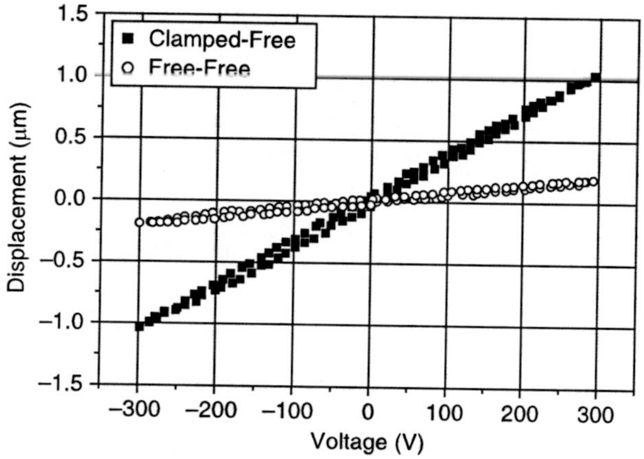

boundary condition were calculated using the following Eqs. (9.1) and (9.2), respectively [1]

$$\varsigma = -\frac{4d_{31}V\cos\beta L^2}{\pi\left(r_o{}^2 + r_i{}^2\right)\ln\frac{r_o}{r_i}} \tag{9.1}$$

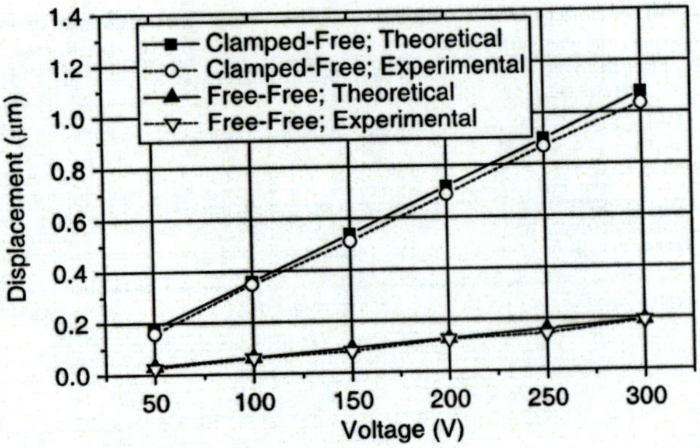

Fig. 9.14 Comparison between calculated and measured displacement of the piezo tube

$$\varsigma = -\frac{4d_{31} V \cos \beta l(L-l)}{\pi \left(r_o^2 + r_i^2\right) \ln \frac{r_o}{r_i}} \tag{9.2}$$

where V is the applied voltage in the wall-thickness direction, r_o and r_i are the outer and inner radius, respectively, β is a constant angle, d_{31} is the piezoelectric constant, L is the length, and l is the distance from end to the supporting point. Figure 9.14 shows both the experimental and theoretical data as a function of applied voltage. The two results clearly are in good agreement. The results also demonstrate that EPD is capable of fabricating piezo actuators with attractive and predictable electromechanical properties.

9.1.5.2 Helical Actuator

Figure 9.15 shows the finite element simulation of the piezoelectric helical actuator. The actuator could generate both rotational displacement Δl (the tip moves from point A to B) and radial displacement ΔR upon the application of electric field. Figure 9.16 shows the measurement results of such two displacements. Figure 9.16a

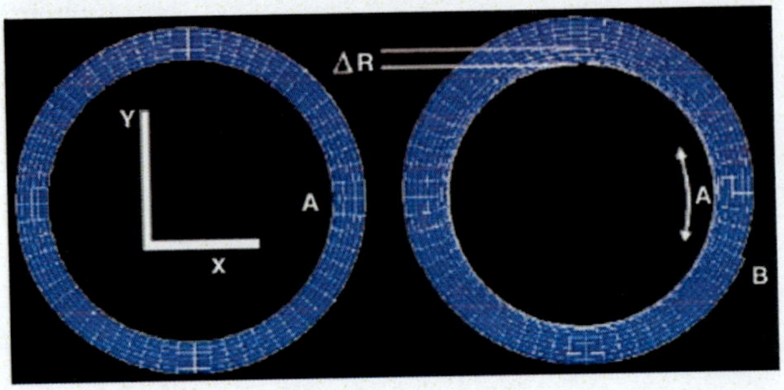

Fig. 9.15 Finite element simulation of the piezo helix showing the motion of the actuator

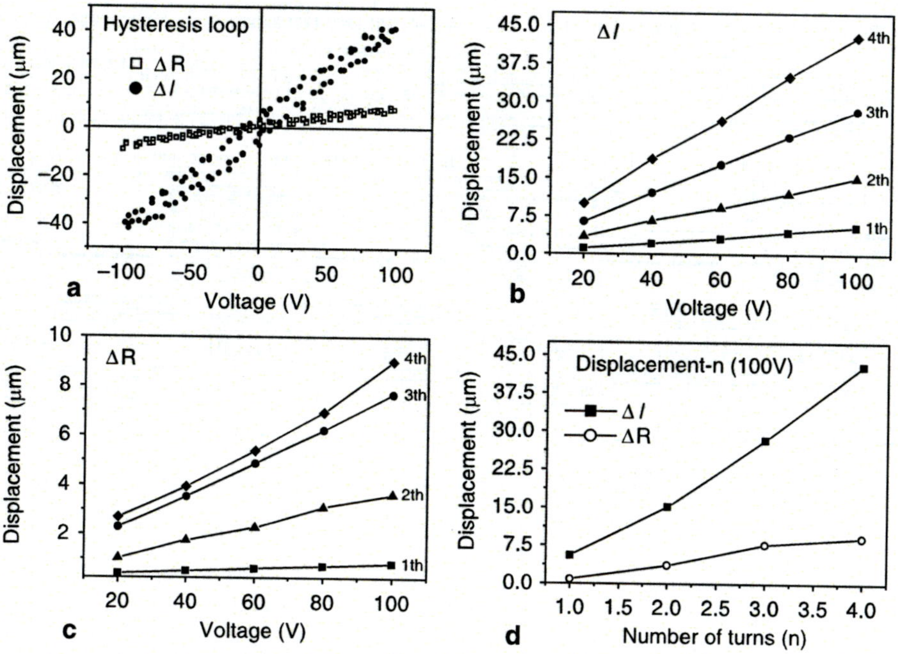

Fig. 9.16 Electromechanical properties of the helical actuator

depicts the field-induced displacement hysteresis loop of the rotational motion Δl and radial motion ΔR. The variation of displacement Δl and ΔR as a function of applied voltage at various turns for the same length of the actuator is plotted in Figs. 9.16b and c. The plots show that maximum displacements of 43 and 9 μm were obtained for Δl and ΔR respectively, which are much larger than that obtained in piezo tubular actuators. Figure 9.16d presents the relationship between displacement and number of helical turns n. The results showed that Δl and ΔR increase linearly with the applied voltage and number of turns.

9.1.5.3 FGM monomorph

The bending displacements of actuators with different dimensions were measured under clamped-free boundary condition as shown in Fig. 9.17. The bending displacement linearly increases with the applied voltage, and the thinner and longer actuators yield larger displacements. This is because the bending displacement of a bending type actuator parabolically increases with the length to thickness ratio (l/h). The thickness of the EPD fabricated FGM monomorph can be very small (<0.3 mm); this confirms the advantage that device miniaturization has in providing larger displacements. Compared to EPD, conventional technique to fabricate bending actuator of such a thickness will be expensive, time consuming and of low yield.

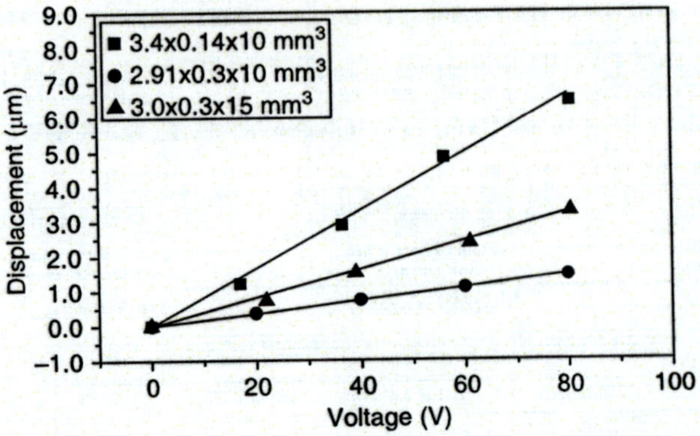

Fig. 9.17 The displacement of monomorph with different dimensions (width × thickness × length)

9.1.6 *Applications*

9.1.6.1 Piezo Tube and Motor

One of the important applications of a piezo tube is the cylindrical-type ultrasonic micromotor as shown in Figs. 9.18 and 9.19. Figure 9.18 shows the working principle of the motor. This consists of two main components, rotor and stator. The stator is the piezo tube working at its first bending vibration mode. Two of such a

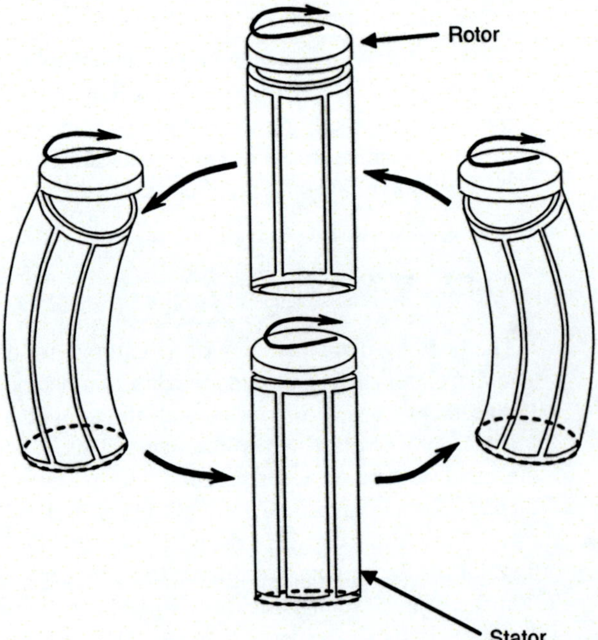

Fig. 9.18 Principle of cylindrical-type ultrasonic micromotor using first bending vibration modes

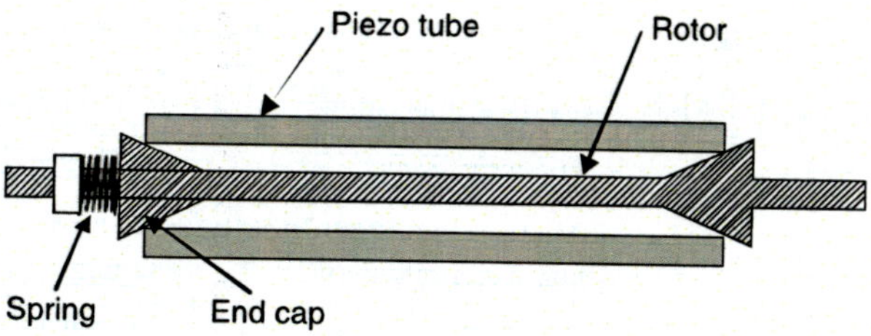

Fig. 9.19 The schematic structure of the piezoelectric cylindrical-type motor

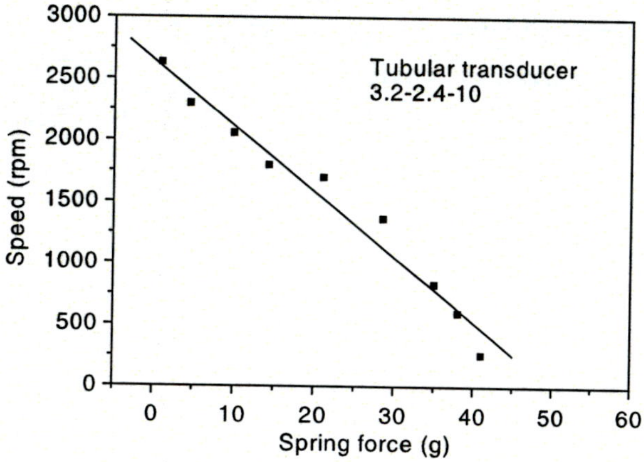

Fig. 9.20 Rotational speed of the motor as a function of the spring force

mode with 90° phase difference will produce the rotation of the stator transducer, which is then able to be used for pushing the rotor to rotate as shown in the figure. The advantages of this kind of motor are good ability for miniaturization, fast rotational speed and simple structure. Figure 9.19 shows the design concept of such a motor, where the rotor is simply clamped at the end of the stator by two end caps and a spring force. Figure 9.20 shows the rotation speed of the motor as a function of the spring force. The present device can achieve a speed up to 2000 rpm, which is superior than that of the traditional ultrasonic motors. The ability to achieve such a high speed is attributed to the small size, good vibration displacement, and also the robust tube obtained from the EPD technique.

9.1.6.2 FGM monomorph and Impedance Pump

One of the novel applications that can be facilitated by the piezoelectric FGM monomorph is an impedance pump. An impedance pump functions most efficiently with a low frequency and high displacement actuator, and the FGM monomorph satisfies this requirement well. Figure 9.21 shows the displacement of an actuator with a dimension of 33.5 mm in effective length, 4.21 mm in width and 0.34 mm in thickness. The displacement is noted to be frequency dependent and reaches a maximum at the resonant frequency around 195 Hz. The maximum displacement

Fig. 9.21 Frequency response of actuator with a dimension 33.5-4.21-0.34 mm^3

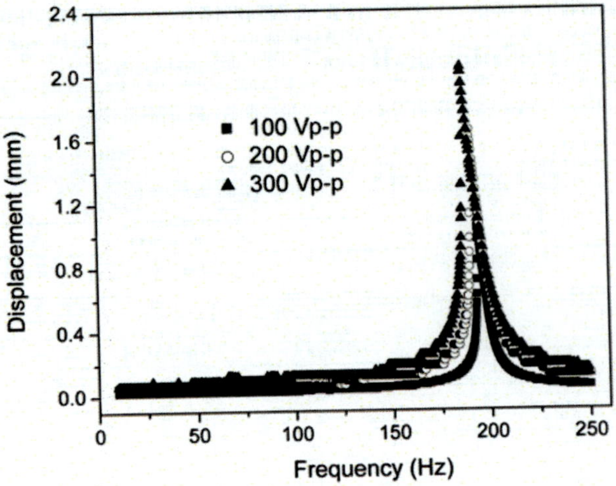

achieved at 300 Vp-p is nearly 2 mm. Both the resonant frequency and displacement obtained are ideal for the impedance pump applications.

The impedance pump, shown in Fig. 9.22, contains a glass flow circuit and a silicone rubber tube. The glass circuit has an inner diameter about 2 mm and perimeter about 34 cm. The rubber tube is 20 mm in length and 50 μm in wall thickness. The two ends of the rubber tube were glued to the glass circuit. The actuator was fixed above the rubber tube with a L-shaped tool head attached at the free tip of the actuator. Upon excitation, the movement of the actuator induced the pinching of the rubber tube at the position x=0.24 of the total length of the rubber tube. The frequency can be tuned to the resonance of the actuator to obtain the largest displacement and hence induce flow rate. To observe the flow and measure the flow rate, fluorescence particles were added into the water in the circuit and observed under UV light. From the product of the traveling velocity of the fluorescence particles and the cross section area of the glass circuit, the flow rate was estimated to be about 9 ml/min. The FGM piezo monomorph provides a neat and simple miniaturized impedance pumping mechanism compared to its electromagnetic counterparts.

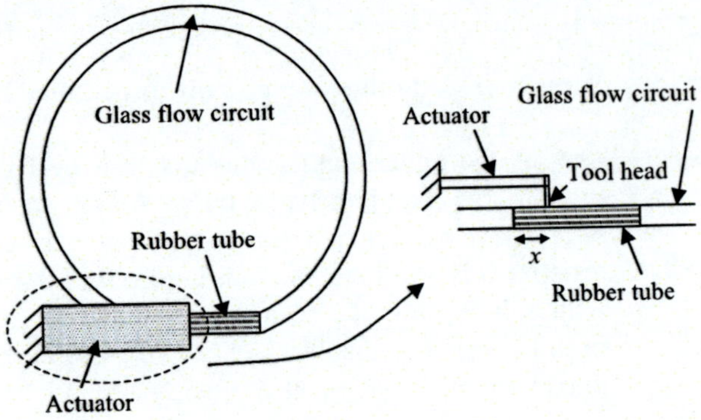

Fig. 9.22 Flow circuit of the impedance pump

9.2 Electrophoretic Deposition of WO_3 Nanorods

9.2.1 *Introduction*

Tungsten oxide (WO_3) is one of the widely studied transition metal oxides for many attractive applications such as energy efficient windows (smart windows), antiglare automobile rear-view mirrors, sunroofs and sensors [21]. WO_3 allows an optical modulation between transparent and blue color upon ion-electron double injection/extraction, which is also known as electrochromism. The reaction can be expressed as:

$$WO_3(transparent) + xe^- + xM^+ \Leftrightarrow M_xWO_3(blue) \tag{9.3}$$

where M^+ stands for cations such as H^+, Li^+ Na^+ or K^+.

Various methods such as sol–gel, electrodeposition, hydrothermal process, evaporation and sputtering have been applied in the formation of crystalline and amorphous WO_3. Among these methods, hydrothermal process has received huge attention for its advantages such as simple operation, low-cost, potential for large scale production and forming crystalline structure with good stability. However, two of the drawbacks of the process are that the hydrothermally synthesized products are dispersed as a suspension and that additional assembly steps are required to adhere the dispersed products onto the substrate. Electrophoretic deposition, which is efficient in forming a film or coating on a conductive surface (even a complex surface morphology) with good control in film thickness and porosity through the electric field, pH value and deposition duration, provides flexibilities in designing electrochromic devices with different degrees of optical modulation (film thickness) and switching time (film porosity) and, hence, widens their application in advanced technologies.

EPD has been applied in surface coating, ceramics processing and forming composite materials, in which particles are usually used as the starting material in the deposition. In this part of the work, 1D nanorods of WO_3 were deposited onto a transparent conductive substrate (ITO glass) using both constant voltage EPD and constant current EPD. Optimum deposition conditions were identified with a discussion in the deposition mechanism. The resultant electrochromic performance of the WO_3 nanorod coated substrates, such as optical modulation and switching time, also were studied.

9.2.2 *Experimental Procedures*

9.2.2.1 WO_3 Synthesis

Tungsten oxide (WO_3) nanorods were prepared through hydrothermal process: 0.825 g of $Na_2WO_4 \cdot 2H_2O$ (Aldrich, 99%) and 0.290 g of NaCl (Sigma Ultra, 99.5%) were dissolved in 20 ml of de-ionized water with a mole ratio of 1:2. The solution was adjusted with 3.0 M HCl until a pH of 2, then transferred into the Teflon liner of the stainless steel autoclave, and then heated at 180°C for 24 h. A white precipitate of nanorods was obtained and centrifuged with ethanol and de-ionized water for several times to remove any left over precursors. The WO_3 nanorod was found to be a hexagonal single crystal with a growth direction of (0002). The diameter and length of the nanorods were around 100 nm and 2 μm, respectively.

9.2.2.2 Electrophoretic Deposition (EPD) of Nanorods

The nanorod suspension was prepared by dispersing the WO_3 nanorods in de-ionized water with a concentration of 10 mg/cm^3. The pH of the suspension was adjusted to 6.77 to obtain an optimal zeta potential of −40 mV. A transparent conductive substrate (ITO coated glass, Delta Technologies, Rs = 15–25 Ω) was used as the working electrode, while a Pt wire was used as the counter electrode. The working electrode was kept at positive voltage or current with distance between the two electrodes remained as 1 cm. The deposition was carried out using an Autolab Potentiostat (PGSTAT302) with a maximum voltage limit of 10 V. Constant voltage and current EPD were carried out with an electric field ranged from 3 to 8 V/cm and current density ranged from 0.2 to 1.4 mA/cm^2, respectively. The current and voltage responses were recorded with respect to the deposition duration for further analysis. The coated substrates were air-dried overnight before morphology and electrochromic studies. Surface morphology of the coated ITO was characterized using field emission scanning electron microscopy (FESEM, JOEL6340F) and the oxide layer thickness was checked by Alpha-step Surface Profiler.

9.2.2.3 Electrochromic Characterization

Optical modulation and switching studies were carried out in a two-electrode system for ±3 V. The coated ITO was used as the working electrode, and a Pt wire was used as the counter and reference electrode. All electrochemical studies were carried out in an electrolyte of 1.0 M $LiClO_4$ in propylene carbonate (PC). Transmittance variation and color changes were detected through UV–Vis spectrometer (Shimadzu UV-2501PC) at visible wavelength, ranging from 300 to 900 nm. Optical modulation and switching time were extracted from the transmittance spectra.

9.2.3 Results and Discussion

9.2.3.1 Electrophoretic Deposition (EPD) Study

WO_3 nanorods suspension of 10 mg/cm^3 was used in the constant voltage EPD. The applied electric field ranged from 3 to 8 V/cm for 900 s deposition and the I–t

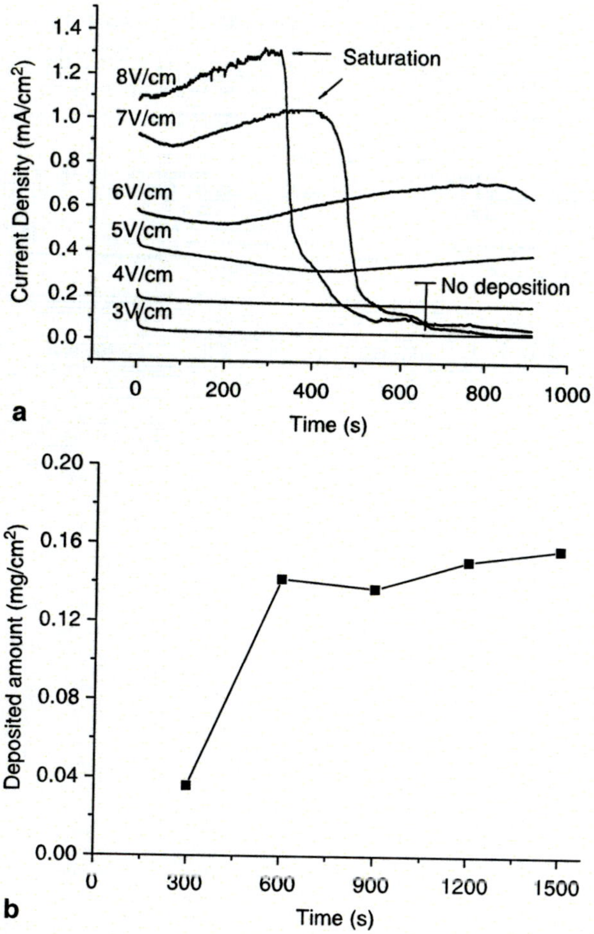

Fig. 9.23 **a** Constant voltage EPD of 10 mg/cm^3 WO_3 nanorods suspension for 900 s deposition, **b** WO_3 nanorods deposited amount at 5 V/cm for 300 to 1500 s deposition duration

behaviors are recorded in Fig. 9.23a. With the various electric field strength studied, the EPD of WO_3 nanorods can be characterized into three regions: no deposition (<4 V/cm), uniform deposition (5–6 V/cm) and saturated deposition (>7 V/cm). For the electric field <4 V/cm, the electric field induced attraction force is not strong enough to drive the WO_3 nanorods onto the ITO glass. Hence, only small amount of nanorods were observed under the FESEM (Fig. 9.24a). When the electric field was increased to beyond 5 V/cm, uniform WO_3 nanorods deposition was observed as a porous oxide layer on top of the ITO (Fig. 9.24b). However, for the electric field >7 V/cm, the deposition ceases after a period of time as showed in Fig. 9.23a with a sharp drop in the current density nearly to zero. This is attributed to the build-up of electrical resistance on the working electrode, which arose from the oxide layer thickness and, hence, from the decrease in the electric field induced electrophoresis in the suspension. The nearly zero current density also indicated that the electric field induced driving force is insufficient to drive the nanorods onto the electrode

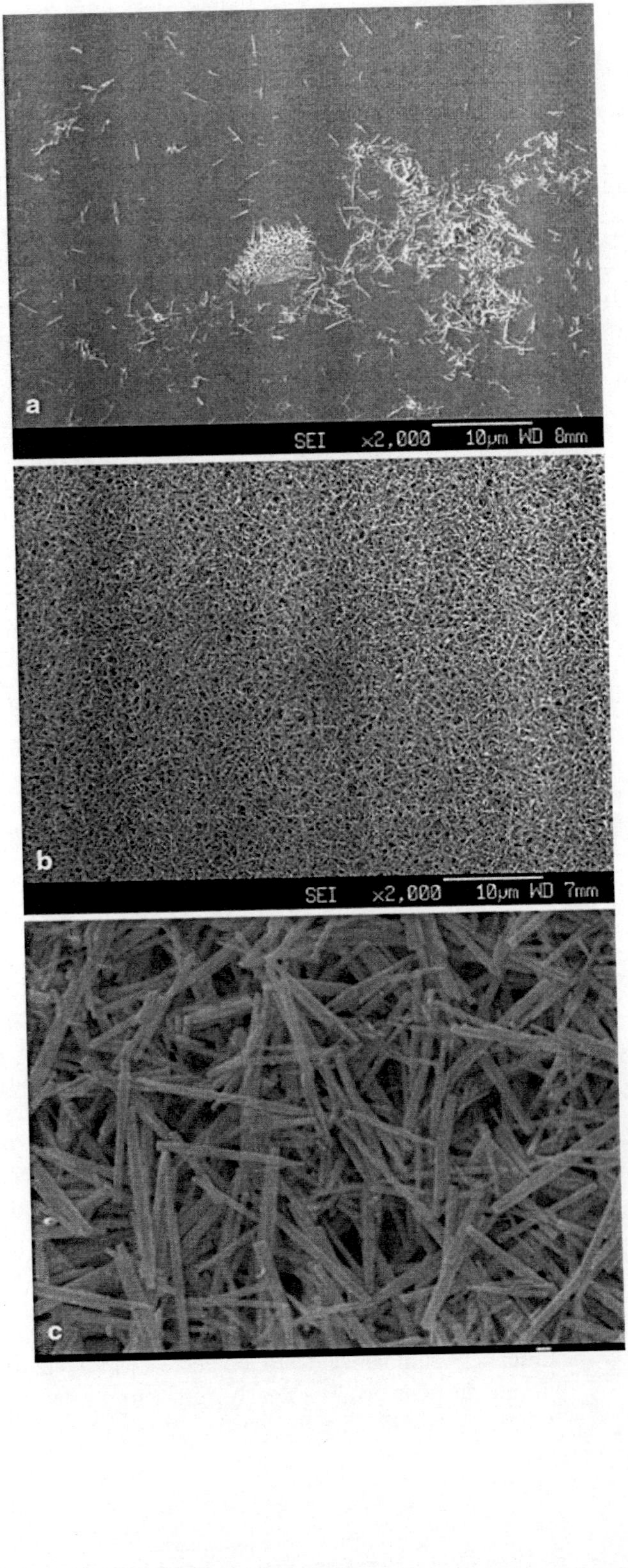

Fig. 9.24 FESEM images of the deposited WO_3 nanorods on ITO glass under **a** 4 V/cm and **b** 6 V/cm and **c** higher magnification of 6 V/cm deposited WO_3 nanorods

after the build-up of the oxide layer. In addition, oxide layer peeling was observed at higher electric field during the substrate removal, which causes non-uniform coverage on the ITO. Thus, it is reckoned that 5–6 V/cm gives the optimal deposition electric field for the present nanorod suspension. Figure 9.23b shows the deposited amount with respect to the deposition duration using 5 V/cm. The relationship turns non-linear after 900 s with a nearly constant deposited amount (corresponds to a thickness of 1.2 μm). This result further indicates that EPD at a constant voltage will cease after the build-up of the non-conductive layer, and longer deposition duration does not significantly improve the deposited amount.

Constant current density EPD was carried out using 0.6–1.4 mA/cm² (Fig. 9.25a). The voltage increased up to the maximum voltage of the potentiostat (10 V) after a period of deposition while optimum deposition was observed to be approximately 0.8 mA/cm². The deposited amount of WO_3 is plotted with respect to the current density in Fig. 9.25b. The deposited amount is found to increase with the increase

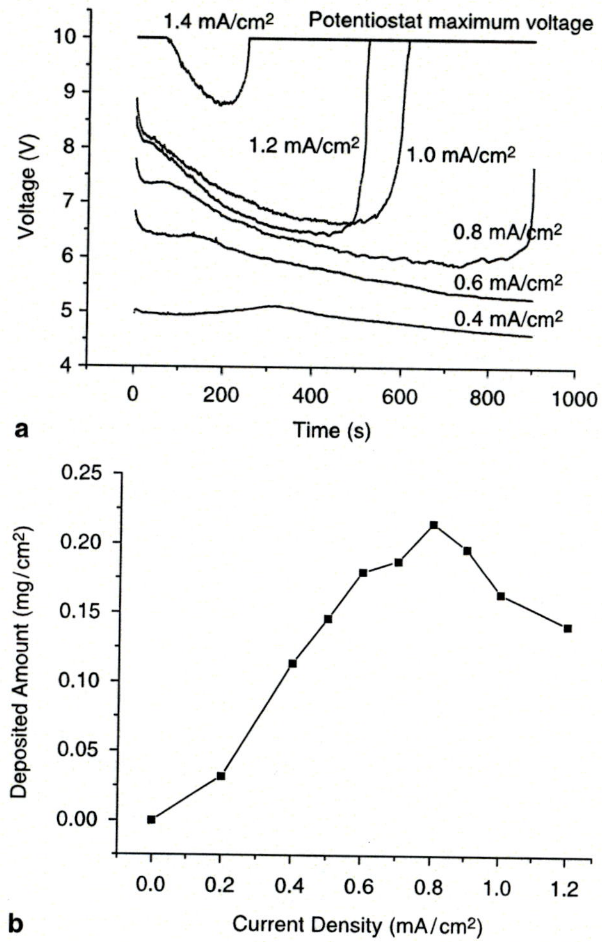

Fig. 9.25 **a** Constant current EPD of 10 mg/cm³ WO_3 nanorods suspension for 900 s deposition, **b** Deposited amount of WO_3 nanorods on ITO glass for 900 s deposition duration at various current densities

of the current density, which implies that the passed charges (current) are proportional to the electrophoretically deposited nanorods. However, at even higher current densities (>0.8 mA/cm^2), the deposited amount is noted to have reduced, which is attributed to the ITO substrate degradation, or poor adhesion of the nanorods to the substrate. Further, ITO glass has been demonstrated as the working electrode for aqueous EPD up to 40 V/cm^2 and 2.5 mA/cm^2 for which no ITO degradation was observed [22, 23]. Thus, ITO electrochemical degradation at high current density should not be the reason for the decrease in the deposited amount. The decrease of deposited mass is hence attributed to the accumulation of large amount of WO_3 nanorods at the electrode within a short duration under a high driving force, which results in the formation of a very porous, inhomogeneous film with reduced adhesion at the electrode surface [24]. As a result, peel-off was observed during the substrate removal.

In summary, a moderate electric field of 5–6 V/cm and current density of 0.8 mA/cm^2 are reckoned to be the optimum deposition conditions for uniform and well adhered nanorod film. An oxide layer shielding effect is observed, but could be overcome by applying constant current with further increased deposition voltage and a longer deposition duration.

Theoretically, Cottrell has proposed the following relationship for a diffusion controlled EPD process:

$$i = KC/(\pi Dt)^{1/2} \tag{9.4}$$

where i is the current, K is the kinetic constant, D is the diffusion coefficient and t is the deposition duration [25]. The equation can be simplified to $i = kt^{-1/2}$, when K, C and D are kept as constants. Figures 9.26a and 9.26b show the experimental data of 5 and 6 V/cm deposition separately. Both cases follow the Cottrell equation where the $i \propto t^{-1/2}$ curve shows an almost linear relationship with a similar slope, which suggests the deposition presented in this work is a diffusion-controlled process. Deviation from the slope is observed at long deposition time, which is attributed to the building-up of the oxide layer which decreased the nanorods electrophoresis.

9.2.3.2 Electrochromic Study

The EPD coated WO_3 nanorods were tested for UV–Vis transmittance under a switching voltage of ±3 V (Fig. 9.27a). The WO_3 nanorods were bleached at +3 V with around 60% of transmittance at 700 nm and showed strong absorption of red-end light at −3 V biasing. The transmittance modulation at 700 nm (ΔT700 nm) is found to be around 40%, which is comparable to other crystalline films [26, 27]. Switching time of the WO_3 nanorods is extracted from the transmittance data in Fig. 9.27b by applying a square wave voltage of ±3 V to the oxide layer (simulating the switching condition). The switching time, namely bleaching (tb) and coloration

Fig. 9.26 Currents density versus inverse square root of deposition duration plot under **a** 5 V/cm and **b** 6 V/cm deposition

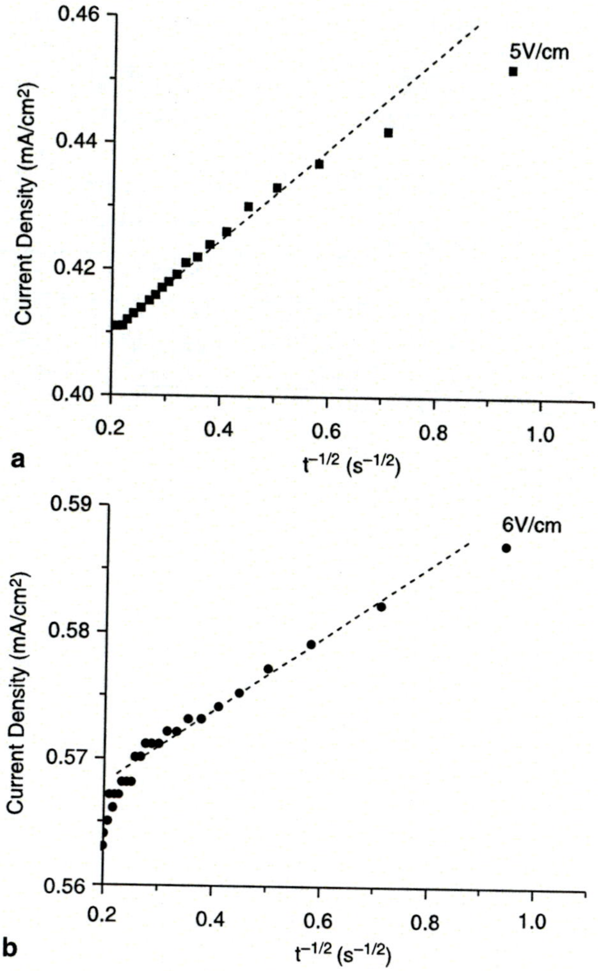

(tc) time, are calculated based on the standard time required for 70% optical modulation. Coloration time, tc70%, is found to be 28.8 s, which is comparable to the reported values (~25 s) [28]. Nevertheless, it should be noted that the bleaching time, tb70%, of 4.5 s is much superior to the crystalline (~22 s) and amorphous WO_3 (~18–36 s) reported [28, 29]. This improvement is attributed to the porous nature of the oxide layer, which consists of nanorods. The porous oxide layer allows electrolyte penetration and shortens the ionic diffusion length from the electrolyte to the centre of the oxide. On the other hand, the high surface to volume ratio of the nanorods facilitates the Li^+ ion movement across the oxide-electrolyte interface by providing a large amount of reaction sites. Hence, shorter switching time is observed as the result of these two factors.

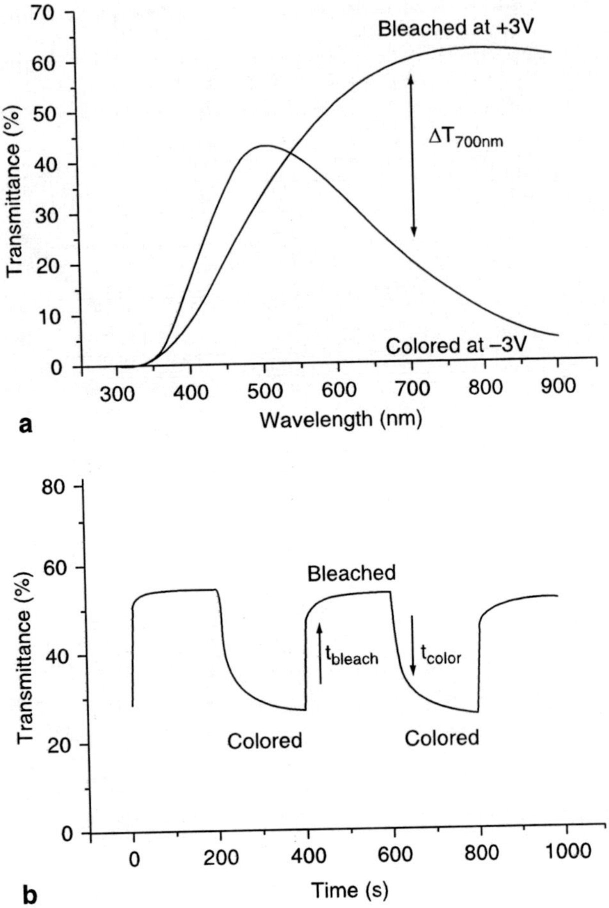

Fig. 9.27 **a** UV-Vis transmittance of WO_3 nanorods at bleaching (+3 V) and coloring (−3 V); **b** Switching study by applying a square wave signal ±3 V at 632.8 nm. All electrochemical studies were carried out in 1 M $LiClO_4$/PC electrolyte

9.3 Conclusions

This paper reviews and discusses the application of EPD technique in the fabrication of functional devices, namely piezoelectric actuators, including piezo tube, piezo helix and FGM monomorph, and nanorod coating for electrochromic applications. In the report on piezo actuator configurations, the fabrication, characterization and applications of the actuators were discussed. Depending on the complicity of the actuator, single, double and multiple deposition techniques were applied respectively. The electromechanical properties were measured and good displacement properties have indicated the success of the EPD process in the forming of these devices. The WO_3 nanorods electrophoretic deposition has also been studied. It is also demonstrated that a thin coating of ~1.2 μm by EPD could achieve good electrochromic properties. In conclusion, the present work confirms that EPD is a good technique to fabricate functional devices with miniaturized dimension and complicated structures.

Reference

1. Li, T.: Development of piezoelectric tubes for micromotor. PhD thesis, Nanyang Technological University (2004)
2. Li, T., Chen, Y.H., Ma, J.: Frequency dependence of piezoelectric vibration velocity. Sensors. Actuat. A. **138**, 404–410 (2007)
3. Li, T., Chen, Y.H., Ma, J.: Factors affecting the performance of piezoelectric bending actuators for advanced applications: an overview. J. Mater. Sci. **44**, 5393–5407 (2009)
4. Li, T., Chen, Y.H., Ma, J.: Development of a miniaturized piezoelectric ultrasonic transducer. IEEE Trans. Ultrson. Ferroelectr. Freq. Control. **56**, 649–659 (2009)
5. Chen, Y.H.: Electrophoretic deposition of advanced ceramic actuator. PhD thesis, Nanyang Technological University (2005)
6. Cheng, Wen: Electrophoretic deposition of advanced ceramics. Master thesis, Nanyang Technological University (2000)
7. Li, T., Ma, J.: Study of piezoelectric tubular transducers. Mater. Sci. Forum. **437–438**, 491–494 (2003)
8. Li, T., Ma, J., Chen, Y.H.: A piezoelectric tube with a double-layer configuration. Ceram. Int. **30**, 1803–1805 (2004)
9. Li, T., Ma, J., Chen, Y.H.: Fabrication and performance of piezoelectric tubes for cylindrical ultrasonic micromotor. Ferroelectrics **315**, 111–121 (2005)
10. Li, T., Chen, Y.H., Ma, J. et al.: Metal-PZT composite piezoelectric transducers and ultrasonic motors. Key. Eng. Mater. **334–335**, 1073–1076 (2007)
11. Ma, J., Li, T., Chen, Y.H. et al.: Piezoelectric materials for biomedical applications. Key. Eng. Mater. **334–335**, 1117–1120 (2007)
12. Chen, Y.H., Li, T., Ma, J. et al.: Electrophoretic deposition and characterization of helical piezoelectric actuator. Ceram. Int. **34**, 1–6 (2008)
13. Chen, Y.H., Li, T., Ma, J.: Investigation on the electrophoretic deposition of a FGM piezoelectric monomorph actuator. J. M. Sci. **38**, 2803–2807 (2003)
14. Chen, Y.H., Ma, J.: Electrophoretic deposition and characterization of a FGM piezoelectric monomorph actuator. Mater. Sci. Forum. **437–438**, 487–490 (2003)
15. Chen, Y.H., Ma, J., Li, T.: A functional gradient ceramic monomorph actuator fabricated using electrophoretic deposition. Ceram. Int. **30**, 683–687 (2004)
16. Chen, Y.H., Ma, J., Li, T.: Electrophoretic deposition and characterization of a piezoelectric FGM monomorph actuator. Ceram. Int. **30**, 1807–1809 (2004)
17. Chen, Y.H., Li, T., Ma, J.: Development of piezoelectric monomorph actuator using electrophoretic deposition. J. M. Sci. **41**, 8079–8085 (2006)
18. Chen, Y.H., Li, T., Ma, J.: Electrophoretic deposition of functionally graded monomorph. Key. Eng. Mater. **314**, 89–93 (2006)
19. Chen, Y.H., Li, T., Ma, J. et al.: Development of FGM monomorph actuator for impedance pump application. Key. Eng. Mater. **334–335**, 1077–1080 (2007)
20. Li, T., Chen, Y.H., Ma, J.: Characterization of FGM monomorph actuators fabricated using EPD. J. M. Sci **40**, 3601–3605 (2005)
21. Khoo, E., Lee, P.S., Ma, J.: Electrophoretic deposition (EPD) of WO_3 nanorods for electrochromic application. J. Eur. Ceram. Soc. **30**, 1139–1144 (2010)
22. Jung, S.M., Jung, H.Y., Suh, J.S.: Horizontally aligned carbon nanotube field emitters fabricated on ITO glass substrates. Carbon **46**, 1973–1977 (2008)
23. Yui, T., Mori, T., Takagi, K.: Synthesis of photofunctional titania nonosheets by electrophoretic deposition. Chem. Mater. **17**, 206–211 (2005)
24. Ma, J., Cheng, W.: Electrophoretic deposition of PZT ceramics. J. Am. Ceram. Soc. **85**, 1735–1737 (2002)
25. Koura, N., Tsukamoto, T., Hotta, T.: Preparation of various oxide films by an electrophoretic deposition method: a study of the mechanism. Jpn. J. Appl. Phys. **34**, 1643–1647 (1995)
26. Liao, C.C., Chen, F.R., Kai, J.J.: WO_{3-x} nanowires based electrochromic devices. Sol. Energy. Mater. Sol. Cells. **90**, 1147–1155 (2006)

27. Joraid, A.A.: Comparison of electrochromic amorphous and crystalline electron beam deposited WO_3 thin film. Curr. Appl. Phys. **9**, 73–79 (2009)
28. Sallard, S., Brezesinski, T., Smarsly, B.M.: Electrochromic stability of WO_3 thin films with nanometer-scale periodicity and varying degrees of crystallinity. J. Phys. Chem. C. **111**, 7200–7206 (2007)
29. Deepa, M., Singh, D.P., Agnihotry, S.A.: A comparison of electrochromic properties of sol-gel derived amorphous and nanocrystalline tungsten oxide films. Curr. Appl. Phys. **7**, 220–229 (2007)

Index

J. H. Dickerson, A. R. Boccaccini (eds.), Electrophoretic Deposition of Nanomaterials, Nanostructure Science and Technology,
DOI 10.1007/978-1-4419-9730-2, © Springer Science+Business Media, LLC 2012